Communications
in Computer and Information Science   2676

Rationale

The CCIS series is devoted to the publication of proceedings of computer science conferences. Its aim is to efficiently disseminate original research results in informatics in printed and electronic form. While the focus is on publication of peer-reviewed full papers presenting mature work, inclusion of reviewed short papers reporting on work in progress is welcome, too. Besides globally relevant meetings with internationally representative program committees guaranteeing a strict peer-reviewing and paper selection process, conferences run by societies or of high regional or national relevance are also considered for publication.

Topics

The topical scope of CCIS spans the entire spectrum of informatics ranging from foundational topics in the theory of computing to information and communications science and technology and a broad variety of interdisciplinary application fields.

Information for Volume Editors and Authors

Publication in CCIS is free of charge. No royalties are paid, however, we offer registered conference participants temporary free access to the online version of the conference proceedings on SpringerLink (http://link.springer.com) by means of an http referrer from the conference website and/or a number of complimentary printed copies, as specified in the official acceptance email of the event.

CCIS proceedings can be published in time for distribution at conferences or as post-proceedings, and delivered in the form of printed books and/or electronically as USBs and/or e-content licenses for accessing proceedings at SpringerLink. Furthermore, CCIS proceedings are included in the CCIS electronic book series hosted in the SpringerLink digital library at http://link.springer.com/bookseries/7899. Conferences publishing in CCIS are allowed to use Online Conference Service (OCS) for managing the whole proceedings lifecycle (from submission and reviewing to preparing for publication) free of charge.

Publication process

The language of publication is exclusively English. Authors publishing in CCIS have to sign the Springer CCIS copyright transfer form, however, they are free to use their material published in CCIS for substantially changed, more elaborate subsequent publications elsewhere. For the preparation of the camera-ready papers/files, authors have to strictly adhere to the Springer CCIS Authors' Instructions and are strongly encouraged to use the CCIS LaTeX style files or templates.

Abstracting/Indexing

CCIS is abstracted/indexed in DBLP, Google Scholar, EI-Compendex, Mathematical Reviews, SCImago, Scopus. CCIS volumes are also submitted for the inclusion in ISI Proceedings.

How to start

To start the evaluation of your proposal for inclusion in the CCIS series, please send an e-mail to ccis@springer.com.

Panos K. Chrysanthis · Kjetil Nørvåg ·
Kostas Stefanidis · Zheying Zhang ·
Elisa Quintarelli · Ester Zumpano
Editors

New Trends in Database and Information Systems

ADBIS 2025 Short Papers, Workshops, Doctoral Consortium and Tutorials
Tampere, Finland, September 23–26, 2025
Proceedings

 Springer

Editors
Panos K. Chrysanthis ⓘ
University of Pittsburgh
Pittsburgh, PA, USA

Kostas Stefanidis ⓘ
Tampere University
Tampere, Finland

Elisa Quintarelli ⓘ
University of Verona
Verona, Italy

Kjetil Nørvåg ⓘ
Norwegian University of Science
and Technology
Trondheim, Norway

Zheying Zhang ⓘ
Tampere University
Tampere, Finland

Ester Zumpano ⓘ
University of Calabria
Rende, Italy

ISSN 1865-0929   ISSN 1865-0937 (electronic)
Communications in Computer and Information Science
ISBN 978-3-032-05726-6   ISBN 978-3-032-05727-3 (eBook)
https://doi.org/10.1007/978-3-032-05727-3

This Springer imprint is published by the registered company Springer Nature Switzerland AG
The registered company address is: Gewerbestrasse 11, 6330 Cham, Switzerland

If disposing of this product, please recycle the paper.

Preface

This CCIS volume includes research and demonstration papers presented at the 29th European Conference on Advances in Databases and Information Systems (ADBIS), research papers from the workshops accompanying ADBIS, and invited papers from the Doctoral Consortium School. The 29th ADBIS conference was held in Tampere, Finland, on September 23–26, 2025, as a fully on-site event.

The first ADBIS Conference was held in Saint Petersburg, Russia (1997). Since then, ADBIS has been continuously organized as an annual event. Its previous editions were held in Bayonne, France (2024); Barcelona, Spain (2023); Turin, Italy (2022); Tartu, Estonia (2021); Lyon, France (2020); Bled, Slovenia (2019); Budapest, Hungary (2018); Nicosia, Cyprus (2017); Prague, Czech Republic (2016); Poitiers, France (2015); Ohrid, North Macedonia (2014); Genoa, Italy (2013); Poznań, Poland (2012); Vienna, Austria (2011); Novi Sad, Serbia (2010); Riga, Latvia (2009); Pori, Finland (2008); Varna, Bulgaria (2007); Thessaloniki, Greece (2006); Tallinn, Estonia (2005); Budapest, Hungary (2004); Dresden, Germany (2003); Bratislava, Slovakia (2002); Vilnius, Lithuania (2001); Prague, Czech Republic (2000); Maribor, Slovenia (1999); and Poznań, Poland (1998). The official ADBIS portal at http://adbis.eu provides up-to-date information on all ADBIS Conferences, persons in charge, publications, and issues related to the ADBIS community.

The program of ADBIS 2025 included keynote talks, research papers, a panel, tutorials, demos, workshops, a doctoral consortium school (DC), as well as a diversity, equity, and inclusion (DEI) panel and keynote. The main conference attracted 66 paper submissions (61 papers and 5 demos). All the papers went through a process of rigorous single-blind reviewing by at least three reviewers. Eventually, the Program Committee selected 14 submissions as full contributions, appearing in this Springer LNCS proceedings volume, and 14 short papers and 3 demo papers, appearing in Springer's Communications in Computer and Information Science (CCIS) series. The selected papers span a large spectrum of topics in the broader field of data management. We have organized the papers in six sessions: (1) Query Optimization, (2) Spatio-Temporal and Graph Data, (3) Data Sharing & Synthesis, (4) Entity Resolution & Integration, (5) Explainable AI, and (6) Data & Machine Learning.

Following a successful tradition, selected best papers of ADBIS 2025 will be invited for a special issue of Information Systems (Elsevier). The PC chairs would like to express their sincere gratitude to the Editors-in-Chief of Information Systems for their approval regarding this special issue.

For this ADBIS event we had keynote talks by experts in different fields of the broader area of data management. In particular (in alphabetical order):

- Eero Hyvönen (Aalto University, Finland) delivered a talk on "Digital Humanities on the Semantic Web: from Infrastructure to Practical Applications, AI-based Knowledge Discovery, and Web of Wisdom"

- Evaggelia Pitoura (University of Ioannina, Greece) delivered a talk on "Explainability, Fairness and Their Interplay"
- Felix Naumann (University of Potsdam, Germany) delivered a talk on "Data Quality in the Age of AI"
- Yannis Ioannidis (National and Kapodistrian University of Athens, Greece) delivered a talk on "Interactive Digital Storytelling"

In addition, four excellent tutorials were also included in the program, which were delivered by:

- Alejandra Josiowicz (State University of Rio de Janeiro, Brazil) and Genoveva Vargas-Solar (CNRS, LIRIS Laboratory, France) who presented a tutorial entitled "Graph Analytics for Bridging Human and Data Sciences",
- Henrietta Dombrovskaya (DRW Holdings, USA) who presented a tutorial entitled "Data Warehousing: The Industrial Perspective",
- Toni Taipalus (Tampere University, Finland) and Jiaheng Lu (University of Helsinki, Finland) who presented a tutorial entitled "Vector Representations of Multi-Modal Data", and
- Valter Uotila (University of Helsinki, Finland), Soror Sahri (Université Paris Cité, France), and Sven Groppe (University of Lübeck, Germany) who presented a tutorial entitled "Utilizing Quantum Computing to Improve the Quality of Data".

ADBIS 2025 hosted a panel titled "Do LLMs Solve the Data Integration Problem Once and for All?" The panel moderator was Matthias Boehm (TU Berlin & BIFOLD, Germany), and our distinguished panelists were:

- Yannis Ioannidis (University of Athens, Greece)
- Oscar Romero Moral (UPC Barcelona, Spain)
- Genoveva Vargas-Solar (CNRS, France)
- Robert Wrembel (Poznań University of Technology, Poland)

The ADBIS 2025 conference participated in the Database Diversity, Equity and Inclusion (DEI) initiative aiming to guide researchers in our community to adopt a more inclusive mindset. Specifically, authors and participants were encouraged to follow DEI-aware communication guidelines when preparing papers and presentations to ensure inclusive communication. The DEI stream offered a keynote address by:

- Maija Hirvonen (Tampere University, Finland) who delivered a talk on "From restriction to resource: Fostering (computing) workplace with diverse abilities".

DEI also hosted a panel titled "Repositioning DEI: The Urgent Reimagining of DEI in an Era of Retrenchment". The panel moderators were Panagiota Fatourou (University of Crete, Greece) and Genoveva Vargas-Solar (CNRS, France), and our renowned panelists were:

- Maija Hirvonen (Tampere University, Finland)
- Yannis Ioannidis (National and Kapodistrian University of Athens, Greece)
- Kaisa Väänänen (Tampere University, Finland)

This CCIS volume includes 17 papers (14 short papers and 3 demos) from the main ADBIS Conference, at an acceptance rate of 27% (17/66), as well as 4 papers from the invited talks of the doctoral consortium school.

This volume also includes the papers accepted at seven workshops which were co-located with ADBIS 2025. Each workshop had its own international program committee, whose members served as the reviewers of the workshop papers included in this volume. The maximum paper acceptance rate at each of these events did not exceed 50%. In total, 48 papers were submitted to these workshops, out of which 24 were selected for presentation at the conference and publication in this volume, giving an overall acceptance rate of 50%.

The following workshops were run at the ADBIS 2025 conference.

MADEISD 2025: 7th Workshop on Modern Approaches in Data Engineering and Information System Design, chaired by Ivan Luković, Slavica Kordić, and Sonja Ristić.

For decades, in many particularly complex organization systems, there has been an open issue of how to support information management processes so as to produce useful knowledge and tangible business value from data being collected. Database and information systems still play one of the central roles in addressing the aforementioned issue. Today we have a huge selection of various technologies, tools, and methods in data engineering as a discipline that helps in support of the whole data life cycle in organization systems, as well as in information system design that supports the software process in data engineering. Despite this, one of the hot issues in practice is still how to effectively transform large amounts of daily collected operational data into useful knowledge from the perspective of declared company goals, and how to set up the information design process aimed at production of effective software services in companies. The main goal of this workshop was to address open questions and real potentials for various applications of modern approaches and technologies in data engineering and information system design so as to develop and implement effective software services in support of information management in various organization systems. We addressed the interdisciplinary character of a set of theories, methodologies, processes, architectures, and technologies in disciplines such as Data Engineering, Information System Design, Big Data, NoSQL Systems, Data Streams, Internet of Things, Cloud Systems, and Model-Driven Approaches in development of effective software services. We invited researchers from all over the world to present their contributions, interdisciplinary approaches or case studies related to modern approaches in Data Engineering and Information System Design. We expressed an interest in gathering scientists and practitioners interested in applying these disciplines in industry, as well as in public and government sectors such as healthcare, education, public administration, or security services. Experts from all sectors and application domains were welcomed.

DOING 2025: 6th Workshop on Intelligent Data - From Data to Knowledge, chaired by Cristina D. Aguiar, Mirian Halfeld-Ferrari, and Carmem S. Hara. The DOING workshop focused on transforming data into information and then into knowledge. The idea was to gather researchers in NLP (Natural Language Processing), DB (Databases), and AI (Artificial Intelligence) to discuss two main problems: (1) how to extract information from textual data and represent it in knowledge bases; (2) how to propose intelligent methods for handling and maintaining these databases with new forms of requests, including efficient, flexible, and secure analysis mechanisms, adapted to the user, and with quality and privacy preservation guarantees. This workshop focused on all aspects concerning these modern infrastructures, giving particular attention (but not only) to data related to health and environmental domains.

K-GALS 2025: 4th Workshop on Knowledge Graphs Analysis on a Large Scale, chaired by Simona Rombo and Ylenia Galluzzo. Knowledge graphs are powerful models to represent networks of real-world entities, such as objects, events, situations, or concepts, by illustrating the relationships between them. Information encoded by knowledge graphs is usually stored in graph databases and visualized as graph structures. Although these models were introduced in the Semantic Web context, recently they have also found successful applications in other contexts, e.g., the analysis of financial, social, geospatial, and biomedical data. Knowledge graphs often integrate datasets from various sources, which frequently differ in their structure. This, together with the increasing volumes of structured and unstructured data stored in a distributed manner, brings to light new problems related to data/knowledge representation and integration, data querying, business analysis, and knowledge discovery. The ultimate goal of this workshop was to provide participants with the opportunity to introduce and discuss new methods, theoretical approaches, algorithms, and software tools that are relevant to Knowledge Graphs-based research, especially when it is focused on a large scale. To this regard, interesting open issues include how Knowledge Graphs may be used to represent knowledge, how systems managing Knowledge Graphs work, and which applications may be provided on top of a Knowledge Graph.

CAIMA 2025: 1st Workshop on Cooperative AI Models and Applications, chaired by Fabrizio Angiulli, Luca Ferragina, Davide Mario Longo, and Simona Nisticò. The CAIMA Workshop centered on Cooperative AI, an approach grounded in Martin Hollis's concept of we-rationality, which suggests that in complex ecosystems, individuals should not act solely based on personal benefit but rather as contributors to a collective good. This idea is particularly relevant in AI, where different types of agents must interact—whether through human-AI collaboration, AI-AI coordination, or AI-environment adaptation—to achieve shared goals. Topics included Active Learning (expert-guided model training), eXplainable AI (enhancing user understanding), Neuro-symbolic AI (hybrid reasoning systems), Federated Learning (coordinated agent evolution), and Reinforcement Learning (reward-based adaptation). A key focus was sustainability, aligning AI development with the UN's Sustainable Development Goals by promoting energy-efficient models and reducing bias. The workshop provided a forum to explore novel theories, methods, tools, and applications for interaction-driven AI systems.

ERGA 2025: 1st Workshop on Entity Resolution and Graph Alignment, chaired by Georgia Koloniari and Alexandros Karakasidis. In the era of a connected world and abundant data, entity resolution and graph alignment have come to play an increasingly important role for ensuring data consistency, accuracy, and quality. Entity resolution identifies and links records referring to the same real-world entities (e.g., people, organizations) despite inconsistencies and has been a cornerstone of data integration with applications in finance, healthcare, and various other domains. Graph alignment maps nodes and edges across different graphs taking into account the relationships and structures between the entities with applications such as social network analysis, bioinformatics, and knowledge graphs. The two tasks share common challenges, particularly in handling large-scale data, requiring efficient solutions and deploying techniques from a great variety of research areas such as database management, big data and parallel processing, machine learning and AI-based techniques, and others. Recognizing the common ground between entity resolution and graph alignment, the goal of this workshop was to bring together researchers, practitioners, and experts in both fields to introduce new methods and techniques, discuss open challenges, and explore future research directions. The workshop aimed to facilitate the exchange of new ideas in each of the two fields, while also highlighting their shared goals and providing new opportunities for collaborations.

FEHDA 2025: 1st Workshop on Fairness Exploration in Heterogeneous Data and Algorithms, chaired by Beatrice Amico, Anna Dalla Vecchia, and Maria Stratigi. Fairness is a complex and multifaceted concept that has garnered increasing attention in many computer science research areas. Defining and measuring fairness is often context-dependent, as different applications may require tailored approaches to address specific ethical concerns. Understanding the relationships between involved groups can be facilitated by defining an ontology or identifying key properties. However, achieving fairness demands ongoing evaluations, transparent decision-making, and adaptability to evolving contexts. A key challenge in achieving fairness is managing heterogeneous data from diverse sources, formats, and structures. Another important aspect is ensuring fairness through the careful design and evaluation of algorithms at all stages—pre-processing, modeling, and post-processing—to mitigate biases and promote equity in data-driven systems. We welcomed articles with a particular focus on descriptive ontologies, fairness metrics, and properties, as well as frameworks for managing data heterogeneity to ensure fairness, and fairness-aware algorithms. We were also open to discussing other fair-related topics that were not explicitly mentioned.

IT4TOCI 2025: 1st Workshop on Information Technology for Tourism and Culture Industries, chaired by Alberto Belussi and Sara Migliorini. IT4TOCI was an interdisciplinary laboratory connecting the research and industry communities operating in tourism and cultural industries. It aimed to explore the role of new technologies and systems in the development and evolution of such sectors. Particular attention was also devoted to aspects related to innovations in the travel and hospitality domains, as well as the promotion of specific forms of tourism, like cultural and agricultural ones.

Acknowledgements. We would like to wholeheartedly thank all participants, authors, PC members, workshop organizers, session chairs, doctoral consortium chairs, workshop chairs, volunteers, and co-organizers for their contributions to making ADBIS

2025 a great success. We would also like to thank the ADBIS Steering Committee and all sponsors.

September 2025

Panos K. Chrysanthis
Kjetil Nørvåg
Kostas Stefanidis
Zheying Zhang
Elisa Quintarelli
Ester Zumpano

Organization

General Chairs

Kostas Stefanidis   Tampere University, Finland
Zheying Zhang   Tampere University, Finland

Program Chairs

Panos Chrysanthis   University of Pittsburgh, USA
Kjetil Nørvåg   NTNU, Norway

Demonstration Chairs

Johann Eder   University of Klagenfurt, Austria
Dimitris Sacharidis   Université Libre de Bruxelles, Belgium

Tutorial Chairs

Georgia Koloniari   University of Macedonia, Greece
Robert Wrembel   Poznań University of Technology, Poland

Panel Chair

Matthias Böhm   Technische Universität Berlin, Germany

Workshop Chairs

Elisa Quintarelli   Università degli Studi di Verona, Italy
Ester Zumpano   University of Calabria, Italy

Doctoral Consortium Chairs

Enrico Gallinucci   University of Bologna, Italy
Genoveva Vargas-Solar   French Council of Scientific Research, France

Diversity and Inclusion Chairs

Panagiota Fatourou   University of Crete, Greece
Genoveva Vargas-Solar   French Council of Scientific Research, France

Publicity Chairs

Barbara Catania   Università degli studi di Genova, Italy
Javier A. Espinosa Oviedo   University of Lyon, France

Sponsorship Chairs

Tony Lindgren   Stockholm University, Sweden
George Papadakis   National and Kapodistrian University of Athens,
                                 Greece
Jaakko Peltonen   Tampere University, Finland

Special Issues Chair

Mirjana Ivanovic   University of Novi Sad, Serbia

Proceedings Chair

Vasilis Efthymiou   Harokopio University of Athens, Greece

Web Chair

Emilia Lenzi   Politecnico di Milano, Italy

Volunteers Chairs

Maria Stratigi	Tampere University, Finland
Toni Taipalus	Tampere University, Finland

Program Committee

Alberto Abella	Universitat Politècnica de Catalunya, Spain
Maribel Acosta	Université Libre de Bruxelles, Belgium
Cristina D. Aguiar	Universidade de São Paulo, Brazil
Bernd Amann	Sorbonne University, CNRS, France
Witold Andrzejewski	Poznań University of Technology, Poland
Sylvio Barbon Junior	University of Trieste, Italy
Nelly Barret	University of Novi Sad, Serbia
Andreas Behrend	University of Bonn, Germany
Ladjel Bellatreche	Universidad de Alicante, Spain
Sandro Bimonte	INRAE, France
Paweł Boiński	University of Orléans, France
Drazen R. Brdjanin	University of Banja Luka, Bosnia and Herzegovina
Richard Chbeir	Université de Pau et des Pays de l'Adour, France
Silvia Chiusano	Politecnico di Torino, Italy
Antonio Corral	University of Almería, Spain
Jerome Darmont	Université Lumière Lyon 2, France
Anton Dignös	Free University of Bozen-Bolzano, Italy
Christos Doulkeridis	University of Piraeus, Greece
Johann Eder	Université Lumière Lyon 2, France
Markus Endres	Technische Universität München, Germany
Javier A. Espinosa-Oviedo	Charles University, Czech Republic
Georgios Evangelidis	University of Macedonia, Greece
Bentayeb Fadila	Université Lumière Lyon 2, France
George Fakas	Uppsala University, Sweden
Johann Gamper	Universität Klagenfurt, Austria
Matteo Golfarelli	University of Augsburg, Germany
Vincenzo Gulisano	Chalmers University of Technology, Sweden
Herodotos Herodotou	Cyprus University of Technology, Cyprus
Tomas Horvath	Eötvös Loránd University, Hungary
Mirjana Ivanovic	Université Paris-Dauphine – PSL, France
Stefan Jablonski	University of Bayreuth, Germany
Pokorny Jaroslav	Universitat Politècnica de Catalunya, Spain
Petar Jovanovic	Poznań University of Technology, Poland

Alexandros Karakasidis	University of Macedonia, Greece
Zoubida Kedad	University of Versailles, France
Georgia Koloniari	University of Macedonia, Greece
Georgia Koutrika	Athena Research Center, Greece
Yannis Manolopoulos	University of Nicosia, Cyprus
Maude Manouvrier	University of Bologna, Italy
Patrick Marcel	University of Ioannina, Greece
Jose-Norberto Mazon	Free University of Bozen-Bolzano, Italy
Sara Migliorini	University of Verona, Italy
Alex Mircoli	Università Politecnica delle Marche, Italy
Angelo Montanari	University of Udine, Italy
Boris Novikov	Unaffiliated, Finland
Boussaid Omar	Université Lumière Lyon 2, France
George Palis	University of Cyprus, Cyprus
George Papastefanatos	Athena Research Center, Greece
Veronika Peralta	University of Tours, France
Giuseppe Polese	University of Salerno, Italy
Elisa Quintarelli	Università degli Studi di Verona, Italy
Franck Ravat	Institut de Recherche en Informatique de Toulouse, France
Gunter Saake	University of Magdeburg, Germany
Mahmoud A. Sakr	ISAE-ENSMA, France
Ingo Schmitt	Brandenburg University of Technology, Germany
Sergey A. Stupnikov	Russian Academy of Sciences, Russia
Bernhard Thalheim	Christian Albrecht University of Kiel, Germany
Goce Trajcevski	Iowa State University, USA
Panos Vassiliadis	Politecnico di Milano, Italy
Yannis Velegrakis	Utrecht University, Netherlands
Goran Velinov	Ss. Cyril and Methodius University in Skopje, North Macedonia
Szymon Wilk	Poznań University of Technology, Poland
Robert Wrembel	Poznań University of Technology, Poland
Demetrios Zeinalipour-Yazti	University of Cyprus, Cyprus
Yongluan Zhou	University of Copenhagen, Denmark
Ester Zumpano	University of Calabria, Italy

Steering Committee Chair

| Yannis Manolopoulos | University of Nicosia, Cyprus |

Steering Committee

Andreas Behrend	TH Köln, Germany
Ladjel Bellatreche	ENSMA Poitiers, France
Maria Bielikova	Kempelen Institute of Intelligent Technologies, Slovakia
Barbara Catania	University of Genoa, Italy
Tania Cerquitelli	Politecnico di Torino, Italy
Richard Chbeir	IUT de Bayonne et du Pays Basque, France
Silvia Chiusano	Politecnico di Torino, Italy
Jérôme Darmont	University of Lyon 2, France
Johann Eder	Alpen-Adria-Universität Klagenfurt, Austria
Johann Gamper	Free University of Bozen-Bolzano, Italy
Tomáš Horváth	Pavol Jozef Šafárik University, Slovakia
Mirjana Ivanović	University of Novi Sad, Serbia
Manuk Manukyan	Yerevan State University, Armenia
Kjetil Nørvåg	Norwegian University of Science and Technology, Norway
Boris Novikov	Unaffiliated, Finland
George Papadopoulos	University of Cyprus, Cyprus
Jaroslav Pokorny	Charles University in Prague, Czech Republic
Oscar Romero	Universitat Politècnica de Catalunya - BarcelonaTech, Spain
Sergey Stupnikov	Russian Academy of Sciences, Russia
Bernhard Thalheim	Christian Albrechts University of Kiel, Germany
Goce Trajcevski	Iowa State University, USA
Valentino Vranić	Pan-European University, Slovakia
Tatjana Welzer	University of Maribor, Slovenia
Genoveva Vargas-Solar	French Council of Scientific Research, France
Robert Wrembel	Poznań University of Technology, Poland
Ester Zumpano	University of Calabria, Italy

Contents

**MADEISD 2025: 7th Workshop on Modern Approaches in Data
Engineering and Information System Design**

ERGA 2025: 1st Workshop on Entity Resolution and Graph Alignment

FEHDA 2025: 1st Workshop on Fairness Exploration in Heterogeneous Data and Algorithms

IT4TOCI 2025: 1st Workshop on Information Technology for Tourism and Culture Industries

Query Optimization

iSearch: Seek Acceleration Through Interpolation in Smart Storage Settings

Christian Knödler[1]([✉]), Arthur Bernhardt[1], Naeem Ramzan[2], and Ilia Petrov[1]

[1] Reutlingen University, Reutlingen, Germany
{christian.knoedler,arthur.bernhardt,
ilia.petrov}@reutlingen-university.de
[2] University of the West of Scotland, Paisley, Scotland
naeem.ramzan@uws.ac.uk

Abstract. Modern database management systems (DBMS) face significant challenges when executing analytical tasks on exponentially growing datasets, often relying on search operation for key lookup. Traditional optimization methods focus on minimizing execution times in host-based systems. In contrast, smart storage devices enable offloading of query plan execution on-device, presenting opportunities for new optimizations. However, these devices operate under strict computational constraints and necessitate efficient resource management. Prevailing DBMS implementations predominantly employ binary search, because of its performance and robustness. In contrast, interpolation search algorithms yield considerable computational savings in smart storage settings, however they are not always robust. In this paper, we propose a novel adaptive search algorithm iSearch, which combines a configurable number of interpolation search iterations with a fallback to binary search. This hybrid approach ensures robust and predictable runtime performance, regardless of the underlying data distribution. We further demonstrate that commodity consumer devices benefit more from adaptive search approaches than traditional host systems, highlighting the potential for improved performance and efficiency in both contexts.

Keywords: Near Data Processing · Binary Search · Interpolation Search · iSearch · Robustness

1 Introduction

Motivation. Modern data-intensive systems face a growing number of queries, many targeting specific values in large datasets. As such applications proliferate, point queries for individual records have increased significantly. At the same time, volumes managed by DBMS are growing exponentially [3,27]. These datasets often lack data locality [10,13], causing excessive data transfers during query execution. These trends are pronounced in modern persistent (LSM-tree-based [18]) key/value (KV) stores like RocksDB [6] or LevelDB [9].

© The Author(s), under exclusive license to Springer Nature Switzerland AG 2026
P. K. Chrysanthis et al. (Eds.): ADBIS 2025, CCIS 2676, pp. 3–13, 2026.
https://doi.org/10.1007/978-3-032-05727-3_1

Combined with the growing complexity of analytical workload, DBMS must perform millions of individual key searches per query. Despite advances in storage and processing technologies, modern DBMS still rely on traditional algorithms and data structures, increasingly limiting their ability to fully exploit new hardware capabilities. One such innovation are intelligent memories [4] and computational/smart storage devices [19–23,34], which enable near-data processing (NDP), allowing operation offloading [2,8,11,13,14,26,29,30,32,33], and partial [14] or full execution of query plans [5,13,14] on the device, close to physical storage. However, intelligent memories/storage employ weak computational capabilities, to keep the price/GB low and align with the commodity model under which memory and storage are sold. As a result, efficient on-device processing becomes essential in such resource-constrained environments. Pre-filtering of data and execution of complex operations on the storage device are now technically feasible. However, this requires the ability to perform low-overhead *search operations*. While adaptive search algorithms [1,12,16,17,28] have shown performance benefits in host-based - especially memory-based - environments, their adoption in LSM-based [18] DBMS remains limited. In the context of smart storage, where compute and memory resources are scarce, adaptive searches offer a compelling solution. Their ability to dynamically adjust to access patterns and data distributions makes them particularly well-suited for on-device processing. Efficient key lookup is a central operation in many query execution plans, and improving this aspect can significantly enhance the overall system performance.

Brief State of the Art Overview. *Binary search* is a widely used algorithm for locating a target key in sorted data. It is fast and robust, with average and worst-case complexity of $\mathcal{O}(log_2(N))$, requiring no knowledge of data distribution. However, this can lead to more comparisons than necessary, where informed methods could terminate earlier. *Interpolation Search,* unlike binary search, improves performance on uniformly distributed data by calculating the probe position (eq. (1)) based on the key's relative position within the dataset. This yields a best-case complexity of $\mathcal{O}(1)$ and an average-case of $\mathcal{O}(log_2(log_2(N)))$, potentially requiring only a single iteration and significantly reducing CPU time and memory accesses.

$$idx = \frac{(key_{target} - key_{left}) \cdot (idx_{right} - idx_{left})}{(key_{right} - key_{left})} + idx_{left} \qquad (1)$$

With non-uniform key distributions, interpolation search may degrade to sequential traversal, leading to $\mathcal{O}(N)$ runtime. While each interpolation step incurs higher CPU overhead than binary search, this is offset under favorable conditions by fewer iterations and comparisons, improving overall efficiency.

Problems. While adaptive search strategies offer notable performance gains for frequent search operations, they are mainly optimized for in-memory workloads and often do not generalize well to persistent storage systems, such as LSM-tree-based architectures. Their effectiveness heavily depends on data characteristics and key distribution, making it difficult to select an optimal strategy

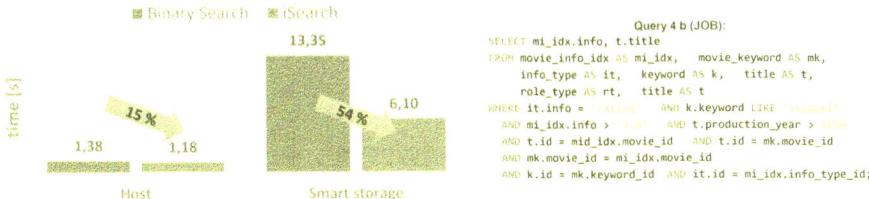

Fig. 1. Utilizing adaptive search strategies in host (left) compared to smart-storage settings (right) for query 4b (JOB). Interestingly, iSearch is more effective on weak smart storage compute elements, compared against binary search.

across varying workloads. In practice, doing so requires insights that may not always be available. Though host-based systems benefit from adaptive search, smart storage environments stand to gain even more due to limited compute and high I/O throughput. However, existing work has largely overlooked the use of adaptive search in near-data processing (NDP). This paper addresses that gap by exploring the application of adaptive search on smart storage devices, where performance benefits are even more significant.

Introductory Experiment. We demonstrate the benefits of adaptive search algorithms on smart storage devices versus host execution in a preliminary experiment. Query 4b of the Join Order Benchmark [15] was run on both host (Fig. 1, left) and NDP (Fig. 1, right) using binary search as a baseline and the proposed iSearch(Sect. 3). Results show a larger performance gain on smart storage (1.9×), though iSearch outperforms the baseline in both environments.

Our **contributions** are: (a) We introduce a configurable adaptive search algorithm for interpolation-search – iSearch; (b) We highlight the efficiency of adaptive search algorithms in smart-storage settings; (c) We demonstrate their usability and stable performance independent of the underlying key distribution.

2   Background and Related Work

Before introducing iSearch, we briefly review the data organization in LSM-based KV-stores, which are ideal candidates for iSearch and NDP due to their large use of sorted data.

nKV. In this paper, we use nKV [29–31], a key value store designed for near data processing (NDP) built on top of RocksDB and integrated as a MySQL storage engine via MyRocks [7]. MyRocks is a MySQL storage engine that combines RocksDB's LSM-tree-based architecture with MySQL's SQL capabilities. In RocksDB, data is stored as key-value pairs. Updates are first written to in-memory buffers (MemTables) and later merged into Sorted String Table (SST) files, consolidating many small random writes into fewer large sequential writes.

LSM Trees. Log-Structured Merge-Trees (LSM-Trees) [18] consist of an in-memory buffer (MemTable) and a set of immutable on-disk files (SST-files).

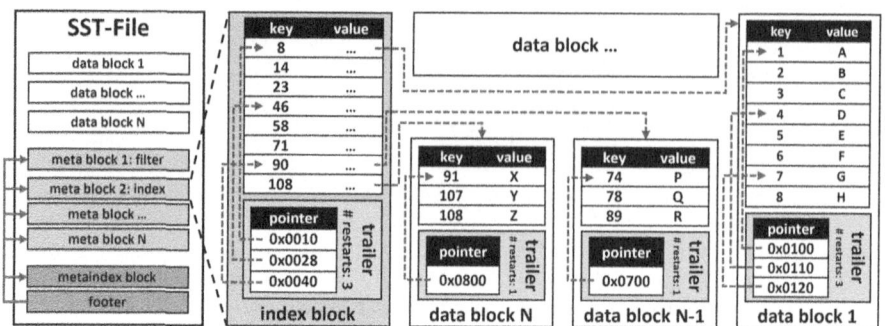

Fig. 2. The internal structure of an SSTable consists of immutable data blocks and accompanying meta-blocks. Meta-blocks include auxiliary structures such as Bloom filters for fast exclusion of absent keys and an index block that maps key ranges to data blocks. Both index and data blocks contain arrays of restart points–periodic offsets that facilitate efficient intra-block navigation.

Inserts and updates are first written to memory and flushed to storage once a threshold is reached. SST-files are organized hierarchically, with each level storing larger, merged datasets from the previous levels. This merging maintains sorted data and reduces the number of files accessed during reads.

Sorted String Tables (SST) are immutable on-disk structures used to store key-value entries in sorted order. Each SST file records its key range (minimum and maximum), allowing efficient exclusion during lookups. SSTs have a configurable size (typically 32 MiB) and consist of three main block types: data blocks, meta blocks, and metaindex blocks. A fixed-size footer at the end of the file contains pointers to the metaindex and index blocks, enabling efficient navigation within the file. Meta blocks store auxiliary data such as file properties, Bloom filters (enable fast checks for key absence, avoiding unnecessary accesses and filter for non-existent keys), and indexes. Data blocks, located near the beginning of the SSTable, contain the actual key-value pairs. Keys are stored in a *memcmp*-compatible format, enabling efficient byte-wise comparisons without conversion overhead during binary search. To accelerate intra-block navigation, data and index blocks include arrays of restart points–offsets serving as sparse indexes within the block. These are placed at configurable intervals to enable efficient access to individual entries. Figure 2 illustrates the internal structure of an SSTable, highlighting the key components essential for efficient key lookups.

Key lookup path on smart storage. During a key lookup (Fig. 3. ①), the active and immutable in-memory MemTables (Fig. 3. ②, ③) are traversed using a skiplist. If the key is not present, Bloom filters of the SST-files (Fig. 3. ④) are probed, excluding non-matching files. The remaining SSTables – along with their physical address ranges – are then dispatched to the computational device (Fig. 3. ⑤). On-device the fence-pointers (minimum and maximum key) are probed and the search algorithm filters for files possibly containing the key. The target SST file is read from flash (Fig. 3. ⑦) and the search algorithm probes the the restart array of the index block searching for a suitable data block (Fig. 3. ⑧). The data block is then loaded from flash (Fig. 3. ⑨), and the search algorithm is applied on the actual data within that block to locate the exact key-value pair. Finally, the position of the matching entry is returned. Whenever an interpolation-based search method is employed (outside of the in-memory skiplist), keys must be converted to the correct byte order prior to any arithmetic to ensure accurate position estimates.

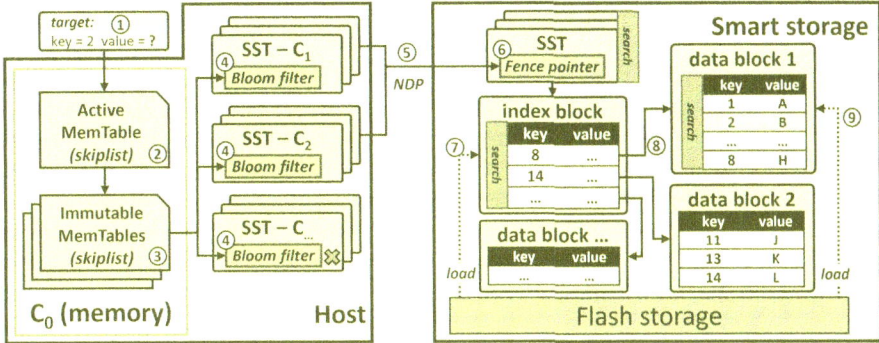

Fig. 3. Searching a key in a multilevel LSM tree using NDP. A lookup in the in-memory structures is performed before probing the SSTs on smart storage.

2.1 Adaptive Algorithms

We now briefly overview the existing interpolation search approaches.
SIP (Slope-reuse Interpolation) [28] enhances classical interpolation search by precomputing and reusing the slope, eliminating redundant calculations and reducing CPU overhead. It uses fixed-point arithmetic to minimize costly integer-to-float conversions and divisions. Additionally, SIP detects when interpolation degrades to sequential scanning and switches to linear search to maintain efficiency. **TIP** (Three Point Interpolation) [28] addresses expensive sequential interpolation by switching to linear search when needed. Unlike SIP, TIP fits non-uniform datasets by estimating the target index using three points, trading higher CPU cost for fewer repeated searches. Both are optimized for *specific* key distributions and require an *a priori* choice of the appropriate algorithm.

3 iSearch

We now present a simple yet effective method that combines traditional binary search with interpolation search advantages in smart storage, outperforming conventional host execution.

Algorithm 1. iSearch-Algorithm (iSearch)

1: **procedure** SEEK($data, left, right, X, rep$)
2:     **data** denotes the array of keys in which the search is performed
3:     **left, right** are the index of the left- and rightmost element of the array
4:     **X** is the key to search for
5:     **rep** is the number of interpolation repetitions

6:     **repeat**
7:         $estimate \leftarrow \left(\frac{(x - data[left]) \cdot (right - left)}{data[right] - data[left]} \right) + left$
8:         **if** $data[estimate] = X$ **then return** $data[estimate]$
9:         **else if** $data[estimate] < X$ **then** $left \leftarrow estimate + 1$
10:        **else** $right \leftarrow estimate - 1$
11:        **end if**
12:        $rep \leftarrow rep - 1$
13:     **until** $rep > 0$ and $left < right$
14:     **while** $left < right$ **do**
15:        $estimate \leftarrow \frac{left + right + 1}{2}$
16:        **if** $data[estimate] < X$ **then** $left \leftarrow estimate$
17:        **else if** $data[estimate] > X$ **then** $right \leftarrow estimate - 1$
18:        **else**
19:            $right \leftarrow estimate$
20:            $left \leftarrow right$
21:        **end if**
22:     **end while**
23:     **return** $data[left]$
24: **end procedure**

Reducing memory accesses and CPU cycles is crucial for search performance. Interpolation search can minimize both but may incur longer runtimes and higher compute costs in some cases. Even adaptive algorithms risk long executions without fine-tuning. We introduce iSearch, a configurable interpolation search with a fallback to binary search. It performs a set number of interpolation steps before switching to binary search if the key remains unfound. Algorithm 1 shows the pseudocode, with an example in Fig. 4. Initially, a standard interpolation search runs (lines 613). To limit attempts, a parameter *rep* sets the maximum interpolations and is decremented each step. If the key isn't found within *rep* tries, the algorithm switches to a conventional binary search (lines 1422), continuing with the current search boundaries (*left* and *right*). Ultimately, the desired key is returned. The algorithm features a configurable parameter *rep* controlling the number of interpolation iterations (lines 5, 6 and 13). Fewer repetitions make it behave more like binary search, while more emphasize interpolation. This tunable balance allows runtime adaptation to data distribution. Setting *rep* to zero yields a binary-only mode; setting it to the maximum results in interpolation-only mode. This flexibility suits environments where dataset characteristics can be approximated by simple statistical moments.

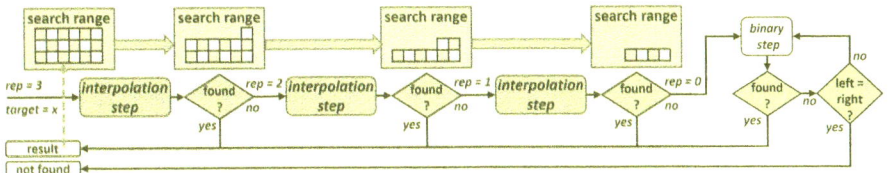

Fig. 4. iSearchperforms three successive interpolation steps over the search range before reverting to binary search, progressively narrowing the interval until the target key is located.

4   Experimental Evaluation

Experimental Setup. The experiments run on a Debian 4.9 server with a quad-core Intel i5 at 3.4 GHz, 6 MB L3 cache, and 4 GB RAM. For smart storage, we use the consumer-grade *COSMOS+* platform [21] via PCIe 2.0 ×8. It features a Xilinx Zynq 7045 SoC with FPGA logic and dual ARM Cortex-A9 cores at 667 MHz. Storage is a 1 TB MLC flash module configured in SLC mode. Testing occurs in two environments: in-memory (using the Revenge benchmark [28] with synthetic data) and LSM-treebased analytical workloads via the Join Order Benchmark (JOB) on MySQL 5.6 over nKV. Host code is in C++, compiled with GCC 6.3.0. Smart storage experiments run on *COSMOS+*, which supports full query execution and isolated operations like seek on embedded cores. All adaptive algorithms leverage index-block restart points; SIP and TIP perform sequential scans after these, while iSearchaccesses fixed-size data blocks directly, eliminating sequential scans.

Baselines. Our evaluation compares iSearchagainst conventional binary search (baseline) and two advanced interpolation methods – SIP and TIP by Van Sandt et al. [28]. All algorithms are tested in the in-memory benchmark and the LSM-based JOB benchmark, under both host and smart-storage (NDP) settings.

Workloads. We evaluate search algorithms using a subset of Join Order Benchmark (JOB) queries on both host and smart-storage devices. Due to implementation constraints, character attributes are adapted to fixed-length byte representations via padding or truncation to meet the *COSMOS+* board's 4-byte alignment requirements. JOB performance is measured by end-to-end query execution time; in-memory tests follow Van Sandt et al. [28] using internal timing mechanisms. The in-memory experiment uses datasets of 10^5 records with uniform (uniform and gap with 0.4 sparsity) and non-uniform (fal, Zipfian-like, and cfal, its cumulative variant) distributions, controlled by a shape parameter of 1.05. All datasets use a fixed seed (42) for reproducibility. Payload sizes of 8, 32, and 128 bytes were tested. In-memory benchmarks pre-access each key to ensure data is fully loaded into memory before measurement.

Experiment 1: Performance comparison of binary search and adaptive algorithms using the Join Order Benchmark. In our first experiment, we compare binary search baseline against iSearch, SIP, and TIP to quantify performance gains from offloading adaptive search to smart storage versus host

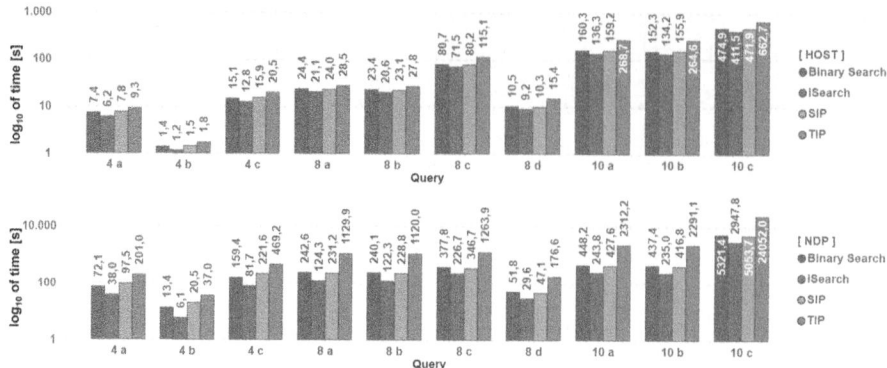

Fig. 5. Comparison of SIP, TIP, iSearchagainst binary search in host (top) and smart storage (bottom) environments using the JOB dataset. iSearchconsistently outperforms both baseline and adaptive methods, delivering better host performance and notably higher efficiency on smart storage.

execution. Figure 5 presents execution times for selected JOB queries across algorithms. The top panel shows host results; the bottom, smart-storage results. In both cases, iSearchoutperforms binary search, SIP, and TIP, with its advantage especially pronounced on smart storage.

On the host system, iSearchachieves an average speedup of 1.16× over conventional binary search across all considered queries (JOB 1a17e) with per-query gains from 0.87× to 2.14×. On the smart storage device, iSearchattains an average speedup of 1.42× versus binary search, ranging from 1.00× to 2.19× per query. SIP and TIP instead perform slightly worse than binary search and iSearchacross all queries. Insight: The results highlight the advantages of adaptive search strategies in host and near-data processing, where resource constraints and data locality amplify the need for efficient algorithms. Offloading to smart storage with iSearchyields significantly greater gains than host execution alone.

Experiment 2: Determining the optimal number of interpolation repeats for iSearch. In our second experiment, we determine the optimal number of interpolation repeats for iSearch. Using the Revenge benchmark [28], we test iSearchacross synthetic datasets (uniform, gap, fal, cfal), varying the number of repeats (Fig. 6) and measuring performance. Results show uniform data favors 45 repeats, while non-uniform datasets perform best with a single interpolation repeat. Overall, the optimal balance between interpolation and binary search is reached at about three interpolation steps.

Insight: Opting up to three interpolation iterations before falling back to binary search provides a robust performance trade-off across different data distributions.

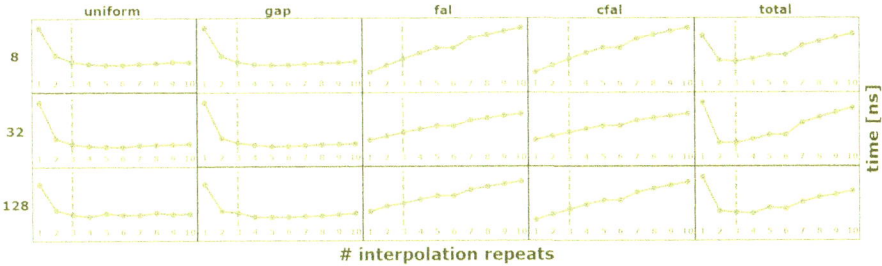

Fig. 6. Performance of iSearch(y-axis) in-memory versus interpolation repeats (x-axis). Columns show key distributions (*uniform, gap, fal, cfal,* and total); rows vary value sizes in bytes. Uniform distributions peak around five repeats, while skewed ones (fal, cfal) favor fewer repeats (typically one). Three repeats offer a balanced trade-off across all distributions.

5   Discussion

With many specialized search algorithms targeting specific workloads, key distributions, and systems, no universal solution exists, making algorithm selection critical. iSearchtackles this by providing a fixed overall structure with dynamically adjustable internal parameters. Dataset characteristics can be captured via lightweight metrics like mean and standard deviation of keys $\sqrt{variance}/average$, computed incrementally during inserts or updates. Using these, thresholds enable selection between algorithms or, for iSearch, dynamic adjustment of interpolation steps – ranging from 0 (binary search) to a set maximum (interpolation-only), or any hybrid in between – without altering the algorithm.

6   Conclusion

This paper presents iSearch, a configurable adaptive algorithm designed to improve search performance over ordered key ranges. iSearchcombines interpolation steps with a binary search fallback, maintaining consistent performance across various LSM-tree workloads on both host and smart storage, without modifying its core structure. By default, it performs three interpolation steps. Our evaluation shows adaptive search algorithms yield greater gains on smart storage than on host CPUs. For JOB queries, iSearchachieves an average 1.42× speedup on smart storage versus a more modest 1.16× on the host.

Acknowledgements. We thank Daniel Stiefel and Benjamin Simonis for providing access to and insights from their Master's theses [24,25], which contributed valuable groundwork for this study. This work has been partially supported by *DFG neoDBMS.2 – 419942270* and *BMBF PANDAS – 01IS18081C/D*.

References

1. Bonasera, B., et al.: Adaptive search over sorted sets. J. Discrete Algorithms **30** (2015)
2. Cao, W., et al.: POLARDB meets computational storage: efficiently support analytical workloads in cloud-native relational database. In: Proceedings of FAST (2020)
3. Corporation C.: Cisco global cloud index: forecast and methodology (2016)
4. Devaux, F.: The true processing in memory accelerator. In: IEEE Hot Chips (2019)
5. Do, J., et al.: Query processing on smart SSDs. In: Proceedings of the SIGMOD (2013)
6. Facebook: Rocksdb a persistent key-value store (2022). http://rocksdb.org
7. Facebook Inc., MyRocks: transaction isolation in myrocks. https://github.com/facebook/mysql-5.6/wiki (2021)
8. Francisco, P.: The Netezza data appliance architecture: a platform for high performance data warehousing and analytics. IBM Redbooks (2011)
9. Ghemawat, S., Dean, J.: LevelDB. Open-Source Implementation (2022). https://github.com/google/leveldb
10. Huang, G., et al.: X-Engine: an optimized storage engine for large-scale e-commerce transaction processing. In: SIGMOD (2019)
11. István, Z., Sidler, D., Alonso, G.: Caribou: Intelligent Distributed Storage. In: Proceedings of the VLDB (2017)
12. Kaporis, A., et al.: Dynamic interpolation search revisited. Inf. Comput. **270**, 104465 (2020)
13. Kim, S., et al.: In-storage processing of database scans and joins. Inf. Sci. (2016)
14. Knöodler, C., Ramzan, N., Petrov, I.: hybridNDP: dynamic operation offloading and cooperative query execution in smart storage settings. In: Proceedings of the EDBT (2025)
15. Leis, V., et al.: How good are query optimizers, really?. In: Proceedings of the VLDB, vol. 9, issue (3) (2015)
16. Mehlhorn, K., Tsakalidis, A.: Dynamic interpolation search. J. ACM **40**(3) (1993)
17. Mohammed, A.S., et al.: Interpolated binary search: an efficient hybrid search algorithm on ordered datasets. IJEST **24**(5) (2021)
18. O'Neil, P., Cheng, E., Gawlick, D., O'Neil, E.: The log-structured merge-tree (LSM-tree). Acta Inform. (1996)
19. OpenSSD: Daisy OpenSSD FPGA platform (2023). https://www.crz-tech.com/crz/article/daisy/
20. OpenSSD: Daisy plus OpenSSD FPGA platform (2023). https://www.crz-tech.com/crz/article/DaisyPlus/
21. OpenSSD Project: COSMOS Project Documentation (2019). http://www.openssd-project.org/wiki/Cosmos_OpenSSD_Technical_Resources
22. Pitchumani, R., Kee, Y.S.: Hybrid data reliability for emerging key-value storage devices. In: Proceedings of the FAST, pp. 309–322. FAST'20 (2020)
23. ProdesignGmbH: prodesign HAWK Versal VC1902 Acceleration Card (2023). https://www.prodesign-fpga-acceleration.com/products/prodesign-hawk-vc1902-acceleration-card
24. Simonis, B.: Suchoptimierung in indexbasierten datenbanksystemen auf basis von apache derby. Reutlingen University, Master Theses (2017)
25. Stiefel, D.: Optimierung der indexsuche in datenbanksystemen. Reutlingen University, Master Theses (2014)

26. Subramaniam, M., et al.: A technical overview of the Oracle Exadata database machine and Exadata storage server (2013). https://www.oracle.com/technetwork/database/exadata/exadata-dbmachine-x4-twp-2076451.pdf
27. Szalay, A., Gray, J.: 2020 computing: science in an exponential world. Nature **440**, 413–414 (2006)
28. Van Sandt, P., Chronis, Y., Patel, J.M.: Efficiently searching in-memory sorted arrays: revenge of the interpolation search?. In: SIGMOD '19, Association for Computing Machinery, pp. 36–53. New York, NY, USA (2019)
29. Vinçon, T., et al.: Near-data processing in database systems on native computational storage under HTAP workloads. Proc. VLDB Endow. **15**(10) (2022)
30. Vincon, T., Weber, L., Bernhardt, A., Koch, A., Petrov, I.: nKV: near-data processing with KV-stores on native computational storage. In: Proceedings of the DaMoN (2020)
31. Vincon, T., et al.: nKV in Action: accelerating KV-stores on native computational storage with near-data processing. PVLDB **12** (2020)
32. Woods, L., István, Z., Alonso, G.: Ibex: an intelligent storage engine with support for advanced SQL offloading. Proc. VLDB (2014)
33. Woods, L., Teubner, J., Alonso, G.: Less watts, more performance: an intelligent storage engine for data appliances. In: Proceedings of the SIGMOD (2013)
34. Xilinx: SmartSSD computational storage drive (2021). https://www.xilinx.com/publications/product-briefs/xilinx-smartssd-computational-storage-drive-product-brief.pdf

Extending the Applicability of Bloom Filters by Relaxing Their Parameter Constraints

Paul Walther$^{(\boxtimes)}$ (ID), Wejdene Mansour (ID), Johann Maximilian Zollner (ID), and Martin Werner (ID)

TUM School of Engineering and Design, Technical University of Munich, Munich, Germany

{paul.walther,wejdene.mansour,maximilian.zollner,martin.werner}@tum.de

Abstract. Bloom filters serve as an efficient probabilistic data structure for representing sets of keys. They allow for set membership queries with no false negatives and with the right choice of the main parameters – length of the Bloom filter (BF), number of hash functions used to map an element to the array's indices, and the number of elements inserted – the false positive rate is optimized. However, the number of hash functions is constrained to integer values, and the length of a BF is usually chosen to be a power of two to allow for efficient modulo operations using binary arithmetic. In this paper, we relax these constraints by proposing the Rational Bloom filter, which allows for non-integer numbers of hash functions. This results in optimized fraction-of-zero values for a known number of elements to be inserted. We further enhance this with the Variably-Sized Block BF to allow for a flexible filter length, especially for large filters, with efficient computation.

Keywords: Bloom Filter · Key-Value-Store · Filter Length · Hash Function

1 Introduction

Key-value stores proved themselves as an efficient storage model to support all steps in the data life cycle [9,27]. Since large, sparse, and low-cardinality data can be modelled as a set and the access is often random, the requirements for an efficient in-memory representation are similar to those of a Key-Value Store. In this context, the Bloom filter (BF) was proposed as a data structure since it allows for a trade-off between its memory footprint and its error rate [19,26].

The BF is a data structure for the efficient probabilistic storage of sets [4]. It is a binary array with methods for storing information in the filter and querying for set membership. An empty BF is an all-zero bit array of length m. Due to its structure, the BF guarantees `true` for inserted items (no false negatives) but may falsely report uninserted items to be `true` (false positives) [7,26]. To *insert* an element x, the element is hashed with k uniformly distributed pairwise

P. K. Chrysanthis et al. (Eds.): ADBIS 2025, CCIS 2676, pp. 14–23, 2026.
https://doi.org/10.1007/978-3-032-05727-3_2

independent hash functions H_i, with $i \in \{1, \ldots, k\}$. The hashing maps from the input space of all elements (universe) to the integer range $\{0, \ldots, m-1\}$. The BF is then set to 1 at the locations denoted by the same hash functions H_i. If the given value is already 1, it remains unchanged. For the *membership query*, the BF checks the k indices computed by H_i. If any of the denoted values is 0, the element is not part of the stored set. Otherwise, if all values are 1, the element is present, or it is a false positive error [14]. Given n elements to store, the false positive rate p_{FP} of the BF can be calculated as [21]:

$$p_{FP} = \left(1 - \left(1 - \frac{1}{m}\right)^{kn}\right)^k \approx \left(1 - e^{-kn/m}\right)^k. \tag{1}$$

The optimal number of hash functions k^* minimizes p_{FP} with a fraction of zeros $foz = \frac{1}{2}$. This maximizes the entropy of the filter and holds the highest information density, yielding

$$k^* = \frac{m}{n} \ln 2. \tag{2}$$

Nayak and Patgiri explain five main challenges with existing BF approaches: to *reduce the* p_{FP}, the *length adaption* of BFs to initially unknown dataset sizes, the *deletion of elements* without recalculation of the whole index, to implement an *efficient hashing* method without negatively influencing the performance, and the *correct determination* of k [22]. An approach to these challenges is the relaxation of the constraints on the parameters m, k, and n.

In the literature, k is restricted to integers. This reduces flexibility in constructing the optimal filter for given m and n, and especially for small k, this may result in a non-optimal p_{FP}. Figure 1 illustrates how an unconstrained, optimal choice of k can lead to improved p_{FP}. Furthermore, p_{FP} is unevenly distributed across elements in multi-hash BFs [1]. As for hash collisions, similar slots denoted by different hash functions for one element increase false positives, and to avoid this effect, using fewer hash functions is beneficial.

General hash functions h_i have to be modified to specific hash functions H_i, mapping the input space to indices of the BF $\{0, \ldots, m-1\}$. The simplest method is to take the modulo $h_i \% m$. It is most efficient for $m = 2^c, c \in \mathbb{N}_0$ as it simplifies to a binary AND ($\wedge$) operation [10]:

$$H_i(x) = h_i(x) \mod m = h_i(x) \wedge (2^c - 1) \tag{3}$$

Still, in computer systems, which allocate memory in power of two increments, this restriction may have downsides if, for example, some space is already used by the operating system. Then, the BF may have at most half of the available memory size. Consequently, freely choosing m might improve the scalability and while keeping the efficient modulo operations [1,10].

In this paper, we propose methods to extend the applicability of Bloom filters based on visions from [25]. First, we introduce a method for a **rational number instead of only an integer number of hash functions** k, which decreases false positive rates if the set is immutable, as it allows for more tailored BFs. Further, we propose a method to choose **non-power-of-two sizes** m **for the BF**

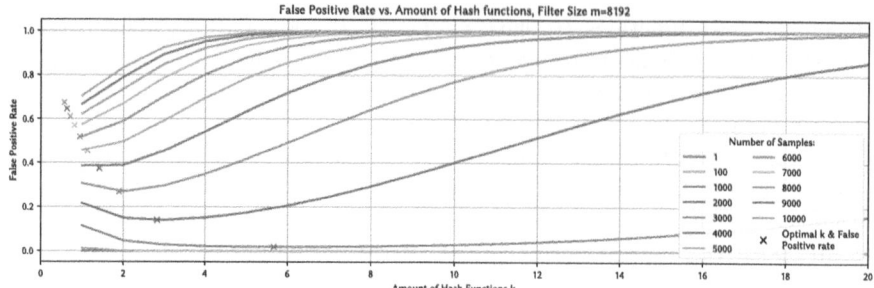

Fig. 1. Dependency of the false positive rate for a fixed filter size of $m = 8192$ on varying numbers of hash functions k and number of samples. The calculated optimal configuration for each number of samples is marked with a cross.

without performance degradation (by non-uniformity of access) and still without explicit modulo computations.

2   Rational Bloom Filter

For BFs, the theoretically optimal number of hash functions k^*, denoted by Eq. 2, is generally not an integer but a rational number. Still, traditional BFs allow for an integer number of hash functions only, requiring an approximation of the optimal k^*. Allowing a non-integer k^* for a BF with known n and a constrained m (e.g., by hardware) poses the advantage to improve the false positive rate and allows for more flexibility in selecting other BF parameters. Therefore, we propose a probabilistic approach to realize rational k^*.

Definition 1. *A hash function is* **probabilistically activated** *if it is not applied to every sample, but instead only activated with a probability of activation* $0 \leq p_{activation} \leq 1, p_{activation} \in \mathbb{R}$.

Based on this and the knowledge that the applied hash functions H_r with $r < \lfloor k \rfloor$ give pseudo-random information which is still deterministic for a given input sample, we develop the Rational Bloom filter (RBF), which is visualized in Fig. 2:

Definition 2. *A* **Rational Bloom filter (RBF)** *is a BF with a non-integer number* $k \in \mathbb{R}^+$ *of hash functions, consisting of standard BF procedures for* $\lfloor k \rfloor$ *hash functions, while a hash function with probabilistic activation represents the non-integer part* $k - \lfloor k \rfloor$.

This can be implemented as described in Algorithm 1, including the Hashing Trick [18].

Lemma 1. *The RBF has no false negatives.*

Algorithm 1: Application of Hash Functions in the RBF

Data: Element x, BF BF with length m, set of always-applied hash functions $\{H_1, \ldots, H_{\lfloor k \rfloor}\}$, probabilistically activated hash function $H_{\lfloor k \rfloor + 1}$, rational number of hash functions k;

Result: Set BF bits for indices denoted by the rational number of hash values of input element x

1 **for** *each H_j in* $\{H_1, \ldots, H_{\lfloor k \rfloor}\}$ **do**
2 $\quad$ Set $BF[H_j(x)] \leftarrow 1$;
3 Set $p_{activation} = k - \lfloor k \rfloor$;
4 Random hash value $H_r(x) \in [0, m]$; // No additional calculation needed if we, e.g., choose $H_r(x) = H_{\lfloor k \rfloor}(x)$;
5 **if** $H_r(x) < (p_{activation} \cdot m)$ **then**
6 $\quad$ Set $BF[H_{\lfloor k \rfloor + 1}(x)] \leftarrow 1$;

Proof. The given RBF with k normal and one probabilistically activated hash function $H_{\lfloor k \rfloor + 1}$ has at least as many 1-bits in the filter as the normal BF with $\lfloor k \rfloor$ hash functions since $H_{\lfloor k \rfloor + 1}$ can only add 1's. A false negative would require a query to expect a 1 in the BF where none exists. For H_1 to $H_{\lfloor k \rfloor}$, this cannot be the case by definition of the standard BF. For $H_{\lfloor k \rfloor + 1}$, activation is deterministic per input, which makes an activation during query without activation during insertion impossible.

Theorem 1. *The false positive rate p_{FP}^{RBF} of the RBF is smaller or equal to the false positive rate p_{FP}^{BF} of a normal BF: $p_{FP}^{RBF} \leq p_{FP}^{BF}$*

Proof. As described in Eq. 2, k^* is solely determined by n and m with the assumption that the highest information can be stored for a fraction of zero $foz_{opt} = \frac{1}{2}$. For one standard BF with k_{BF} hash functions and one RBF, the RBF's k_{RBF} is chosen to be optimal, $k_{RBF} = k^*$. Based on the construction of k^* as a minima of the false positive rate p_{FP} and the monotonicity of this function we can conclude that $p_{FP}(k^*) < p_{FP}(k_{RBF}) \forall k_{RBF} \neq k^*$. For $k_{BF} = k_{RBF}$ the optimal k^* is an integer. Therefore, $BF = RBF$ and consequently $p_{FP}^{RBF} = p_{FP}^{BF}$.

Consequently, RBFs serve as a new possibility to allow for a more flexible choice of the number of hash functions. In the following we extend this idea to variable filter lengths.

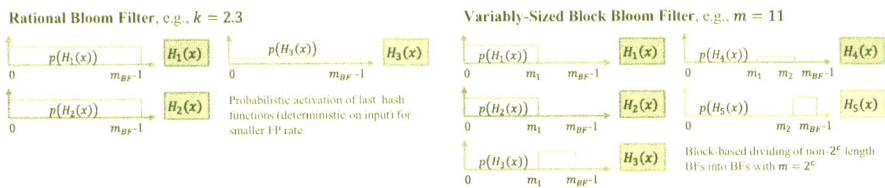

Fig. 2. Rational Bloom filter and Variably-Sized Block Bloom filter

3   Variably-Sized Block Bloom Filters

This can be used to efficiently compute hash functions for $m \neq 2^c, c \in \mathbb{N}_0$.

Definition 3. *The **Variably-Sized Block Bloom filter (VSBBF)** is a BF of length m_{BF}, with $2^c < m_{BF} < 2^{c+1}, c \in \mathbb{N}_0$, where the actual filter is subdivided into J blocks of sizes $m_j, j \in \{1, \ldots, J\}$ with $m_j > m_{j+1}$ and $m_j = 2^c, c \in \mathbb{N}$. Each block is denoted by a set of indices I^j and is filled by k_j subset hash functions H_j^i. $k_j \in \mathbb{R}$ is thereby the optimal k_j^* for the respective filter block.*

Unlike a standard BF, the VSBBF, visualized in Fig. 2, maps hashes to blocks of length 2^c using efficient modulo operations as described in Algorithm 2.

Algorithm 2: Inserting an Element in the VSBBF

> **Data**: Element x, total elements to insert n, VSBBF BF of total length m_{BF},
> uniform, pairwise independent hash functions h_1, h_2
> **Result**: Inserted element x into the VSBBF
> 1   binary $\leftarrow$ BinaryRepresentation(m_{BF});
> 2   length $\leftarrow$ Length(binary);
> 3   hvs $\leftarrow h_1(x), h_2(x)$;
> 4   offset $\leftarrow 0$;
> 5   **for** $j \leftarrow 0$ **to** *length - 1* **do**
> 6   **if** *binary[j] == 1* **then**
> 7 $m_j \leftarrow 2^{\text{length}-j-1}$;
> 8 $k_j \leftarrow \frac{m_j}{n} \cdot \ln 2$;
> 9   **for** $i \leftarrow 0$ **to** k_j **do**
> 10 $H_i(x) = ((hvs_1 + (i + m_j) \cdot hvs_2)\&(m_j - 1)) + \text{offset}$;
> 11 $BF[H_i(x)] \leftarrow 1$;
> 12   offset $+= m_j$;

Lemma 2. *The VSBBF has no false negatives.*

Proof. By definition, the filter blocks are all either normal or RBFs, and thus, it follows from Lemma 1 that the Block BF has no false negatives.

Theorem 2. *The false positive rate of VSBBF is calculated with the chain rule* $p_{FP}^{Block} = \prod_j p_{FP}^j \approx \prod_j \left(1 - e^{-k_j n / m_j}\right)^{k_j}$.

Proof. For several BFs representing the same set, which is similar to hash functions only mapping to subsets of the BF, deciding whether a queried sample is in the desired set is always requires checking all applied hash functions. An item can only be false positive if these all point wrongly to a 1 element. This is equal to all subset BFs wrongly denoting the element to be in the set.

Corollary 1. *For the optimal choice of $k_j = k_j^*$, the false positive rate of the combined filters of length $m_j = 2^{c_j}$ stays the same as the single BF of size m_{BF}:*

$$p_{FP}^{Block} \approx \prod_j \left(1 - e^{-k_j n/m_j}\right)^{k_j} = \left(1 - e^{-kn/m_{BF}}\right)^k \approx p_{FP}^{BF}. \tag{4}$$

Proof. With Eq. 2 for optimal $k_j^* = \frac{m_j}{n} \ln 2$ and $k^* = \frac{m_{BF}}{n} \ln 2$:

$$\prod_j \left(1 - e^{-\ln 2}\right)^{\frac{m_j}{n} \ln 2} = \left(1 - e^{-\ln 2}\right)^{\frac{m_{BF}}{n} \ln 2} \tag{5}$$

and as $\sum_j m_j = m_{BF}$ by construction, for the choice of optimal k_j^* it holds $p_{FP}^{Block} = p_{FP}^{BF}$ as we can assume optimality from RBF paradigms.

Theorem 3. *The Block BF improves the equal distribution of false positives over the to-be-tested elements compared to a standard BF of the same size.*

Proof. An element is more likely to be a false positive if its footprint has fewer 1's due to clashing hash values, increasing the risk of overlap with existing elements in the BF. For a fully filled filter with $foz = 0.5$ the false positive rate is $p_{FP} = 0.5^k$ and for o clashing hash functions it increases to $p_{FP}^{clash} = 0.5^{(k-o)} > p_{FP}$.

For a given Block BF, the sum of applied hash functions $\sum_j k_j^*$ is equal to the optimal number k^* for a filter of the summed lengths of the blocks $m = \sum_j m_j$.

The probabilities for one element x being inserted into a filter BF with k^* hash functions having less than k^* hash values are

$$p_{clash} = \sum_{i=1}^{k-1} \frac{i}{m_{BF}}, \qquad p_{clash}^{Block} = \sum_j \sum_{i=1}^{k_j-1} \frac{i}{m_j}. \tag{6}$$

After reformulation, we can show for all $J > 1$ that $p_{clash} > p_{clash}^{Block}$.

Corollary 2. *The proposed solution requires no storage overhead for the description of the block sizes m_j and the number of hash functions k_j.*

Proof. This information is contained in the non-blocked representation's properties, the assumption that blocks are sorted descendingly with $m_{j+1} < m_j$ and the rule that the optimal number of hash functions k_j^* is always chosen per block.

The proposed solution reduces processing time: $t(BF) \geq t(BF^{Block})$ for insertion and querying, when $m_{BF} \neq 2^c, c \in \mathbb{N}_0$ and $\sum_j m_j^{Block} = m_{BF}$ and the runtime of one modulo calculation by $m_{BF} \neq 2^c$ is computationally more expensive than the computation of J modulo operations with the binary bit trick [24].

A further advantage of Block BF is the easy and meaningful compression by simply taking subsets of the filter, which comprise a certain number of blocks. In tendency, this enables more compression steps than the standard BF of size $m_{BF} = 2^c$ as not only halving the size is possible but also any combination of block sizes applied in the BF. Available compression ratios are $\frac{\sum_i m_i}{m_{BF}}$, with $i \subseteq j$.

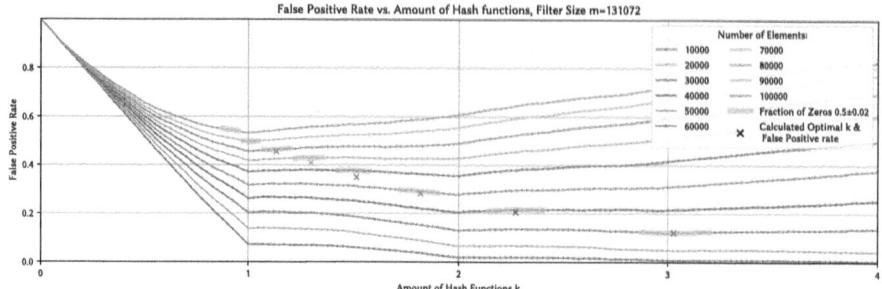

Fig. 3. False positive rate of RBFs with filter size 131,072 for different k and n, queried on 10,000 unseen elements. Experiments with $foz = 0.5 \pm 0.02$ are shadowed grey. Theoretically optimal configurations are denoted by a cross.

4 Implementation Artifacts

To validate our theories, we implemented RBF and VSBBF in C++ with a Murmur Hash using the hashing trick and efficient modulo calculation with binary arithmetic. As a baseline, we also implemented a standard BF with the same optimizations. Interestingly, *RBF* showed unexpected artifacts: The false positive rate has local minima not at the theoretically optimal rational k^*, but near integer values instead, as shown in Fig. 3. For example, with $m = 131,072$ and $n = 60,000$, the global minimum occurs at $k = 2$ and another local minimum at $k = 1$, although the optimal value lies in between.

This was counterintuitive, and we cannot yet fully explain it. We suspect this is due to the non-uniform behavior of the hashing algorithm. However, similar artifacts were also observed with SHA256, which should lead to a more uniform distribution of hash values. Additionally, larger filter sizes and numbers of elements inserted did not mitigate this, though logically, this should reduce the impact of non-uniform hash functions. Comparing the *foz* of the near-optimal RBF and BFs with integer k shows that RBF achieves better *foz* rates. For *VSBBF*, the false positive rates are similar to a standard BF, but no insertion time improvements were observed. Although our practical results fall short of the theoretical gains, we still believe our methods are a valuable contribution to BF theory and open new application possibilities, especially for learned BF approaches and input-dependent hash functions.

5 Related Work

To the best of our knowledge, no prior work addresses rational numbers of hash functions, though many improve the hashing: *Perfect hashing* uniformly maps inputs to m buckets [14]. Common hash functions like Murmur Hash [2] are used with consecutive modulo operations. For $k > 2$, *double hashing* calculates two hashes and combines them $h_i(x) = h_1(x) + f(i)h_2(x)$ [14,18]. *Partitioned hashing* [14] assigns each function disjoint ranges of length m/k, which may increase false positives due to more 1s. Hao et al. group inputs and apply *different uniform hash functions*, optimizing for the highest *foz* [17]. Bruck et al. allocates hash functions based on query likelihood [7]. In [27], *non-uniform hash functions* and float-based representations on GPUs come at the cost of higher memory and complexity; unlike our RBF, which only relaxes parameter constraints. Several BFs adapt their size to the number of elements. *Incremental BF* [15] sets a fill threshold for a minimum *foz* and adds new filters when reached. Similarly, [1] proposes adding *plain BFs* of increasing size $m_i = m_0 \cdot s^{l-1}$, applying a geometric progression on error bounds to keep false positives and elements-per-filter constant. *Slicing* assigns distinct BF regions per hash function to avoid overlap and achieves more uniform false positives [1,6,8]. *Dynamic BF* [13] supports deletions and is adopted from Counting BFs [5,12] and *Block BF* [23], which consists of multiple small cache-line-sized BFs and inserts each element in just one. Furthermore, *Combinatorial BF* and *Partitioned Combinatorial BF* [16] use a set of BFs when one element belongs to several sets and partition a BF into smaller ones of similar size. These approaches split the BF into smaller pieces for scalability or memory locality, but still use power-of-two sizes and similarly-sized filter subparts. Last, *learned BFs* mimic BFs with a learned function, allowing false negatives and using a small BF to filter them out [20]. Beyond direct improvements to BFs, *Quotient* [3] and *Cuckoo Filters* [11] enhance data locality and space efficiency, and support deletions without recalculations, often outperforming BFs. Still, BFs variants remain popular in data management systems for their simple architecture. Our approaches allow for improvements in these domains without a general change in system architecture.

6 Conclusion

By relaxing the classic constraints on Bloom filter parameters, such as requiring a natural number of hash functions and filter sizes that are powers of two, we present two new BF designs. The RBF uses a probabilistic activation of hash functions, which is still deterministic with respect to the inputs to mimic the theoretically optimal number of hash functions. We prove that this theoretically achieves superior false positive rates compared to standard BFs of the same size. Further, VSBBFs allow for variable filter sizes and efficient modulo operation, enabling a better distribution of false positives over all elements in the universe, as a clash of hash functions is less probable. The probabilistic activation of hash functions in RBFs opens new research directions, e.g., how learned approaches might use rational numbers of hash functions. This extends existing approaches that only rely on learned pre-filters to increase learned BF's efficiency.

Acknowledgements. This work is funded by the Deutsche Forschungsgemeinschaft (DFG, German Research Foundation) - 507196470.

References

1. Almeida, P.S., Baquero, C., Preguiça, N., Hutchison, D.: Scalable bloom filters. Inf. Process. Lett. **101**(6), 255–261 (2007). https://doi.org/10.1016/j.ipl.2006.10.007

2. Appleby, A.: aappleby/smhasher: MurmurHash: GitHub Repository (2008). https://github.com/aappleby/smhasher

3. Bender, M.A., et al.: Don't Thrash: how to Cache your Hash on Flash. In: Proceedings of the VLDB Endowment, vol. 5, issue 11, pp. 1627–1637 (2012). https://doi.org/10.14778/2350229.2350275

4. Bloom, B.H.: Space/time trade-offs in hash coding with allowable errors. Commun. ACM **13**(7), 422–426 (1970). https://doi.org/10.1145/362686.362692

5. Bonomi, F., Mitzenmacher, M., Panigrahy, R., Singh, S., Varghese, G.: An improved construction for counting bloom filters. In: Azar, Y., Erlebach, T. (eds.) ESA 2006. LNCS, vol. 4168, pp. 684–695. Springer, Heidelberg (2006). https://doi.org/10.1007/11841036_61

6. Bose, P., et al.: On the false-positive rate of Bloom filters. Inf. Process. Lett. **108**(4), 210–213 (2008). https://doi.org/10.1016/j.ipl.2008.05.018

7. Bruck, J., Gao, J., Jiang, A.: Weighted bloom filter. In: 2006 IEEE International Symposium on Information Theory, pp. 2304–2308. John Wiley (2006). https://doi.org/10.1109/ISIT.2006.261978

8. Chang, F., Feng, W.-C., Li, K.: Approximate caches for packet classification. In: IEEE INFOCOM 2004, pp. 2196–2207. IEEE (2004). https://doi.org/10.1109/infcom.2004.1354643

9. DeCandia, G., et al.: Dynamo: Amazon's highly available key-value store. ACM SIGOPS Operating Syst. Rev. **41**(6), 205–220 (2007). https://doi.org/10.1145/1323293.1294281

10. Estébanez, C., Saez, Y., Recio, G., Isasi, P.: Performance of the most common non-cryptographic hash functions. Softw. Pract. Exper. **44**(6), 681–698 (2014). https://doi.org/10.1002/spe.2179

11. Fan, B., Andersen, D.G., Kaminsky, M., Mitzenmacher, M.D.: Cuckoo filter. In: Proceedings of the 2014 CoNEXT: December 2-5, 2014, Sydney, Australia, pp. 75–88. ACM (2014). https://doi.org/10.1145/2674005.2674994

12. Fan, L., Cao, P., Almeida, J., Broder, A.Z.: Summary cache. ACM SIGCOMM Comput. Commun. Rev. **28**(4), 254–265 (1998). https://doi.org/10.1145/285243.285287

13. Guo, D., Wu, J., Chen, H., Yuan, Y., Luo, X.: The dynamic bloom filters. IEEE Trans. Knowl. Data Eng. **22**(1), 120–133 (2010). https://doi.org/10.1109/TKDE.2009.57

14. Gupta, D., Batra, S.: A short survey on bloom filter and its variants. In: 2017 International Conference on Computing, Communication and Automation (ICCCA), pp. 1086–1092 (2017). https://doi.org/10.1109/CCAA.2017.8229957

15. Hao, F., Kodialam, M., Lakshman, T.V.: Incremental Bloom Filters. In: INFOCOM 2008. The 27th Conference on Computer Communications. IEEE, pp. 1067–1075. IEEE Computer Society (2008). https://doi.org/10.1109/INFOCOM.2008.161

16. Hao, F., Kodialam, M., Lakshman, T.V., Song, H.: Fast multiset membership testing using combinatorial bloom filters. In: IEEE INFOCOM 2009, pp. 513–521 (2009). https://doi.org/10.1109/INFCOM.2009.5061957

17. Hao, F., Kodialam, M., Lakshman, T.V.: Building high accuracy bloom filters using partitioned hashing. In: Proceedings of the 2007 ACM SIGMETRICS, pp. 277–288. ACM Digital Library (2007). https://doi.org/10.1145/1254882.1254916

18. Kirsch, A., Mitzenmacher, M.: Less hashing, same performance: building a better bloom filter. In: Azar, Y., Erlebach, T. (eds.) ESA 2006. LNCS, vol. 4168, pp. 456–467. Springer, Heidelberg (2006). https://doi.org/10.1007/11841036_42

19. Lu, G., Nam, Y.J., Du, D.H.C.: BloomStore: bloom-filter based memory-efficient key-value store for indexing of data deduplication on flash. In: 2012 IEEE 28th Symposium on Mass Storage Systems and Technologies, pp. 1–11. IEEE (2012). https://doi.org/10.1109/MSST.2012.6232390

20. Mitzenmacher, M.: A model for learned bloom filters and optimizing by sandwiching. In: Advances in Neural Information Processing Systems, vol. 31. Curran Associates (2018). https://proceedings.neurips.cc/paper_files/paper/2018/file/0f49c89d1e7298bb9930789c8ed59d48-Paper.pdf

21. Mullin, J.K.: A second look at bloom filters. Commun. ACM **26**(8), 570–571 (1983). https://doi.org/10.1145/358161.358167

22. Nayak, S., Patgiri, R.: A review on role of bloom filter on DNA assembly. IEEE Access **7**, 66939–66954 (2019). https://doi.org/10.1109/ACCESS.2019.2910180

23. Putze, F., Sanders, P., Singler, J.: Cache-, hash-, and space-efficient bloom filters. ACM J. Exper. Algorithmics **14** (2009). https://doi.org/10.1145/1498698.1594230

24. Reed, I.: A class of multiple-error-correcting codes and the decoding scheme. Trans. IRE Prof. Group Inf. Theory **4**(4), 38–49 (1954). https://doi.org/10.1109/tit.1954.1057465

25. Walther, P.: Advancements of randomized data structures for geospatial data. In: Proceedings of the Workshops of the EDBT/ICDT 2024 (2024). https://ceur-ws.org/Vol-3651/PhDW-1.pdf

26. Werner, M.: GloBiMapsAI: an AI-enhanced probabilistic data structure for global raster datasets. ACM Trans. Spatial Algorithms Syst. **7**(4), 1–24 (2021). https://doi.org/10.1145/3453184

27. Werner, M., Schönfeld, M.: The Gaussian bloom filter. In: Renz, M., Shahabi, C., Zhou, X., Cheema, M.A. (eds.) DASFAA 2015. LNCS, vol. 9049, pp. 191–206. Springer, Cham (2015). https://doi.org/10.1007/978-3-319-18120-2_12

CoDD: A Constraint-Based Dataset Discovery Tool for Open Data Lakes

Tim Otto$^{(\boxtimes)}$, Christopher Rawald, and Stefan Deßloch

Heterogeneous Information Systems, RPTU Kaiserslautern-Landau, Kaiserslautern,
Germany
{tim.otto,stefan.dessloch}@cs.rptu.de, christopher.rawald@edu.rptu.de

Abstract. Data lakes offer the flexibility to store large volumes of heterogeneous data with minimal curation. However, this flexibility comes at a cost: traditional keyword-based dataset discovery methods require reliable metadata such as table names or column headers, and become ineffective when this metadata is either missing or incomplete. This issue is especially pronounced in open or poorly maintained data lakes, where the quality of metadata cannot be guaranteed. In this paper, we present CoDD, a system for constraint-based dataset discovery in open data lakes. Instead of querying metadata (query-by-metadata), CoDD profiles datasets by extracting structured facts directly from the data using modular, user-definable components. Users can perform query-by-constraint searches by specifying constraints over the profiled facts in an interactive, question-driven interface. Early results from our user study show that CoDD enables users to find relevant datasets when traditional keyword-based approaches fail due to insufficient or misleading metadata. Furthermore, CoDD performs comparably well even when accurate metadata is available, demonstrating that query-by-constraint is a robust and scalable alternative for dataset discovery in open data lake environments.

Keywords: Dataset Discovery · Open Data Lakes · Constraints

1 Introduction

The term *dataset discovery* in (open) data lakes broadly refers to the process of identifying relevant datasets from a repository that may contain millions of uncurated and unmaintained datasets. Based on the nature of the input, two primary approaches can be distinguished: keyword-based search, which relies on single keywords, and table-based search, which uses an example table to find similar datasets.

Keyword-based search can be further categorized into query-by-metadata and query-by-data. Query-by-metadata methods, such as those that search table or column names, are widely used. However, in many real-world scenarios, especially in open data lakes, metadata is often missing, inconsistent, or misleading [2]. Query-by-data approaches, which bypass metadata, use inverted indexes over

P. K. Chrysanthis et al. (Eds.): ADBIS 2025, CCIS 2676, pp. 24–31, 2026.
https://doi.org/10.1007/978-3-032-05727-3_3

raw data values to identify datasets containing specific entities. While more robust to metadata in open data lakes, these methods face serious scalability challenges in large-scale repositories. Open data lakes tend to accumulate large numbers of uncurated datasets, often in structured format and sourced from various contributors. For example, Data.gov [6], which began in 2009 with only 47 datasets, now hosts over 300,000. In such environments, metadata-based searches often fail, and raw data value search becomes computationally expensive.

Recent table-based search approaches use automatic annotations or embedding-based representations derived from ontologies or large language models (LLMs) to match relevant datasets to an input table [1]. While promising, these methods depend on accurate ontologies or well-trained models, raising concerns about domain coverage, generalizability, and robustness. They also perform poorly with datasets that primarily consist of numerical values, where ontologies are inadequate and sentence-based models less effective, as illustrated in Table 1. Furthermore, table-based search is only applicable when an example table is available which is often not the case in practice.

Table 1. Sample dataset containing technical specifications of the NVIDIA RTX 5090, 5080, and 5070 graphics cards.

col1	col2	col3	col4	col5	col6	col7	col8
2025	28	448	900	1200	8192	256	128
2025	16	256	900	1215	6912	432	192
2025	12	192	1825	2000	5120	320	128

To overcome the challenges of dataset discovery in open data lakes, we introduce a new keyword-based search tool called Constraint-based Dataset Discovery (CoDD), which establishes a novel *query-by-constraint* paradigm. This approach addresses the shortcomings of query-by-metadata and query-by-data by summarizing and transforming data into adhoc-metadata in a reproducible and structured way. In CoDD, datasets are profiled by extracting structured facts directly from the data itself, without relying on existing metadata or predefined schemas. These facts are generated by modular components called *FactProviders* and stored as consistent, semi-structured JSON profiles.

This fact-based profiling offers several advantages. All datasets are described using a uniform structure, making them easy to compare and reason about. Furthermore, unlike query-by-data approaches, CoDD's profiles are more scalable and storage-efficient, and in contrast to query-by-metadata, query-by-constraint enables users to formulate absolute rules on table profiles. CoDD's user interface enables an interactive, question-driven search process where users can dynamically add, refine, or remove constraints at any point. It suggests possible constraints based on the profiled facts, guiding users through the discovery process. The result is a ranked list of top-k datasets that match the specified constraints. These resulting datasets can then be used in downstream tasks like Table Union

Search or Table Join Search [1]. By grounding dataset discovery in actual data-derived constraints rather than uncertain metadata, CoDD improves the accuracy and reliability of data-driven workflows.

Our contributions are as follows:

- We introduce CoDD, and its query-by-constraint strategy as a robust alternative to traditional query-by-metadata and query-by-data methods for dataset discovery in open data lakes.
- We demonstrate CoDD, a modular plugin-based system developed in Python, featuring a user-friendly web interface. CoDD's plugin interface and WebUI enable users to easily create custom *FactProviders* and perform effective and efficient, constraint-based searches without requiring prior knowledge of the dataset corpus. For the demonstration, we use the Wiki-Union dataset [5].

2   Related Work

Since query-by-constraint is intended as a substitute for query-by-metadata and query-by-data, and not directly comparable to table-based search, we focus exclusively on related work addressing keyword-based search approaches for open data lakes, considering only those that are freely available. A tool that meets these criteria and offers a superset of the functionalities found in comparable systems is OpenMetadata (OM) [3].

OpenMetadata supports data discovery through three ingestion pipelines: Metadata, Profile, and Sample. The Metadata and Sample pipelines enable keyword-based searches on table and column names, and allow users to view sample data values directly in the WebUI. In contrast, the Profile pipeline generates a limited, non-searchable set of dataset statistics such as null counts per column.

3   System Architecture and Demonstration Outline

The architecture of CoDD consists of three main components, as illustrated in Fig. 1. The first component is the CoDD Profiler, which scans data lakes and uses an active set of FactProviders to generate semi-structured table profiles. These profiles contain data-derived facts that characterize each dataset. The second component is the document store, which acts as a central repository for all generated profiles. The profiles serve as structured representations of datasets, capturing key statistical and structural facts directly from the data, independent of any metadata. The third component is the CoDD WebUI, a web-based user interface that enables users to interactively search for datasets in (open) data lakes by formulating constraints over the profiled facts. Designed as a guided, question-and-answer system, the interface helps users to define constraints without requiring prior knowledge of specific FactProviders or datasets in the data lakes.

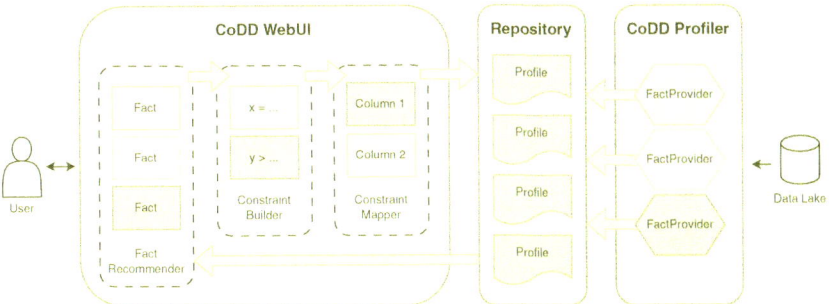

Fig. 1. CoDD's main components showing a human-in-the-loop frontend with the WebUI and the profiling of open data lakes in the backend.

3.1   CoDD Profiler

The core principle of query-by-constraint is the availability of reliable and interpretable facts. To support this, CoDD employs a plugin-based architecture implemented in Python, allowing user-defined FactProviders to be dynamically loaded at runtime from a designated plugin directory. Each FactProvider processes structured documents from a specified data lake and generates a table profile that includes facts for each column as well as for the table as a whole. These facts may include statistics such as minimum and maximum values, the number of null entries, or detected value patterns such as the presence of "@." in columns containing email addresses. The resulting profiles are stored as semi-structured JSON documents in a document store, where each FactProvider's ID serves as a key and the corresponding generated fact as its value. Indexes can be created for selected facts as needed to improve query performance. FactProviders can be activated or deactivated as required, helping to reduce runtime overhead and storage usage. Importantly, they are managed entirely through the plugin directory, without requiring modifications to CoDD's core codebase.

CoDD's admin interface provides an overview of available data lakes and displays the processing status of documents within each lake. Only the Fact-Providers present in the plugin folder at the time profiling begins are considered during the profiling process. Profiling can be repeated at any time to include newly added or updated FactProviders. Furthermore, the system enables users to begin formulating query-by-constraint searches using the CoDD WebUI even while the profiling process is still ongoing, enabling early and efficient interaction with the data.

3.2   CoDD WebUI

The CoDD WebUI is the central component of the system, as it combines an intuitive user interface with the powerful functionality of formulating query-by-constraint queries on table profiles. Unlike traditional query-by-data approaches, where users search for specific values within datasets, CoDD allows users to search based on dataset descriptions generated from previously profiled facts by the CoDD Profiler. In this way, queries are formulated in terms of adhoc-metadata grounded in the actual content of the data, rather than relying on often unreliable or missing metadata. The WebUI is divided into six interactive panels, which we explain in pairs. To demonstrate CoDD's capabilities, we use a search example for the dataset shown in Table 1, which has been added to the Wiki-Union data lake [5].

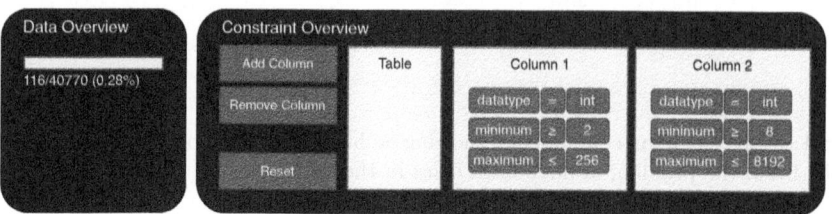

Fig. 2. The top row of the CoDD WebUI displays the query's selectivity on the left and the active constraints, along with their assignment to table and columns, on the right.

The first row of panels, as shown in Fig. 2, consists of two key components. The right panel supports the graphical construction of query-by-constraint queries. To simplify interaction, constraints are visually represented as boxes. We distinguish between table boxes and column boxes to separate constraints at the table level from those at the column level. By default, a table box is provided. Users can add any number of column boxes, with each corresponding to a set of constraints for a single column. The number of column boxes also acts as an implicit filter: only tables with at least that many columns are considered. By clicking on a box, users can select relevant table- or column-level facts to define their constraints. CoDD then maps the table box to a table profile and column boxes to specific columns using a best-match procedure, based on similarity between the specified constraints and the facts available in the profiles. For this purpose, a recursion tree is constructed by assigning each set of constraints to a single column. If a complete assignment cannot be achieved along a given path, that path is pruned, and the process continues with an alternative branch. The left panel displays the proportion of datasets remaining after applying the current query-by-constraint, serving as an indicator of the query's selectivity. A lower proportion indicates a more selective query and helps guide users toward more expressive constraints. The total number refers to the documents already profiled across all data lakes.

For our search for the GPU dataset presented in Table 1, whose structure and content are unknown at this point, we assume that a GPU should include at least two specifications: Memory and Memory Bus Width. Both are expected to be integer values, likely within the ranges of 2 to 256 and 8 to 8192, respectively. Applying these constraints alone prunes 99.72 percent of the over 40,000 datasets, leaving only 116 candidates remaining.

Looking at the middle row, as shown in Fig. 3, the left panel displays a list of recommended FactProviders, hereafter referred to as Facts, once a user selects a table or column box. Only Facts that have already been computed by the CoDD Profiler and are applicable to the selected box are shown. For example, Facts related to numerical data will not be displayed for columns containing only strings. When a user selects a Fact, a guided question-and-answer dialogue begins in the right-hand panel. This panel presents a brief description of the selected Fact and offers suitable comparison operators specific to the respective data types. After the user selects an operator, they are prompted to enter a comparison value. Once the constraint is fully defined, it is visually added to the corresponding table or column box, and the proportion of matching datasets is updated in real-time. Constraints can be removed by right-clicking on them. Users can also delete individual boxes or reset the entire query at any time.

Continuing with the example above, the fact *maximum* is selected for each of the column boxes, using 256 or 8192 as the comparison value with the less than or equal to operator. Similarly, the *datatype* and *minimum* facts were selected and configured in the same way.

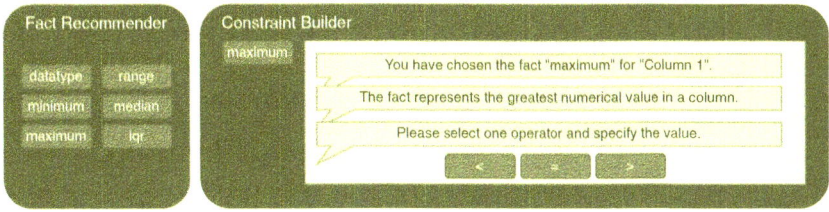

Fig. 3. The second row of the CoDD WebUI displays the available facts for selection on the left. When a fact is selected, its explanation and configuration options for building the constraint appear on the right.

As with other tools in this domain, it is essential that users can validate both their query and the resulting datasets. To support this, the bottom row of panels, shown in Fig. 4, provides two key views. The left panel lists the names of all datasets that currently match the active query-by-constraint. Each entry includes the name of the data lake and the dataset identifier. By clicking on a dataset in this list, users can view a preview of its contents in the right panel, enabling direct inspection and validation of how well the dataset matches the specified constraints.

To conclude the example above, we determine that the selectivity is sufficient and choose not to add further constraints. We continue to review and inspect the list of resulting datasets until the matching datasets are found.

Of course, additional constraints can be added to increase selectivity. This involves iteratively creating new boxes, selecting relevant facts, and configuring them until the desired level of precision is achieved.

3.3   Demo Setup

For the demonstration, we extend the existing Wiki-Union data lake [5] by adding custom datasets that serve as search targets. Search tasks provide a brief textual

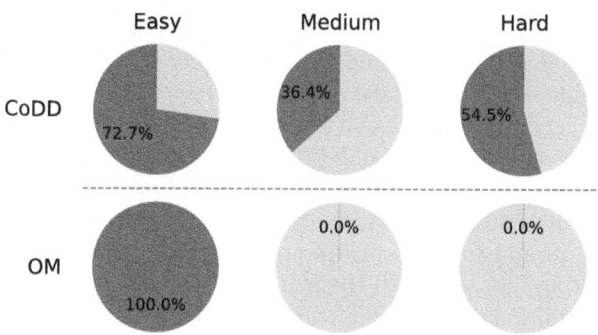

	col1	col2	col3	col4	col5	col6	col7	col8
Remaining Files								
datasetA/a.csv	2025	28	448	900	1200	8192	256	128
datasetA/b.csv	2025	16	256	900	1215	6912	432	192
datasetA/c.csv	2025	12	192	1825	2000	5120	320	128

Fig. 4. The third row of the CoDD WebUI displays the names of the matching datasets on the left and a preview of their contents on the right.

description of a dataset to be found, requiring users to discover relevant datasets based solely on that description. The task of finding the associated datasets using CoDD can be classified into three levels of difficulty: Easy, Medium, and Hard. Easy datasets have clear and unambiguous metadata. Medium datasets contain metadata with challenges such as abbreviations, synonyms, or homonyms. Hard datasets either lack metadata entirely or contain misleading or incorrect metadata. In the demo, participants are invited to try search tasks of varying difficulty, beginning with either CoDD or OpenMetadata (OM) at random. They complete the first set of tasks using their chosen tool before switching to the other for the remaining tasks. This sequential setup allows participants to directly experience and compare the search effectiveness of CoDD and OM.

We expect results similar to a previous user study conducted in our research group, as illustrated in Fig. 5. In that study, users are unable to find any datasets in the Medium and Hard categories using OpenMetadata. In contrast, CoDD consistently enables users to discover a substantial number of relevant datasets, demonstrating a clear advantage in scenarios where metadata is limited or unreliable.

Fig. 5. Proportion of Easy, Medium, and Hard datasets successfully identified in the user study, depending on the tool used - CoDD or OpenMetadata (OM).

4   Conclusion

The vast size of open data lakes makes dataset discovery based on raw data values inefficient, while the lack of maintenance of such lakes causes keyword-based search on metadata, known as query-by-metadata, to be ineffective when table or column names are missing, misleading, or incorrect. In this demonstration, we present CoDD, a Constraint-based Dataset Discovery Tool for Open Data Lakes that introduces a new paradigm: *query-by-constraint*. CoDD bridges the gap between data and metadata by allowing user-defined facts to be extracted directly from the data, treating them as first-class searchable entities. These facts are profiled into adhoc-metadata and used to formulate queries as constraints through an intuitive graphical interface. The demonstration highlights CoDD's capabilities by presenting participants with search tasks of varying difficulty levels. To showcase not only CoDD's effectiveness but also its superiority over conventional tools, the same queries are executed in OpenMetadata for comparison. Results from an internal user study showcase what can be observed during the demonstration: CoDD's query-by-constraint approach outperforms traditional query-by-metadata tools like OpenMetadata in terms of precision, especially in the presence of poor or missing metadata. By grounding dataset discovery in factual data rather than speculative metadata, CoDD also enables deeper, follow-up searches such as Table Union Search and Table Join Search. The source code for CoDD is publicly available [4].

References

1. Miller, R.J., Nargesian, F., Zhu, E., Christodoulakis, C., Pu, K.Q., Andritsos, P.: Making open data transparent: data discovery on open data. IEEE Data Eng. Bull. **41**(2), 59–70 (2018)
2. Nargesian, F., Zhu, E., Miller, R.J., Pu, K.Q., Arocena, P.C.: Data lake management: challenges and opportunities. Proc. VLDB Endow. **12**(12), 1986–1989 (2019)
3. OpenMetadata: https://github.com/open-metadata/OpenMetadata
4. Otto, T., Rawald, C.: Codd. https://sci-git.cs.rptu.de/t_otto18/codd
5. Srinivas, K., et al.: LakeBench: benchmarks for data discovery over data lakes. CoRR abs/2307.04217 (2023)
6. U.S. General Services Administration: data.gov (2025). https://data.gov/

Spatio-Temporal and Graph Data

Could More Be Less: The Case of Location(s) Awareness in Time Series Prediction

Bala Sai Sathwick Reddy Mora🆔 and Goce Trajcevski[✉]🆔

Iowa State University, Ames, IA 50010, USA
{sathwick,gocet25}@iastate.edu

Abstract. Time series data is essential in many applications of high societal relevance and the problem of efficient and effective *prediction* is at the heart of many recent research efforts tackling domains like energy demand prediction, weather forecasting, traffic density prediction, market predictions (to name but a few). In the recent years, multiple deep learning approaches have been proposed, each presenting different architectures/models. One universally common theme is "the more, the merrier" in terms of the (training) data. In this paper, we focus on domains in which the time series data items are bound to *locations* – specifically, energy demand forecasting and weather forecasting, where each data source (e.g., a power transformer units or a weather station) pertains to a particular location. While many recent works have considered the incorporation of spatio-temporal dependencies in the model to improve the effectiveness – as it turns out, the inclusion of multiple sources may have adverse impacts. We present the results of two heuristics that we developed for the purpose of determining when is it that "more is less" in such settings.

Keywords: Time series · Location awareness · Prediction

1 Introduction and Motivation

Owing to their ubiquity, time series data have been one of the most extensivelyy investigated domains of forecasting, with a rich history spanning from statistical and traditional machine learning (ML) approaches to more contemporary deep learning (DL) ones [10]. The objective is to use past/historic data to make predictions about the values of the phenomenon of interest in the future – which is at the heart of multiple applications of extreme societal interests (finance, traffic management, healthcare, meteorology, astrophysic, etc.) [18].

In the recent years, plethora of works emerged addressing different variations of the problem of time series forecasting (TSF) and proposing variety of architectures based on different paradigms (e.g., recurrent neural networks (RNN), Long-Short Term Memor (LSTM); graph neural networks (GNN), etc.) [2,13]. Taxonmies have been proposed according to various contexts like, for example

© The Author(s), under exclusive license to Springer Nature Switzerland AG 2026
P. K. Chrysanthis et al. (Eds.): ADBIS 2025, CCIS 2676, pp. 35–45, 2026.
https://doi.org/10.1007/978-3-032-05727-3_4

long-term and short-term forecasting [12]; augmentation [20] of time series data, etc. One particular aspect that has gained a recent popularity and is part of the motivation for this work, is the forecasting in *spatio-temporal* context. Namely, in many forecasting applications such as energy demand [16], weather [5] and traffic [8], the data representing the dynamics of the evolution corresponds to measurements in particular locations (e.g., roadside/inductive sensors, power stations, weather stations). This, in turn, implies that the changes of the phenomena at one location can influence the evolution in another location (and vice-versa), with a note that more distant locations are likely to exert the influence further in the future.

When it comes to deep learning methodologies for prediction in the realm of time series data, typically "the more, the merrier" spirit prevails. As a complementary observation, to improve the quality of the model often times researchers apply data imputation methods [21]. At the heart of the motivation for this work is that, unlike traffic-based scenarios where one could expect some cascade-like impacts over neighboring zones and road segments, in applications such as weather forecasting and energy demand forecasting the distance may play an adverse effect. We postulate that the geo-distance between the locations of the (multivariate) time series data sources should not be neglected – i.e., it may actually have an adverse effect on the quality of prediction of the time series values at a particular location. Towards this, the main contributions of the work can be summarized as follows: (1) We investigate whether including additional data sources may adversely affect the quality of the prediction at specific location. (2) We propose two strategies for investigating the impact of including more locations: nearest-neighbor based and distance based. (3) We conducted experimental evaluation over three datasets with varying density and mutual distances between the data sources confirming out hypothesis.

2   Related Works

We now review the evolution of TSF methodologies, highlight key advancements, and identify the gaps addressed in this work.

Traditional Statistical Models: These have long been the foundation of time series analysis [1]. Techniques like ARIMA (Autoregressive Integrated Moving Average) and Exponential Smoothing are well-known for their simplicity and ease of interpretation. ARIMA, for instance, breaks down a time series into autoregressive (AR), integrated (I), and moving average (MA) components, making it effective for capturing linear trends in univariate data. Although these models are still widely used as benchmarks in research [17,25], they often struggle to capture the complex, non-linear patterns and multivariate relationships found in many real-world datasets.

Deep Learning for Sequential Data: Fully recognizing the impact of traditionsl ML models (e.g., Support Vector Machines (SVMs) [14]), here we focus on models leveraging deep learning to capture non-linear relationships and intricate temporal patterns [1].

– *Recurrent Neural Networks (RNNs)*: LSTMs (Long Short-Term Memory) [11] and GRUs (Gated Recurrent Units) [7] became popular for sequence modeling. Their gating mechanisms allow them to capture long-term dependencies better than vanilla RNNs, mitigating the problem of vanishing gradients. However, their sequential nature can make training slow and parallelization difficult.

– *Convolutional Neural Networks (CNNs/TCNs)*: Initially designed for image processing, CNNs, particularly 1D-CNNs or Temporal Convolutional Networks (TCNs) [15], have been adapted for TSF. TCNs use causal convolutions (preventing information leakage from the future) often combined with dilated convolutions (to increase receptive field exponentially) and gating mechanisms (Gated-TCN) [6] to capture temporal patterns efficiently through parallel computation.

– *MLP-based Models*: Recently, simpler architectures based entirely on Multi-Layer Perceptrons (MLPs) have shown competitive performance. Models like NLinear and DLinear [24] use simple linear layers, potentially with normalization or decomposition schemes, challenging the necessity of complex recurrent or attention structures for some TSF tasks. TSMixer [4] employs MLPs to mix information across both time and feature dimensions, proving effective in multivariate forecasting.

– *Transformer Architecture in Time Series*: Initially successful in Natural Language Processing (NLP) [19], transformers have been widely adapted for TSF, marking a third generation of models [1]. The core innovation is the self-attention mechanism, which allows the model to weigh the importance of different past time steps when predicting a future value. This enables direct modeling of long-range dependencies, overcoming a key limitation of RNNs. In TSF, Transformers can be viewed as learning a message passing graph among time steps.

Spatio-Temporal Forecasting: Integrating Space and Time: There are several complementary aspects:

– *Hybrid Architectures for Spatial and Temporal Integration*: Traditional forecasting models often treated spatial and temporal components independently, leading to oversimplified representations of real-world systems. Hybrid architectures, such as Graph Neural Networks (GNNs) combined with Transformers or Long Short-Term Memory (LSTM) networks, have demonstrated superior performance by explicitly modeling spatial adjacency and temporal sequences.

– *Graph Neural Networks (GNNs)*: Graph Neural Networks (GNNs) are deep learning models designed to operate directly on graph-structured data [22]. They work via a message passing paradigm. Each node aggregates information from its neighbors (and potentially itself) to update its own representation (embedding). This process is typically repeated over several layers, allowing information to propagate across the graph.

Spatio-temporal forecasting requires models that can simultaneously capture dependencies across both space and time. STGNNs have emerged as a dominant approach, aiming to combine the spatial modeling capabilities of GNNs with sequence modeling techniques for the temporal dimension [3]. Most STGNNs follow a framework involving modules for spatial learning (often GNN-based) and temporal learning (often RNN or TCN-based), integrated via some fusion mechanism. The goal is to learn hidden representations that encode both spatial context and temporal dynamics.

– *Transformers in Spatio-Temporal Modeling*: Given their success in TSF, Transformers are increasingly being applied to spatio-temporal problems, leveraging attention to capture dependencies across both dimensions. Models may apply attention mechanisms across spatial locations, time steps, or jointly across space-time. Models like Spacetimeformer [9] aim to apply attention directly across a combined spatio-temporal domain, potentially using efficient attention mechanisms to handle the scale. To overcome quadratic complexity, various efficient attention approximations have been proposed, aiming for linear or near-linear complexity ($O(L \log L)$ or $O(L)$).

What separates our work from the related literature is that, to our knowledge, we are the first to investigate the impact of the distance/neighborhoods to the quality of the predictions on datasets of different types, scale and density.

3   Datasets and Methodology

We selected three publicly available datasets, and we aimed at using different densities for the locations/sources of the time series:

D1 – Australian Energy Market Operator (AEMO): this dataset is primarily used for energy forecasting tasks. It captures detailed information such as electricity demand and pricing across various time intervals, typically every 5 or 30 min. This multivariate dataset is particularly valuable for building models that predict future energy usage or grid behavior, as it reflects both temporal patterns and regional interdependencies within Australia's power system. Geographically, the dataset encompasses regions covered by the National Electricity Market (NEM), including New South Wales (NSW), Victoria (VIC), Queensland (QLD), South Australia (SA), and Tasmania (TAS). These regions serve as spatial nodes in forecasting models, with data collected from substations, transmission zones, or demand regions. The spatial relationships between these locations based on either physical grid connections or learned dependencies allow for advanced modeling techniques such as graph-based neural networks. The dataset is publicly available at: https://www.aemo.com.au/.

D2 – New York Independent System Operator (NYISO): The dataset used in this study contains load data from several regions within the NYISO grid, which manages electricity distribution across New York State. The regions included in the dataset (referred to as CAPITL, CENTRL, DUNWOD, GENESE, HUD VL, LONGIL, MHK VL, MILLWD, N.Y.C., NORTH, and WEST) represent a wide range of areas, each with unique electricity demand characteristics. The inclusion of both urban and rural areas in the dataset provides a comprehensive overview of New York's electricity consumption patterns. Each of these regions has its own distinctive load data, which reflects local energy demands influenced by various factors like population density, industrial activity, and weather conditions. For instance, N.Y.C. shows high demand due to the large population and commercial activities, while more rural regions like NORTH and WEST may experience lower but more seasonal variations in electricity usage. The dataset is publicly available at: https://www.nyiso.com/energy-market-operational-data.

D3 – Automated Surface Observing System (ASOS): This dataset includes weather data collected from the ASOS network, which is operated by the National Weather Service (NWS) in the United States. In this study, specific ASOS data has been utilized from the stations in Iowa and its adjacent states (Nebraska, South Dakota, Minnesota, Wisconsin, Illinois, Missouri, and Kansas). The weather variables captured by ASOS play a critical role in spatio-temporal forecasting as the stations are often much closer when compared to the locations in the previous two datasets. The dataset is publicly available at: https://mesonet.agron.iastate.edu/request/download.phtml.

The raw data from each dataset was initially pivoted so that each row corresponds to a specific time step T, and each column represents a measured attribute (such as load or temperature) at a specific location. We considered two time ranges for experimental analysis: 2015âÅŞ2025 and 2020âÅŞ2025. To regularize the temporal resolution, the data was aggregated into hourly intervals. Locations with a substantial proportion of missing data were excluded from the analysis to prevent them from negatively impacting model performance. For the remaining data, rather than applying conventional imputation techniques which might introduce bias or artificial patterns, missing values were handled using the null value embedding functionality of the Spacetimeformer model [9]. It explicitly informs the model of missing entries without distorting the temporal or spatial characteristics of the original data. Finally, the cleaned and transformed datasets were segmented into batches. Each batch was constructed using a context window y_c, representing the input sequence provided to the model, and a target window y_t, which the model is trained to predict.

For fair comparisons between regions of varying geographic sizes, we computed a Normalized Distance metric that considers not just the physical distance between regions, but also the spatial extent (area) of each. This was done in two main steps: first, by calculating the effective radius of each region based on its area, and second, by using the Haversine formula to compute the true geographic distance between centroids, which is then normalized by the regions' sizes.

Step 1: Effective Radius – The effective radius provides a single value that approximates the "spread" of a region, assuming it's roughly circular. It is computed as: $r = \sqrt{\frac{area}{\pi}}$. This step helps in capturing the fact that a larger region naturally spans more physical space and therefore should tolerate greater distances from other areas before being considered "too far".

Step   2:   Haversine   Distance   –   to   calculate   it,   we   used:
$R = 6371.0\,\text{km}$ (Earth's radius)

$\Delta\phi = \phi_2 - \phi_1$ (latitude difference in radians)

$\Delta\lambda = \lambda_2 - \lambda_1$ (longitude difference in radians)

$$a = \sin^2\left(\frac{\Delta\phi}{2}\right) + \cos(\phi_1) \cdot \cos(\phi_2) \cdot \sin^2\left(\frac{\Delta\lambda}{2}\right)$$

$$c = 2 \cdot \arctan 2\left(\sqrt{a}, \sqrt{1-a}\right)$$

$d = R \cdot c$ (physical distance in kilometers)

Step 3: Normalized Distance – Finally, to make distances comparable across differently sized regions, Normalized Distance was defined as: $D_{\text{norm}} = \frac{D_{\text{phys}}}{r_1 + r_2}$, Where: D_{phys} is the Haversine distance between the two region centroids; and r_1 and r_2 are the effective radii of the two regions. This normalized metric allows for a more consistent measure of "closeness" that adjusts for how large the regions are. To improve the model's handling of time series data, we used two key preprocessing settings: (1) RevIN (Reversible Instance Normalization), which adaptively normalizes each input time series instance based on its own statistical properties. Unlike global normalization methods, RevIN ensures that the model learns patterns that are invariant to the scale and distribution of individual series, while also allowing for accurate inversion of the normalization during inference; (2) Seasonal decomposition which separates the original time series into trend, seasonal and residual components. By removing the seasonality during training, the model is able to focus on learning the more complex or non-repetitive aspects of the data, and the seasonal signal can be reintroduced during reconstruction. This is especially beneficial for datasets where regular periodic behaviors can mask other underlying dynamics. We used two basic strategies to test the hypothesis.

S1: For both the AEMO and NYISO datasets, a progressive spatial modeling strategy was employed in order to assess the impact of the spatial context on forecasting performance. Specifically, for each location in the respective datasets, the model was trained using only the historical time series data of that single location in isolation. This provided a kind of a baseline – i.e., understanding how well the model could learn temporal patterns without any spatial dependencies.

Subsequently, spatial nearest neighbors were added incrementaly. At each step of this process, the nearest unconsidered location (according to the normalized distance) was added to the input set, and the model was *retrained* with this expanded context. This allowed us to observe the marginal benefit of adding each neighboring region in terms of forecasting accuracy, and to examine the impact of spatial information from nearby locations on the performance.

S2: For this strategy, we used the ASOS dataset (the densest one) and we divided the region around each of the reference weather station location into 50KM concentric circles. Once again, we first trained the model on a single location and subsequently considered adding the neighbors from each successive ring between consecutive circles. We limited the experimental evaluation to at most 10 rings from the reference location.

4 Experimental Results

For reproducibility, all the datasets and the code used for pre-processing and as well as the implementation of the model are publicly available at: https://github.com/Sathwick-Reddy-M/improving-temporal-forecasting. Due to a lack of space, we cannot present all the experiments here – however, they are available in an expanded form in a document that is part of the public repository. We used four measures: – Mean Absolute Error (MAE); – Mean Squared Error (MSE); – Normalized MAE; – Normalized MSE (the last two to account for potentially different scales).

We firste present the results obtained by using Spacetimeformer [9], which achieves competitive results on benchmarks from traffic forecasting to electricity demand and weather prediction and is publicly accessible via https://spacetimeformer.readthedocs.io/en/latest/. Based on the amount of training data considered we have included the architecture with an encoder and decoder layer, and incorporated Multi-Head Attention. The shapes of the K, Q, and V matrices (cf. [9]) and the hyperparameters including the learning rate, number of context points, target points and FFN dimensions have been experiemented with a few number of possible values with each of the datasets. We used two key preprocessing settings for improving the model's handling of time series data: `-use_revin` for normalization using RevIN (Reversible Instance Normalization), and `-use_seasonal_decomp` for seasonal decomposition of the input data, explained in the public repository.

Results Based on S1: We use abbreviations that correspond to a particular location from the original datasets. We firstly present a global evaluation of two smaller subsets (each with three locations) from the ASOS dataset and compare it with the evaluation on the combined triplets into a dataset of six locations. The respective SpaceTimeFormer models were trained using NY-TX weather data from 2016-01-01 to 2025-01-01. (1) **Triplet 1 (NY-Only)**: Using stations ALB, JFK, LGA. (2) **Triplet 2 (TX-Only)**: Using stations ABI, ACT, AMA. (2) **Combined**: Using all six (NY + TX) stations.

Table 1. Performance of two triplets vs. sixtuple

	MAE	MSE	Norm MAE	Norm MSE
TX-Only	1.85	6.2	0.18	0.07
NY-Only	1.93	7.5	0.21	0.09
Combined	2.75	13.5	0.26	0.12

As shown in Table 1, the combined model performs worse – which is to be expected, given that the locations in each of the triplets (NY and TX) are relatively close to each other, whereas in the sixtuple there is a huge distance between the two pairs of three locations each.

Table 2. ASOS hops results (MSE)

Loc-ID	MSE	Loc-ID	MSE	Loc-ID	MSE	Loc-ID	MSE	Loc-ID	MSE
ADU-1	1.73	ADU-2	1.95	ADU-3	1.41	ADU-4	1.14	ADU-5	2.05
ADU-6	2.37	ADU-7	2.51	ADU-8	2.54	ADU-9	2.58	ADU-10	2.76
AIO-1	6.22	AIO-2	5.69	AIO-3	5.92	AIO-4	5.62	AIO-5	6.30
AIO-6	7.99	AIO-7	8.25	AIO-8	8.55	AIO-9	8.66	AIO-10	8.82
AMW-1	16.35	AMW-2	16.68	AMW-3	16.96	AMW-4	16.84	AMW-5	18.17
AMW-6	17.23	AMW-7	18.22	AMW-8	17.12	AMW-9	19.22	AMW-10	18.94
AWG-1	10.32	AWG-2	8.23	AWG-3	9.98	AWG-4	11.22	AWG-5	10.67
AWG-6	12.34	AWG-7	12.11	AWG-8	13.78	AWG-9	12.08	AWG-10	13.59
DSM-1	1.29	DSM-2	1.90	DSM-3	2.31	DSM-4	2.22	DSM-5	2.58
DSM-6	3.13	DSM-7	2.99	DSM-8	3.12	DSM-9	3.87	DSM-10	3.89
DVN-1	19.11	DVN-2	17.21	DVN-3	18.83	DVN-4	19.91	DVN-5	20.88
DVN-6	20.99	DVN-7	19.81	DVN-8	20.87	DVN-9	20.33	DVN-10	19.99

Results Based on S2: Recall that the sequence of rings around a given location progressively increases the respective radii by 50km, and then iteratively executes the Spacetimeformer [9]. We report the results for the densest datasets – ASOS – of time series for weather data from multiple locations. When it comes to the notation used in this subsection, we use the following: (1) once again, each string corresponds to a given location as used in the original (publicly available) ASOS dataset; (2) to indicate the number of iteration, we add a subscript to each location. Thus, as an example, in Table 2: – ADU corresponds to a given location from the ASOS dataset; – ADU-1 corresponds to a set consisting of (the locations of) all the weather stations within the first ring/disk of radius 50km around ADU; – ADU-2 adds the locations of all the datasets within the second 50km ring to ADu-1; ...; – Finally, ADU-10 adds the locations from within the 10th ring of 50km centered at ADU to all the previously added locations.

Table 2 shows the result for five consecutive locations and successive inclusion of their ring of neighbors in terms of MSE. We observe that: (1) In some cases – e.g., AMW and DSM, the best results are obtained when using the very first ring (i.e., disk centered at the respective location). (2) In other cases it is the 2nd ring that yields the optimal value (e.g., AWG and DVN; or the 4th ring (e.g., ADU and AIO); or even the 6th ring (e.g., DSM). (3) However, what is consistent is that in each case, past certain iteration the results begin to deteriorate.

We close this section by presenting the results from the ongoing investigation of evaluating our hypothesis – except now we used the MTGNN model [23] – which is highly cited in the literature. We used RMSE as a measure and PJM regions from NYISO dataset. Table 3 shows the results of the evaluation.

Table 3. RMSE results for PJM regions (Hops 1–10)

Loc-ID	RMSE	Loc-ID	RMSE	Loc-ID	RMSE	Loc-ID	RMSE	Loc-ID	RMSE
AE-1	0.0813	AE-2	0.0745	AE-3	0.0924	AE-4	0.1321	AE-5	0.1701
AE-6	0.2176	AE-7	0.0758	AE-8	0.0851	AE-9	0.1919	AE-10	0.1784
ATSI-1	0.0975	ATSI-2	0.0889	ATSI-3	0.0847	ATSI-4	0.1055	ATSI-5	0.0918
ATSI-6	0.0994	ATSI-7	0.14	ATSI-8	0.161	ATSI-9	0.168	ATSI-10	0.1218
DUQ-1	0.0793	DUQ-2	0.0855	DUQ-3	0.1198	DUQ-4	0.0776	DUQ-5	0.0768
DUQ-6	0.1182	DUQ-7	0.0787	DUQ-8	0.1324	DUQ-9	0.0818	DUQ-10	0.1295
JC-1	0.0989	JC-2	0.0718	JC-3	0.1362	JC-4	0.1847	JC-5	0.1412
JC-6	0.1018	JC-7	0.0789	JC-8	0.1014	JC-9	0.0886	JC-10	0.0753
PL-1	0.1095	PL-2	0.1588	PL-3	0.1283	PL-4	0.1439	PL-5	0.1434
PL-6	0.1528	PL-7	0.0831	PL-8	0.1347	PL-9	0.1732	PL-10	0.1072
PN-1	0.1293	PN-2	0.1181	PN-3	0.134	PN-4	0.1304	PN-5	0.1196
PN-6	0.1201	PN-7	0.1353	PN-8	0.2721	PN-9	0.1344	PN-10	0.1199

5   Concluding Remarks

In this work, we investigated the impact of location-awareness to the problem of quality of prediction in time series data. We used publicly available time series dataset from the domains of energy load and weather used in forecasting tasks, and proceeded with including more locations in the model, and considered two strategies for adding locations: (1) based on nearest neighbors; (2) based on spatial distances – and iteratively included more and more of them. In each strategy, past certain threshold, the inclusion of more data sources began causing a deterioration in the performance, thus confirming our hypothesis that *more* (data) could mean *less* quality of the prediction. We reported a subset of our study – with a reminder that the source code and datasets are publicly available – mostly focusing on the Spacetimeformer [9] model, and also presented part of our ongoing work using GNN based models (MTGNN [23]). We believe that our results will generate an interesting follow-up research that will improve the quality of the models addressing the forecasting problem when time series datasets are bound to different geo-locations.

Acknowledgments. Research partially supported by the Kingland foundation. We are grateful to Beryl Caleb Shelton for the help with the MTGNN results.

References

1. Andreai, A.V., Velev, G., Toma, F.M., Pele, D.T., Lessman, S.: Energy price modelling- a comparative evaluation of four generations of forecasting methods
2. Benidis, K., Rangapuram, S.S., Flunkert, V.e.a.: Deep learning for time series forecasting: Tutorial and literature survey **55**(6) (2022)
3. Bui, K.-H.N., Cho, J., Yi, H.: Spatial-temporal graph neural network for traffic forecasting: an overview and open research issues. Appl. Intell. , 1–12 (2021). https://doi.org/10.1007/s10489-021-02587-w
4. Chen, S.A., Li, C.L., Yoder, N., Arik, S.O., Pfister, T.: Tsmixer: An all-MLP architecture for time series forecasting (2023). arXiv preprint arXiv:2303.06053
5. Ciszynski, M., Chrominski, K.: Applying machine learning techniques in weather forecast modeling. In: KES-2024. Procedia Computer Science, vol. 246, pp. 4133–4141. Elsevier (2024)
6. Dauphin, Y.N., Fan, A., Auli, M., Grangier, D.: Language modeling with gated convolutional networks. In: International Conference on Machine Learning, pp. 933–941. PMLR (2017)
7. Dey, R., Salem, F.M.: Gate-variants of gated recurrent unit (GRU) neural networks. In: 2017 IEEE 60th International Midwest Symposium on Circuits and Systems (MWSCAS), pp. 1597–1600. IEEE (2017)
8. Gao, Q., Wang, Z., Huang, L., Trajcevski, G., Zhang, K., Chen, X.: Enhancing dependency dynamics in traffic flow forecasting via graph risk bootstrap. In: ACM SIGSPATIAL (2024)
9. Grigsby, J., Wang, Z., Nguyen, N., Qi, Y.: Long-range transformers for dynamic spatiotemporal forecasting (2023). https://arxiv.org/abs/2109.12218
10. Hendikawati, P., Subanar, Abdurakhman, Tarno: A survey of time series forecasting from stochastic method to soft computing **1613**(1), 012019 (2020)
11. Hochreiter, S., Schmidhuber, J.: Long short-term memory. Neural Comput. **9**(8), 1735–1780 (1997)
12. Li, Y., et al.: Towards long-term time-series forecasting: Feature, pattern, and distribution. In: 39th IEEE International Conference on Data Engineering, ICDE 2023, Anaheim, CA, USA, April 3-7, 2023, pp. 1611–1624. IEEE (2023)
13. Liu, X., Wenmin, W.: Deep time series forecasting models: a comprehensive survey. Mathematics **12**(10) (2024)
14. Noble, W.: What is a support vector machine? Nat Biotechnol **24** (2006)
15. van den Oord, A., et al.: Wavenet: A generative model for raw audio (2016). arXiv preprint arXiv:1609.03499
16. Özen, S., Yazici, A., Atalay, V.: Hybrid deep learning models with data fusion approach for electricity load forecasting. Expert Syst. J. Knowl. Eng. **42**(2) (2025)
17. Santos, M.L., García-Santiago, X., Camarero, F.E., Gil, G.B., Ortega, P.C.: Application of temporal fusion transformer for day-ahead PV power forecasting. Energies **15**(14) (2022)
18. Shumway, R.H., Stoffer, D.S.: Time Series Analysis and Its Applications with R Examples. Springer (2017)
19. Vaswani, A., et al.: Attention is all you need. Adv. Neural Inf. Process. Syst. pp. 5998–6008 (2017)
20. Victor, A.O., Ali, M.I.: Enhancing time series data predictions: a survey of augmentation techniques and model performances. In: ACSW (2024)
21. Wang, J., et al.: Deep learning for multivariate time series imputation: a survey. CoRR **abs/2402.04059** (2024). https://doi.org/10.48550/arXiv.2402.04059

22. Wu, Z., et al.: A comprehensive survey on graph neural networks. IEEE Trans. Neural Netw. Learn. Syst. **32**(1), 4–24 (2021)
23. Wu, Z., Pan, S., Long, G., Jiang, J., Zhang, C.: Connecting the dots: Multivariate time series forecasting with graph neural networks (2020). arXiv preprint arXiv:2005.11650
24. Zeng, A., Chen, M., Zhang, L., Xu, Q.: Are transformers effective for time series forecasting? arXiv preprint arXiv:2205.13504 (2022)
25. Zhang, B., Song, C., Jiang, X., Li, Y.: Electricity price forecast based on the STL-TCN-NBEATS model. Heliyon **9**(1), e13029 (2023)

Spatio-Temporal Data and Molecular Dynamics: Challenges and Opportunities (Vision Paper)

Goce Trajcevski[1]([✉])(ID), Ashfaq Khokhar[1,2,3], Sohail Murad[2], and Cynthia Jameson[3]

[1] Iowa State University, Ames, IA 50010, USA
{gocet25,ashfaq}@iastate.edu
[2] Illinois Institute of Technology, Chicago, IL 60616, USA
murad@iit.edu
[3] University of Illinois at Chicago, Chicago, IL 60607, USA
cjjames@uic.edu

Abstract. The key challenge in plethora of natural sciences – chemistry, biology, biophysics, material science, etc. – is how to understand the behavior of evolving systems consisting of vast number of atoms belonging to different molecules that engage in various interactions. In addition to understanding the fundamental nature of the underlying processes, understanding this behavior is key to applications of societal relevance, such as drug design and discovery, modeling and prediction of protein structures, grain size evolution, etc. Due to safety concerns, as well as (prohibitive) costs, instead of experimental studies – especially in the preliminary research stages – Molecular Dynamics (MD) is an attractive alternative for generating trajectories of atoms that move in a 3D space and are subject to different laws of quantum physics and analytical chemistry. However, analytical expressions cannot be used to characterize every particle, nor can they be used with a confidence for predicting occurrence(s) of events of interest. In the recent years, multiple data analytics methods have been developed to analyze the vast datasets obtained via simulation. In this paper, we address questions related to a possible "push-pull" entaglement between: (1) how and to what extent can spatio-temporal data management and analytics improve the process of knowledge generation in the micro-world of molecular reactions? (2) are there issues from this realm (and using MD data) that can initiate novel challenges for spatio-temporal data management and analytics?

Keywords: Spatio-temporal data · Molecular Dynamics

1 Introduction and Motivation

Since the late 1990s the research related to spatio-temporal phenomena has generated a large body of results – as an example, a search on DBLP with

P. K. Chrysanthis et al. (Eds.): ADBIS 2025, CCIS 2676, pp. 46–55, 2026.
https://doi.org/10.1007/978-3-032-05727-3_5

'spatio-temporal' keyphrase yields (at the moment of preparing this submission) 12,541 matches. These results have had a significant practical impact on multiple applications of high societal relevance such as urban computing and smart cities [51] (with various facets – e.g., traffic management [48], crime [52]), ecology [35], wildlife [10], diseases modeling/tracking [31] – to name but a few.

As the field of spatio-temporal data management was getting shaped, multiple formalization efforts have been undertaken since its early days – from formalizing data types and operators to query constructs and processing algorithms along with indexing structures [20]. Plethora of research results targeting specific problems followed over the last two+ decades. Following the technological advances that enabled smart sensing capacities and Internet of Things (IoT) [15] and Big Data [44], many new challenges related to spatio-temporal data analytics emerged [3,21]. These, in turn, spurred many efforts in applying novel Machine Learning (ML) and deep learning techniques to address multiple problems related to spatio-temporal data enabling effective models for various classification and prediction tasks, offering novel challenges from data science perspective [27] and even yielding a well-established paradigm of Spatio-Temporal Graph Neural Networks (STGNN) [38].

What motivates this work is the observation that with all these advancements, to date, there has been no significant collaboration between the research tackling various facets of spatio-temporal data (management and querying; classification and prediction) and the field of chemistry studying evolution of molecular interactions. We postulate that effective management and querying of big spatio-temporal data [16] may yield significant contributions, as:

– There are large volumes of spatio-temporal data corresponding to motion of atoms participating in reactions of interest in many societaly critical applications – from drug manufacturing to new materials development.
– There are multiple constraints on that motion – from atoms belonging to molecules, to various forces/laws governing the interactions between atoms in different molecules.
– There are unique semantic constraints on reactions, and their detection (events) and tracking pose significant computational challenges.

In the rest of this paper, after presenting some background information and setting the context in Sect. 2, in Sect. 3 we will make a case that there is a significant potential for a "push-pull" kind of entaglement between multiple topics from spatio-temporal data management and the dynamics of molecular interaction. Specifically, we will demonstrate that: (a) several well-established topics in spatio-temporal data management are likely to enable novel findings of interest to dynamic chemical interactions; (b) the existing solutions cannot be readily applied and they need to overcome distinct challenges. We present conclusions in Sect. 4

2   Preliminaries

Typically, a chemical reaction between two mollecules occurs when certain conditions are met like, for example, proximity between specific atoms and the angular position between the molecular structures to which those atoms belong. In such cases, the existing molecular structures are disrupted and new kinds of structures are formed. A common category of such reactions is one of forming various *bonds* (e.g., Hydrogen Bond (HB), ionic bond, π-π, etc.).

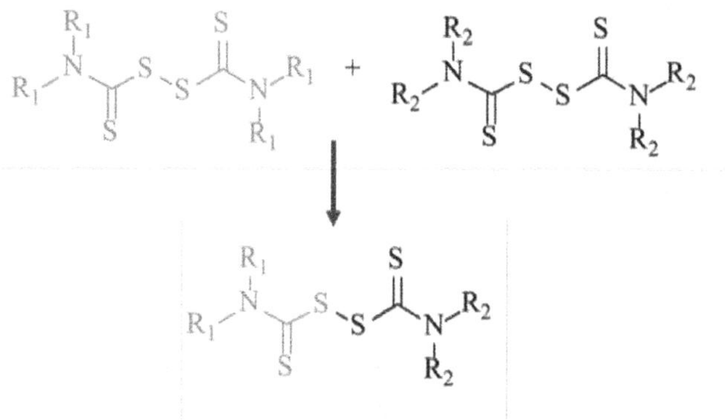

Fig. 1. Typical illustration of bonds formation

Most often, when it comes to modeling, such reactions are illustrated with 2D schematics as shown in Fig. 1 (adapted from [34]) which illustrates a case of dynamic disulfide exchange reaction of a molecular model containing an okuranum disulfide bond. However, while such structural graphs may convey certain information, the reality is rather different. Firstly, upon starting a particular reaction by mixing specific compounds, there are a lot of atoms and molecules involved and the motion is in 3D (cf. Fig. 2(a)) – e.g., a mixture of 10 flavanone drug molecules and 4 polymer substrates amounts to over 6000 atoms, and reactions last for over 200,000 timestamps. Secondly, it is not only the proximity of the molecules that matters. Another criterion is the relative spatial orientation of the corresponding structure (aka rings – which also varies over time) that is an important criterion for forming particular bond, as shown schematically in Fig. 2(b) (cf. [47]). Clearly, there are other laws (i.e., domain-based semantics) that govern the physical properties, which, in turn, provide yet another kind of constraints on the occurrence of reactions.

Among the key concerns of in many practical applications are: (a) generating real-time data via dynamic system measurements using actual sensors is extremely costly; (b) the exploratory stages (e.g., for drug discovery) are extremely slow, costly and sometimes involve risks. Complementary to these, any detection, classification and prediction tasks based on data science and ML that would aid the process, require large volumes of data [16]. To alleviate

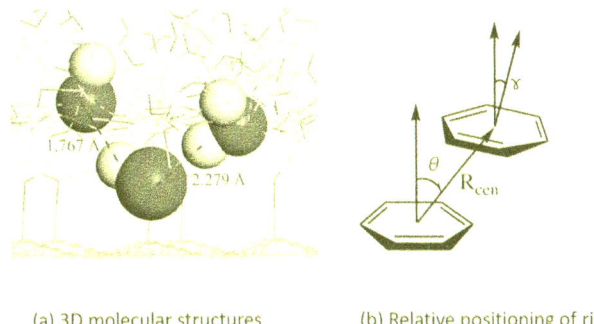

(a) 3D molecular structures   (b) Relative positioning of rings

Fig. 2. Reactions and bonds in 3D

these issues, domain scientists often resort to simulation methods. While traditional (e.g., Monte Carlo based) ones provide insights, the most popular tool used in drug discovery and protein activity simulation is *Molecular Dynamics* (MD) [9]. Using corresponding models for capturing the internal forces dictating interactions (e.g., Hooke's law; Lennard-Jones potential) MD simulations (MDS) typically solve numeric equations corresponding to Newtonian motion, enabling structural fluctuations to be observed over time [1]. Such MDS datasets capturing the evolution of participating chemical entities have proven to provide important insights into DNA structures [6], protein interactions binding [11], cancerous cell membranes and drug delivery [18], polymer formation [29] and chiral separation [47] – to name but a few applications[1].

However, despite its high utility in terms of enabling insights in multiple case-studies, MDS faces several challenges:

1. The detection of "interesting events" is often done by scanning the entire output dataset and based on domain expertise/experience.
2. The computations are resource-heavy and generate large volumes of data, which is often stored as plain csv files containing *(x, y, z, t)* motion data, augmented with certain annotations (e.g., atom type, corresponding molecule, etc.).
3. The data is also accompanied by structural constrains, in terms of (evolving) spatial orientation of the molecular rings.

As a complementary observation, we note that in the recent years multiple works have tried to capitalize on ML techniques to use MDS data for specific purposes [5,16,25,32,37,39,42]. However, the models developed mostly pertain to classifying/predicting specific states/motifs – not tackling evolving and/or persistent events.

---

[1] Once again, for illustration, a DBLP search with 'molecular dynamics' yields 3071 different publications (with a note that DBLP need not be the most relevant source, as it does not index some journals from chemistry).

3 Spatio-Temporal Data Challenges

We now discuss several existing categories of research in spatio-temporal data and identify challenges that, if overcome, are likely to increase the efficiency and effectiveness of discoveries in the domain of chemical interactions. We respectfully (and with apologies to multiple researchers in the field of spatio-temporal data) note that due to space constraints we confine to citing survey papers or some representative works for illustrating a particular context – as we assume a relative familiarity of the topics by the targeted audience.

3.1 Query Processing

The results in processing spatio-temporal queries (e.g., range, k-Nearest Neighbor, join, etc.) abound [3], relying on the paradigm of *filtering* (using index structures [24,30]) to eliminate unqualified data) and *refinement* (using algorithms to process the post-filtering data). Both range and k-NN queries would be helpful in analyzing MDS data – for example, for fast extraction of potential zones of interest, or for detecting potential formation of a bond. As a simple example, consider the query: *"Retrieve all the molecules containing Hydrogen atoms which are within 6 Angstroms from molecules containing Nitrogen atoms for at least 20 consecutive time-steps"*. However, the state of the art (SoA) falls short of addressing such queries because: (1) The motion of the data is in 3D, for which a straightforward adaptation of data-oriented and space-oriented indexes may be inadequate (e.g., dead-space; too many false positives); (2) MDS data is annotated with semantic descriptors. One may be tempted to adapt the works from *semantic trajectories* [36] – however, an additional challenge that needs to be tackled is that the data also has structural constraints (i.e., the rings and their orientation), for which the IG-tree based indexes are inadequate.

Two particular categories of queries that may be of interest are the *pattern* and the *co-moving* (i.e., clustered motion such as flocks, platoons, convoys, etc.) queries [23]. The reason is that the chemists are often interested in transitions among (sequence of) states, as well as persistency of certain events (e.g., a bond). However, in addition to the reasons identified above (3D motion, semantics not only in types but also in structures) there a peculiar tollerances that may need to be incorporated – e.g., a particular bond may break and re-form, for as long as the gap in time is below certain threshold.

3.2 Spatio-Temporal Data Compression

Many works have addressed the problems of spatio-temporal (and, in particular, trajectories) data compression as well as reduction and simplification [13]. Multiple results have been generated focusing on subtle properties such as direction awareness and stay-points preservation, as well as properties of the (source) application domain (e.g., maritime, air vessels, pedestrians, GPS vs. vibration sensors tracking, etc.). In addition, some global properties such as error-bounds have been considered [22].

In addition to the data-related issue outlined in Sect. 3.1, there are a few more challenges related to MDS data reduction: (1) Different kinds of *generalization* techniques will need to be employed to preserve semantic properties during the compression; (2) New *distance functions* and *interestingness* measures [49] will be needed.

While the issues/challenges above persist – a specific application of spatio-temporal data reduction is the *online* variant [46]. However, in addition to the sheer reduction of volume, of particular interest in MDS setting is to reduce the overall simulation process, due to the heavy computational demands [2]. In part, this calls for novel ML methods to effectively recommend how to steer (or, eventually, early-terminate) the simulation.

3.3   Visual Analytics and Simulation

The objective of visual analytics is to optimize the use of human vision in the processes of understanding and analysis of interesting phenomena in targeted domains. In the context of chemistry, several tools have recently been developed for that purpose – e.g., interactive exploration of chemical spaces [17]; graph-based recognition of chemical structures [28]; and visualization of interesting events in MDS [14]. However, in our assessment, these tools do not yet possess the maturity and robustness of the existing tools from spatio-temporal visual analytics [4] in terms of well-defined workflows that can be ported with minimal overheads. More importantly, they lack the formalization of querying capabilities (such as TimeMask [4]) for speeding up the retrieval of the data. However, the TimeMask paradigm cannot be straighforwardly carried to the domains of MDS data due to its peculiarities in terms of semantic features and 3D motion – which present other research challenges.

While it may seem a bit like a paradox (given how well-established is the MDS and its ability to provide various insights), it is likely that some aspects of spatio-temporal simulations may bring benefits for investigating chemical interactions. Namely, most of the MDS in its core have numerical simulations that observe the constraints of physical laws. However, taking the perspective of treating atoms and molecules like "smart mobile agents" [19] may yield different insights, opening the room for further benefits of experiences from spatio-temporal visualizations in terms of social networks [4].

3.4   Machine Learning and MDS

As indicated (cf. Sect. 2), recent research has investigated the use of ML techniques for MDS data (see [33] for a more recent survey). However, we postulate that many of the paradigms used in the existing results in spatio-temporal data settings are still to be fully exploited to improve the efficiency and effectiveness in the realm of MDS. For example, GNN based solutions have been recently used to improve the explainability of formation of molecular structures in terms of influential components [43,50]. However, the underlying models are not properly considering the impact of the temporal evolution – and, as recently demonstrated in [40], an important aspect is not only the formation of a particular bond, but also its persistence (i.e., how long does that bond last). As a complementary example – Metalearning has been used (e.g., for fine-tuning simulations [8]) in MDS. However, the different aspects such as data sparsity and combined classification and prediction [53] or explainability [45] that have been investigated in spatio-temporal settings may yield potential improvements in MDS context. Similarly, federated learning has also been used in MDS research, enabling a privacy-preserving collaborative sharing [12]. But, once again, the context is focused on topological/spatial structures, without proper exploitation of the federated learning in spatio-temporal evolution domains [7]. We believe that one of the main obstacles is that MDS data typically has various semantic constraints (e.g., only certain atoms from certain molecules can spur a particular reaction/bond formation), which is yet to be addressed by the spatio-temporal ML community.

We close this section with an observation that an emerging trend in computational chemistry is quantum computing [26], and its potentials in spatial data and localization have only recently been explored [41] – leaving opportunities for spatio-temporal quantum computing.

4   Concluding Remarks

We presented a vision of potentially advancing the research of evolving chemical interactions and MD with efficient and effective methods rooted in spatio-temporal data management and analytics (e.g., queries processing, data reduction, ML for spatio-temporal data, visual analytics, ML). Conversely, we also demonstrated that the MD types of problems will pose novel challenges in the existing areas of spatio-temporal data management and analytics. We hope that this coupling will also contribute to similar opportunities for spatio-temporal data techniques being identified and tackled for other scientific domains (e.g., astrophysics, biophysics, etc.).

References

1. Adcock, S.A., McCammon, J.A.: Molecular dynamics: survey of methods for simulating the activity of proteins. Chem. Rev. **106**(5) (2006)
2. Ahmadian, M., Zhuang, Y., Hase, W.L., Chen, Y.: Data reduction through increased data utilization in chemical dynamics simulations. Big Data Res. **9**, 57–66 (2017)
3. Alam, M.M., Torgo, L., Bifet, A.: A survey on spatio-temporal data analytics systems. ACM Comput. Surv. **54**(10s), 219:1–219:38 (2022)
4. Andrienko, N.V., Andrienko, G.L.: Spatio-temporal visual analytics: a vision for 2020s. J. Spatial Inf. Sci. **20**(1), 87–95 (2020)
5. Astero, M., Rousu, J.: Learning symmetry-aware atom mapping in chemical reactions through deep graph matching. J. Cheminform. **16**(46) (2024)
6. Barsky, D., Foloppe, N., Ahmadia, S., III, D.M.W., Jr., A.D.M.: New insights into the structure of ABASIC DNA from molecular dynamics simulations. Nucleic Acids Res. **28**(13) (2000)

7. Belal, Y., Mokhtar, S.B., Haddadi, H., Wang, J., Mashhadi, A.: Survey of federated learning models for spatial-temporal mobility applications. ACM Trans. Spatial Algorithms Syst. **10**(3), 18:1–18:39 (2024). https://doi.org/10.1145/3666089, https://doi.org/10.1145/3666089

8. Chen, J., et al.: Mixup-augmented meta-learning for sample-efficient fine-tuning of protein simulators (2023)

9. Durrant, J., McCammon, J.: Molecular dynamics simulations and drug discovery. BMC Biol. **9**(71) (2011)

10. Focardi, S., Cagnacci, F.: Animal movement. In: Mobility Data: Modeling, Management, and Understanding. Cambridge University Press (2013)

11. Fu, Y., Zhao, J., Chen, Z.: Insights into the molecular mechanisms of protein-ligand interactions by molecular docking and molecular dynamics simulation: a case of oligopeptide binding protein. Comput. Math. Methods Medicine **2018** (2018)

12. Hanser, T.: Federated learning for molecular discovery. Curr. Opin. Struct. Biol. **79** (2023). https://doi.org/10.1016/j.sbi.2023.102545

13. Herrero, D.A., Pedroche, D.S., García, J., Molina, J.M.: Review and classification of trajectory summarisation algorithms: From compression to segmentation. Int. J. Distributed Sens. Networks **17**(10) (2021)

14. Jaeger-Honz, S., Klein, K., Schreiber, F.: Systematic analysis, aggregation and visualisation of interaction fingerprints for molecular dynamics simulation data. J. Cheminformatics **16**(1), 28 (2024)

15. Jin, B., Zhuo, W., Hu, J., Chen, H., Yang, Y.: Specifying and detecting spatio-temporal events in the internet of things. Decis. Support Syst. **55**(1), 256–269 (2013)

16. Rodrigues Jr., J.F., Florea, L., de Oliveira, M.C.F., Diamond, D., Jr., O.N.O.: A survey on big data and machine learning for chemistry. CoRR abs/1904.10370 (2019). http://arxiv.org/abs/1904.10370

17. Kale, B., Clyde, A., Sun, M., Ramanathan, A., Stevens, R., Papka, M.E.: Chemograph: interactive visual exploration of the chemical space. Comput. Graph. Forum **42**(3), 13–24 (2023)

18. Kordzadeh, A., Zarif, M., Amjad-Iranagh, S.: Molecular dynamics insight of interaction between the functionalized-carbon nanotube and cancerous cell membrane in doxorubicin delivery. Comput. Methods Programs Biomed. **230**, 107332 (2023)

19. Kotnana, S., Han, D., Anderson, T., Züfle, A., Kavak, H.: Using generative adversarial networks to assist synthetic population creation for simulations. In: ANNSIM, pp. 1–12. IEEE (2022)

20. Koubarakis, M., et al. (eds.): Spatio-Temporal Databases: The CHOROCHRONOS Approach. Lecture Notes in Computer Science. Springer (2003). https://doi.org/10.1007/b83622

21. Liang, H., Zhang, Z., Hu, C., Gong, Y., Cheng, D.: A survey on spatio-temporal big data analytics ecosystem: resource management, processing platform, and applications. IEEE Trans. Big Data **10**(2), 174–193 (2024). https://doi.org/10.1109/TBDATA.2023.3342619

22. Lin, X., Ma, S., Jiang, J., Hou, Y., Wo, T.: Error bounded line simplification algorithms for trajectory compression: an experimental evaluation. ACM Trans. Database Syst. **46**(3), 11:1–11:44 (2021)

23. Loglisci, C.: Using interactions and dynamics for mining groups of moving objects from trajectory data. Int. J. Geogr. Inf. Sci. **32**(7), 1436–1468 (2018). https://doi.org/10.1080/13658816.2017.1416473

24. Mahmood, A.R., Punni, S., Aref, W.G.: Spatio-temporal access methods: a survey (2010–2017). GeoInformatica **23**(1), 1–36 (2019)

25. Majumdar, S., Palma, F.D., Spyrakis, F., Decherchi, S., Cavalli, A.: Molecular dynamics and machine learning give insights on the flexibility-activity relationships in tyrosine KINOME. J. Chem. Inf. Model. **63**(15), 4814–4826 (2023)

26. Metcalf, M., Bauman, N.P., Kowalski, K., de Jong, W.A.: Resource-efficient chemistry on quantum computers with the variational quantum eigensolver and the double unitary coupled-cluster approach. J. Chem. Theory Comput. **16**(10) (2020)

27. Mokbel, M., et al.: Mobility data science: Perspectives and challenges. ACM Trans. Spatial Algorithms Syst. (2024). just Accepted

28. Morin, L., et al.: Molgrapher: graph-based visual recognition of chemical structures. In: ICCV (2023)

29. Nedyalkova, M., Russo, G., Loche, P., Lattuada, M.: Revealing the formation dynamics of Janus polymer particles: Insights from experiments and molecular dynamics. J. Chem. Inf. Model. **63**(23), 7453–7463 (2023). https://doi.org/10.1021/ACS.JCIM.3C01547

30. Nguyen-Dinh, L., Aref, W.G., Mokbel, M.F.: Spatio-temporal access methods: Part 2 (2003–2010). IEEE Data Eng. Bull. **33**(2), 46–55 (2010)

31. Orozco-Acosta, E., Adin, A., Ugarte, M.D.: Big problems in spatio-temporal disease mapping: methods and software. Comput. Methods Programs Biomed. **231**, 107403 (2023). https://doi.org/10.1016/J.CMPB.2023.107403

32. Prasnikar, E., Ljubic, M., Perdih, A., Borisek, J.: Machine learning heralding a new development phase in molecular dynamics simulations. Artif. Intell. Rev. **57**(4), 102 (2024)

33. Prasnikar, E., Ljubic, M., Perdih, A., Borisek, J.: Machine learning heralding a new development phase in molecular dynamics simulations. Artif. Intell. Rev. **57**(4), 102 (2024). https://doi.org/10.1007/S10462-024-10731-4

34. Qingjie, M., Liang, L., Yanfeng, Z., Yunpeng, G., Jun, L.: Progress in sulfur-containing dynamic polymers. J. Funct. Polym. **37**(1) (2024)

35. Raffaetà, A., et al.: An application of advanced spatio-temporal formalisms to behavioural ecology. GeoInformatica **12**(1), 37–72 (2008)

36. Renso, C., Bogorny, V., Tserpes, K., Matwin, S., de Macêdo, J.A.F.: Multiple-aspect analysis of semantic trajectories(master). Int. J. Geogr. Inf. Sci. **35**(4), 763–766 (2021)

37. Ru, Z., Wu, Y., Shao, J., Yin, J., Qian, L., Miao, X.: A dual-modal graph learning framework for identifying interaction events among chemical and biotech drugs. Briefings Bioinform. **24**(5) (2023)

38. Sahili, Z.A., Awad, M.: Spatio-temporal graph neural networks: a survey. CoRR abs/2301.10569 (2023)

39. Schütte, C., Klus, S., Hartmann, C.: Overcoming the timescale barrier in molecular dynamics: transfer operators, variational principles and machine learning. Acta Numer **32**, 517–673 (2023)

40. Shamail, A., Anowar, M.H., Trajcevski, G., Khokhar, A., Murad, S., Jameson, C.J.: Bond-aware moving clusters of atomic trajectories with relaxed persistency. In: Proceedings of the 32nd ACM SIGSPATIAL, pp. 605–608. ACM (2024). https://doi.org/10.1145/3678717.3691298

41. Shokry, A., Youssef, M.: Towards quantum computing for location tracking and spatial systems. In: SIGSPATIAL '21: 29th International Conference on Advances in Geographic Information Systems (2021)

42. Stavrogiannis, C., Sofos, F., Sagri, M., Vavougios, D., Karakasidis, T.E.: Twofold machine-learning and molecular dynamics: a computational framework. Comput. **13**(1), 2 (2024)

43. Tai, W., Zhong, T., Trajcevski, G., Zhou, F.: Redundancy undermines the trust-worthiness of self-interpretable GNNs. In: International Conference on Machine Learning (ICML) (2025). (accepted, to appear)
44. Tan, H., Luo, W., Ni, L.M.: Clost: a hadoop-based storage system for big spatio-temporal data analytics. In: CIKM (2012)
45. Tang, J., Xia, L., Huang, C.: Explainable spatio-temporal graph neural networks. In: Proceedings of the 32nd ACM International Conference on Information and Knowledge Management, CIKM (2023)
46. Trajcevski, G., Cao, H., Scheuermann, P., Wolfson, O., Vaccaro, D.: On-line data reduction and the quality of history in moving objects databases. In: Fifth ACM International MobiDE Workshop, pp. 19–26. ACM (2006)
47. Wang, X., Jameson, C.J., , Murad, S.: Molecular dynamics simulations of chiral recognition of drugs by amylose polymers coated on amorphous silica. Mol. Phys. **119**(1920) (2021)
48. Yang, Y., Xu, Y., Han, J., Wang, E., Chen, W., Yue, L.: Efficient traffic congestion estimation using multiple spatio-temporal properties. Neurocomputing **267**, 344–353 (2017)
49. Zhang, Y., Paquette, L.: An effect-size-based temporal interestingness metric for sequential pattern mining. In: Proceedings of EDM (2020)
50. Zhao, T., Luo, D., Zhang, X., Wang, S.: Towards faithful and consistent explanations for graph neural networks. In: International Conference on Web Search and Data Mining, pp. 634–642 (2023)
51. Zheng, Y., Capra, L., Wolfson, O., Yang, H.: Urban computing: concepts, methodologies, and applications. ACM Trans. Intell. Syst. Technol. **5**(3), 38:1–38:55 (2014). https://doi.org/10.1145/2629592
52. Zhou, B., Chen, L., Zhao, S., Zhou, F., Li, S., Pan, G.: Spatio-temporal analysis of urban crime leveraging multisource crowdsensed data. Pers. Ubiquitous Comput. **27**(3), 599–612 (2023)
53. Zhou, F., Liu, X., Zhong, T., Trajcevski, G.: Metamove: on improving human mobility classification and prediction via metalearning. IEEE Trans. Cybern. **52**(8), 8128–8141 (2022)

From ER Conceptual Models
to Document-Based NoSQL Logical
Models

Alberto Belussi and Sara Migliorini(✉)

Department of Computer Science, University of Verona (Italy), Verona, Italy
{alberto.belussi,sara.migliorini}@univr.it

Abstract. NoSQL systems have gained popularity in recent years due
to their capability to deal with huge amounts of information, even in an
unstructured way. The schemaless nature of NoSQL databases leads to
the widespread opinion that no modeling activity is required for them.
However, in the literature, many researchers recognized the need for
the classical three levels of modeling also in this kind of system. This
paper proposes a solution for the conceptual and logical modeling of a
document-based database, which starts from the well-known ER conceptual model and automatically translates it into a set of document
collections through predefined labeling and translation rules. The usage
of a standard conceptual model, like the ER one, and the JSON Schema
language as a document-based logical model, promotes the usage of the
proposed rules, as well as the modeling activity in the NoSQL world.

Keywords: NoSQL · document-based · modeling · translation rules

1 Introduction

Relational models have been the standard logical model for database construction for a long time, starting back in the 1980 s till the early 21st century. However, in recent years, NoSQL systems have gained popularity over the former due
to their ability to manage big data properly. In general, NoSQL systems provide greater flexibility in organizing data since they relax the rigidity provided
by relational models, at the detriment of other properties like full consistency,
which is replaced by eventual consistency. Several different families of NoSQL
databases are available, each one tailored for a specific task or need. Accordingly to [7], we can identify the following classification: graph databases, column-
oriented databases, key-value databases, and document-oriented databases. Each
of these families is particularly tailored to solve a different limitation of relational
databases for a given purpose. In particular, the usefulness of NoSQL databases
is recognized when the data of interest do not fit well in the rigid structure of
flat relational tables, where instead document-based systems allow more complex and rich nesting of objects, or if the access to data consists in only simple

P. K. Chrysanthis et al. (Eds.): ADBIS 2025, CCIS 2676, pp. 56–66, 2026.
https://doi.org/10.1007/978-3-032-05727-3_6

read-write operations, like in big data parallel computations for which column-oriented or key-value databases provides better support and scalability [6,9].

Regardless of the specific family, all NoSQL databases are considered schemaless in nature, which leads to the opinion that the modeling phase is irrelevant for them. Therefore, NoSQL databases are mostly conceived only at the physical design level, following a set of storage and structural rules regulated by each specific database family. However, several authors have observed that the development of high-level methodologies and tools supporting NoSQL database design is needed [2,3,8]. Concluding that also the design of such kind of databases can be divided into the classical three levels of abstraction: conceptual, logical, and physical [10]. In particular, there is still the need for a conceptual model to define how data will be structured in the database from a higher-level point of view [1]. Despite the fact that several authors highlighted the need for NoSQL database modeling, only some of them provide a real solution to the problem. More specifically, many works only concentrate on the definition of a unified logical model for all families of NoSQL databases [5] or, eventually, in the extension of existing conceptual models, taking UML class diagrams as a reference base model, to properly capture the abstractions underlying a specific NoSQL family [4].

In this paper, we take a different perspective, which starts from the original definition of conceptual model: a formalism that is useful to define the structure of data of interest independently from the underlying system and implementation, and so the operations to be performed. In this light, a conceptual model cannot depend on the choice of a NoSQL system rather than a relational one. As a specific formalism, we consider the Entity-Relationship (ER) model due to its traditional widespread adoption for conceptual database modeling. Conversely, relative to the logical model, we refer again to the original definition, which talks about a model strictly related to the data model implemented by the chosen system, even without considering the specific physical details. Therefore, in our opinion, each family of NoSQL databases can be considered a different target data model which can lead to different logical schemas. In particular, in this paper, we concentrate on the document-based family due to its common usage in application contexts that use data in a way very similar to relational models. Given that, we propose a set of standard translation rules similar to the one defined in literature for the translation from the ER to the relation model [11]. In this case, the translation procedure is a two-step process: a labeling step, in which a set of labels is added to a standard ER conceptual model to identify the encapsulation choices, and a translation step, in which the labeled conceptual schema can be directly translated into a set of document collections. The main contribution of the paper is the identification of a set of labeling and translation rules that allow the translation of a standard ER conceptual model into a document-based logical model. The advantage is clearly the possibility of exploiting previous knowledge and experience about the ER conceptual modeling or even starting from an existing ER conceptual model and translating it into a set of document collections instead of into a set of relational tables. The use of a standard modeling language extends the applicability of the approach.

2   Labeling and Translation Rules

This section introduces the proposed set of labeling and translation rules from an ER conceptual model to a document-based relational model described using the JSON Schema syntax [12]. The formalization of such rules requires starting from the definition of the conceptual schema.

Definition 1 (ER conceptual schema). *A database ER conceptual schema S is a tuple $\langle \mathcal{E}, \mathcal{R} \rangle$, where $\mathcal{E}$ is the set of entities, while $\mathcal{R}$ is the set of relations.*

Each entity $E_i \in \mathcal{E}$ can be represented as a tuple $\langle \ell_i, A_i = \{a_1, \ldots, a_{n_i}\}\rangle$, where ℓ_i is the name of the entity (or label) and A_i is the set of attributes of E_i. Each attribute $a = \langle \ell_a, \tau_a, c_a = (min, max)\rangle \in A$ is characterized by a name ℓ_a, a type τ_a, and a cardinality c_a denoting the minimum (i.e., $c_a.min$) and maximum (i.e., $c_a.max$) number of attribute values for such attribute for each instance of E_i.

Each relation $R \in \mathcal{R}$ is a tuple $\langle E_h, E_k, c_h, c_k, A_r \rangle$ where E_h and E_k are the two entities involved in the relation, c_h and c_k are the cardinalities on the E_h and E_k side, respectively, and A_r is the set of attributes of R whose representation is the same used for entity attributes.

Notice that in an ER model, relationships typically do not have a direction, so $\langle E_h, E_k, c_h, c_k, A_r \rangle = \langle E_k, E_h, c_k, c_h, A_r \rangle$ are the same relation and are represented only once. In the following, we will indifferently reference a particular direction depending on the specific needs.

Relatively to the NoSQL logical model, we reference the document-based one provided by MongoDB in which a database is a set of collections, and each collection is a named group of documents. Each document is a complex value represented as a set of attribute-value pairs, where the type of each attribute could be a simple value, an array (i.e., a list), or a nested document. A main document is a top-level document with a unique identifier represented by a special attribute _id associated with a special type ObjectID.

Definition 2 (JSON logical schema). *A database JSON schema is a logical schema in which the information is organized into a set of documents that are gathered together in collections. Such schema could be represented as a set of collections $\mathcal{C} = \{C_1, \ldots, C_n\}$, where each collection contains a set of documents $C_i = \{D_1, \ldots, D_m\}$ representing a homogeneous piece of information.*

Given a database conceptual schema S, its translation into a set of object collections $\mathcal{C}$ is obtained through the application of the following rules, whose application assumes the preliminary satisfaction of the following hypothesis.

Hypothesis 1 (Preliminary hypothesis). Let us consider a database conceptual schema $S = \langle \mathcal{E}, \mathcal{R} \rangle$, the following hypotheses need to be satisfied before performing the labeling and translation phases towards a JSON schema $\mathcal{C}$:

(i) S has already been restructured: namely, it does not contain generalization, and all the relations between more than two entities (n-th relations) have been removed and replaced with the appropriate set of binary relations.

(ii) The set $\mathcal{C} = \{C_1, \ldots, C_n\}$ of document collections have been identified based on the operations to be performed on such data. In particular, there is at most one collection $C \in \mathcal{C}$ for each entity $E \in \mathcal{E}$, and each collection $C \in \mathcal{C}$ corresponds to one and only one entity $E \in \mathcal{E}$, which is called *main entity*. The set of all identified main entities is denoted as $\mathcal{P} \subseteq \mathcal{E}$.

Given such preliminary hypotheses, the first step in the generation of the JSON logical schema consists of labeling the conceptual schema $\mathcal{S}$ through the rules LR 1-4 formalized below. The labeling essentially consists in the addition inside $\mathcal{S}$ of: a label for each entity corresponding to the name of a main document or of an object property inside another document, an encapsulation arrow (i.e., thick gray arrow) from each entity to be encapsulated towards its encapsulating entity, and a reference arrow (i.e., thin black arrow) from each entity to be referenced towards its referencing entity.

The labeling starts with the labeling of entities identified as the main ones.

Labeling Rule 1 (Main entity). Given an ER conceptual schema $\mathcal{S} = \langle \mathcal{E}, \mathcal{R} \rangle$ and a desired set of document collections $\mathcal{C} = \{C_1, \ldots C_n\}$ to be obtained, each entity $P_i \in \mathcal{P} \subseteq \mathcal{E}$, which has been identified as a main entity in $\mathcal{S}$, is labeled as in Fig. 1.a with a label C_i representing the name of the corresponding main document collection in $\mathcal{C}$.

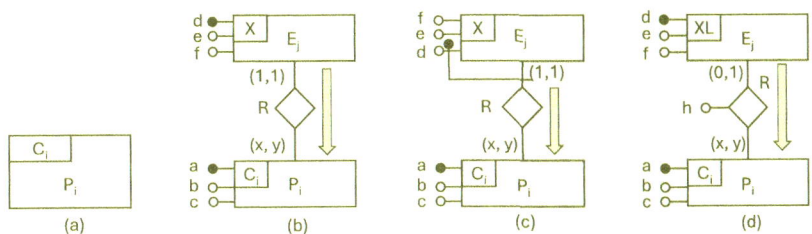

Fig. 1. (a) LR 1 for a main entity $P_i \in \mathcal{E}$. (b)-(d) First two cases of LR 2.

Labeled the set of main entities, we can then continue with the entities that have not been identified as main ones, but are connected to one of them.

Labeling Rule 2 (Entity connected to a unique main entity). Given an ER conceptual schema $\mathcal{S} = \langle \mathcal{E}, \mathcal{R} \rangle$ and a desired set of document collections $\mathcal{C} = \{C_1, \ldots C_n\}$ to be obtained, each entity $E_j \in \mathcal{E} \setminus \mathcal{P}$, which has not been labeled as a main one but is connected through a relation $R \in \mathcal{R}$ to a unique main entity $P_i \in \mathcal{P}$, will be encapsulated into P_i. Figure 1.a-c and Fig. 2.a-b illustrate the different possible configurations which depend on the cardinality of the relation $R = \langle E_j, P_i, c_j, c_i, A \rangle \in \mathcal{R}$ connecting E_j with P_i:

1. R *has cardinality one-to-one or one-to-many, with* $c_j = (1,1)$ (see Fig. 1.a and Fig. 1.b). Entity E_j is labeled as X, and an encapsulation arrow is added from E_j to P_i.

2. *R has cardinality one-to-one or one-to-many with* $c_j = (0,1)$ (see Fig. 1.c). Entity E_j is labeled as XL, and an encapsulation arrow is added from E_j to P_i. The instances of E_j that are not connected with any instance P_i will be lost. This is represented by the L inside the label of E_j.
3. *R has cardinality many-to-many with* $c_j = (1,N)$ (see Fig. 2.a). Entity E_j is labeled as XR, and an encapsulation arrow is added from E_j to P_i. Each instance of E_j that is connected to multiple instances of P_i will be represented multiple times. This is represented by the R inside the label of E_j.
4. *R has cardinality many-to-many with* $c_j = (0,N)$ (see Fig. 2.b). Entity E_j is labeled as XRL, and an encapsulation arrow is added from E_j to P_i. In this case, each instance of E_j that is connected to multiple instances of P_i will be represented multiple times, and the instances of E_j that are not connected to any instance of P_i will be lost. This is represented by the presence of both the R and the L inside the label of E_j.

The remaining entities are the ones that are neither main ones nor connected with one of them, but are connected to another non-main entity.

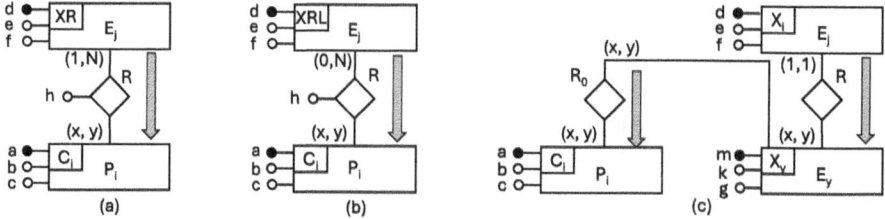

Fig. 2. (a)-(b) Last two cases of LR 2. (c) Application of LR 3.

Labeling Rule 3 (Entity connected to another Entity). Given an ER conceptual schema $S = \langle \mathcal{E}, \mathcal{R} \rangle$ and a desired set of document collections $\mathcal{C} = \{C_1, \ldots C_n\}$ to be obtained, each entity $E_j \in \mathcal{E} \setminus \mathcal{P}$, which is not connected to a main entity but is connected to another entity $E_y \in \mathcal{E} \setminus \mathcal{P}$ connected to a main entity $P_i \in \mathcal{P}$, will be encapsulated into E_y. This situation is depicted through the encapsulation arrow in Fig. 2.c. In this case, the label X_i associated with E_i depends on the cardinality c_y of the relation $R = \langle E_j, E_y, c_j, c_y, A \rangle \in \mathcal{R}$ connecting E_j with E_y:

- If $c_y = (1,1)$, then the label X_i is followed by the same suffix L, R or LR, following X_y, if present.
- If $c_y \neq (1,1)$, then the label X_i becomes: X_iL if $c_y = (0,1)$, X_iR if $c_y = (1,N)$, or X_iLR if $c_y = (0,N)$.

The final labeling step regards the relations between two main entities.

Labeling Rule 4 (Relation between two main entities). Given an ER conceptual schema $S = \langle \mathcal{E}, \mathcal{R} \rangle$ and a desired set of document collections $\mathcal{C} = \{C_1, \ldots C_n\}$ to be obtained, for each binary relation $R \in \mathcal{R}$ between two main entities $P_i, P_j \in \mathcal{P} \subseteq \mathcal{E}$, it is necessary to decide in which of the two entities the encapsulation is performed. The possible choices are:

- R is encapsulated in one of the two main entities P_i or P_j, as illustrated in Fig. 3.a and Fig. 3.b, respectively. In this case, a reference arrow is added with the label *id* from P_i to P_j, or vice versa.
- R is encapsulated in both entities P_i and P_j as in Fig. 3.c. This choice generates redundancy and is depicted by the reference arrow with a double direction labeled as R.

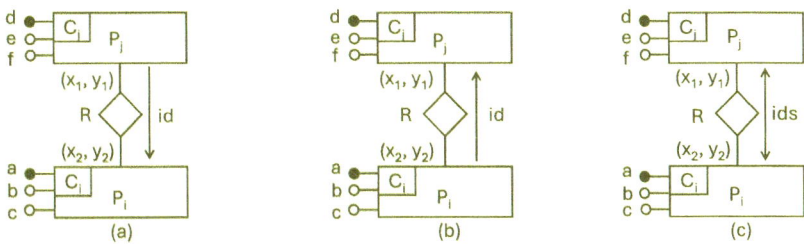

Fig. 3. Application of LR 4.

The same labeling rule also applies in the presence of entities that are not main entities but have already been encapsulated by using one of the previous rules. In this case, a reference arrow is added to the conceptual schema S to denote where the identifier of this entity will be referenced.

The application of the labeling rules LR 1–4 produces a so-called labeled conceptual schema defined in the following way.

Definition 3 (Labeled conceptual schema). *Given a database conceptual schema $S = \langle \mathcal{E}, \mathcal{R} \rangle$ and a desired set of document collections $\mathcal{C} = \{C_1, \ldots C_n\}$ to be obtained, the labeled conceptual S^ℓ, obtained from S by applying the rules LR 1–4, is a tuple $\langle \mathcal{E}, \mathcal{R}, \lambda, \alpha, \rho \rangle$ where:*

- *$\lambda : \mathcal{E} \to \Lambda$ is the function assigning to each entity its corresponding label. The set Λ of possible entity labels includes the name of the collections in $\mathcal{C}$ together with the set of property labels defined for the entities that are not main, with the corresponding suffix L or R or both.*
- *$\alpha : \mathcal{R} \to \Delta$ is the function associating to each relation $R \in \mathcal{R}$ the corresponding encapsulation arrow, if assigned.*
- *$\rho : \mathcal{R} \to \mathcal{F}$: is the function associating to each relation $R \in \mathcal{R}$ the corresponding reference arrow, if assigned.*

Given a labeled conceptual schema $\mathcal{S}^\ell$ obtained by applying the rules LR 1–4, its translation towards the desired set of document collections $\mathcal{C} = \{C_1, \ldots C_n\}$ is obtained through the rules TR 1–4 formalized below. The first step creates a document schema (i.e., collection) corresponding to each identified main entity.

Transaction Rule 1 (Translation of a main entity). Given a labeled conceptual schema $\mathcal{S}^\ell = \langle \mathcal{E}, \mathcal{R}, \lambda, \alpha, \rho \rangle$, a desired set of document collections $\mathcal{C} = \{C_1, \ldots C_n\}$ to be obtained, and a main entity $P_h = \langle \ell, A = \{a_1, \ldots, a_n\} \rangle \in \mathcal{P} \subseteq \mathcal{E}$ that has been labeled as C_h, each instance i of P_h will be translated into a document $D_i \in C_h$ comply with the following JSON schema:

```
Dᵢ = { "type": "object",
       "properties": { "_id": {"type": ObjectID},
             "a₁": {"type": <type>}, ..., "aₕ": {"type": <type>},
             (*1) },
       "required": [...]
}
```

where the attribute `_id` is automatically added to each document and its value is autogenerated by the system (i.e., see the keyword `ObjectID` coming from the MongoDB terminology), while the type `<type>` associated with each attribute $a_i \in A$ is the JSON type corresponding to τ_i. Symbol (*1) denotes the part of the document that will be populated through the following rules TR 2–4.

According to the JSON Schema syntax [12], the set of possible types for each attribute of an entity $E \in \mathcal{E}$ is `<type>` $\in \{string, number, boolean, array\}$. Notice that JSON does not include specific types for date and time values, but it is possible to use the property `format` to indicate that a string value could be interpreted as a date. For instance, an attribute a_k representing a date could be represented as a_k: `{"type": string, format: "date"}`. Possible formats for this purpose are: `"date-time"`, `"time"`, `"date"`. Conversely, the type `array` is used in the presence of an attribute a_h with maximum cardinality greater than one (i.e., $c_h.max > 1$) to denote an ordered collection of values of type `<type>` separated by commas. Finally, the property `required` will contain the list of attributes $a_h \in A$ such that $c_h.min > 0$.

Besides properties with simple types, a document could also contain properties that are nested documents. This is mainly obtained in the presence of relations that will be translated as described by the following three rules.

Transaction Rule 2 (Entity connected to a unique main entity). Given a labeled conceptual schema $\mathcal{S}^\ell = \langle \mathcal{E}, \mathcal{R}, \lambda, \alpha, \rho \rangle$, a desired set of document collections $\mathcal{C} = \{C_1, \ldots C_n\}$ to be obtained, and an entity $E_j = \langle \ell, A = \{a_1, \ldots, a_m\} \rangle \in \mathcal{E} \setminus \mathcal{P}$ which is connected to a main entity P_i through a relation $R = \langle E_j, P_i, c_j, c_i, A_r \rangle \in \mathcal{R}$. Each instance of E_j will be encapsulated into the corresponding instance of P_i enriching the document schema of D_i as follows:

```
(*1) := "R": { <rType> }, ...
```

where the structure of `<rType>` depends on the cardinality of R and of the set of attributes A_r, as described in detail below, while ... denotes that other documents are nested in the presence of other relations encapsulated in P_i.

1. General structure of R:
   (a) If $A = \{a_0, \ldots, a_k\}$, then

   ```
   "R": {"type": object,
         "properties": { <E_j>,
             "a_0": { "type": <type>" }, ..., "a_k: { "type": <type>" }}}
   ```

   where $\texttt{<E_j>}$ represents the translation of the instances E_j participating to the relation as discussed in point 2.
   (b) If $A = \emptyset$, then "R": $\texttt{<E_j>}$.
2. Structure of $\texttt{<E_j>}$:
   (a) If R *has cardinality one-to-one with* $c_j = (1, 1)$ *and* $c_i = (x, 1)$, then E_j has been labeled as X and there exists only one instance of E_j that is connected with the same instance of P_i. This instance is translated as:

   ```
   <E_j> := {"type": "object",
             "properties": {a_1: {"type": <type>}, ...,
                            a_m: {"type": <type>}, (*2)}}
   ```

   (b) If R *has cardinality one-to-one with* $c_j = (0, 1)$ *and* $c_i = (x, 1)$, then E_j has been labeled as XL and $\texttt{<E_j>}$ is translated as in the previous point 2a, but in this case some instances of E_j could be loss.
   (c) If R *has cardinality one-to-many with* $c_j = (1, N)$ *and* $c_i = (x, 1)$, then E_j has been labeled as XR and R is translated as in case 2a, but in this case some instances of E_j could be repeated inside different main objects, generating redundancy.
   (d) If R *has cardinality one-to-many with* $c_j = (0, N)$ *and* $c_i = (x, 1)$, then E_j has been labeled as XRL and R is translated as in case 2a, but in this case some instance of E_j could be repeated generating redundancy, while others could be lost.
   (e) If R *has cardinality many-to-many with* $c_j = (x, y)$ *and* $c_i = (w, N)$, then E_j has been labeled as X, XR, XL or XRL depending on the values of x and y. Independently from this, due to the cardinality c_i, the same instance of P_i needs to encapsulate multiple instances of E_j. Therefore, the translation of R becomes:

   ```
   <E_j>: {"type": "array",
           "items": {"type": "object",
                     "properties": {a_1: {"type": <type>}, ...,
                                    a_m: {"type": <type>}, (*2)}}}
   ```

   In other words, the relation R is translated into an array of nested objects, each one representing an instance of E_j connected to the same instance P_i. If the label of E_j contains the suffix R, then some instances could appear inside multiple documents, while if it contains L, then some instances will be lost in the encapsulation.

   Similarly to TR 2, the following rule deals with an encapsulated entity, but this time the encapsulating entity is not a main entity.

Transaction Rule 3 (Entity connected to another entity). Given a labeled conceptual schema $S^\ell = \langle \mathcal{E}, \mathcal{R}, \lambda, \alpha, \rho \rangle$, a desired set of document collections $\mathcal{C} = \{C_1, \ldots C_n\}$ to be obtained, and an entity $E_j = \langle \ell, A = \{a_1, \ldots, a_m\} \rangle \in \mathcal{E} \setminus \mathcal{P}$, which is connected through a relation $R = \langle E_j, E_y, c_i, c'_y, A_r \rangle \in \mathcal{R}$ to another entity $E_y \in \mathcal{E} \setminus \mathcal{P}$, that in turn is connected to a main entity $P_i \in \mathcal{P}$ through a relation $R_0 = \langle E_y, P_i, c_y, c_i, A_{r_0} \rangle \in \mathcal{R}$. Each instance of E_j will be encapsulated into the corresponding instance of E_y by enriching the schema of document D_i corresponding to the main entity P_i.

1. Starting from the appropriate case of TR 2 for translating the relation R_0 between E_y and P_i, generate the preliminary structure of the document D_i.
2. Inside the definition of `<E`$_y$`>`, after the translation of the properties $a_1, \ldots, a_m$, in correspondence to the expansion point (*2), we will add another property for representing the relation R. Based on the cardinalities c_j and c'_y the relation R will be translated as described in the second point of TR 2:
   - Given $R_0 = \langle E_y, P_i, c_y, c_i, A_{r_0} \rangle$, if $c_i = (x, 1)$ then E_y will be translated as a property of P_i of type object representing the single connected instance of E_y. Conversely, if $c_i = (x, N)$, then E_y will be translated as a property of P_i of type array, whose items are objects each one representing an instance of E_y.
   - Given $R = \langle E_j, E_y, c_i, c'_y, A_r \rangle$, if $c'_y = (x, 1)$ then E_j will be translated as a property of E_y of type object representing the single connected instance of E_j. Conversely, if $c'_y = (x, N)$, then E_j will be translated as a property of E_y of type array, whose items are objects each one representing an instance of E_j.

The final translation rule regards the relations between two main entities obtained through the use of reference properties inside one or both documents.

Transaction Rule 4 (Relation between two entities). Given a labeled conceptual schema $S^\ell = \langle \mathcal{E}, \mathcal{R}, \lambda, \alpha, \rho \rangle$, a desired set of document collections $\mathcal{C} = \{C_1, \ldots C_n\}$ to be obtained, for each binary relation $R = \langle P_i, P_j, c_i, c_j, A_r \rangle \in \mathcal{R}$ between two main entities $P_i, P_j \in \mathcal{P} \subseteq \mathcal{E}$, the translation is performed in different ways based on the encapsulation strategy given by the reference arrow. The same translation rule also holds in the presence of a relation R between entities, which are not main but have already been encapsulated into another entity and for which a reference arrow has been added in the labeled schema S^ℓ.

1. R has cardinality one-to-one
   (a) If R is encapsulated into P_i, then a property is added to the document C_i as follows: (*1) := "P_j"_id : {"type": "numeric"}.
   (b) If R is encapsulated in both main entities, then besides the addition of the property into C_i, also C_j will be extended with a similar property (*1) := "P_i"_id : {"type": "numeric"}.
2. R has cardinality one-to-many with $c_j = (x, 1)$

    (a) If R is encapsulated into P_i, then a property is added to the document C_i as follows: (*1) := "P_j"_id : {"type": "numeric"}.

    (b) If R is encapsulated into P_j, then a property is added to C_j as follows: (*1) := "P_i_ids":{"type":"array", "items":{type:"numeric"}}.

    (c) If R is encapsulated in both entities, then both 2a-2b are applied.

3. *R has cardinality many-to-many*

    (a) If R is encapsulated into P_i, then a property is added to the document C_i as follows: (*1) := "P_j_ids":{"type":"array", "items":{"type":"numeric"}}.

    (b) If R is encapsulated in both main entities, then besides the addition of the property into C_i, also C_j will be extended with a similar property (*1) := "P_i_ids":{"type":"array", "items":{"type":"numeric"}}.

3 Conclusion

This paper proposes a set of labeling and translation rules to map a standard ER conceptual model into a document-based logical model. The aim is to promote the use of the standard three-level modeling approach also in the NoSQL world, where the schemaless nature of the model typically induces the idea that no data modeling is needed at all. With respect to existing works, the approach proposed in this paper exploits the well-known ER conceptual model without introducing any extension or changes, while proposing a set of labeling and translation rules that mainly resemble those developed in the relational database field.

Acknowledgments. This study was carried out within the Interconnected Nord-Est Innovation Ecosystem (iNEST) and received funding from the European Union Next-GenerationEU (Piano Nazionale di Ripresa e Resilienza (PNRR) – Missione 4 Componente 2, Investimento 1.5 – D.D. 1058 23/06/2022, ECS00000043). This manuscript reflects only the authors' views and opinions, neither the European Union nor the European Commission can be considered responsible for them.

References

1. Asaad, C., Baïna, K.: NoSQL databases - seek for a design methodology. In: MEDI, pp. 25–40 (2018). https://doi.org/10.1007/978-3-030-00856-7_2
2. Atzeni, P., Jensen, C.S., Orsi, G., Ram, S., Tanca, L., Torlone, R.: The relational model is dead, SQL is dead, and i don't feel so good myself. SIGMOD Rec. **42**(2), 64–68 (2013). https://doi.org/10.1145/2503792.2503808
3. Badia, A., Lemire, D.: A call to arms: revisiting database design. SIGMOD Rec. **40**(3), 61–69 (2011). https://doi.org/10.1145/2070736.2070750
4. Banerjee, S., Sarkar, A.: Logical level design of NoSQL databases. In: 2016 IEEE Region 10 Conference (TENCON), pp. 2360–2365 (2016). https://doi.org/10.1109/TENCON.2016.7848452
5. Bugiotti, F., Cabibbo, L., Atzeni, P., Torlone, R.: Database design for NoSQL systems. In: Conceptual Modeling (ER), pp. 223–231 (2014)

6. Cattell, R.: Scalable SQL and NoSQL data stores. SIGMOD Rec. **39**(4), 12–27 (2011). https://doi.org/10.1145/1978915.1978919

7. Davoudian, A., Chen, L., Liu, M.: A survey on NoSQL stores. ACM Comput. Surv. **51**(2) (2018). https://doi.org/10.1145/3158661

8. Mohan, C.: History repeats itself: sensible and NonsenSQL aspects of the NoSQL hoopla. In: Proceedings of the 16th International Conference on Extending Database Technology, pp. 11–16. EDBT '13 (2013). https://doi.org/10.1145/2452376.2452378

9. Sadalage, P.J., Fowler, M.: NoSQL Distilled: a brief guide to the emerging world of polyglot persistence. Addison-Wesley Professional, 1st edn. (2012)

10. Shin, K., Hwang, C., Jung, H.: NoSQL database design using UML conceptual data model based on Peter Chen's framework. Int. J. Appl. Eng. Res. **12**(5), 632–636 (2017)

11. Storey, V.C.: Relational database design based on the entity-relationship model. Data Knowl. Eng. **7**(1), 47–83 (1991). https://doi.org/10.1016/0169-023X(91)90033-T

12. Wright, A., Andrews, H., Hutton, B., Dennis, G.: JSON Schema: a media type for describing JSON documents. Tech. rep. (2020). https://json-schema.org/draft/2020-12/json-schema-core.html

Querying Property Graphs with XPath

Marko Junkkari[✉] ⓘ, Sami-Santeri Svensk ⓘ, and Jyrki Nummenmaa ⓘ

Tampere University, Tampere, Finland
{marko.junkkari,jyrki.nummenmaa}@tuni.fi,
samisanterisvensk@gmail.com

Abstract. XPath has been established as the de facto standard for searching data items from hierarchical XML structures. Due to its popularity and compact path expressions, XPath has also been recognized as a query language candidate for graph databases where the structure does not follow a hierarchical order. Graph databases are based on graph theory and the data are organized accordingly. Among different types of graphs, property graphs have gained special interest because they allow data associated with edges as well as vertices, reflecting that edges represent relationships and relationships are generally allowed to have properties, just like entities. Earlier proposals to apply XPath to graph databases do not allow manipulation of the properties of edges in a property graph. The present study focuses on this issue. We show how XPath can be applied to full-scale property graphs. This requires a novel mapping of XPath primitives to the primitives of property graphs. Based on this mapping, we define graph-based semantics for XPath by regular path queries, an established logical approach for querying vertices and edges.

Keywords: Property graph · Graph database · XPath · Regular path query

1 Introduction and Related Work

Graph databases [2, 6, 17, 26] are in growing demand for analyzing linked data in various domains [3, 26]. In graph databases, the data is organized using graph structures, emerging from heavily studied graph theory [5], giving graph databases a strong theoretical foundation [20]. There is, however, no common data model for all graph databases [34]. Common features exist, though, like for instance index-free adjacency [34]. Retrieving data from graph databases can be performed using the graph operations defined in the graph theory [2, 12]. Like other NoSQL databases, graph databases store semi-structured data containing the schema within the data [13].

There is a consensus on modeling entities as vertices and their relationships as edges. A special type of graph data model, called property graphs [33], has emerged both in the theoretical context, e.g. [18, 26], and in practical implementations [29]. A property graph relies on vertices and edges, both of which can be labeled. The most popular graph database Neo4J [29] is based on the property graph model. The graph database community has recently seen the rise of commercial query languages such as Cypher [28], PGQL [32], SQL/PGQ [16] and Gremlin [7]. In 2024, Graph Query Language (GQL) became the standard query language for property graphs [21].

© The Author(s), under exclusive license to Springer Nature Switzerland AG 2026
P. K. Chrysanthis et al. (Eds.): ADBIS 2025, CCIS 2676, pp. 67–76, 2026.
https://doi.org/10.1007/978-3-032-05727-3_7

The history of graph query languages starts from the 1980s, when the language G was proposed in [15] to query edge-labeled graph with regular expressions. Mendelzon and others [27] created the semantics based on language G to find a simple path between two vertices. This approach describing paths is nowadays known in literature as regular path queries (RPQ). RPQ:s have influenced the research from then on and have had a huge impact for query language design, and the navigation can be found from many of those languages. Since then, RPQs have been extended with various extensions to enhance expressivity. Such extensions include RPQ with inverse [14], conjunctive RPQ [14], extended conjunctive RPQ [9] and RPQ with data tests [25]. Path queries bring an expressive way to query databases [4]. In our opinion, RPQ forms a similar basis to graph query languages as relational algebra or calculus formed to the relational query languages.

Pattern matching, finding nodes connected by paths, and aggregating the results are important features for graph query languages [4, 38]. These features have also been identified to form the core of XPath [35], a thoroughly researched e.g. [10, 11] query language to address parts in XML [36] documents. Even though XPath is designed to operate on tree-like-structured XML documents, Buneman [13] sees these structures essentially as rooted graphs. Angles and others [2] describe XML as restricted type of graph and see essential theoretical basis, graph theory influencing both graph databases and XML documents. Despite the fact that XPath has been developed originally for addressing parts of XML documents, Libkin and others [24] see XPath as probably overlooked as a candidate language for graph databases, as its goal seems very similar to querying graph databases.

Libkin and others [24, 25] have noted the proximity of XPath to first order logic or modal logic and van Rest and others [32] compare XPath to Tarski's algebra, which has similarities to the basics of many graph query languages. XPath and RPQ have been compared in the context of graph databases [8, 19, 23, 24]. Oltenau and others [30] describe the relationship between XPath and RPQ as an abstraction of the navigational features of XPath, where support for XPath axes child, descendant, parent and ancestor is provided. Despite the similarities, XPath cannot fully be subsumed by RPQ or its various extensions [24]. In some scenarios, XPath goes beyond the path queries by defining patterns that cannot be captured by paths [25]. Libkin and others [25] study the capabilities of potential languages, including XPath, to query graph databases combining topology and data. According to them, XPath succeeds by describing the properties of paths and patterns, considering both their purely navigational aspects and the data contained in the database. Their analysis is on a theoretical level, and they do not provide an implementation or an approach for implementation. Barceló [8] has mentioned XPath's branching operator providing good expressive power in graphs. In general, there is a wide consensus that XPath is a promising querying approach to graph data model.

Libkin and others [24] have created semantics for XPath in graphs and studied the expressiveness and complexity of various XPath formalisms called GXPath. Their semantics were expressed with respect to a graph structure called data graph. Research of XPath in graphs has focused on simple graph structures like data graphs [24, 25], where the data is contained in nodes as single values. There has not yet been a direct link between the formalism of XPath and the increasingly common property graph data model, even though both are data models designed for semi-structured data and have similar capabilities to store data into properties of the elements building the structure of the model. We define how the primitives of XPath are mapped to the primitives of the property graph. The path steps of XPath are not labeled, and they do not have properties while the edges of a property graph are labeled and may possess properties. We manipulate both vertices and edges as XPath nodes. We match selected fragments of XPath to the features of property graphs and define compilation from XPath to RPQ by attribute grammars [22].

The rest of the paper is organized as follows. Section 2 introduces the graph structure and notations for regular path queries. In Sect. 3, we define the mapping between XPath primitives and graph primitives. In this context, we give an informal introduction for using XPath in property graphs. In Sect. 4, we define PRQ based semantics and use them to evaluate an example query. We give our conclusions in Sect. 5.

2 Property Graphs and Regular Path Queries

Property graphs are directed multigraphs, where vertices represent entities and edges relationships between them [5, 12]. Figure 1 illustrates this conceptualization without properties. Customer, Order, Product, Supplier and Category are entities, modeled as vertices. Arrows represent relationships, modeled as directed edges.

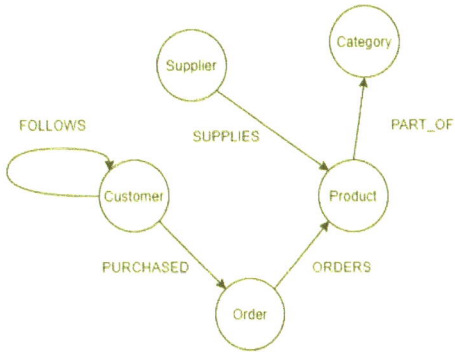

Fig. 1. Example graph

We follow the definitions of Angles and others [5], however, we do not allow multi-labeling of vertices and edges nor multivalued properties. Let L be a set of labels, P a set of properties and PV a set of property values. Then, the property graph can formally be defined as the tuple G = (V, E, ρ, λ, σ), where.

1. V is a finite set of the vertices.
2. E is a finite set of edges.
3. ρ: E → (V × V) assigns each edge of E to a pair of nodes in V.
4. λ: (V ∪ E) → L labels edges and vertices with the set L.
5. σ: (V ∪ E)(V ∪ E) × P → PV assigns a value to a property of a vertex or edge.

Following this notation, the example graph used through the paper can be expressed as follows:

- V = {v1, v2, v3, v4, v5, v6}
- E = {e1, e2, e3, e4, e5}
- P = {type, name, quantity, country}
- L = {Customer, Supplier, Category, Product, Order, SUPPLIES, PURCHASED, ORDERS, PART_OF, FOLLOWS}
- λ = {⟨v1, *Customer*⟩, ⟨v2, *Customer*⟩, ⟨v3, *Order*⟩, ⟨v4, *Supplier*⟩, ⟨v5, Pr *oduct*⟩, ⟨v6, *Category*⟩, ⟨e1, *FOLLOWS*⟩, ⟨e2, *PURCHASED*⟩, ⟨e3, *ORDERS*⟩, ⟨e4, *SUPPLIES*⟩, ⟨e5, *PART_OF*⟩}
- ρ = {⟨e1, ⟨v1, v2⟩⟩, ⟨e2, ⟨v1, v3⟩⟩, ⟨e3, ⟨v3, v5⟩⟩, ⟨e5, ⟨v5, v6⟩⟩, ⟨e5, ⟨v4, v6⟩⟩}
- σ = {⟨⟨v6, type⟩, junk⟩, ⟨⟨v3, name⟩, Smith⟩, ⟨⟨e3, quantity⟩, 10⟩, ⟨⟨v4, Country⟩, UF⟩, ⟨⟨v5, Name⟩, toy⟩}

We can allow L to assign similar labels to vertices and edges. Our example does not do that, so we can use labels when referring to vertices and edges.

Regular path queries are based on regular expressions determining one or several paths in a graph. Formally, a regular path query can be expressed by a triple (X, RE, Y), where X and Y are variables that refer to the end and start vertices of the underlying path expression, and RE is a regular expression over the vocabulary of edge labels. Operations, such as repetitions (+, *), can be used and a regular expression may contain a complex clause with patterns including various edges. However, for the purpose of the present study, complex regular expressions are not needed. RE can be represented in the form -label- > where the label may involve the + postfix for denoting one or several occurrences of the edge. In RE, it is also possible to express that no label is specified. This is denoted by --- >. Thus, any path can be denoted by -- + − >. Property value restrictions for an edge can be represented within parentheses. For example, -ORDERS(Quantity = 10)- > refers to the order edges that have the quantity property with value 10.

The labels of vertices can be restricted with additional facts of the form (X, is, label), where X is a variable. For example, (X −· + − > Y) ∧ (X, is, Customer) ∧ (Y, is, Order) refers to any path from a customer vertex to an order vertex. Property value restrictions of vertices can be represented by additional facts. For example, (Z, is, Category) ∧ (Z, has, type = 'junk') means that Category has an attribute with name 'type' and value 'junk'.

3 XPath for Graph Database

XPath consists of path steps Axis::node_test[predicate] separated by slashes (/). An axis determines the relationships between a context node and the connected processing nodes. The most common axes are self, child, parent and descendant-or-self. Element label and attribute are the most common node types. A predicate can determine conditions for both nodes and paths. Using predicates, a search tree structure can be represented in a serialized form, where each branch can have individual predicates. For the most common axes, abbreviators are established: the child axis is the default value, and it is typically not expressed, double dash (//) denotes descendants, dot (.) corresponds to the self axis, two dots (..) determines the parent axis and @ is refers to an attribute.

The simplest way to apply XPath to graphs would be by mapping nodes to vertices and child relationships to directed edges. For example, the XPath expression Customer/Product would be mapped to the edge Customer → Product. The expression Customer//Category would correspond to any directed path from Customer to Category. Following this approach, attributes can be used in referring to the properties of vertices. For example, Category[@type ='junk'] could denote the type property having value 'junk' in a Category node.

The problem of this trivial approach is that it is not suitable for property graphs where edges are labelled and may have properties. The child relationships in XML are not labelled, and they have no properties. Therefore, we propose that also edges are manipulated as XPath nodes. In other words, we map a path in a graph to an XPath path. A path in a graph is started from a vertex, and every other member is an edge, and every other a vertex [15]. For example, in the expression a/b/c/d/e, the nodes a, c and e refer to vertices and b and d refer to edges. In other words, a/b returns the edge b and a/b/c returns the vertex c. It is possible to refer the properties of vertices and edges as our example in Sect. 4 will demonstrate.

Like in XPath, the asterisk refers to an unlabeled node, for both vertices and edges. For example, Order/*/Product means a step from an order to a product via any edge whereas Order/ORDERS/* means any successor of an order through an ORDERS edge. The descendant notation can also be used in the context of graphs. The path Customer//Category determines all paths from a customer to a category. It is worth noting that the descendant relationship may refer to either vertices or edges if the asterisk is used. For example, in the expression Order//* the asterisks may refer to ORDERS, Product, PART_OF or Category. As such this kind of expression is hardly useful but this allows powerful expressions in queries containing uncertain aspects. For example, it is possible to determine a path from a node to another and give a value restriction to an attribute in an edge or a vertex in the path. In the path Customer//*[unitPrice > 10 000]//Category, the unit price can be in any node between Customer and Category.

The branches can be expressed between square brackets in serialized expressions. For example, the path Order[ORDERS/Product[@name = 'toy']] refers to an order that contains a toy. The double dot denotes a parent relationship. In the context of graphs, we interpret this as traversal to the inverse direction in a directed edge. For example, the path Product/../Supplier denotes the inverse direction from a product to a supplier. The parent notation can also be used in serialized expressions. For example in the path Product[../Supplier[@Country ='UK']/*/Category, the fragment../Supplier[@Country ='UK'] means that a product vertex must have an inverse path to a supplier whose country is UK.

Above, the abbreviations of XPath are used for navigation. Basically, navigation is based on axes that determine the displacement from a context node to its relatives. The relationship between XPath axes and their interpretation in graph databases is as follows, with the arrow from axis to interpretation: *child → immediate successor; parent → immediate predecessor; descendant → successor; ancestor → predecessor; self → self; descendant-or-self → successor or self; ancestor-or-self → predecessor or self;*

attribute → *property*. The axes *following, preceding, following-sibling,* and *preceding-sibling* do not have an interpretation in a graph.

4 Regular Path Query Based Semantics

Attribute grammars (AGs), widely used in compilers [1, 37], are used to define both the syntax and semantics of a formal language [22, 31]. We introduce only such notational conventions for attribute grammars that are applied in this study.

Let AG = (G, A, R), be a triple where G is a context free grammar, A is a finite set of attributes and R is a finite set of semantic rules associated with the attributes. A context free grammar G = (NT, T, P, D) defines a syntax for a formal language. In G, NT is a finite set of non-terminals and T is a finite set of terminals such that NT ∩ T = Ø. Elements in NT and T are called grammar symbols. P is a finite set of productions. Each production is of the form X → α, where X ∈ N and α ∈ (NT ∪ N)*. P may contain alternative productions e.g. X → α1 and X→ α2 so that α1 ≠ α2. D (∈ N) is a start symbol. Each attribute (∈ A) is associated with one or several non-terminals and if X ∈ N, then the attribute set of X is denoted by A(X). A(X) is partitioned into two exclusive sets: inherited attributes I(X) and synthesized attributes S(X) so that I(X) ∩ S(X) = Ø and I(X) ∪ S(X) = A(X). Usually there are two ways to denote the selection of an attribute a of symbol X. One follows a record style, X.a; the other one follows a functional style a(X). In this paper we adopt the second one. Each semantic rule (∈ R) is associated with a production p (∈ P) to define the evaluation of a synthesized attribute of the symbol in the left hand side of p, or the evaluation of an inherited attribute of a symbol on the right-hand side of p.

We define the context free grammar G_{XRPQ} as the tuple (NT, T, P, Q) where NT = {Q, P, E, V, VN, EN}, T = {/, //,.., *, [,], @, =, ≠, <, >} ∪ E_names ∪ V_names ∪ A_names ∪ A_values. Q is the starting symbol, and P is the set of productions represented in the second column of Table 1. E_names and V_names are the names of edges and vertices, respectively. A_names and A_values are the sets of property names and values. We define the attribute grammar AG_{XRPQ} to be the triple (G_{XRPQ}, A, R) where A = {*v, result, ret, name, var, first_var*} and R is the semantic rules associated with the P. The attributes in A have the following intention:

- *v* is an inherited attribute for the variable associated with a vertex or an edge.
- *result* is a synthesized attribute that describes the final conjunctive query of the regular path queries.
- *res* is a synthesized attribute that contains a set of regular path queries associated within parsing the XPath query.
- *var* is a synthesized attribute that contains the variable associated with a vertex.
- *first_var* is a synthesized attribute that contains the variable associated with the first vertex of a sub-path.
- The attributes are associated with the grammar symbols as follows:
- I(V) = I(V) = {*v*}
- S(Q) = {*result*}
- S(P) = {ret, first_var}
- S(E) = S(EN) = {*name*}

- $S(VN) = \{name, var\}$
- $S(V) = \{var, ret\}$

The rules of R are represented in the third and fourth columns of Table 1. In the third column, the function *new*() generates a new variable.

Table 1. An attribute grammar for compiling XPath to regular query queries

Id	Production	Inherited	Synthesized
1	Q → P		$result(Q) = \bigwedge_{x \in ret(P)} x$
2	P1 → V / E / P2	$v(V)$ = new()	$ret(P1) = \{(var(V)$ -name(E)-> $first\_var(P2))\} \cup ret(V) \cup ret(P2) \; first\_var(P1) = v(V)$
3	P1 → V /../ P2	$v(V)$ = new()	$ret(P1) = \{first\_var(P2)$ -•-> $(var(V))\} \cup ret(V) \cup ret(P2)$
			$first\_var(P1) = v(V)$
4	P → V	$v(V)$ = new()	$ret(P) = ret(V)$
			$first\_var(V) = v(V)$
5	P1 → V//P2	$v(V)$ = new()	$ret(P1) = \{(v(V)$ -.+-> $first\_var(P1))\} \cup ret(V) \cup ret(P2)$
			$first\_var(P1) = v(V)$
6	V → VN[E/P]	$v(NV) = v(V)$	$ret(V) = \{(v(VN),$ **is**, $name(VN)),$
			$\quad (v(VN)$ -name(E)-> $first\_var(P))\} \cup ret(P)$
			$var(V) = v(VN)$
7	V → VN	$v(NV) = v(V)$	$ret(V) = \{(v(VN),$ **is**, $name(VN))\}$, if $name(VN) \neq$ 'null'
			$\qquad\qquad \varnothing \qquad\qquad\qquad$, otherwise
			$var(V) = v(VN)$
8	E → EN		$name(E) = name(EN)$
9	V1 → V2[@name o d]	$v(V2) = v(V1)$	$ret(V1) = ret(V2) \cup \{(v(V2),$ has, name), (name o d)\},
			$\quad$ where name ∈ names and d ∈ values and o ∈ {=, ≠, <, >}
			$var(V1) = v(V2)$
10	E1 → E2[@name o d]		$name(E1) = name(E2) \oplus$ (name o d) ,
			$\quad$ where name ∈ names and d ∈ values and o ∈ {=, ≠, <, >}
11	VN → name		$name(VN) = name$, where name ∈ V-names
12	EN → name		$name(EN) = name$, where name ∈ E-names
13	VN → *		$name(VN) =$ 'null'
14	EN → *		$name(EN) = •$

Related to the graph of Fig. 1, an example evaluation for an XPath query

```
Customer/*/Order/ORDERS[@quantity=10]/Product[PART_OF/
Category[@type = 'junk']]/../Supplier
```

is given in Fig. 2. The query associates customers and suppliers in a case where a customer has bought 10 pieces of products whose category are 'junk', and the products have the same supplier. Variables are labeled X1, X2, X3, etc. following their creating order.

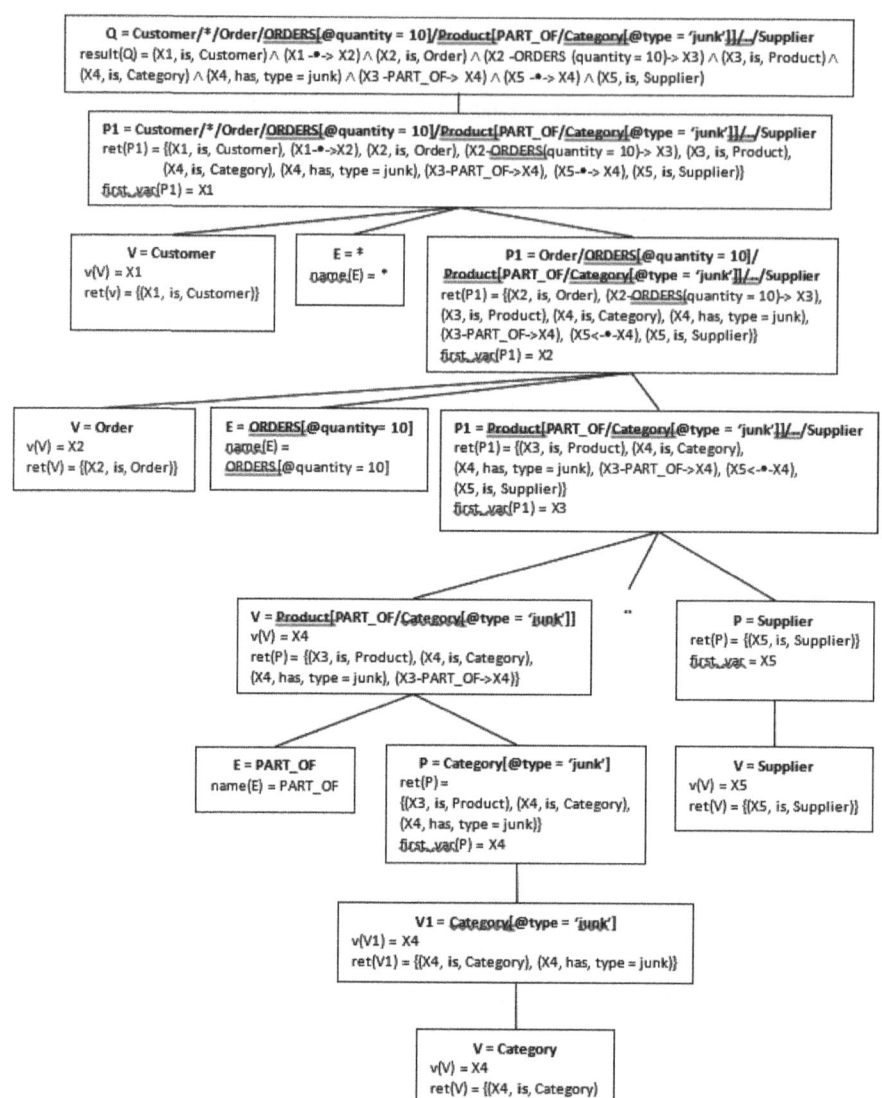

Fig. 2. Parsing an XPath Query.

5 Conclusions

The likely reason why XPath has not been applied to property graphs is that XML data model does not contain the properties of relationships. We solve the problem by mapping XPath nodes to both vertices and edges in the graph data model, thus enabling consistent handling of vertices and edges. Using attribute grammars, we compile XPath expressions to regular path queries in a natural way, giving a framework to compile XPath to other graph query languages. Expressing XPath's nested paths in the predicates treats structural and data filtering as equal, which can be seen as a non-conventional approach.

References

1. Aho, A.V., Sethi, R., Ullman, J.D.: Compilers Principles, Techniques and Tools. Addison-Wesley, Reading (1986)
2. Angles, R., Gutierrez, C.: Survey of graph database models. ACM Comput. Surv. **40**(1), 1–39 (2008)
3. Angles, R., Prat-Perez, A., Dominguez-Sal, D., Larruba-Pey, J.L.: Benchmarking database systems for social network. In: First International Workshop on Graph Data Management Experiences and Systems, pp. 1–7 (2013)
4. Angles, R., Arenas, M., Barceló, P., Hogan, A., Reutter, J., Vrgoč, D.: Foundations of modern query languages for graph databases. ACM Comput. Surv. **50**(5), 1–40 (2017)
5. Angles, R.: The Property Graph Database Model. Universidad de Talca, Department of Computer Science (2018)
6. Angles, R., et al.: G-CORE: A core for future graph query language. In: Proceedings of the 2018 International Conference on Management of Data, pp. 1421–1432 (2018)
7. Apache TinkerPop, Gremlin Query Language. https://tinkerpop.apache.org/gremlin.html. Accessed 16 Oct 2024
8. Barceló, P.: Querying graph databases. In: Proceedings of the 32nd ACM SIGACT-SIGMOD-SIGART Symposium on Principles of Database Systems, pp. 175–187 (2013)
9. Barceló, P., Libkin, L., Lin, A.W., Wood, P.T.: Expressive languages for path queries over graph-structured data. ACM Trans. Database Syst. **37**(4), 1–46 (2012)
10. Benedikt, M., Koch, C.: XPath leashed. ACM Comput. Surv. **41**(1), 1–54 (2009)
11. Benedikt, M., Wenfei, F., Kuper, G.: Structural properties of XPath fragments. Theoret. Comput. Sci. **336**, 3–31 (2005)
12. van Bruggen, R.: Learning Neo4j: Run Blazingly Fast Queries on Complex Graph Datasets with the Power of the Neo4j Graph Database. Packt Publishing (2014)
13. Buneman, P.: Semistructured data. In: Proceedings of the Sixteenth ACM SIGACT-SIGMOD-SIGART Symposium on Principles of Database Systems, pp. 117–121. (1997)
14. Calvanese, D., De Giacomo, G., Lenzerini, M., Vardi, M.Y.: Containment of conjunctive regular path queries with inverse. In KR **2000**, 176–185 (2000)
15. Cruz, I., Mendelzon, A., Wood, P.: A graphical query language supporting recursion. SIGMOD Record **16**(3), 323–330 (1987)
16. Deutsch, A., et al.: Graph pattern matching in GQL and SQL/PGQ. In: Proceedings of the 2022 International Conference on Management of Data, pp. 2246–2258 (2022)
17. Foulds, L.R.: Graph Theory Applications. Springer, New York (1992)
18. Francis, N., et al.: Cypher: An evolving query language for property graphs. In: Proceedings of the 2018 International Conference on Management of Data, pp. 1433–1445. (2018)
19. Furche, T., Linse, B., Bry, F., Plexouakis, D., Gottlob, G.: RDF querying: Language constructs and evaluation methods compared. In: Barahona, P., Bry, F., Franconi, E., Henze, N., Sattler, U. (eds), Reasoning Web. Reasoning Web 2006. LNCS, vol. 4126, pp. 1–52, Springer, Heidelberg (2006)
20. Harrison, G.: Next Generation Databases NoSQL and Big Data. Apress (2015)
21. ISO, ISO/IEC 39075:2024. https://www.iso.org/standard/76120.html. Accessed 16 Mar 2025
22. Knuth, D.: Semantics of context-free languages. Math. Syst. Theory **2**(2), 127–145 (1968)
23. Kostylev, E., Reutter, J., Vrgoč, D.: Static analysis of navigational XPath over graph databases. Inf. Process. Lett. **116**(7), 467–474 (2016)

24. Libkin, L., Martens, W., Vrgoč, D.: Querying graph databases with XPath. In: Proceedings of the 16th International Conference on Database Theory, pp. 129–140 (2013)
25. Libkin, L., Martens, W., Vrgoč, D.: Querying graphs with data. J. ACM **63**(2), 1–53 (2016)
26. Maiolo, S., Etcheverry, L., Marotta, A.: Data profiling in property graph databases. ACM J. Data Inf. Qual. **12**(4), 1–27 (2020)
27. Mendelzon, A., Wood, P.T.: Finding regular simple paths in graph databases. SIAM J. Comput. **24**(6), 1235–1258 (1995)
28. Neo4J, Cypher Query Language. https://neo4j.com/developer/cypher/. Accessed 28 Oct 2024
29. Neo4J. https://neo4j.com/. Accessed 16 Oct 2024
30. Oltenau, D., Furche, T., Bry, F.: Evaluating complex queries against XML streams with polynomial combined complexity. Key Technol. Data Manage. **3112**, 31–44 (2004)
31. Paakki, J.: Attribute grammar paradigms - a high-level methodology in language implementation. ACM Comput. Surv. **27**(2), 196–255 (1995)
32. van Rest, O., Hong, S., Jinha, K., Meng, X., Chafi, H.: PGQL: a property graph query language. In: Proceedings of the Fourth International Workshop on Graph Data Management Experiences and Systems, pp. 1–6 (2016)
33. Rodriquez, M., Neubauer, P.: Constructions from dots and lines. Bull. Am. Soc. Inf. Sci. Technol. **36**(6), 35–41 (2010)
34. de Virgilio, R., Maccioni, A., Torlone, R.: Converting relational to graph databases. In: First International Workshop on Graph Data Management Experiences and Systems, pp. 1–6 (2013)
35. W3C, XML Path Language (XPath). Accessed 16 Oct 2024
36. W3C, Resource Description Framework (RDF) Concepts and Abstract Syntax. Accessed 16 Oct 2024
37. Waite, W., Goos, G.: Compiler Construction. Springer, New York (1983)
38. Wood, P.: Query languages for graph databases. ACM SIGMOD Rec. **41**(1), 50–60 (2012)

TimeVizBench: An Interactive Platform for Evaluating Techniques for Efficient Large Time Series Visualization

Vassilis Stamatopoulos[1,2](✉) ⓘ, Stavros Maroulis[1] ⓘ, Christos Pantoleon[3] ⓘ,
George Papastefanatos[1] ⓘ, and Panos Vassiliadis[2] ⓘ

[1] ATHENA Research Center, Athens, Greece
{bstam,stavmars,gpapas}@athenarc.gr
[2] University of Ioannina, Ioannina, Greece
panos.vassiliadis@cs.uoi.gr
[3] Athens University of Economics and Business, Athens, Greece
chr.pantoleon@aueb.gr

Abstract. Interactive time series visualization is essential in domains like IoT monitoring but is often constrained by latency and scalability challenges. Various methods have been proposed to address these issues, each with different trade-offs between efficiency, interactivity, and visualization accuracy, making systematic evaluation crucial. Traditional benchmarking approaches, however, fail to capture user-perceived responsiveness and accuracy in real-world exploration scenarios. To bridge this gap, we introduce *TimeVizBench*, an interactive evaluation platform for scalable time series visualization methods across performance and accuracy dimensions. *TimeVizBench* enables users to configure different methods, explore visual outputs interactively, and dynamically assess performance and accuracy. It also provides a standardized interface for integrating and comparing additional methods.

Keywords: Interactive Visualization · Time Series · Approximate Visualization

1 Introduction

Visualizing large-scale, multivariate time series data presents unique challenges due to the sheer volume, high dimensionality, and dynamic nature of the data. In domains such as IoT monitoring, financial analysis, and anomaly detection, users rely on interactive visual exploration performing operations like panning, zooming, and pattern highlighting. Achieving interactive response times is essential for effective real-time analysis, while ensuring visualization accuracy is critical to avoid misleading or incomplete insights. However, the latency associated with fetching, processing, and rendering such data often hinders interactivity.

Approaches for Scalable Time Series Visualization. Various methods have been proposed to address the challenges of large-scale time series visualization.

© The Author(s), under exclusive license to Springer Nature Switzerland AG 2026
P. K. Chrysanthis et al. (Eds.): ADBIS 2025, CCIS 2676, pp. 77–84, 2026.
https://doi.org/10.1007/978-3-032-05727-3_8

Traditional techniques, such as sampling and aggregation, reduce data volume to improve performance but often distort visual representation and affect visualization accuracy. Visualization accuracy is typically assessed at the pixel level rather than in the data domain, measuring how closely the rendered visualization aligns with the expected output from raw data. Metrics like the Structural Similarity Index Measure (SSIM) [12] quantify visual differences by evaluating structural and perceptual similarities, making them suitable for assessing visualization accuracy.

Visualization-aware methods, such as M4 [6], account for visualization parameters (e.g., width and height of the chart canvas) to ensure accurate representations. M4 aggregates data into pixel-wide intervals, preserving the visualization's shape with 100% accuracy. However, it requires querying all relevant data for each interaction, increasing latency and reducing interactivity.

Progressive approaches, like OM3 [11], incrementally refine visualizations through a precomputed multi-level representation, eventually achieving an error-free visualization. However, OM3 lacks error guarantees for intermediate visualizations, requiring users to wait until full convergence for accuracy assurances.

To improve interactivity, caching-based methods have emerged. MinMax-Cache [9] reduces latency in time series visual exploration while ensuring accuracy. It dynamically aggregates and caches data at granularities optimized for visualization, considering user-defined accuracy constraints, enabling efficient reuse of cached results during panning and zooming. This approach minimizes redundant data fetching while maintaining pixel error-bound guarantees, ensuring accurate visual representations with low latency. Unlike precomputation-based methods, MinMaxCache adapts dynamically to user exploration and supports streaming data, making it well-suited for interactive scenarios.

Other methods, such as MinMaxLTTB [3], integrate min-max preselection with downsampling techniques like Largest Triangle Three Buckets (LTTB) to improve scalability while maintaining visual quality. However, they lack explicit error quantification relative to fully accurate references.

Challenge and Main Beneficiaries. Comparing and systematically evaluating these methods remains challenging. Existing evaluation practices often rely on controlled experiments or algorithm reimplementations, which are labor-intensive and fail to capture the dynamic, user-driven nature of modern visualization tools. Interactive visualization scenarios also demand careful consideration of user experiences—particularly perceptions of latency and accuracy—which directly influence both utility and usability. Addressing these needs requires an interactive experimentation platform that enables the evaluation of diverse methods across various metrics, facilitating the addition of new ones. Such a platform should enable its key beneficiaries, i.e., researchers, and software engineers to accomplish real-time exploration of trade-offs, systematically evaluate visualization techniques, and provide built-in support for measuring key metrics through a well-defined interface, ensuring ease of use and extensibility.

Related Work. While many techniques support interactive, low-latency, scalable visualization, the lack of standardized evaluation frameworks limits systematic comparison. The need for benchmarking in visualization has been well established [1], highlighting the importance of performance evaluation in interactive data systems. Generic benchmarks like TPC-H focus on backend performance metrics, while domain-specific benchmarks, such as those for time series data [7], provide specialized evaluation frameworks. While effective for assessing generic query performance and scalability, these benchmarks do not address the interactivity and accuracy requirements of visualization systems. Visualization-specific benchmarks, such as the Visual Analytics Benchmark Repository [10], focus on analytical effectiveness but offer limited support for evaluating system performance during interactive operations. More recent efforts include interactive data exploration benchmarks [2,4], which provide systematic approaches for gaining insights into query execution and system-level performance for interactive exploration and visualization. While these benchmarks enable standardized evaluation of system performance, they do not explicitly target timeseries visualization methods and often fail to capture the real-time user experience, including user-perceived latency, visual accuracy, and the critical trade-offs between usability and performance essential in interactive scenarios.

Contribution. In this work, we introduce *TimeVizBench*, an extensible platform for evaluating methods designed to enhance the interactivity and scalability of time series visual exploration, focusing on key dimensions such as performance and visualization accuracy. *TimeVizBench* supports state-of-the-art approaches like MinMaxCache [9] and M4 [6], while offering a flexible framework for integrating alternative algorithms. Method-specific parameters are declaratively defined and translated into user interface controls, enabling effortless experimentation. Users can interactively explore multivariate time series data through operations like pan and zoom, while comparing the efficiency and effectiveness of different methods in real time. The platform measures key performance metrics, including query time, network transfer, and rendering time, while also assessing visualization accuracy via SSIM and tracking data reduction efficiency based on the amount of retrieved data.

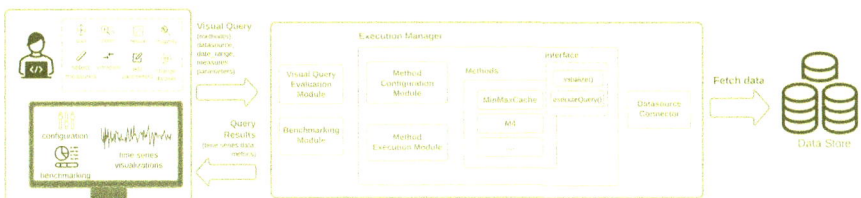

Fig. 1. *TimeVizBench* Platform Architecture

2 *Time VizBench* Architecture

Figure 1 presents the architecture of the *Time VizBench* system, designed for evaluating methods that support interactive, large-scale time series visualization. The platform is composed of a web-based UI and a backend API responsible for executing the evaluated methods and interacting with the database.

The UI allows users to select datasets, configure various parameters, and compare the visualizations generated by different methods. Each query specifies the method to execute, the datasource to query, the date range, measures to visualize, and any required parameters for the selected method. Additionally, many visualization-aware methods require parameters such as the width and height of the visualization canvas, which are automatically included. Query results include the data points (e.g., timestamps and values) to visualize for one or more variables, performance metrics (e.g., query time and total response time), and visualization quality metrics (e.g., SSIM).

The backend is orchestrated by the **Execution Manager**, which manages query evaluation, method execution, and benchmarking. To evaluate a new method, developers must implement a standardized interface and annotate it with @VisualMethod, giving it a name and a description. This annotation mechanism automatically registers the method with the system, making it available to both backend and frontend components.

Parameters for each method are defined using the @Parameter annotation. A built-in boolean field in the annotation setup, distinguishes **initialization parameters**, which are set once during method instantiation and **query-time parameters**, which are specified at runtime for each query (e.g., the accuracy constraint in MinMaxCache). When a method is initialized, any necessary pre-computed data can be loaded within the initialize() routine.

When a user selects a method in the UI, the frontend dynamically queries the backend for its parameter definitions. These are extracted by introspecting the annotated Java fields, so the UI can render the appropriate configuration options without manual intervention. Outputs are always returned using standardized classes to ensure compatibility with frontend requirements.

The **Method Execution Module** processes the configured methods, generating and executing queries through the **Datasource Connector**. This connector retrieves data as needed from the specified datasource. Processed results, including time series data and computed metrics, are returned to the **Benchmarking Module**, which evaluates method performance across multiple dimensions. *Time VizBench* reports metrics for three key evaluation dimensions:

- **Performance** is measured by *query time*, which captures the time taken by the database to evaluate the query, *network time* for data transfer, and *rendering time* for visualizing the results.
- **Accuracy** is assessed using the *Structural Similarity Index Measure (SSIM)*, quantifying how closely a method's visualization matches the reference visualization.

– **Data reduction efficiency** is evaluated by measuring the *amount of data retrieved from the database* for each method. This metric helps assess how effectively a method reduces the data volume needed to fetch from the data store, impacting both performance and visualization latency.

These metrics allow users to systematically compare methods and explore trade-offs between interactivity, accuracy, and data reduction efficiency.

Currently, the platform supports SQL databases and InfluxDB, with the **Datasource Connector** enabling metadata retrieval for available datasets and time series measures. The system is designed for extensibility, allowing new data-sources to be easily added.

To ensure fair benchmarking, an optional cleaning method is invoked in the **Datasource Connector** after executing each query, which clears any cached results from the underlying database, minimizes the impact of database opti-mizations like caching and ensuring consistent evaluation conditions across meth-ods. The processed results and benchmarking metrics are sent to the frontend, enabling users to analyze and compare methods across latency, accuracy, and interactivity dimensions.

Implementation Details. *TimeVizBench* features a Java 17 backend built with Spring Boot and a React-Redux frontend. Time series rendering is handled using D3.js. The source code is available under the MIT license[1].

3 *TimeVizBench* User Interface

This section presents the *TimeVizBench* interface (Fig. 2), which comprises three components: the *Configuration Panel*, *Visualization Panel*, and *Performance Metrics Panel*.

The *Control Panel* ((A_1)), enables users to set up the parameters for explo-ration. Users can select a time range through interactive date and time pickers, though this interval can also be adjusted dynamically via pan and zoom on the charts. Additionally, users can choose from available datasets and specify one or more measures to visualize. The interface supports multi-value selection, enabling analysis of multiple variables simultaneously.

Users can also select different methods for evaluation. Upon selection of a method, users are prompted to configure the *initialization parameters*, which remain fixed for that method instance throughout the session ((A_2)). In contrast, *query-time parameters*, such as accuracy thresholds, can be modified dynamically at any time, even after instantiation. The platform also allows for comparing multiple instances of the same method by configuring different values for either the initialization or query-time parameters. This flexibility enables users to sys-tematically compare methods or explore variations of a single method under different configurations.

---

[1] The source code is available at https://github.com/athenarc/TimeVizBench.

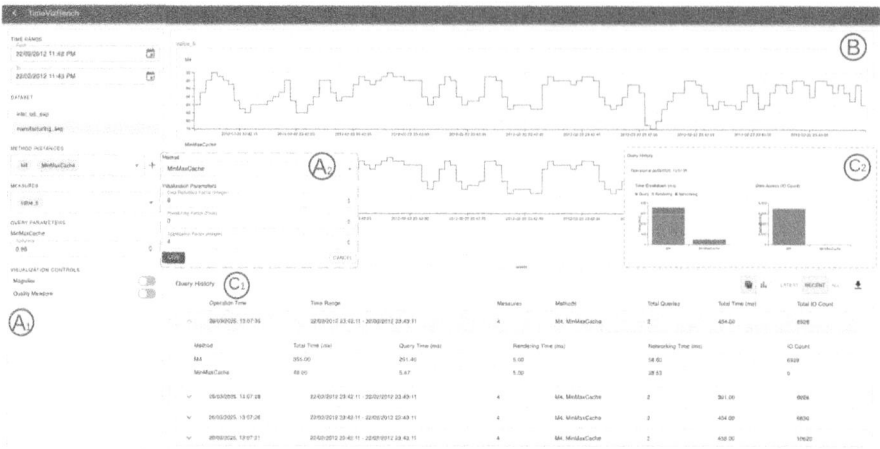

Fig. 2. *TimeVizBench* User Interface for the interactive evaluation of methods for large-scale time series exploration. The *Control Panel* (A₁), enables users to set up the parameters for exploration. Upon selection of a method, users are prompted to configure the *initialization parameters* (A₂). The *Visualization Panel* (B) displays time series line charts for the selected measures and method instances. The *Performance Metrics Panel* (C) *provides insights into the performance of each configured method and is organized into two complementary views. A* table view (C₁) itand a chart view (C₂).

Furthermore, this panel provides controls for the generated visualization. When enabled, the *Magnifier*, displays a circled segment of the time series in an enlarged overlay for closer inspection. Meanwhile, the *Quality Measure* switch overlays an error-free reference series on each chart for comparison against method-generated visualizations and displays the corresponding Structural Similarity Index Measure (SSIM) [12]. Due to the large-scale nature of the data, fetching raw data to verify accuracy would be prohibitively expensive; however, M4 has been verified as error-free (SSIM = 1) relative to the actual raw data [6], making it an ideal reference for measuring how closely each method's output matches the baseline.

The *Visualization Panel* (B) displays time series line charts for the selected measures and method instances. Each measure is shown in a separate panel, with visualizations for all configured methods stacked vertically for direct comparison. The visualizations are synchronized, ensuring that user interactions such as panning or zooming on one chart automatically update all others.

The *Performance Metrics Panel* (C), provides insights into the performance of each configured method and is organized into two complementary views. A *table view* (C₁) lists each query with its time interval, chosen dataset, requested measures, and all methods used, grouping together the performance metrics (e.g., query time, network time, rendering time, and data I/O counts). A *chart view* (C₂) displays the *same* metrics in bar-chart form, with each bar broken down to show the contributions of querying, networking, and rendering, alongside a visualization of data retrieval counts.

Additionally, *TimeVizBench* provides an export functionality, allowing users to download query history and corresponding performance metrics in CSV format. Each exported file includes details such as the time interval, selected dataset, measures requested, and performance metrics recorded for each method. This feature facilitates further offline analysis, enabling users to systematically compare different configurations, track performance trends over time, and reproduce evaluations for consistency.

Availability. The tool and its functionalities can be accessed online[2]. A video demonstration is also available[3].

4   Demonstration Outline

In this section, we describe the demonstration scenario for *TimeVizBench*, where attendees will explore methods for scalable, low-latency time series visualization. The demonstration showcases *TimeVizBench*'s ability to benchmark visualization methods across performance and accuracy dimensions using real-world datasets. Attendees will be asked to perform the following tasks:

– Add an example method for benchmarking, implemented via *TimeVizBench*'s interface (e.g., a simple averaging method per pixel column).
– Select a dataset from preloaded time series data (e.g., sensor readings from [8] or electrical power measurements from [5]) and choose one or more measures for visualization.
– Instantiate and configure visualization methods, with support for multiple configurations of the same method.
– Interactively explore time series visualizations, with synchronized panning and zooming across selected methods for direct comparison.
– Analyze performance metrics, including query time, total response time, and the amount of data retrieved from the database per method.
– Compare visualization accuracy using SSIM to assess differences between method-generated visualizations and a reference visualization.
– Export session results, including query history, method configurations, and performance metrics, for further analysis.

By engaging with these tasks, attendees will gain hands-on experience with *TimeVizBench*, understanding how different methods balance interactivity, performance, and accuracy in large-scale time series visualization.

Acknowledgments. This work was supported by the ExtremeXP project (EU Horizon program, GA 101093164).

---

[2] http://timevizbench.imsi.athenarc.gr/.
[3] https://vimeo.com/1070287190.

References

1. Battle, L., Chang, R., Heer, J., Stonebraker, M.: Position statement: the case for a visualization performance benchmark. In: 2017 IEEE Workshop on Data Systems for Interactive Analysis (DSIA), pp. 1–5 (2017). https://doi.org/10.1109/DSIA. 2017.8339089

2. Battle, L., et al.: Database benchmarking for supporting real-time interactive querying of large data. In: Proceedings of the 2020 ACM SIGMOD International Conference on Management of Data, SIGMOD 2020 pp. 1571–1587. Association for Computing Machinery, New York (2020). https://doi.org/10.1145/3318464. 3389732

3. Donckt, J.V.D., Donckt, J.V.D., Rademaker, M., Hoecke, S.V.: Minmaxlttb: leveraging minmax-preselection to scale lttb (2023). https://arxiv.org/abs/2305.00332

4. Eichmann, P., Zgraggen, E., Binnig, C., Kraska, T.: Idebench: a benchmark for interactive data exploration. In: Proceedings of the 2020 ACM SIGMOD International Conference on Management of Data, SIGMOD 2020, pp. 1555–1569. Association for Computing Machinery, New York (2020). https://doi.org/10.1145/ 3318464.3380574

5. Jerzak, Z., Heinze, T., Fehr, M., Grober, D., Hartung, R., Stojanovic, N.: The debs 2012 grand challenge. DEBS, pp. 393–398 (2012). https://debs.org/grand-challenges/2012/

6. Jugel, U., Jerzak, Z., Hackenbroich, G., Markl, V.: M4: a visualization-oriented time series data aggregation. Proc. VLDB Endowment **7**(10), 797–808 (2014)

7. Khelifati, A., Khayati, M., Dignös, A., Difallah, D., Cudré-Mauroux, P.: Tsmbench: benchmarking time series database systems for monitoring applications. Proc. VLDB Endow. **16**(11), 3363–3376 (2023). https://doi.org/10.14778/3611479. 3611532

8. Lab IBR: Intel lab dataset (2004). http://db.csail.mit.edu/labdata/labdata.html

9. Maroulis, S., Stamatopoulos, V., Papastefanatos, G., Terrovitis, M.: Visualization-aware time series min-max caching with error bound guarantees. Proc. VLDB Endow. **17**(8), 2091–2103 (2024). https://doi.org/10.14778/3659437.3659460

10. Plaisant, C., Fekete, J.D., Grinstein, G.: Promoting insight-based evaluation of visualizations: from contest to benchmark repository. IEEE Trans. Visual Comput. Graph. **14**(1), 120–134 (2008). https://doi.org/10.1109/TVCG.2007.70412

11. Wang, Y., et al.: Om3: An ordered multi-level min-max representation for interactive progressive visualization of time series. Proc. ACM Manag. Data **1**(2), 1–24 (2023)

12. Wang, Z., Bovik, A.C., Sheikh, H.R., Simoncelli, E.P.: Image quality assessment: from error visibility to structural similarity. IEEE Trans. Image Process. **13**(4), 600–612 (2004)

Data Sharing and Synthesis

Decentralized Research Data Sharing Management Using Blockchain Technology

Otto Hylli[✉][iD], David Hästbacka[iD], and Kari Systä[iD]

Computing Sciences, Tampere University, Tampere, Finland
{otto.hylli,david.hastbacka,kari.systa}@tuni.fi

Abstract. Sharing of data between researchers can benefit data owners, data users, and overall the scientific community. However, sharing presents challenges such as data owners losing control of their data. In this paper we explore a blockchain-based tool for managed data sharing. We present a concrete demonstration where data owners can control access to their data. We also discuss the possibilities and challenges of such an approach. Possibilities include flexible data storage and data access tracing without relying on a trusted third party. Challenges include identity verification, ease of use and blockchain transaction costs.

Keywords: blockchain · data sharing · Decentralized Data Management · Scientific Research · Data Ownership · Access Control

1 Introduction

Data is a valuable asset and companies see data as an opportunity for new business. That has led to needs and solutions for data sharing. Similarly, data sets are important assets in many fields of research. Research groups with good data sets can be more productive. Researchers are asked – and often agree – to share research data openly in a similar way to open access publications [12].

This study is motivated by a desire for a computer-aided service to support sharing of research data. We explore the potential of blockchain technology to manage sharing of the research data in a non-centralized way. In our proof-of-concept we use data trading infrastructure Ocean Protocol [2] that we extend to validate and demonstrate our proposal. We selected Ocean Protocol, since it already includes a suitable user interface and the trading functionality resembles conditions and negotiations required for research data sharing, but on a conceptual level some other blockchain approach could have been used as well.

In this paper we explore the issues and solutions for sharing research data in a decentralized way. In particular, we want to explore the applicability of blockchain-based solutions. Thus, the research questions can be expressed as:

RQ1 Can a blockchain-based solution be used in research data sharing?
RQ2 What kind of additional solutions and practices would be needed for blockchain-based data sharing?

P. K. Chrysanthis et al. (Eds.): ADBIS 2025, CCIS 2676, pp. 87–97, 2026.
https://doi.org/10.1007/978-3-032-05727-3_9

The paper is organized as follows. In Sect. 2 we discuss the needs, i.e., the issues of data sharing in the research context, introduce the blockchain technology, and present related work. Sections 3 and 4 present our blockchain-based concept and demonstration findings. Finally, conclusions are given in Sect. 5.

2   Background

2.1   Data Sharing in Research Context

The lack of suitable data can prevent scientific advancement. For example, in one survey 67 % of respondents agreed that lack of access to data generated by other researchers has restricted their ability to answer scientific questions [15]. Also, 85 % of respondents were interested in using data from other researchers. Researchers are generally willing to share their data. In another survey only 18 % of respondents were not interested in sharing their data [10].

Researchers also have concerns about sharing their data, and often do not want to share it without restrictions. Concerns include privacy of data subjects, desire to protect future publication opportunities, and a general desire to retain exclusive rights to data that has required effort to collect [13]. Another major issue is the effort required in sharing the data [7]. Legal questions about ownership of the data [9] and intellectual property rights related to sharing of research data.

Researchers are more likely to share their data, if they can set restrictions on access, get information about who uses their data, and for what purpose [15]. In their literature survey about academic data sharing Fecher et al. call this degree of control [10]. In that survey 56 % of respondents were willing to share only on request or demanded a context with access control. Another notable requirement for data sharing is recognition or other kind of returns. [10]. There are various forms for these returns that both data providers and data users see as fair. In the survey by Tenopir et al., options considered acceptable by a significant portion of respondents included, co-authorship, formal citation of the data provider in publications using the data, and providing the data provider an opportunity to review and comment on works based on their data [15].

2.2   Blockchains and Smart Contracts

A data sharing system can be either centralized or decentralized and there is no clear consensus on which is better [10]. For a decentralized solution blockchain can be one approach, since decentralization and lack of centralized authorities are core elements. Blockchain is a distributed ledger of transactions where every participant in the network has a copy of the ledger. Various cryptographical techniques such as private-public key cryptography and hashing are then used to ensure the authenticity and immutability of the transactions.

Smart contracts enable various uses for blockchains. Smart contracts are small applications whose code and state are stored on the blockchain. End-users can then interact with these smart contracts, e.g., via a web application and a blockchain wallet. The wallet manages the users cryptographic keys and uses them to sign blockchain transactions that the application constructs.

One popular platform for smart contract applications is the Ethereum blockchain [6]. Its own cryptocurrency Ether is used to pay for transaction costs. One common use for smart contracts is to manage assets and their ownership with tokens. An asset could be for example a currency, stock in a company or voting rights in a blockchain based organization. These tokens can be owned and traded by users. The token contract, i.e., the smart contract implementing a particular token keeps a record of token balances for each user. It also handles transfer of tokens between users, and can also allow specific users to create, i.e., mint new tokens. A more specialized version of a blockchain token is a non-fungible token (NFT). An NFT represents ownership of an unique asset such as a digital art piece. This is in contrast to fungible tokens, where the individual units of the asset the token represents are interchangeable with each other.

Blockchain and smart contracts provide the backend for decentralized applications (dApp). Execution is shared across the network, so no single party can alter the logic once deployed. Users interact pseudonymously: each action is a signed blockchain transaction sent from their client program A transfer of the dApp's own token, for instance, simply calls the token contract, which verifies the sender's balance and then atomically updates both accounts. Every accepted transaction is immutably recorded and publicly auditable. Which makes dApps attractive for research data sharing, where trust and traceability are important.

2.3   Related Work

CESSDA[1] and the UK data service[2] focus on open data sharing. CESSDA (Consortium of European Social Science Data Archives) federates various national social sciences data archives in Europe offering a unified catalog of its members' data offerings. The UK data service hosts large amounts of data about economics, population and social issues in the UK. B2share is an open source platform for sharing scientific data, focusing on data discoverability, integrity and reliability. It can be used for open data publication but it also offers access control features for restricted access. It can be customized for the needs of different communities [4]. DataONE is federating data repositories of biological and environmental data. It aims to engage the relevant stakeholder communities, enabling easy, secure, and persistent storage of data, and development of tools for data discovery, analysis, visualization, and decision-making. [11]

Blockchains have been used as part of various data sharing solutions both for general purpose data sharing and sharing data related to a specific domain. As an example of general data sharing Wang et al. [16] propose a data sharing system which combines the peer-to-peer distributed file storage system IPFS (interplanetary file system), the Ethereum blockchain, and attribute based encryption. Users can store their encrypted files to IPFS and then via smart contracts apply

---

[1] https://www.cessda.eu/.
[2] https://ukdataservice.ac.uk/.

fine-grained access control to their data. Smart contracts are also used for keyword based search of data for approved users.

The medical and healthcare field is an example of a specific domain blockchain based data sharing solutions have been proposed for. For example, Cheng et al. [8] propose a blockchain and cloud based system for patient data sharing between hospitals. Data is stored in the cloud and data hashes are stored on the blockchain to ensure the data is not tampered with. Blockchain is also used in authenticating access to the data, avoiding the need for a trusted third party.

There is also some existing work on sharing of research data via blockchain. Shrestha and Vassileva [14] propose a system to incentify the data owners with crypto payments or acknowledgements. In addition, smart contracts are also used to set access permissions and keep track of data accesses thus giving the data owner control over their data. They propose a hybrid blockchain approach where part of the functionality, like the crypto incentives, are done on public Ethereum, and part like the data access are handled on a private, permissioned blockchain. The Molecule protocol proposes a system where intellectual property and data rights for a research project are tokenized as NFTs [3]. The main purpose is to create a new system for funding research projects by connecting the legal agreements related to research funding to NFTs and allowing community management of these NFTs. However, the protocol also has some features related to data sharing with a possibility to store encrypted research data linked to the NFT in a decentralized storage and then granting access to it to desired users.

3   Proposed Concept

We investigate the applicability of blockchain by building a concrete proof-of-concept and demonstrator. This allows us to gain a concrete understanding of the issues and opportunities, and will also enable collection of concrete feedback.

Ocean Protocol, which already provides basic functionalities of a data marketplace, was chosen as the demonstration platform because building on plain blockchain infrastructure would have required substantial implementation effort.

3.1   Decentralized Research Data Sharing System Requirements

Our target is a decentralized data sharing system with the following features:

1. Data owner can publish the availability of their data sets
2. Data owner can store the data itself, flexibly where they want and still make it available via the system
3. Data owner can determine if their data is available as open data or if it is only available on request (restricted access)
4. Data consumer can discover available data sets
5. Data consumer can download open access data sets independently without the involvement of the data owner

6. Data consumer can send access requests for restricted data
7. Data owner can manage access requests by accepting or denying them
8. Data consumer gets restricted data only if the request has been accepted

3.2   Concept Demonstration Based on Ocean Protocol

Ocean Protocol was developed for blockchain-based data markets, where data owners could profit by selling their data to interested buyers. The data owner keeps control of their data and can store it anywhere as long as it is accessible via the internet. Information about available data is stored on the blockchain making the system decentralized. Getting access to the data is also handled via blockchain. Thus, no centralized trusted third party is needed. End-users, such as data owners and consumers, interact via the Ocean market web application and a blockchain wallet such as Metamask[3]. Ocean protocol has three layers:

1. The app layer: The end-user facing applications such as the market and data management platforms.
2. Middleware layer: Consists of the metadata cache Aquarius, the Ocean Provider used to access the data and various software libraries used to interact with Ocean Protocol.
3. Smart contracts layer: The actual blockchain part that has information about data ownership, metadata and handles data access rights via data tokens. Works on Ethereum based blockchains.

Since each data set is unique, Ocean uses non-fungible tokens for managing ownership. To publish data the data owner deploys a data NFT smart contract instance on the blockchain. This NFT represents their ownership of the data. Access to the data is handled with fungible data tokens whose token contracts are linked to the data NFT. In a simple case only one data token contract can be deployed. Its tokens can, for example, grant perpetual access to the data. The URL to access data is encrypted and stored as part of the NFT smart contract. Metadata about the data set such as name and description are also stored as part of the NFT contract. Only the Ocean provider service can decrypt the URL to the data. The Ocean provider is responsible for relaying the data to users who have been given access to the data.

The Ocean protocol is intended for commercial data markets, and it does not offer all the features required in research data sharing. However, it can offer a basis for such a solution with its existing features such as NFTs representing data ownership and data tokens used to manage access. Notably, Ocean Protocol does not allow data owners to manage access on the level of individual users. For some data owners this might not give enough control on their data.

3.3   Proof-of-Concept Implementation

We modified how Ocean handles free data sets. The *dispenser* smart contract, which usually gives data tokens on request, we modified to manage access

---

[3] https://metamask.io/.

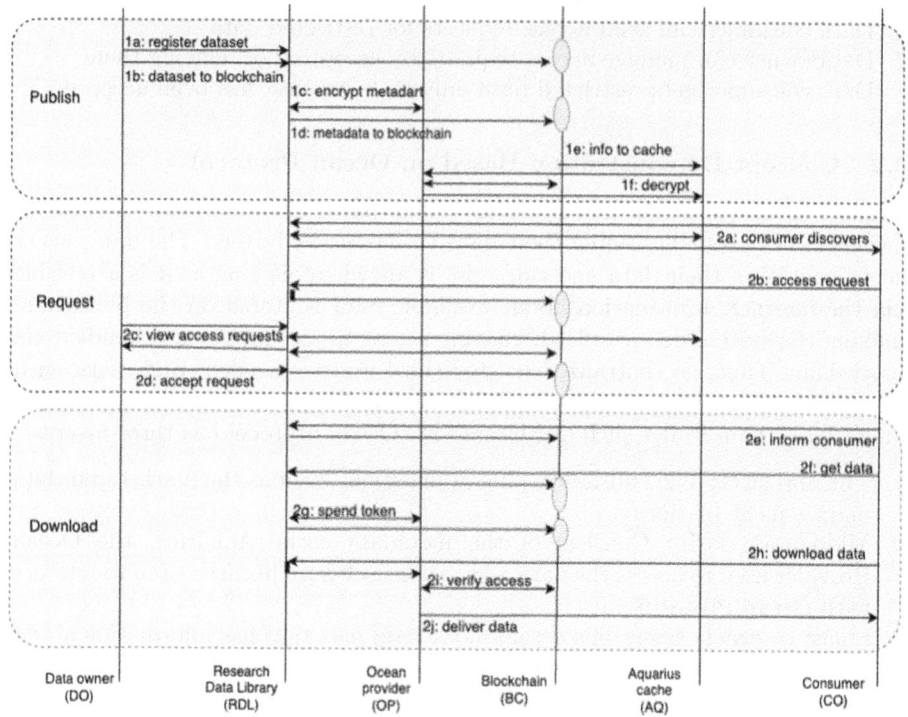

Fig. 1. Component interactions with ovals highlighting blockchain transactions.

requests to the data set. We then modified the Ocean market web application to support these access requests and renamed it to Research Data Library. The normal Ocean flow for publishing and downloading of free data was transformed into: publish data set, request access and download data set. This flow, shown in Fig. 1, has six actors:

1. *Data Owner (DO)* publishes the data and accepts the usage requests.
2. *Consumer (CO)* is a party that uses data after getting rights to do so.
3. *Ocean Provider (OP)* is a service that accesses data and known by the DO.
4. *Blockchain (BC)* is the distributed ledger storing all contracts.
5. *Aquarius (AQ)* is a metadata cache for all available data items.
6. *Research Data Library (RDL)* is a web user interface towards DO and CO.

In the publish phase, the CO uses the RDL to deploy a data NFT and associated data token smart contracts. The data NFT contains the data set metadata the CO defined. This metadata is encrypted by the OP and also includes the data access URL. See steps 1*a* - 1*d* in Fig. 1. If the CO sets the data set as restricted access, the NFT deployment also sets up the *dispenser* smart contract to manage access requests to the data set. The metadata cache AQ detects the creation of a new data set and caches its metadata for further queries. This includes requesting the OP to decrypt the metadata apart from the data access URL which stays encrypted. See steps $1c$ and $1f$.

In the request phase, the CO uses the RDL to browse available data sets. RDL queries data sets and their metadata from AQ (step $2a$). When the CO finds an interesting data set, they use the RDL to make an access request which includes a message for the DO ($2b$). The access requests are managed by the *dispenser* smart contract. The RDL offers a view for the DO for managing access requests to the DO's data sets ($2c$). From there the DO can accept or reject access requests. This decision is then registered and stored by the *dispenser* contract ($2d$).

In the download phase, the CO can check the status of their access request on the RDL, which queries it from the *dispenser* contract ($2e$). When the request has been accepted, the CO can download the dataset. This download process, managed by the RDL, first uses the *dispenser* to get a data token for the CO ($2f$). However, the *dispenser* gives a data token only, if the DO has accepted the access request. Next, the data token is immediately spend, granting the CO access to the data ($2g$). CO can then use the RDL to download or later redownload the data. The download is facilitated by the OP, which first verifies the spending of the data token ($2h$). Then the OP internally decrypts the data access URL and delivers the data to the CO ($2j$).

4   Findings and Discussion

In Subsect. 2.1 we discussed the expectations and the benefits of research data sharing. However, we also noted challenges such as privacy of research subjects, intellectual property rights, required publishing effort, need for control of the data, and lack of added value for the owner who shares the data. Our design and proof of concept show that blockchain technology could be used for sharing of research data and that it can solve some of the challenges without the need of centralized authorities. Data owners can control their data and track the data access. This helps in ensuring proper returns or recognition such as citations. Though, some challenges such as intellectual property issues cannot be solved via technological means alone. Also blockchain technology can present its own challenges including difficulty of use and transaction costs. These observations among other issues are discussed further in this section.

We have expressed usage conditions as free form text and stored on the blockchain. Similarly data accesses are recorded on the blockchain. Thus, there is an immutable record of what conditions applied when the user was granted access to the data. We assume that the conditions are similar for several data sets and a predefined condition terminology could be used. The system should propose some often used conditions that the research community recognizes. The definition of these conditions is a topic for future research.

The data owner might want to know the identity of the data user. This would be necessary for monitoring of the agreed usage conditions. However, blockchains are typically pseudonymous, and the users are represented just by their public addresses that are not connected to any identity information. Due to the transparency of blockchains, the actions done by the account can be tracked based on the blockchain transactions related to it. If data sharing use case needs some kind of accountability from the data consumer, the real identity of the user has to be connected to their blockchain identity.

Handling of the real identity of data users is also a topic for future research. One way of implementing identity in a decentralized way is the use of W3C verifiable credentials as used in the European Blockchain Service Infrastructure (EBSI) [1]. Verifiable credentials are machine verifiable and readable documents that contain statements about the credential subject stated by an issuer [5]. Integration to the widely used ORCID[4] could be one possible future development.

Additional features for research data sharing could be added to the system. For instance, many researchers might not need to grant individual access permissions for their data, but they still want to be aware who has been using their data and might want to check if the data has been used according to the conditions they have set. The real identity verification together with the ability to track transactions could give the data set owner a view on how their data is used. The data set owner could require that a smart contract allows only users whose real identity is verifiable to access the data.

In this research we had a simple model of users and roles – just data owners and data. In future research we should consider issues, such as shared ownership and different usage roles, e.g., future research, external validation, reviewers of the related publications. Blockchain technology could be used to implement shared ownership in the form of a distributed autonomous organization (DAO). DAO is a smart contract where actions require approval from multiple users.

Ocean protocol was used in our demo mainly as a convenience to make the development faster. An actual production implementation do not need be based on Ocean, it can use some ideas from it such as how data ownership is represented with NFTs and data access with fungible tokens, and how a metadata cache component can improve performance. Some parts might then require changes such as how data access and storage is handled. The current data storage and access solution of Ocean Protocol consists of two parts. The data can be stored anywhere the data owner wants as long as it is publicly reachable by an URL. Access is then restricted by encrypting the URL as part of the data NFT. The URL can be decrypted only by the Ocean provider instance the data owner selected when they published their data set. While this can offer flexibility, it can also be needlessly complex for some data owners, when they need to get both a storage solution and a Ocean provider instance for managing access to the data. A component that would combine both storage and access control could be created to simplify the solution.

The transaction costs of blockchains could be an issue. In public blockchains such as Ethereum the sender has to pay for the processing of the transaction with the network's crypto currency that is to reward the operators and to ensure the network is not overwhelmed by meaningless transactions. The cost of the transaction depends on the required amount of computation used, i.e., the complexity

---

[4] https://orcid.org/.

of the smart contract operations performed. In addition, the amount of storage needed on the blockchain also affects the cost. Especially the latter is an issue in our case since the data set metadata and access request messages are stored on the blockchain, and storing data in blockchain may be expensive. For use cases with no financial gains such as sharing of scientific data, the transaction costs should be minimized. One possible approach is a non-public permissioned blockchain which has other means of limiting the amount of transactions, and could be operated e.g. by a group of research organizations such as universities.

The immutability of blockchain must be considered in the system design. Users make mistakes and information, e.g., data usage conditions can change. It must be possible to update the stored information. Still, the old information stays on the blockchain and can be checked, e.g., to verify what version of the usage conditions applied when a specific access request was accepted.

To encourage data sharing, the system should be easy to use. Since, the use of blockchain based systems are not yet widely used, the related complexities such as using a wallet to sign transactions, could be an impediment for the wide adoption. To alleviate this issue users could be offered an alternative to use a wallet managed by a trusted party such as their own research institution. This way, the user is not responsible for managing and backing up their own crypto-graphic keys, where errors may cause the user to lose access to their blockchain account. In general, in a production version of this kind of system extra attention has to be paid towards making the user experience as smooth as possible.

Integration to data portals, especially with respect to emerging data spaces developments, would require further research. This could solve some of the previously mentioned challenges, and could possibly act as the abstraction layer many end users would like to operate with.

5   Conclusions

Data is important in research, and collection of data sets is often a big investment. Controlled sharing of research data would benefit many stakeholders. Data owners can get credits for their data, see further research based on the data, and establish collaborations. Data users can benefit from existing data, and the research community could see more validation of published research.

Some solutions have been created for trading data in business operations. However, established solutions, which offer flexible access control, do not yet exist for research data. In this paper we explore the opportunities of data sharing mechanisms for research data and propose a tailored trading solution for research data. Our concrete experiment was based on an existing data trading solution based on blockchain – with reasonable modifications to the system we adopted the solution for sharing of the research data.

Our results show that a blockchain-based approach is feasible, there are still a few issues to be resolved. These include: 1) traceability to actual data user, either by adding user identity or by some other mechanisms; 2) formulation of some pre-defined rules and conditions that make agreement on the conditions easier, and; 3) the system should be integrated to other tools and systems, e.g., to platforms that actually store the shared data. We see that this work is just a beginning. The current implementation is just a proof-of-concept and the used technologies namely Ethereum and Ocean protocol are not necessarily the best choices for a proper implementation. In addition, the idea and solutions should be discussed in the research community before going to actual implementations.

Acknowledgments. This work is supported by Business Finland project ECADEC.

Disclosure of Interests. The authors have no competing interests to disclose.

References

1. European blockchain services infrastructure (ebsi). https://ec.europa.eu/digital-building-blocks/wikis/display/EBSI/Home, Accessed 1 Aug 2023
2. Ocean protocol. https://oceanprotocol.com/
3. What is molecule? - molecule docs. https://docs.molecule.to/documentation/introduction/readme, Accessed 11 Sep 2023
4. Ardestani, S.B., et al.: B2share: an open escience data sharing platform. In: 2015 IEEE 11th Intl. Conference on e-Science, pp. 448–453 (2015)
5. Burnett, D., Sporny, M., Zundel, B., Noble, G., Hartog, K.D., Longley, D.: Verifiable credentials data model v1.1. W3C recommendation, W3C (Mar 2022). https://www.w3.org/TR/2022/REC-vc-data-model-20220303/
6. Buterin, V.: A next-generation smart contract and decentralized application platform. https://ethereum.org/content/whitepaper/whitepaper-pdf/Ethereum_Whitepaper_-_Buterin_2014.pdf (2014)
7. Campbell, E.G., Bendavid, E.: Data-sharing and data-withholding in genetics and the life sciences: results of a national survey of technology transfer officers. J. Health Care Law Policy **6**, 241–255 (2002)
8. Cheng, X., Chen, F., Xie, D., Sun, H., Huang, C.: Design of a secure medical data sharing scheme based on blockchain. J. Med. Syst. **44**(2), 52 (2020)
9. Enke, N., Thessen, A., Bach, K., Bendix, J., Seeger, B., Gemeinholzer, B.: The user's view on biodiversity data sharing — investigating facts of acceptance and requirements to realize a sustainable use of research data —. Eco. Inform. **11**, 25–33 (2012)
10. Fecher, B., Friesike, S., Hebing, M.: What drives academic data sharing? PloS one **10**(2) (2015)
11. Michener, W., Vieglais, D., Vision, T., Kunze, J., Cruse, P., Janée, G.: Dataone: data observation network for earth—preserving data and enabling innovation in the biological and environmental sciences. D-Lib Mag. **17**(1/2), 12 (2011)
12. Open Research Europe: Open data, software and code guidelines. https://open-research-europe.ec.europa.eu/for-authors/data-guidelines, Accessed 3 April 2024
13. Savage, C.J., Vickers, A.J.: Empirical study of data sharing by authors publishing in plos journals. PLOS ONE **4**(9), 1–3 (2009)
14. Shrestha, A.K., Vassileva, J.: Blockchain-based research data sharing framework for incentivizing the data owners. In: Chen, S., Wang, H., Zhang, L.-J. (eds.) ICBC 2018. LNCS, vol. 10974, pp. 259–266. Springer, Cham (2018). https://doi.org/10.1007/978-3-319-94478-4_19

15. Tenopir, C., et al.: Data sharing by scientists: practices and perceptions. PLOS ONE **6**(6), 1–21 (2011)
16. Wang, S., Zhang, Y., Zhang, Y.: A blockchain-based framework for data sharing with fine-grained access control in decentralized storage systems. IEEE Access **6**, 38437–38450 (2018)

EXPERIVERSUM: An Environment for Curating Data-Driven Experimental Sciences

Genoveva Vargas-Solar[1]([✉]) [ID], Umberto Costa[2] [ID], Jérôrme Darmont[3] [ID],
Javier A. Espinosa-Oviedo[4] [ID], Carmem Hara[5] [ID], Sabine Loudcher[3] [ID],
Regina Motz[7] [ID], Martin A. Musicante[2] [ID], and José-Luis Zechinelli-Martini[6] [ID]

[1] CNRS, Univ Lyon, INSA Lyon, UCBL, LIRIS, UMR5205, 69221 Villeurbanne,
France
genoveva.vargas-solar@cnrs.fr
[2] Federal University Rio Grande do Norte, DIMAP, Natal, Brazil
{umberto.costa,martin.musicante}@ufrn.br
[3] Université Lumière Lyon 2, ERIC, Lyon, France
{jerome.darmont,sabine.loudcher}@univ-lyon2.fr
[4] Université Claude Bernard Lyon 1, ERIC, Villeurbanne, France
javier.espinosa@univ-lyon1.fr
[5] Federal University of Parana, Curitiba, Brazil
carmemhara@ufpr.br
[6] Fundación Universidad de las Américas Puebla, Puebla, Mexico
joseluis.zechinelli@udlap.mx
[7] Universidad de las República, Montevideo, Uruguay

Abstract. This paper introduces EXPERIVERSUM, a lakehouse-based environment that supports the curation, documentation and reproducibility of data-driven scientific experiments. EXPERIVERSUM enables structured research through iterative data cycles, while capturing metadata and collaborative decisions. Demonstrated through case studies in the earth, life and political sciences, EXPERIVERSUM promotes transparent workflows and multi-perspective result interpretation. EXPERIVERSUM bridges exploratory and reproducible research, encouraging accountable and robust data-driven practices across disciplines.

Keywords: Data and experiment curation · Reproducible research · Lakehouse architecture · Data processing pipelines · Metadata

1 Introduction

Massive data production is becoming increasingly vital in experimental sciences such as the life sciences, earth sciences, social sciences, and humanities, where large-scale and cost-effective data acquisition is now possible. Such fields generate diverse datasets of varying quality, enabling multifaceted analyses. Traditional schema-on-write methods such as ETL (Extraction, Transformation, Loading)

© The Author(s), under exclusive license to Springer Nature Switzerland AG 2026
P. K. Chrysanthis et al. (Eds.): ADBIS 2025, CCIS 2676, pp. 98–107, 2026.
https://doi.org/10.1007/978-3-032-05727-3_10

struggle with such heterogeneity. Data lakes provide a flexible alternative by storing raw data in original formats, but require effective metadata extraction to integrate data and ensure reproducibility.

In the context of open science, it is not enough to simply share data; researchers must also document the experimental context, including the conditions and decisions made during an experiment. This requires detailed metadata that captures both the data itself and the process through which it was produced. The main challenge is twofold: (i) designing metadata models that accurately represent both data and processing workflows, and (ii) implementing ELT (Extraction, Loading, Transformation) pipelines that support experiment curation and track how decisions influence outcomes. In this setting, metadata must serve as an execution guide for ELT processes to ensure transparency and reproducibility.

This paper introduces EXPERIVERSUM, a lakehouse-based environment that applies a meta model to curate and manage data-driven experiments. EXPERIVERSUM enables researchers to explore, analyse and reuse experiments with rich metadata, supporting the principles of open science. Consequently, the remainder of the paper is structured as follows. Section 2 reviews related works on metadata, provenance and reproducibility. Section 3 introduces the EXPERIVERSUM environment. Section 4 details the system's architecture, curation processes and exploration functions. Section 5 presents use cases in life, earth, and social sciences. Section 6 concludes and outlines future work.

2   Related Works

This section reviews key approaches for curating experimental data and processes, covering storage and management systems such as data warehouses, data lakes, lakehouses and dataverses [19,22]. We also compare data lake solutions used in earth, life, and social sciences.

The Evolving Practice of Data Curation. Data curation has evolved from focusing on preservation and quality control [12,16] to a value-added process that includes metadata enrichment and contextualization [18]. In fields of earth sciences and biodiversity, this shift supports reusability and clear provenance [5]. Modern platforms such as dataverse combine automation with expert oversight to support the full research lifecycle [23].

Infrastructure for Modern Research. Managing today's research data, ranging from structured tables to unstructured content, requires flexible systems. Data warehouses are optimized for structured analytics, but lack support for diverse formats [3]. Data lakes address this with schema-on-read flexibility [11]. However, without proper governance, data lakes risk becoming "data swamps" [2]. The lakehouse model combines the strengths of both data warehouses and data lakes [1], while dataverses offer curated, citable storage [6,15]. Our work advocates for integrating a lakehouse and a dataverse approaches in data-driven experimental sciences.

Discipline-Specific Data Challenges. Different disciplines require tailored infrastructures. In natural sciences, for instance, repositories support metadata standards for reproducibility [20]. In social sciences, data is often qualitative and harder to standardize. Data lakes offer needed flexibility [10], but must preserve context and consider ethical issues, especially with personal or indigenous data [4]. Hybrid solutions aim to balance scale and detail [8,22].

Innovations and Remaining Challenges. Emerging technologies such as conversational analytics using Large Language Models (LLMs) are reshaping interaction with data [7,14,17]. While promising, LLMs raise concerns about accuracy and trust [9]. Interoperability remains difficult across disciplines and data types [21]. Ultimately, success depends not only on technical solutions but also on institutional support and user adoption [13].

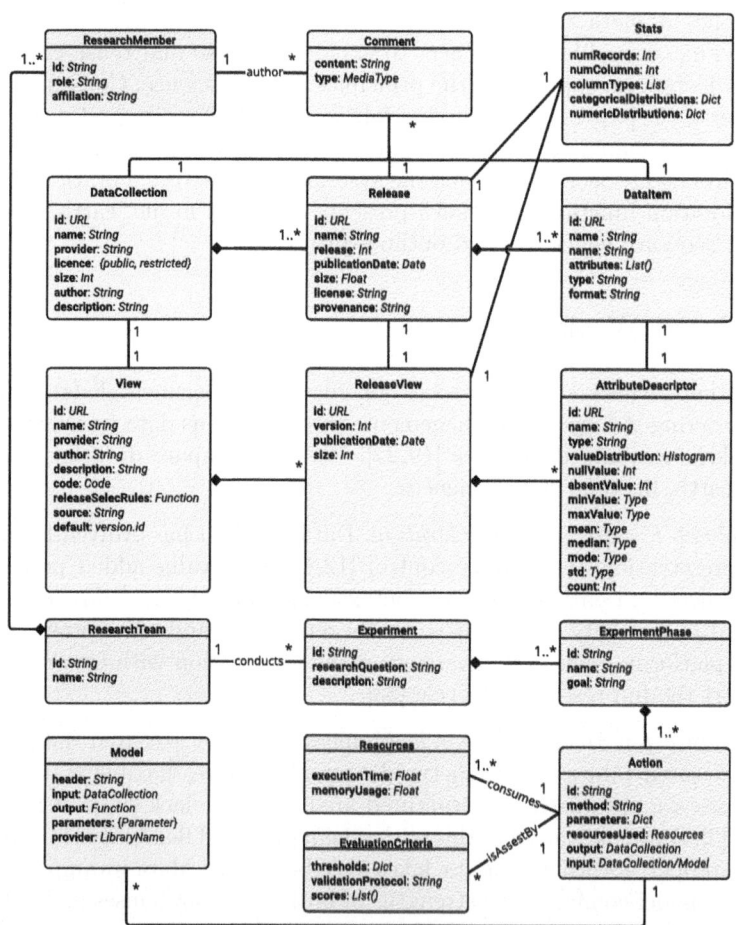

Fig. 1. EXPERIVERSUM Metamodel UML Class Diagram

3   Modeling Data-Driven Experiments

A data-driven experiment consists of three key elements: (1) raw data from empirical sources, (2) the research team responsible for data selection, methods, and validation, and (3) contextual metadata describing collection, processing, and analysis conditions. Curation ensures all components are documented for transparency and reproducibility. We define a metadata model structured around these concepts. Figure 1 illustrates this model.

Level 1: Raw Content. The green classes in Fig. 1 represent data ingested or produced during experiments. Metadata are extracted through automated and manual processes, capturing summaries, distributions and structure (e.g., column types and format). Each data release is profiled (e.g., licensing, size and provenance) and can include tabular, textual or signal data. Data items can also be annotated with multimedia or textual comments to enhance interpretability.

Level 2: Experimental Specifications. The yellow classes represent actions performed on data collections, whether manual or automated. Actions produce artefacts or models, with metadata describing structure, execution and provenance. Parameters, evaluation criteria and validation protocols are recorded to trace how and why actions were performed or repeated.

Level 3: Experiment Context. The blue classes describe the research team's composition (e.g., roles and seniority), responsibilities and the guiding research question. It captures the decision-making context and provides a basis for comparing experiments.

4   EXPERIVERSUM Environment

The EXPERIVERSUM environment ensures preservation, documentation and reproducibility of scientific experiments using a lakehouse infrastructure. Figure 2 shows EXPERIVERSUM's main components.

Extraction and Loading. Raw data such as seismic signals or social media posts are ingested in their original format (e.g., signals, text and media). Data collections are grouped into catalogues based on attributes such as ingestion date, size, format and quantitative traits.

Metadata Management. This component links metadata from raw and processed data to the experimental context, supporting reproducibility. Scientists can navigate datasets using quantitative summaries and relevant descriptors. Metadata is extracted, normalized and transformed following the curation metamodel, then stored in a metadata repository, enabling comparison and exploration across experiments.

Experiment Curation. Researchers can specify experimental parameters such as selection criteria, team roles, questions and performance constraints. They

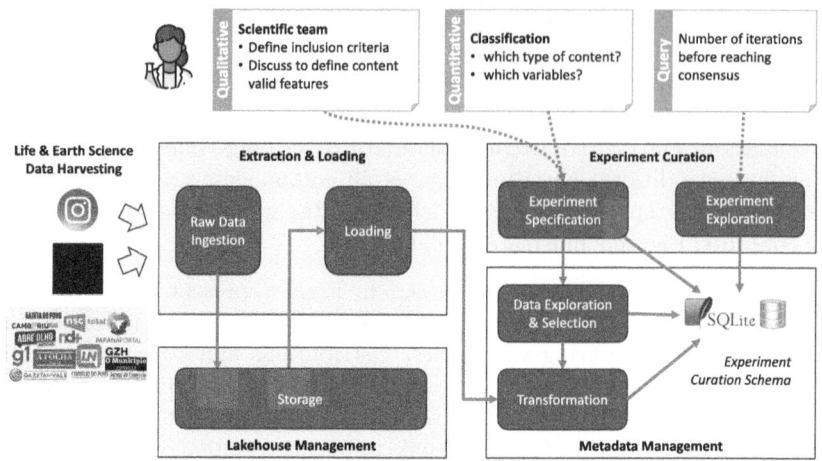

Fig. 2. EXPERIVERSUM Architecture

can explore experiments on similar topics conducted under different conditions, review methods and assess outcomes—supporting comparative analysis and enhancing reproducibility.

EXPERIVERSUM Management. This component orchestrates EXPERIVERSUM pipelines, managing seamless data flow from ingestion to analysis. It ensures consistency, performance, and smooth transitions across the infrastructure.

Extraction and Loading Pipeline. The EL pipeline handles data ingestion, cleaning and transformation before loading it for analysis—crucial for any data workflow. Key steps include (i) data extraction retrieves raw data from sources such as APIs, sensors, databases or unstructured files (e.g., seismic logs and social media); (ii) data cleaning removes duplicates, errors and inconsistencies; (iii) data enrichment adds contextual details such as timestamps or geolocations reliability; (iv) data loading stores data with associated structural and quantitative metadata in organized collections.

Tagging Experimental Processes Pipeline. This pipeline improves reproducibility and collaboration by assigning structured tags to experimental workflows. It consists of three tasks: (i) experiment specification records metadata such as experiment ID, name and date to link processes and data; (ii) tagging applies algorithmic or user-defined tags to annotate datasets and processes; (iii) tag storage maintains tag traceability for reuse and reference.

Transformation Pipeline. This pipeline is responsible of converting raw or semi-structured data into usable formats aligned with the metadata model. It consists of three tasks: (i) structuring maps text, signals and media to metadata entities; (ii) contextual enrichment adds metadata to reflect experimental settings; (iii) preparation formats data for analytics, machine learning or further experimentation.

Exploring and Querying Processes. The exploration and analytics pipeline enables users to query, analyze and visualize curated datasets. It supports exploratory data analysis (EDA), statistical modeling and machine learning to uncover insights. There are five tasks:

1. *Experiment querying and retrieval* to access datasets by filtering parameters such as time, location or experiment settings for efficient data selection.
2. *Filtering and aggregation* to refine data by extracting relevant subsets and aggregating across dimensions (e.g., time and region) to produce summary metrics.
3. *Descriptive and predictive analytics* to perform statistical analysis (averages, correlations, trends) and advanced tasks (classification, regression, clustering, anomaly detection) for pattern discovery and forecasting.
4. *Data visualisation* to display results using graphs, charts and heatmaps to simplify interpretation, trend spotting and anomaly identification.
5. *Collaboration and sharing* to share results, export outputs and integrate findings into reports or publications to support teamwork and dissemination.

5   Use Case-Based Validation

The first prototype of EXPERIVERSUM was implemented using SQLite3 as the metadata storage backend. Data pipelines were developed in Python and Jupyter Notebooks to support three use cases: biodiversity analysis, seismic event detection, and graffiti classification.

Tracking "Caravelas Portuguesas" along the Brazilian Coastline. This use case classifies sightings of the jellyfish *Physalia physalis* along Brazil's coastline[1].

- *Raw data extraction and loading.* Instagram posts tagged with relevant hashtags (e.g., #aguaviva, #caravelaportuguesa) are collected and converted into CSV files containing metadata (e.g., post's ID, source, location, media URL), which are then uploaded into EXPERIVERSUM.
- *Data transformation.* CSV headers are mapped to our metadata model (e.g., experiment, media, content, and tags). Unstructured or imprecise geo-temporal data (e.g., 'last summer' or inaccurate locations) is cleaned and corrected. Each transformation, along with its derived dataset, is recorded in EXPERIVERSUM.
- *Experimental settings.* Two research teams collaborate: data scientists use machine learning models to classify posts, while biologists manually tag and define classification categories. Settings include: inclusion criteria, (e.g., location/time), the gender of the jellyfish's victim, model calibration and performance thresholds.

---

[1] https://es.wikipedia.org/wiki/Physalia_physalis.

– *Exploration and querying.* Users query jellyfish occurrences by time and region, explore ecological associations and compare human and machine learning classifications to study methodological differences.

Classification of Seismic Activity in Northeast Brazil. This use case curates seismic data to distinguish natural from anthropogenic events, producing validated bulletins that summarize seismic activity.

– *Experiment setup.* Participants include seismographs (data generation), data collectors (retrieval), junior analysts (event detection) and senior analysts (review and bulletin publication).
– *Data extraction and loading.* Seismographs produce SAC files, which are uploaded and validated in EXPERIVERSUM. Amplitude values along each axis (X, Y, Z) are extracted from the SAC files and stored together with metadata such as *station_id, channel_id* and *timestamps*.
– *Data transformation and tagging.* Junior analysts plot waveforms to detect and tag events. Each event is annotated by *station, year* and *magnitude*. Analysts identify P and S wave arrivals. A triangulated event list forms the basis of the official bulletin, validated by a senior analyst.
– *Exploration.* Waveforms and results are visualised through a Web interface. Analysts can share or publish curated outcomes.

Graffiti Analysis for Political Messaging. A two-member team (junior + senior) conduct qualitative analysis to classify political graffiti across a city.

– *Research framing.* Over two cycles, the team refines the central research question: *"Can political messages be traced through graffiti?"*. They define inclusion criteria and political graffiti indicators through discussion.
– *Data collection.* The junior researcher photographed 1050 graffiti across districts. After review, 546 photos were validated and shared on Instagram[2].
– *Analysis.* Manual classification is complemented by unsupervised machine learning (k-means and hierarchical clustering). Results from both are iteratively refined and interpreted collaboratively.
– *Results.* Narratives and metadata are compiled through successive review rounds. Final deliverables include classifications, summaries and reproducibility documentation.

Lessons Learned. Developing an experiment curation system reveals key insights into the challenges and benefits of structuring data-driven research. It underscores how data, metadata and decisions intersect, and the importance of systematic curation for transparency, reproducibility and collaboration.

Curation and Reproducibility. EXPERIVERSUM supports the curation of varied data types (seismic signals, social media and multimedia) through ingestion,

---

[2] https://www.instagram.com/graffitis.montevideo.

transformation and tagging pipelines. These pipelines enrich content with contextual metadata, enabling reproducible experiments and traceable results. *Lesson*: Metadata models are crucial for linking data with experiments, ensuring interpretability beyond storage.

Data Transformation and Tagging. While structured data such as seismic signals are easily processed, tagging unstructured content (e.g., social media) prove more difficult, requiring advanced techniques. *Lesson*: Automated tagging suits structured data. Unstructured sources need robust Natural Language Processing (NLP) methods.

Using Metadata to Understand Experiments. Fig. 3 illustrates metadata-driven queries across three use cases. The first chart shows political graffiti labels by annotator, with juniors contributing most tags, indicating their key role in inter-

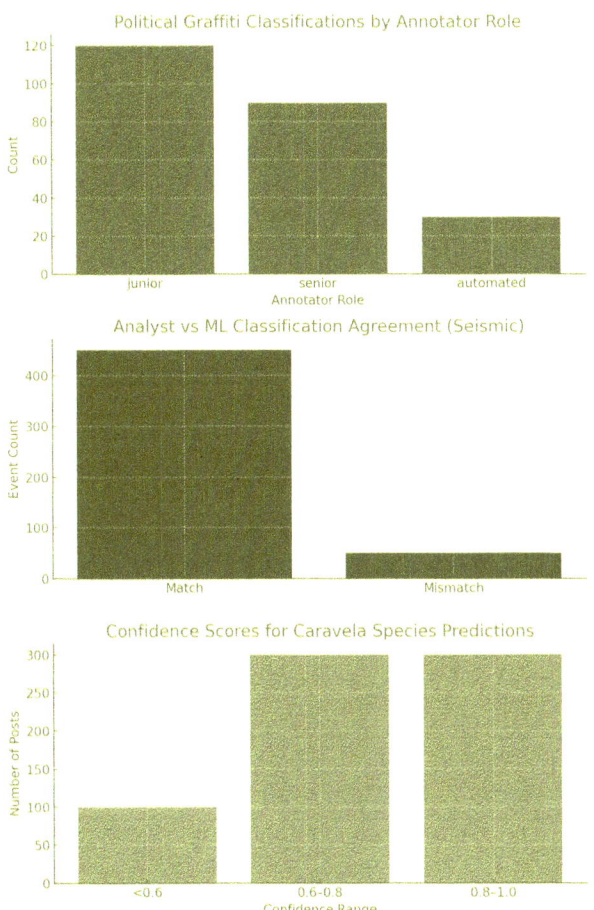

Fig. 3. Query Visualisation in EXPERIVERSUM

pretation. The second chart compares human and machine classifications in seismic monitoring, showing general agreement but highlighting some discrepancies needing expert review. The third chart displays confidence scores for jellyfish classification, while many fall in the 0.8âĂŞ1.0 range, lower-confidence cases (<0.6) point to the need for manual checks. Such visualisations show how curated metadata improves analysis, validation and understanding across complex experiments.

6 Conclusion and Future Work

This paper introduces EXPERIVERSUM, a lakehouse-based platform for curating, exploring and reproducing data-driven experiments. EXPERIVERSUM integrates ELT pipelines with a structured metamodel to link raw data to experimental intent, enabling workflow reuse across disciplines. Case studies with biodiversity, seismic data and graffiti classification highlight its flexibility for interdisciplinary research. The main insight is that reproducibility requires preserving full experimental context, not just raw data. Our metadata model and curated workflows improve traceability and reuse across diverse data types.

Future work includes extending metadata coverage, using NLP and graph techniques for tagging, adding privacy-aware analytics, and deploying the platform in real infrastructures to support open and collaborative science.

Acknowledgments. This work was partially supported by the LETITIA project (http://vargas-solar.com/letitia) through the Lyon Computer Science Federation (FIL), and by the Research Direction program at Universidad de las Américas, Puebla, through the Process Curation special research project.

References

1. Armbrust, M., Ghodsi, A., Xin, R., Zaharia, M.: Lakehouse: a new generation of open platforms that unify data warehousing and advanced analytics. In: Proceedings of the 11th Conference on Innovative Data Systems Research (CIDR) (2021)
2. Becker, C., Genschel, U., Siegfried, T.: From data swamp to data lakehouse: metadata management in interdisciplinary research. J. Inform. Manag. Data Sci. **5**(2) (2022)
3. Bimonte, S., Coulibaly, F.A., Rizzi, S.: An approach to on-demand extension of multidimensional cubes in multi-model settings: application to iot-based agroecology. Data Knowl. Eng. **150**, 102267 (2024)
4. Carroll, S.R., et al.: The care principles for indigenous data governance. Data Sci. J. **19**, 43 (2020)
5. Cheney, J., Chiticariu, L., Tan, W.C., et al.: Provenance in databases: why, how, and where. Foundations Trends Databases **1**(4), 379–474 (2009)
6. Crosas, M., King, G., Honaker, J., Sweeney, L.: Automating open science for big data. Ann. Am. Acad. Pol. Soc. Sci. **659**(1), 260–273 (2015)
7. Dubey, R., Chakraborty, P., Getoor, L.: Chatgpt for data science: promise and pitfalls. Commun. ACM **67**(4), 48–56 (2024)

8. Dunning, A., Erp, M., Skarpelis, C.: Advancing fair data practices in the social sciences: lessons from the sshoc project. Data Sci. J. **20**, 1–9 (2021)
9. Ghodsi, A., Zaharia, M., Xin, R.: Can we trust llms for data tasks? challenges and research directions. IEEE Data Eng. Bull. **46**(2), 9–20 (2023)
10. Hai, R., Geisler, S., Quix, C.: Conquering the data lake: a research agenda. In: Proceedings of the 2016 International Conference on Information Systems (2016)
11. Hegde, M., Smit, C., Pilone, P., Petrenko, M., Pham, L.: Use of schema on read in earth science data archives. In: 2017 American Geophysical Union (AGU) Fall Meeting (2017)
12. Higgins, S.: The dcc curation lifecycle model. Int. J. Digit. Curation **3**(1), 134–140 (2008)
13. Jagadish, H., Abiteboul, S., Buneman, P.E.A.: The big data conundrum in the social sciences. Commun. ACM **64**(3), 68–77 (2021)
14. Kerner, N., Batra, I., Hosanagar, K.: Prompting for insights: how business analysts use llms for data exploration. Proc. ACM Human-Comput. Interact. **7**(CSCW2), 1–27 (2023)
15. King, G.: An introduction to the dataverse network as an infrastructure for data sharing (2007)
16. Lord, P., Macdonald, A.: e-science curation report. Data curation for e-Science in the UK: an audit to establish requirements for future curation and provision (2003)
17. Nguyen, M., Bao, J.: Conversational analytics: a survey of large language models for data analysis. arXiv preprint arXiv:2403.00012 (2024)
18. Palmer, C.L., Renear, A.H., Cragin, M.H.: Purposeful curation: research and education for a future with working data (2008)
19. Rocha, H.F., et al.: Identifying occurrences of the cnidarian physalia physalis in social media data. Comput. Sci. Inf. Syst. **21**(4), 1887–1911 (2024)
20. Russom, P.: Data warehouse modernization. TDWI Best Pract Rep (2016)
21. Sawadogo, P., Darmont, J.: On data lake architectures and metadata management. J. Intell. Inform. Syst. **56**(1), 97–120 (2021)
22. Vargas-Solar, G., et al.: Dataversifying earth sciences: pioneering a data lake architecture for curated data-centric experiments in life and earth sciences. In: Proceedings of the Workshops of the EDBT/ICDT 2024 Joint Conference, Paestum, Italy, 25 March 2024. CEUR Workshop Proceedings, vol. 3651. CEUR-WS.org (2024)
23. Zgolli, A., Collet, C., Madera, C.: Metadata in data lake ecosystems. Data Lakes **2**, 57–96 (2020)

Extension of Data Catalog Vocabulary for Federating Open Datasets in Data Spaces

Adriana Morejón$^{(\boxtimes)}$ (ID), Lucía de Espona(ID), Alberto Berenguer(ID),
David Tomás(ID), and Jose-Norberto Mazón(ID)

University Institute for Computing Research (IUII), University of Alicante, Alicante,
Spain
adriana.morejon@ua.es

Abstract. Data spaces require the federation of open data from diverse providers. To support this, data catalogs play a critical role in managing and exposing metadata that enables dataset discovery from open data portals. However, most existing catalogs rely heavily on high-level metadata at the dataset level (such as title, license, and keywords) often aligned with standards like DCAT (Data Catalog Vocabulary). While useful, this coarse-grained metadata often falls short in supporting discoverability. To address this limitation, this paper proposes a novel extension to the DCAT standard specifically designed to discover relevant open data to be federated in data spaces. The extension enriches dataset descriptions with fine-grained, content-level metadata, including field-level details, descriptions, and representative samples of values. These additional metadata elements provide critical context for understanding the structure and semantics of datasets, enhancing their discoverability.

Keywords: open data · data space · metadata · catalog · DCAT · LLM

1 Introduction

Data spaces represent a paradigm shift in how data is shared across organizations [16]. These federated infrastructures are built upon principles such as data sovereignty, interoperability, and trust, and are designed to enable data collaboration in B2B (Business-to-Business) scenarios [6].

Interestingly, within this context, the role of open data becomes increasingly significant [18]. Open data (i.e., data that is freely available for anyone to use, reuse, and redistribute) has the potential to enrich data spaces. However, this potential is often underutilized due to the lack of robust discoverability mechanisms that would allow open data to be seamlessly federated into data sharing ecosystems such as data spaces [5]. Data catalogs play a critical role in addressing this challenge by enabling effective metadata management, which is essential for enhancing data discoverability across organizations [1].

P. K. Chrysanthis et al. (Eds.): ADBIS 2025, CCIS 2676, pp. 108–117, 2026.
https://doi.org/10.1007/978-3-032-05727-3_11

Metadata can refer either to the dataset as a whole or to its internal data content [15]. When describing the dataset as a whole, metadata typically fall into three categories [17]: (i) descriptive metadata, which supports identification through elements such as title, author, publisher, subject, and description; (ii) structural metadata, which outlines the organisation and scope of the dataset; and (iii) administrative metadata, which provides additional information such as creation date, technical specifications (e.g., file format), and licensing or access rights. In contrast, metadata about the content focuses on individual data elements, such as field names (e.g., the column headers in a spreadsheet) and their definitions [15]. These metadata are essential not only for helping users understand the structure of the datasets and their intended use, but also for enhancing data discoverability.

A widely adopted metadata standard for creating data catalogs is the Data Catalog Vocabulary (DCAT)[1]. DCAT is a W3C recommendation designed to promote interoperability and it is the reference for implementing cataloging mechanisms in various data ecosystems, including open data portals (e.g., through CKAN) or reference architectures for data spaces such as the International Data Spaces Association (IDSA)[2] and Gaia-X[3], implemented through connectors like the Eclipse Dataspace Components (EDC)[4] and the FIWARE Data Space Connector[5]. Given its widespread adoption and alignment with interoperability goals, DCAT would serve as a strong foundation for discovering useful open data to be federated into data spaces. Unfortunately, while DCAT has significantly advanced the harmonization of high-level metadata, its current implementations primarily focus on dataset descriptions (such as title, description, license, publisher, and thematic keywords). Such high-level metadata often fails to meet the data discoverability needs of open datasets [4]. Therefore, this metadata gap introduces friction in the data discovery process and hampers the federation of open datasets into data spaces. To address these limitations, this paper proposes an extension to the DCAT standard focused on incorporating fine-grained, content-level metadata into open data catalogs. The proposed extension specifies additional metadata elements that describe individual dataset fields and provide sample values. By enriching dataset descriptions with this deeper level of detail, our DCAT extension aims to facilitate more effective discovery of relevant open datasets suitable for federation into data spaces. To achieve this, it is also essential to map our extension to the cataloging approaches employed by existing open data portals and data spaces.

The contributions of this paper are twofold. First, it introduces a novel DCAT-compliant extension that incorporates content-based metadata, including new classes and properties for describing dataset content such as fields, types, descriptions, and representative sample values. Second, it provides a mapping

---

[1] https://www.w3.org/TR/vocab-dcat-3/.
[2] https://internationaldataspaces.org/.
[3] https://gaia-x.eu/.
[4] https://projects.eclipse.org/projects/technology.edc.
[5] https://github.com/FIWARE/data-space-connector.

of how this extension can be practically adopted in the context of open data software like CKAN and data space frameworks such as EDC and IDSA.

This paper is structured as follows. Section 2 provides related work on data spaces. Section 3 discusses the design principles and requirements for extending DCAT to include content-level metadata, as well as the mappings of the proposed DCAT extension to catalogs from open data portals and data spaces. Finally, Sect. 4 sketches out conclusions and directions for future research.

2   Related Work

Open data portals are online platforms where governments and organizations make public sector data available for access, use, and reuse. These portals serve as a central repository for datasets, providing a user-friendly way to discover, download, and analyze information. They play a crucial role in promoting transparency, innovation, and civic engagement by enabling citizens, researchers, and businesses to access a wide range of data [7].

CKAN (Comprehensive Knowledge Archive Network)[6] represents one of the most widely adopted open-source data portal platforms in the open data ecosystem [2]. Developed by the Open Knowledge Foundation and maintained by an active community, CKAN has been implemented by numerous governmental bodies, research institutions, and organizations worldwide to publish and share their datasets. The platform implements a metadata schema that partially aligns with DCAT principles, offering a standardized approach to dataset description.

While open data portals have successfully democratized access to public datasets, data spaces offer a complementary paradigm that interconnects components of a broader data sovereignty ecosystem. A data space is a federated data infrastructure that supports trustworthy data sharing among data providers and data consumers [16]. To enable this, data spaces rely on connectors [9], which act as interfaces between participants and the ecosystem. Connectors ensure technical interoperability and enforce access policies, allowing providers to keep data sovereignty when sharing data while granting consumers seamless access without compromising control. Hence, data spaces enable new options for value creation where providers and consumers easily interact and work together to find, access, publish, consume, and reuse data, as well as to stimulate innovation [10].

Numerous research initiatives have resulted in conceptual frameworks and reference architecture models to accelerate the development of data spaces [3].

The i4Trust initiative emerges as a collaboration program, targeting the creation of data spaces by proposing an architecture based on commonly agreed building blocks [11]. They are a combination of components from FIWARE and iSHARE, two data space foundations. The iSHARE Foundation[7] maintains a trust framework for data spaces. The FIWARE Foundation[8] introduces an open

---

[6] https://ckan.org/.

[7] https://ishare.eu/.

[8] https://www.fiware.org.

source framework that promotes the development of interoperable, smart solutions.

Another innovative data space technical framework has been developed by Gaia-X, aiming to establish a trustworthy ecosystem where data is shared, maintaining the user's digital sovereignty of the data. This standard-based framework allows the implementation of distributed data systems in all European countries in a legally secure manner, enabling compliance with GDPR and other data regulations [8].

The International Data Spaces Association (IDSA) is a non-profit organization focusing on establishing and promoting standards for data spaces as trusted environments where organizations can share data while retaining full control over its use. The IDS Reference Architecture Model (IDS-RAM) [13] materializes the standard developed from collecting requirements from various industries and the results gained from the model's implementation. Additionally, the IDS-RAM connectors fit the GAIA-X principles and architecture model [12].

The Eclipse Dataspace Components (EDC) is a comprehensive framework (concept, architecture, code, samples) providing a basic set of features (functional and non-functional) that dataspace implementations can re-use and customize by leveraging the framework's defined APIs and ensure interoperability by design. The EDC project is governed by the Eclipse Foundation and it is powered by the specifications of the Gaia-X AISBL Trust Framework and the IDSA Dataspace protocol.

One core element of both data space architectures and open data portals is the data catalog, which serves as a structured inventory of available datasets and associated metadata and access policies or usage licenses.

The Data Catalog Vocabulary (DCAT), a W3C recommendation, has emerged as the predominant metadata standard for data catalog implementation. Prior research has established DCAT's significant contribution to metadata interoperability across diverse data ecosystems, from open data portals utilizing CKAN to advanced data space architectures like those proposed by IDSA and Gaia-X. These architectures have operationalized DCAT through connector technologies such EDC and the FIWARE Data Connector. Several publications [14] have highlighted DCAT's potential as a foundation for open data discovery in federated data spaces, given its widespread adoption and interoperability-focused design.

Unfortunately, current DCAT-based catalogs are insufficient for effective data discovery, posing two key limitations:

- Limited metadata: existing data catalogs primarily rely on high-level metadata to describe datasets but lack detailed structural information of dataset content, such as field names or descriptions, which are essential for precise dataset discover.
- Content blindness: current data catalogs exclude actual dataset content, relying solely on metadata. This prevents data spaces initiatives from gaining deeper insights into open data relevance before federating the full dataset.

As an attempt to enhance semantic interoperability across European data portals, the extension DCAT-AP[9] emerges. Based on W3C's DCAT, it supports standardized dataset descriptions, enabling efficient data exchange and reuse. The basic use case for DCAT-AP is to enable cross-data portal search for datasets and make public sector data better searchable across borders and sectors. Although this extension improves data sharing among different sectors and includes additional support for access policies and legal aspects, the aforementioned two limitations regarding content description remain unresolved.

To overcome these challenges in data discovery, reference architectures for data spaces and open data portals require enhancing data catalogs with content-based metadata. In response to these requirements, the following section proposes an extension of DCAT that integrates content-based metadata, while it is mapped to current data catalogs for the sake of compatibility.

3   DCAT Extension for Content-Level Metadata

The Data Catalog Vocabulary (DCAT)[10] is a W3C standard that facilitates interoperability between data catalogs published on the Web. While DCAT is well-suited to describing datasets and services, it lacks built-in support for (i) describing sample data, and (ii) representing content-level metadata such as types, descriptions, and example values. It makes DCAT not enough for discovering open data to be federated in data spaces. To address these needs, we have created an extension of DCAT, the CBM (Content-Based Metadata) extension (see Fig. 1) in which three new classes are introduced (cbm:Sample, cbm:DataDictionary, and cbm:Field) along with associated properties.

3.1   Core DCAT Concepts Being Extended

Core DCAT classes to be extended are described next:

- **dcat:Catalog**. It represents a collection of datasets and services. It contains dcat:Resource instances using dcat:resource.
- **dcat:Resource**. Abstract superclass for cataloged resources, such as dcat:Dataset and dcat:DataService.
- **dcat:Dataset**. Collection of data published or curated by an agent. Datasets have dcat:distribution. It links to a downloadable dataset representation.
- **dcat:Distribution** Describes a representation of a dataset (e.g., a CSV file).
- **dcat:DataService**. It describes an endpoint or service providing access to datasets.

---

[9] https://semiceu.github.io/DCAT-AP/releases/3.0.0/.
[10] https://www.w3.org/TR/vocab-dcat-3/.

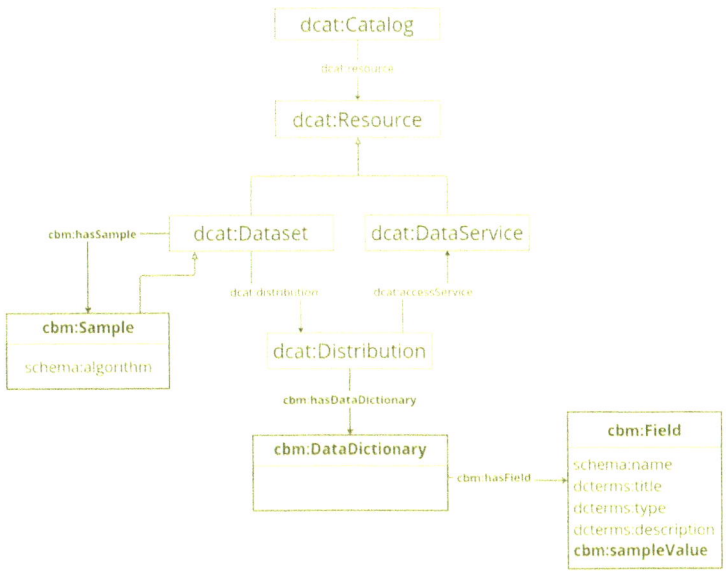

Fig. 1. Schema of the DCAT extension.

3.2   CBM Extension for DCAT

The following classes have been defined to extend DCAT:

- **cbm:Sample**. A demonstration or snippet of dataset use.
- **cbm:DataDictionary**. A structural description of fields.
- **cbm:Field**. Metadata about a dataset variable (e.g., a column in a CSV file).

Corresponding properties of the CBM extension are described in Table 1.

Table 1. Properties of the DCAT extension for Content-Based Metadata.

Property	Domain	Range	Description
cbm:hasSample	dcat:Dataset	cbm:Sample	Relates a dataset to example or sample data.
cbm:hasDataDictionary	dcat:Distribution	cbm:DataDictionary	Links a distribution to its structural metadata.
cbm:hasField	cbm:DataDictionary	cbm:Field	Lists the fields described by the data dictionary.
cbm:sampleValue	cbm:Field	Literal	A representative value of the field.

3.3   Mapping DCAT Extension with Data Ecosystem Catalogs

Our proposed DCAT extension is compatible with catalogs from data space frameworks and open data portals. Next, we will show how our CBM extension can be mapped into two representative data space catalogs (from EDC and IDSA) as well as into the catalog of the CKAN open data portal, thereby facilitating adoption across heterogeneous data ecosystems without requiring significant architectural modifications.

As shown in Fig. 2a, the DCAT extension entities can be mapped to the Eclipse Dataspace Components (EDC) in a straightforward way. An EDC Data Provider has a "catalog"; that corresponds to the *dcat:Catalog* class. This catalog contains multiple "assets" whose access is defined by the policies in the contract. Since this authorization is defined at the asset level, it is necessary to have the original and sample data as separate assets, corresponding to two different DCAT datasets: the original one with class *dcat:Dataset* and the sample one mapped to a *cbm:Sample* instance with open access policies. Since in EDC the data format is defined at the asset level, there is only one *dcat:Distribution* instance per dataset, and we can store the content metadata, equivalent to the *cbm:DataDictionary* within the asset metadata. The "distribution" entity in EDC is equivalent to the *dcat:DataService* and defines the data access methods.

The data model from the International Data Spaces Association (IDSA)[11] data provider and its correspondence to the DCAT extension classes are depicted in Fig. 2b. The *dcat:Catalog* class is equivalent to the "catalog" of the IDSA data provider, which contains the offered data resources. These resources or "offers" correspond to *dcat:Dataset* entities, and the access to them is determined by contract templates and rules. Similarly to the previous data model from EDC, this authorization access is defined at the resource level; thus, in order to keep the access to the sample unrestricted, two datasets are required: the original one corresponding to a *dcat:Dataset* and the one offering the sample data, corresponding to an entity of the *cbm:Sample* subclass. A resource can have multiple "representations," each of them corresponding to a different data format that can be mapped to a *dcat:Distribution* instance. Representations have associated "artifacts" that correspond to the *dcat:DataService* class as they define the access to the data if this is stored remotely or the local data itself. The sample data can be implemented easily as an artifact with locally stored data. To facilitate the retrieval by the data consumer, the content metadata (equivalent to the *cbm:DataDictionary*) is stored together with the sample data at the artifact level.

The case of CKAN, a well-known open data portal technology, allows a different approach due to the lack of access restrictions to the offered data as represented in Fig. 2c. The closest equivalence to a *dcat:Catalog* is an "organization," as each published "dataset" belongs to a single organization. Users have certain privileges such as editing defined at the organization level, but the read access to these datasets is public, which allows both the original and sample data to be hosted within the same dataset as different "resources." The resource entity has its own metadata, defines the data format, and holds the data access that can be hosted locally or remotely. Each resource entity with the original data will inherit the metadata from the dataset and map at the same time to the three DCAT classes: *dcat:Dataset*, *dcat:Distribution*, and *dcat:DataService*. Similarly,

---

[11] https://internationaldataspaces.org/.

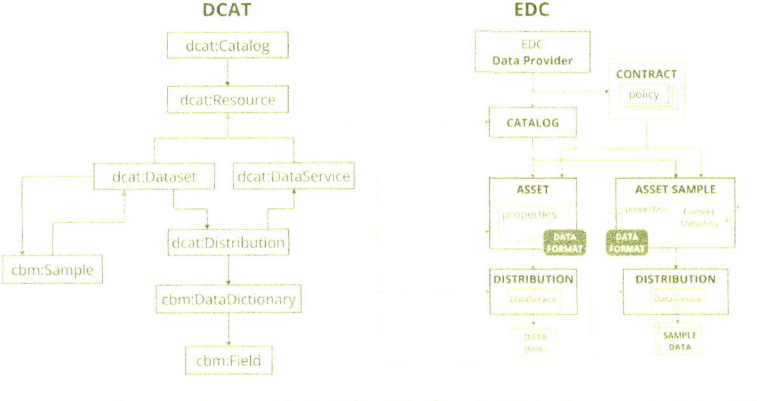

(a) Mapping of the DCAT extension to the EDC data model.

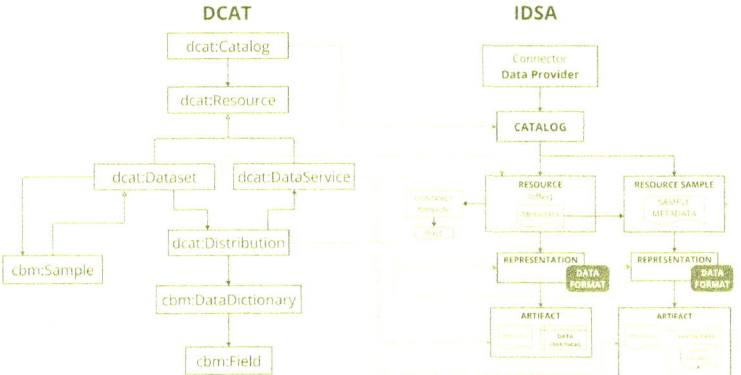

(b) Mapping of the DCAT extension to the IDSA data model.

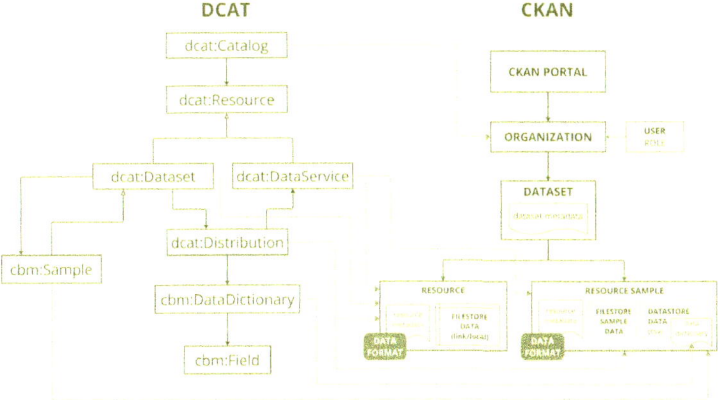

(c) Mapping of the DCAT extension to the CKAN data model.

Fig. 2. Mapping of the DCAT extension EDC, IDSA, and CKAN data models.

the sample resource will correspond to a *cbm:Sample*, *dcat:Distribution*, and *dcat:DataService* class. CKAN has a "filestore" to host local data, and for some specific data formats such as CSV, it also has available a "datastore": a structured storage that offers further functionalities such as preview or filtering. The sample resource could be implemented as a usual resource in the filestore hosted locally and including the content metadata as part of the resource metadata. A better approach, if the data format allows it, is the one depicted in Fig. 2c where the sample data can be additionally imported into the datastore. The content metadata corresponding to the *cbm:DataDictionary* has a more suitable implementation possibility, as the datastore has a data dictionary structure specifically designed to hold content-based metadata such as field name, description, etc.

4   Conclusions and Future Work

This work presents an extension of the DCAT standard aimed at supporting content-level metadata in open data catalogs, particularly within the context of data spaces. By incorporating content-based metadata (including representative data samples), our DCAT extension improves discoverability of datasets to be federated in data spaces thanks to the mapping proposed between our extension and data catalogs of other ecosystems, such as open data portals or data spaces.

Several directions are envisaged for future work. As an immediate next step, we aim to evaluate our DCAT extension in the short term. Furthermore, while this paper focuses on tabular data, a natural extension is to adapt the generation of the proposed content-based metadata and indexing strategies to other data types, such as spatial data or unstructured data. We also aim to collaborate with data space initiatives to assess our approach for data discovery in federated environments beyond open data portals. Lastly, applying this approach to sensitive data would raise new challenges related to privacy, access control, and metadata exposure.

Acknowledgements. This work is part of the HELEADE project (TSI-100121-2024-24), funded by Spanish Ministry of Digital Processing and by the European Union NextGeneration EU. Also, part of the research was partially funded by CIAICO/2022/019 project from Generalitat Valenciana (Spain).

References

1. Ailamaki, A., et al.: The cambridge report on database research. arXiv preprint arXiv:2504.11259 (2025)
2. Attard, J., Orlandi, F., Scerri, S., Auer, S.: A systematic review of open government data initiatives. Gov. Inf. Q. **32**(4), 399–418 (2015)
3. Bacco, M., Kocian, A., Chessa, S., Crivello, A., Barsocchi, P.: What are data spaces? systematic survey and future outlook. Data Brief **57**, 110969 (2024)
4. Bernhauer, D., Nečaský, M., Škoda, P., Klímek, J., Skopal, T.: Open dataset discovery using context-enhanced similarity search. Knowl. Inf. Syst. **64**(12), 3265–3291 (2022)

5. Corcho, O., Simperl, E.: data.europa.eu and the european common data spaces: a report on challenges and opportunities. Report, data. europa. eu—The Official Portal for European Data, Luxembourg (2022)
6. Curry, E., Tuikka, T.: An organizational maturity model fordata spaces: A data sharing wheel approach. In: Data Spaces: Design, Deployment and Future Directions, pp. 21–42. Springer (2022)
7. Enríquez-Reyes, R., Cadena-Vela, S., Fuster-Guilló, A., Mazón, J.N., Ibáñez, L.D., Simperl, E.: Systematic mapping of open data studies: Classification and trends from a technological perspective. IEEE Access **9**, 12968–12988 (2021)
8. Gaia-X European Association for Data and Cloud AISBL: Gaia-X Architecture Document (2024). https://docs.gaia-x.eu/technical-committee/architecture-document/latest/
9. Gieß, A., Hupperz, M.J., Schoormann, T., Möller, F.: What does it take to connect? unveiling characteristics of data space connectors. In: Bui, T.X. (ed.) 57th Hawaii International Conference on System Sciences, HICSS 2024, Hilton Hawaiian Village Waikiki Beach Resort, Hawaii, USA, 3-6 January 2024, pp. 4238–4247. ScholarSpace (2024). https://hdl.handle.net/10125/106895
10. Gieß, A., Möller, F., Schoormann, T., Otto, B.: Design options for data spaces. In: Thirty-first European Conference on Information Systems (ECIS 2023) (June 2023)
11. i4Trust: B2B Data Sharing Playbook - the i4Trust approach to Data Sharing (2021). https://i4trust.org/wp-content/uploads/i4Trust_DataSharingPlaybook.pdf
12. International Data Spaces Association: Position Paper - GAIA-X and IDS (2021). https://internationaldataspaces.org/wp-content/uploads/dlm_uploads/IDSA-Position-Paper-GAIA-X-and-IDS.pdf
13. International Data Spaces Association: IDS Reference Architecture Model, Version 4.0 (2022). https://docs.internationaldataspaces.org/ids-knowledgebase/ids-ram-4/introduction/1_1_goals_of_the_international_data_spaces
14. Kirstein, F., Dutkowski, S., Dittwald, B., Hauswirth, M.: The european data portal: scalable harvesting and management of linked open data. In: ISWC (Satellites), pp. 321–322 (2019)
15. Kitchin, R.: The data revolution: Big data, open data, data infrastructures and their consequences. Sage (2014)
16. Otto, B.: A federated infrastructure for european data spaces. Commun. ACM **65**(4), 44–45 (2022)
17. Riley, J.: Understanding metadata. Washington DC, United States: National Information Standards Organization. http://www.niso.org/publications/press/UnderstandingMetadata.pdf (2017)
18. Scerri, S., Tuikka, T., de Vallejo, I.L., Curry, E.: Common european data spaces: challenges and opportunities. Data Spaces: Design, Deployment and Future Directions, pp. 337–357 (2022)

Enhancing SQL Learning Through Generative AI and Student Error Analysis

Davide Ponzini$^{(\boxtimes)}$ [ID], Barbara Catania [ID], and Giovanna Guerrini [ID]

Università di Genova, Genova, Italy
davide.ponzini@edu.unige.it,
{barbara.catania,giovanna.guerrini}@unige.it

Abstract. Despite the widespread use of SQL in both academic and professional contexts, students often struggle with writing correct queries due to a range of syntactic and semantic misconceptions. In this work, we propose a framework supporting SQL learning centered around errors as central to the learning process. The framework integrates generative AI and automated error categorization to foster metacognitive engagement and support personalization.

1 Introduction

Context. Learning SQL presents difficulties for students, both in terms of syntax and semantics, even in degrees in which database related courses play an essential role. Prior research has shown that student misconceptions in SQL can be persistent and difficult to correct without targeted interventions. Identifying and classifying students' errors in SQL queries [19] can be a first step to enable a data-driven approach [5], provide specific feedback, and develop resources that target specific student misconceptions. SQL learning can be enhanced by approaches and prototypes that identify errors, offering constructive feedback and supporting self-guided improvement [8]. The integration of generative AI in such frameworks can be extremely beneficial, as it offers unprecedented scalability, and, among many opportunities, can provide students with clearer feedback.

Contribution. Grounded on the above mentioned body of data system education work, and along the same lines of approaches for integrating generative AI in programming environments for educational use, such as approaches to fine-tune GPT models to detect and provide feedback on SQL errors [3,11], we designed and are currently developing an SQL tutoring framework. The framework integrates a data-driven approach with interactive learning mechanisms, relying on generative AI, to provide structured feedback and encourage deeper understanding. More specifically, it: (i) logs and analyzes all student queries to identify common mistakes; (ii) provides an interactive AI-powered tutor that guides students through the error resolution process without offering direct solutions; (iii) generates personalized exercises that target specific weaknesses, leveraging students' interests to enhance engagement. These interventions aim to develop students'

© The Author(s), under exclusive license to Springer Nature Switzerland AG 2026
P. K. Chrysanthis et al. (Eds.): ADBIS 2025, CCIS 2676, pp. 118–128, 2026.
https://doi.org/10.1007/978-3-032-05727-3_12

ability to reason about mistakes, while enabling instructors to better understand learning trajectories and provide tailored support.

The main novelty of our approach lies in combining previous work in more traditional areas of error analysis [19] and automatic feedback [7] with a principled integration of generative AI. Our goal is to mitigate the risks associated with the use of AI from two perspectives: minimizing the impact of AI errors and hallucinations, and guiding students' use of AI in a way that fosters genuine learning. Preliminary results on the use of the tool are encouraging and show the effectiveness of the overall proposed methodology.

Related Work. Error analysis as a pedagogical technique is not new, even in the SQL context; e.g., Taipalus et al. [18] and Miedema et al. [12] explore students' reasoning to reveal typical errors and their causes. SQL query errors range from simple syntax issues to deeper semantic and logical misunderstandings and different classifications have been proposed over the years [2,4,14,15]. Our framework adopts the taxonomy by Taipalus et al. [19], which to the best of our knowledge is the most complete and frequently used in the literature. It classifies errors as syntactic, semantic (incorrect regardless of the request), logical (correct but mismatched), or overly complex. Using this framework, we categorized student mistakes from our database courses at the University of Genoa [10,13].

SQL learning benefits from error-aware learning tools [8], whose potential can only be leveraged if they extend beyond grading efficiency by also providing tutoring capabilities to the students. Hint generation is, e.g., the focus of [7,9]. These approaches are at the basis of the error categorization in our framework.

Generative AI integrated in programming environments for educational use is more and more frequently exploited. In the SQL context, their effects on student performance have been analyzed in [16]. Other uses of generative AI in SQL learning are for the automatic exercise generation [1] and for the assessment of authentic learning [17]. The approach we propose is more similar but complementary to the ones proposed in [3,11] to fine-tune GPT models to detect and provide feedback on errors in SQL queries.

Outline of the Paper. The remainder of the paper is organized as follows. Section 2 presents the framework, discusses the main methodologies we plan to follow for each framework component, and provides details about the currently operational components. Preliminary results are presented in Sect. 3. Finally, Sect. 4 concludes the paper and outlines future work.

2   AI-Powered Tutoring Framework

The goal of our framework is to enhance SQL learning by focusing on systematic error detection, AI-assisted feedback, and assignment generation. The framework relies on a multi-faceted strategy: we collect and analyze student queries, classify errors automatically, provide AI-driven assistance, and generate targeted assignments to reinforce learning.

Figure 1 illustrates the framework components. Users interact with the system through a web browser, by which they can submit queries and get the

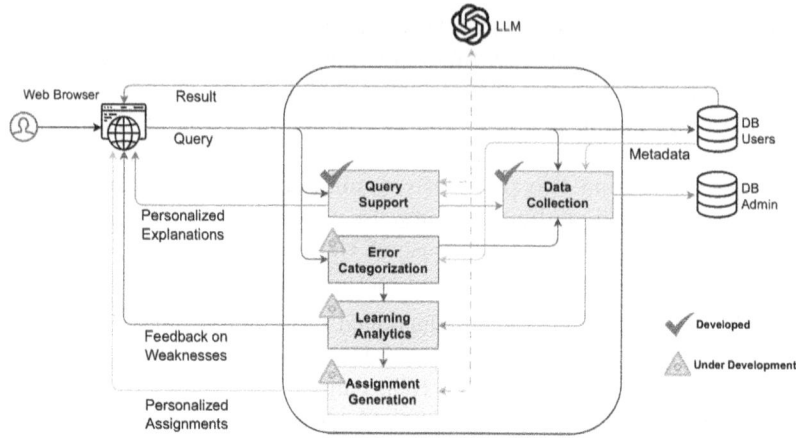

Fig. 1. Architecture of the AI-Powered Tutoring Framework

corresponding results. The framework relies on two separate database instances: DB Users contains a single database for each student, where students can execute any operation; DB Admin is used only to log student queries and perform subsequent analysis. The framework is composed of five modules, described in the following: two have already been implemented (green tick in the figure) and three are currently under development (orange tick).

2.1   Data Collection Module

The Data Collection Module logs all queries executed by students along with their execution outcomes. For each query, the system records: the query text; the execution status (successful or erroneous); the timestamp; the student who submitted it; the corresponding request in natural language; the expected correct SQL solution (provided by the instructor and hidden from students), if available; the active search_path at the time of execution; the type of SQL statement (e.g., SELECT, INSERT, CREATE, etc.); the entire database structure at the time of execution (i.e. available tables, their structure, and their constraints). If the query executes successfully, the resulting output is also stored. In cases where execution fails, the associated error message is logged.

2.2   Query Support Module

The Query Support Module presents users with interactive buttons, providing tailored and meaningful explanations, depending on the query written by the user, its output, and additional metadata (e.g., the database structure). Depending on the query execution status, different options are available.

For *syntactically incorrect queries*, the module provides the following options:

– *Explain Error:* clarifies the DBMS error message in relation to the specific mistake in the query.
– *Provide an Example:* presents a simplified query that produces the same type of error, helping students understand the issue in a broader context.
– *Locate Error:* highlights the specific part of the code responsible for the error, assisting students who have not yet identified the exact cause.
– *Suggest Fix:* offers a possible correction to resolve the error.

These options are progressively made available, to encourage active problem-solving and deeper engagement with error analysis. Initially, students can access explanations and illustrative examples to help them understand the nature of the error. If they still struggle to identify the issue, they can use the error location feature to pinpoint the problematic part of their query. The suggest fix option becomes available only after students have exhausted all other support mechanisms, ensuring that they make a genuine effort to diagnose and understand the error before receiving a direct solution.

For *syntactically correct queries*, the module provides the following options:

– *Describe My Query:* generates a brief natural language summary of the user query, helping to identify potential logical or reasoning errors. If the description does not align with the user intended goal, this may indicate a misunderstanding of the query actual behavior.
– *Explain My Query:* provides a detailed step-by-step breakdown of each clause in natural language, clarifying the function of different query components. This allows students to reflect on the role of each clause and detect possible logical inconsistencies or misinterpretations.

When the user selects one button, a prompt is automatically generated and sent to the ChatGPT API (model `gpt-4o-mini`). The prompt contains only the information necessary to fulfill the selected request. For instance, the *Explain Error* button provides the model with the error message and the original query, and explicitly instructs it to explain the error without offering a correction. Additionally, a response template is embedded in the prompt to maintain stylistic consistency across outputs. To avoid misleading the model, any comment present in the SQL code is removed, as inconsistencies between code and comments can distort the model's interpretation. Figure 2 shows an example of a personalized error explanation for a query returning multiple rows instead of a single value.

These interactive features support students in refining their understanding of SQL query logic. By offering progressively detailed explanations and comparisons, we encourage self-reflection, helping students diagnose errors independently before resorting to external assistance. The proposed features create an interactive dialogue, helping students understand their mistakes and learn from them in real time. The goal is to mimic a human tutor guidance by offering just-in-time explanations and suggestions as students work through SQL problems.

While the integration of AI-driven support can enhance scalability and personalization opportunities, it is important to acknowledge potential risks. The explanations and suggestions generated by the LLM may occasionally contain

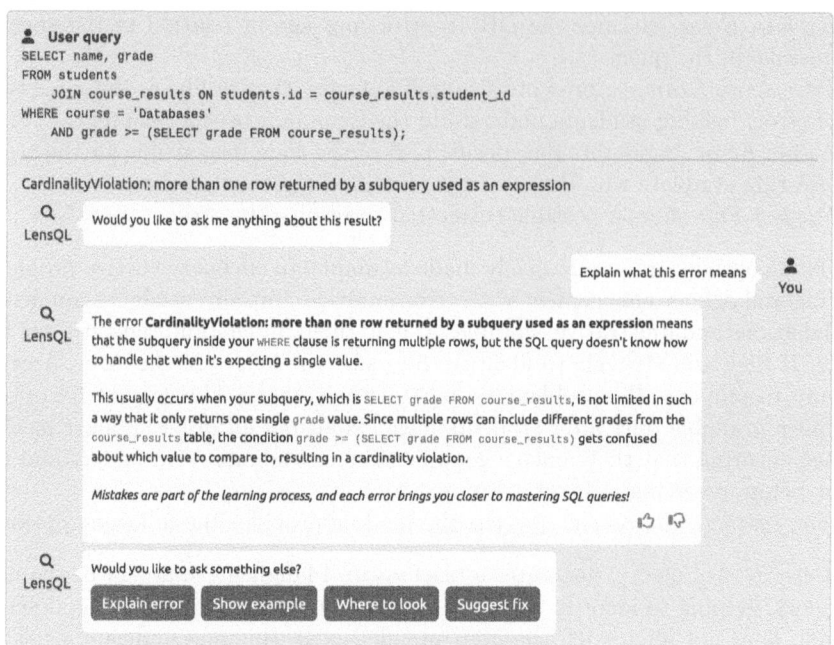

Fig. 2. Example of personalized error explanation for a syntactically-wrong query.

inaccuracies or misinterpret the user's intent, especially in edge cases or ambiguous scenarios. To mitigate this, the system encourages critical engagement by presenting AI-generated feedback only after students have attempted to understand the request or resolve the issue themselves.

2.3   Error Categorization Module

To systematically address mistakes, submitted queries will be processed through an automated error detection module, currently under development. This component will classify errors based on the taxonomy proposed in [19] which delineates errors into four primary categories: syntax errors, semantic errors, logical errors, and complications. Furthermore, this taxonomy encompasses a three-level hierarchy with 105 specific error types, enabling a nuanced analysis of student mistakes. *Syntax errors* cover issues that violate SQL grammar (and thus are caught by the database engine), whereas semantic and logical errors involve queries that execute but return incorrect results. *Semantic errors* are characterized by inherent contradictions or tautologies within the query logic, rendering them incorrect regardless of the specific data demand. *Logical errors*, on the other hand, are context-dependent and arise when a query, though syntactically and semantically correct, fails to retrieve the intended data due to misalignment with the user's specific information need. *Complications* refer to unnecessary or overly complex parts of a query that do not change the result but make the solution harder to understand.

We plan to detect a substantial number of errors through parsing the SQL query structure and running it against a set of rules [4], optionally supplemented by additional metadata, such as the active `search_path` or information about the available schema(s) at the time of execution. For more complex errors, particularly logical errors, it will also be possible to leverage the expected solution provided by the instructor [7], as well as the request in natural language.

As an example, consider the following query, which attempts to retrieve the names of students enrolled in a course named *Databases* who scored above the average:

```
SELECT name FROM students
JOIN course_results ON students.id = course_results.student_id
WHERE course LIKE Databases OR grade > AVG(grade);
```

These are the error categories the module should be able to detect:

- **Syntax Error (Undefined Database Object): omitting quotes around character data**—The string literal `Databases` is not enclosed in single quotes. As a result, it is interpreted as an identifier, leading to a syntax error if no such column or variable exists.
- **Syntax Error (Illegal Aggregate Function Placement): using aggregate function outside SELECT or HAVING**—The aggregate function `AVG(grade)` is used directly in the `WHERE` clause, which violates SQL syntax rules. Aggregate functions must appear in `SELECT`, `HAVING`, or in a scalar subquery.
- **Logical Error (Operator Error): OR instead of AND** – The use of `OR` instead of the intended `AND` changes the logic of the query, possibly returning unintended results and failing to meet the user's information need.
- **Complication: unqualified column references**—The columns `name`, `course`, and `grade` are referenced without table aliases. While syntactically valid, this reduces clarity and increases the risk of ambiguity in multi-table queries.
- **Complication: LIKE without wildcards** – The `LIKE` operator is used without wildcards, which reduces it to a simple equality check. This adds unnecessary complexity and reduces clarity.

By automatically tagging each query with these error categories, we will ensure consistency in how mistakes are identified. This automation makes it easier to spot recurring misconceptions. While the example shown involves moderate complexity, the module is being designed to eventually support more advanced queries, including those with subqueries or grouping conditions.

2.4   Learning Analytics Module

The aim of the Learning Analytics Module is to leverage error classifications and student interaction data to construct detailed learner profiles and support data-driven adaptation of instruction. For each student, the module will be able to continuously track error trends, such as the frequency and types of mistakes (syntax, semantic, or logical) across different exercises. It will also record how long students take and how many attempts they need to resolve each issue, providing information on the learning process and persistence. Based on successful query completions and error patterns, the system will infer concepts' mastery levels across core SQL constructs such as joins, aggregation, and subqueries. Additionally, by analyzing behaviors like repeated queries and use of support features, it will estimate confidence levels, helping to identify cases of overconfidence or uncertainty.

We plan to create interactive dashboards that will provide real-time insights for both students and instructors. Students will have access to a personalized dashboard showing their progress across topics, a heatmap of error types, and a "confidence vs. accuracy" plot to reflect on how well their perceived understanding aligns with actual performance. The system will also track success streaks, defined as sequences of exercises correctly solved on the first attempt, which signal growing proficiency and can boost motivation. Instructors, on the other hand, will be presented with aggregated class-wide analytics that highlight common misconceptions, detect at-risk or high-performing students, and compare cohort performance.

To further enhance diagnostic capabilities, we are also planning to cluster students based on shared error patterns and learning behaviors, allowing instructors to tailor interventions more effectively. This will also surface high-impact misconceptions that are strongly correlated with low performance or slow correction times. Finally, by integrating engagement metrics (e.g., time on task, help interactions) and mapping student performance against predefined learning objectives, the module will produce individualized "SQL skill coverage maps", offering a comprehensive overview of student understanding and instructional needs.

2.5   Assignment Generation Module

Building on the metrics computed by the learning analytics module, this component will allow users to generate personalized SQL assignments that directly target the specific misconceptions identified in each student's queries. To do this effectively, we plan to incorporate metadata from the database, such as table names, schema structure, and attribute types, ensuring that generated exercises are both realistic and pedagogically aligned with the course content.

Using generative AI, the module will be able to produce customized exercises that mirror the context in which the student made the error, but with slight variations designed to challenge and correct the misunderstanding. For example, if a student frequently omits a `GROUP BY` clause when using aggregate functions, the system will generate a new query task requiring correct grouping to produce the intended results. This integration of prior error data and schema-aware content creation will ensure each student receives assignments that are both contextually grounded and instructionally relevant. Each assignment will consist of: (*i*) a relational schema and synthetic dataset generated to match the selected domain; (*ii*) a natural language description of the task to be solved; (*iii*) one or more expected outputs, against which students can compare their query results.

In addition to targeting specific errors, the module will also allow students to choose a topic of personal interest, which is used to generate a synthetic dataset and schema relevant to that context (e.g., sports, music, or books). By grounding exercises in topics that resonate with students' interests, we aim to enhance motivation and engagement. These assignments could be used to complement standard course exercises by offering focused practice that adapts to each student's needs, while also promoting self-paced exploration of SQL topics in personally meaningful contexts.

3   Preliminary Results

The current system was employed in the context of an *"Introduction to Databases"* second year Bachelor course [6], with 115 students. The learning outcomes of the course, like those of typical introductory database courses, are quite broad, covering a wide range of skills and abilities, that for querying includes *expressing queries and manipulating operations in the SQL language.*

To date, we have recorded a total of 82,813 student-submitted SQL queries. Approximately 60% of the queries executed successfully, while the remaining 40% contained at least one error. Notably, only 4.14% of the queries were submitted during scheduled lab sessions, while the vast majority (over 95%) were submitted outside official lab hours. This usage pattern suggests that the tool is actively supporting students beyond the constraints of in-person instruction.

Focusing on `SELECT` queries, which represent the most common command type (about 45% of the executed commands) around 70% were executed without errors. The remaining 30% included one or more errors, reflecting misconceptions and challenges related to syntax and query logic, even in basic scenarios. The most frequent error categories identified by PostgreSQL are the following[1]:

– **Undefined (47.18%)**—Situations in which something referenced in the SQL query is not defined or does not exist. For example, the student might be trying to select from a table that has not been created or spelled correctly.
– **Syntax Errors (29.28%)**—Violations of SQL grammar rules, such as missing keywords or incorrect clause ordering.
– **Grouping Errors (10.55%)**—Misuses of aggregate functions or missing `GROUP BY` clauses, one of the most conceptually challenging aspects of SQL.

Other less frequent categories include ambiguous column references due to unqualified names in multi-table queries (3.21%), data type mismatches such as comparing numeric values with strings (2.77%), invalid operations due, e.g., to datetime format or casting (2.68%), cardinality violations due to subqueries used in expressions but returning more than one row (1.56%), insufficient privileges

---

[1] Note that here we are referring to the error messages raised by PostgreSQL that will be mapped to error types in the *syntactic* category of the reference taxonomy [19].

(1.07%), and other datetime-related issues (0.49%). These results confirm prior findings that conceptual mastery of grouping operations and schema navigation remain key learning challenges for SQL novices.

In terms of AI-assisted support, students submitted a total of 2,784 help requests through the Query Support Module. The most commonly used features were *Suggest Fix* (49% of requests) and *Explain Error* (33% of requests), showing a clear preference for actionable guidance and concise feedback. Temporal analysis of tool usage reveals distinct patterns. Query activity increased steadily over the semester, with noticeable spikes coinciding with more complex lab sessions and immediately prior to exam dates. Help requests mirrored this trend, with *Suggest Fix* and *Explain Error* being particularly prominent during peak periods. A sharp increase in student engagement was observed during weeks dedicated to advanced topics, especially for operational SQL, as well as near assessment deadlines.

4   Conclusions and Future Work

In this paper we presented a data-driven framework designed to support SQL learning by leveraging student errors as opportunities for reflection and conceptual growth. The system combines query logging, AI-generated feedback, error categorization, learning analytics, and personalized assignments to promote a deeper understanding of SQL. Rather than offering direct solutions, it fosters active reasoning and iterative learning, aligning with pedagogical principles that treat errors as a core part of the learning process. Two components—data collection and query support—have been implemented, and preliminary results show the framework is capable of identifying meaningful error patterns and is actively used as a learning aid.

Future developments will focus on completing the framework, by implementing the missing and incomplete components, and on its validation. We plan to assess the effectiveness of our approach through a comparative study using historical data from different editions of our introductory database course [6]. We will examine exam performance, both quantitatively, through grade distributions, and qualitatively, through an analysis of recurring error types, to determine whether the system contributes to more robust learning outcomes. The data collection will also enable a detailed analysis of learning trajectories over time. By analyzing how students move from incorrect to correct solutions, and how they respond to feedback and guidance, we aim to identify which features of the system are most effective in supporting progress. These insights will inform iterative refinements to the framework and lay the groundwork for broader applications. Longer-term directions include the application of the framework to different courses and learning contexts, such as K-12, and the application of the same approach to other areas within computer science education.

References

1. Aerts, W., Fletcher, G., Miedema, D.: A feasibility study on automated SQL exercise generation with ChatGPT-3.5. In: Proceedings of 3rd Int'l Workshop on Data Systems Education, pp. 13–19 (2024)
2. Ahadi, A., Prior, J., Behbood, V., Lister, R.: Students' semantic mistakes in writing seven different types of SQL queries. In: Proceedings of ACM Conference on Innovation and Technology in Computer Science Education, pp. 272–277 (2016)
3. AlRabah, A., Yang, S., Alawini, A.: Optimizing database query learning: a generative AI approach for semantic error feedback. In: Proceedings of ASEE Annual Conference and Exposition (2024)
4. Brass, S., Goldberg, C.: Semantic errors in SQL queries: a quite complete list. J. Syst. Softw. **79**(5), 630–644 (2006)
5. Cagliero, L., De Russis, L., Farinetti, L., Montanaro, T.: Improving the effectiveness of SQL learning practice: a data-driven approach. In: IEEE COMPSAC 2018, vol. 1, pp. 980–989 (2018)
6. Catania, B., Guerrini, G., Traversaro, D.: Collaborative learning in an introductory database course: a study with think-pair-share and team peer review. In: Proceedings of 1st Int'l. Workshop on Data Systems Education, pp. 60–66 (2022)
7. Hu, Y., Gilad, A., Stephens-Martinez, K., Roy, S., Yang, J.: Qr-hint: actionable hints towards correcting wrong sql queries. Proc. ACM Manag. Data **2**(3), 1–27 (2024)
8. Kenny, C., Pahl, C.: Automated tutoring for a database skills training environment. In: Proceedings of 36th SIGCSE Technical Symposium on Computer Science Education, pp. 58–62 (2005)
9. Kleiner, C., Heine, F.: Enhancing feedback generation for autograded SQL statements to improve student learning. In: Proceedings of Conference on Innovation and Technology in Computer Science Education, pp. 248–254 (2024)
10. Livani, A.: Do the errors produced by generative AI in formulating queries reflect students' misconceptions in learning SQL? Master's thesis, MSc in Computer Science, University of Genova (2024)
11. Manikani, K., Chapaneri, R., Shetty, D., Shah, D.: SQL Autograder: web-based LLM-powered autograder for assessment of SQL Queries. Int. J. Artifi. Intell. Educ., 1–31 (2025)
12. Miedema, D., Aivaloglou, E., Fletcher, G.: Identifying SQL misconceptions of novices: findings from a think-aloud study. In: Proceedings of ACM Conference on International Computing Education Research, pp. 355–367 (2021)
13. Ponzini, D., Livani, A., Guerrini, G., Catania, B., Coccoli, M.: Analyzing common student errors in sql query formulation to enhance learning support. In: Proceedings of 33rd Symposium on Advanced Database Systems (2025)
14. Poulsen, S., Butler, L., Alawini, A., Herman, G.L.: Insights from student solutions to SQL homework problems. In: Proceedings of ACM Conference on Innovation and Technology in Computer Science Education, pp. 404–410 (2020)
15. Presler-Marshall, K., Heckman, S., Stolee, K.: SQLRepair: identifying and Repairing Mistakes in Student-authored SQL Queries. In: Proceedings of IEEE/ACM ICSE-SEET, pp. 199–210 (2021)
16. Ramakrishnan, C., Cassidy, S., Bower, M.: Evaluating student performance and interactions in generative AI-integrated SQL practical tests. In: Proceedings of ACM Conference on Innovation and Technology in Computer Science Education (2025)

17. Sooriamurthi, R., Tu, X., Pensky, A.E.C.: A Generative AI Tool to Foster and Assess Authentic Learning: A Case Study in Teaching SQL. In: Proceedings of ACM Conference on Innovation and Technology in Computer Science Education, pp. 465–471 (2025)
18. Taipalus, T.: Explaining causes behind SQL query formulation errors. In: Proceedings of IEEE Frontiers in Education Conference, pp. 1–9 (2020)
19. Taipalus, T., Siponen, M., Vartiainen, T.: Errors and complications in SQL query formulation. ACM Trans. Comput. Educ. **18**(3), 1–29 (2018)

GESONGEN: An Interface for Generating and Visualizing Geosocial Networks

Julius Hoffmann[1] , Panagiotis Bouros[1(✉)] ,
and Theodoros Chondrogiannis[2]

[1] Johannes Gutenberg University Mainz, Mainz, Germany
jhoffm09@students.uni-mainz.de, bouros@uni-mainz.de
[2] Norwegian University of Science and Technology, Trondheim, Norway
theodoros.chondrogiannis@ntnu.no

Abstract. The ubiquity of mobile location-aware devices and the proliferation of social networks gave rise to *geosocial networks*, where users
not only form social connections but also perform geo-referenced actions.
Examples include traditional social networks with geo-annotated posts,
e.g., Twitter, and networks that directly offer geosocial services, e.g.,
Yelp. However, despite the strong interest by both the industry and
academia in geosocial networks, a limited number of datasets are in fact
publicly available. To fill this gap, we present GESONGEN, an interactive interface for generating geosocial networks. We built upon a recently
proposed generation process which combines independently generated
graphs and geospatial data or re-uses existing datasets. GESONGEN can
visualize geosocial networks, with different view options, and to modify
networks.

Keywords: Geosocial networks · generator · visualization

1 Introduction

The proliferation of location-based services and social networks gave rise to
geosocial networks, which model both the social interactions of users and
their geo-referenced actions, e.g., check-ins. Examples are typical social networks extended with geospatial information such as X (formerly, Twitter) and
Facebook, and networks directly offering geosocial services such as Yelp and
Foursquare.

Despite the interest from the research community and the industry on query
processing [3,7,17], indexing [20,21], recommender systems [15,19] and on tasks
such as influence maximization [5,16] and community search [6,9], a limited number of geosocial networks are in fact publicly available. For instance, Yelp[1] offers
an official dump for academic purposes, and SNAP's page[2] offers a Brightkite

---

[1] https://www.yelp.com/dataset/.

[2] http://snap.stanford.edu/data/index.html#locnet.

© The Author(s), under exclusive license to Springer Nature Switzerland AG 2026
P. K. Chrysanthis et al. (Eds.): ADBIS 2025, CCIS 2676, pp. 129–137, 2026.
https://doi.org/10.1007/978-3-032-05727-3_13

and a Gowalla dump. Another option for acquiring geosocial network datasets is to use official APIs e.g., from X[3] and Foursquare[4], similar to [15,19][5]. However, these APIs restrict the number of requests per day; for unlimited downloads, fees are charged. A common practice to deal with the limited availability of real datasets is to generate synthetic ones. Synthetic geosocial networks can be used for benchmarking the efficiency and the robustness of geosocial queries, for hypothesis testing, "what-if" scenarios, and simulations.

Existing Generators. Network and spatial data generation have individually received significant attention in the past. For the first, the goal is to generate synthetic networks whose properties match the ones in real networks. Real-world social networks in particular, typically exhibit a vertex-degree distribution that follows a power law, and a small diameter ("small-world" phenomenon, or "six degrees of separation"). Under this, the majority of the proposed models [1, 8,23,24] use some form of preferential attachment to progressively construct a synthetic network, adopting a "rich get richer" approach. For spatial data, Beckmann and Seeger [4] presented a generator that was later extended to build Spider [12,22].

In contrast, generating geosocial networks has received limited attention. One approach is to leverage geo-simulation by creating a digital twin of a real urban environment and then use co-location between agents to form social connections [11,13,14]. Alizadeh et al. [2] adopted a different approach, modifying the generation process of the Erdős-Rényi [8], Barabási-Albert [1] and Watts-Strogatz [23] models so that geographic information dictates how network vertices are connected. Gallagher et al. [10] extended this approach. The vertices are no longer randomly positioned in space; instead the distribution of real geospatial data is used. Lastly, Sarsour et al. [18] proposed to fully decouple the generation of the two data components. The authors presented three types of synthetic networks for different real-world geosocial networks (see Sect. 2) and a generation process that combines the output of a graph and a spatial data generator.

Contributions. We present GEoSOcial Network GENerator (GESONGEN)[6], an interactive interface for geosocial networks with a threefold mission. First, our system enables users to generate synthetic geosocial networks by building on the generation process in [18]. Second, users can visualize generated geosocial networks; GESONGEN offers three view options, two that prioritize the social or the spatial component, and a third, combined view option. Finally, users can also modify a network by adding, removing, or editing vertices and edges.

2   Synthetic Geosocial Networks

A *geosocial* network is a labeled graph $G = (V, E)$ where every edge $(u, v) \in E \subseteq V \times V$ represents a relationship between the entities modeled by vertices u

---

[3] https://developer.twitter.com/en/products/twitter-api.

[4] https://location.foursquare.com/developer/reference/places-api-overview.

[5] https://archive.org/details/201309_foursquare_dataset_umn.

[6] https://gesongentool.github.io and https://github.com/pbour/geosocialgenerator.

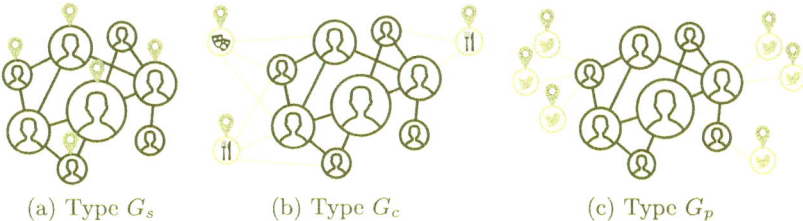

(a) Type G_s (b) Type G_c (c) Type G_p

Fig. 1. Types of synthetic geosocial networks: non-spatial vertices and edges in between them are drawn in black, spatial vertices are marked by a red pin. Without loss of generality, we draw only undirected edges. (Color figure online)

and v. The set $V_s \subseteq V$ contains *spatial vertices*, i.e., those are associated with a geometry *v.geom* in the two or three-dimensional space , e.g., a point or a polygon. We revisit the three types of synthetic geosocial networks introduced by Sarsour et al. [18] which are fully supported by GENSOGEN. Figure 1(a) illustrates the first type, denoted by G_s. All vertices represent the same type of entity, e.g., users of the network, and the edges model relationships between such entities, e.g., *FRIEND_OF*. Some vertices are also spatial (marked by a red pin). An academic geosocial network created on co-authorship is an example of G_s type, where geospatial information models the affiliation of a researcher.

Figure 1(b) depicts the second type of geosocial networks. G_c models two disjoint sets of entities as vertices. Specifically, the network comprises a set of non-spatial vertices, i.e., social vertices, and the edges connecting these vertices that model social relationships. Spatial vertices can be connected to one or more social vertices, but never to each other. The social vertices and the edge between them are in black, spatial vertices in green, also marked by a red pin, and the edges connecting social and spatial vertices in gray. Foursquare is an example of the G_c type with gray edges capturing *CHECKED_IN* actions by the users.

The third type of network G_p, is shown in Fig. 1(c). Similarly to G_c, G_p also models two disjoint sets of entities using social and spatial vertices. But, in contrast to G_c, each spatial vertex is always connected to exactly one social vertex. The social vertices and edges of the network are again in black, and the spatial vertices, in blue with a red pin. Social networks such as X or Facebook are examples of G_p networks, where the spatial vertices represent geo-annotated posts (namely, tweets in X). Naturally, as users make multiple posts, a non-spatial vertex can be connected to multiple spatial ones.

3   The GESONGEN System

GESONGEN adopts a typical client-server architecture, illustrated in Fig. 2. The frontend, a Web-based application developed in React, is responsible for collecting the parameters required to generate a synthetic geosocial network, and for visualizing and modifying networks. Visualization is powered by the

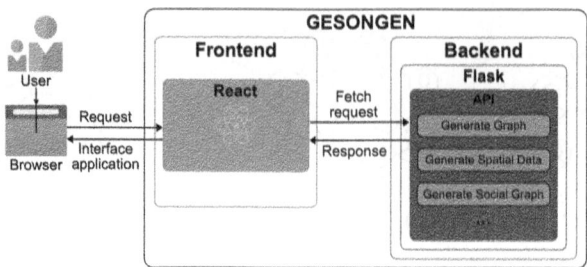

Fig. 2. GESONGEN architecture

D3.js Javascript Library and its d3-force-3d extension.[7] The backend, written in Python, uses Flask to provide a lightweight API that handles the requests between the two system components,[8] and is responsible for generating the networks according to the process in [18]. We use graph generators provided by NetworkX and the Spider spatial generator.[9]

The two system components communicate by exchanging .gr and .co files. Specifically, the backend stores the graph of the generated network inside a .gr file, and its spatial vertices inside a .co file. These files are then sent to the frontend using the Flask-based API for visualization. If the user decides to re-use the graph of an existing social network or an existing collection of geospatial objects, the frontend also sends a .gr or a .co file to the backend, respectively. Figures 3 and 4 detail the two file types. the first lines specify important metadata, i.e., the number of vertices and edges and the number of spatial vertices along with the type of their geospatial information (currently points or boxes). The rest of the lines in the .gr file list the edges with an optional label for each, while in the .co file the geometries of the spatial vertices.

GESONGEN is designed with extendability in mind. We can replace or extend the generators in the backend, by incorporating other graph or spatial data libraries. As long as the generated network is stored inside a .gr and a .co file, the system can operate as described above. We only need small changes in the frontend to modify or include new user parameters for the generation process.

3.1 Frontend

We first elaborate on the frontend. Figure 5 exemplifies GESONGEN's main screen comprising two frames. The left frame is further split into two subframes. The top subframe lists the necessary input for generating a new geosocial network:

(1) the type of the synthetic network, i.e., one of G_s, G_c, G_p;

---

[7] https://d3js.org and https://github.com/vasturiano/d3-force-3d.
[8] https://react.dev and https://flask.palletsprojects.com/.
[9] https://networkx.org and https://spider.cs.ucr.edu.

```
{number of vertices} {number of edges}
{source vertex id} {target vertex id} {label (opt)}
...

5 6
1 2 FRIEND_OF
2 3 FRIEND_OF
1 3 FRIEND_OF
1 4 CHECKED_IN
2 4 CHECKED_IN
2 5 CHECKED_IN
```

Fig. 3. Format (top) and an example (bottom) of a `.gr` file

```
{number of spatial vertices} {type: point or box}
{vertex id} {geometry}
...

2 point
4 0.9366755118414897 0.4555963756560951
5 0.30657038904474687 0.6100456763118138
```

Fig. 4. Format (top) and an example (bottom) of a `.co` file

(2) the nature of the social and the spatial component, i.e., generated or uploaded by the user via a `.gr` and `.co` file - in the first case, the user provides the values for the generation parameters;

(3) the parameters for the combiner to be used (see Sect. 3.2).

The bottom left subframe and the entire right frame visualize a network.

Visualization Options. GESONGEN offers three visualization options, called *views*. The *social* view (as the word suggests) prioritizes the social component of the network. The right frame visualizes the network graph; the user is allowed to select the colors for the vertices (both spatial and non-spatial), and the edges. The left frame (bottom subframe) is used to partially display geospatial information of the network, i.e., for the spatial vertices selected by the user. Figure 5(a) exemplifies the social view. The second visualization option, called the *spatial* view, prioritizes the spatial component of the network. The right frame now plots the geometries of all spatial vertices, while the left subframe is used to partially display the social this time information. Specifically, we draw for the selected spatial vertex its directly connected vertices of the network. Figure 5(b) provides an example of the spatial view. Finally, the third option - the *combined* view, visualizes both components of the geosocial network at the same time, as the name suggests. The combine view defines two layers; the top layer displays the social vertices of the network and their edges, while the bottom layer plots the spatial vertices. Figure 5(c) exemplifies the combined view.

Statistics. GESONGEN provides evidence on how realistic generated geosocial networks are. We consider statistics previously used in [11] for the same purpose, i.e., the vertex degree and its distribution histogram, the diameter of the network and the number of contained triangles. Figure 6 exemplifies the statistics screen.

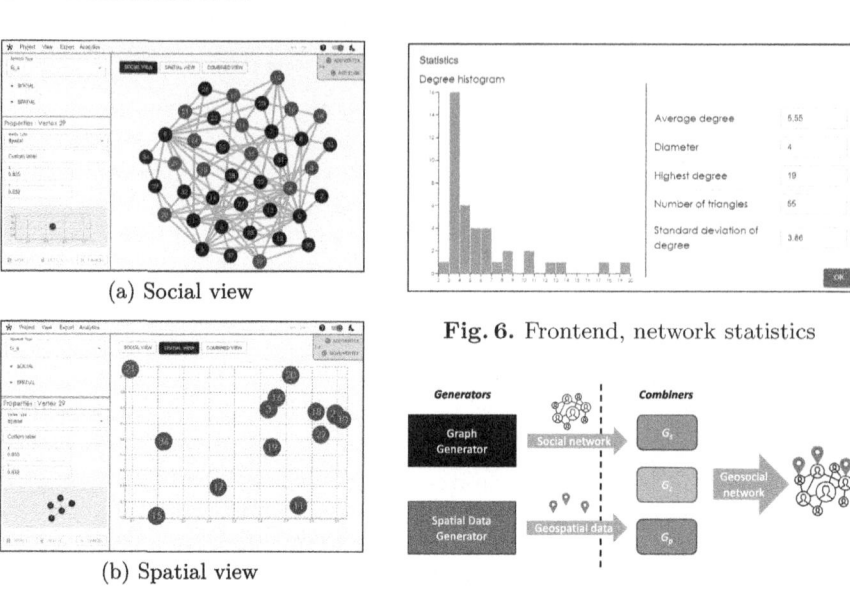

(a) Social view

(b) Spatial view

(c) Combined view

Fig. 5. Frontend, visualization options; spatial vertices in purple (Color figure online)

Fig. 6. Frontend, network statistics

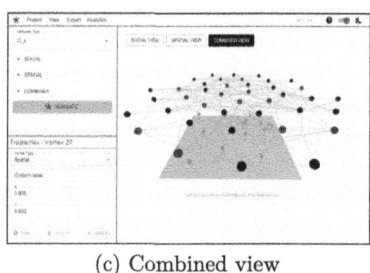

Fig. 7. Backend, generation process

3.2   Backend

To generate a synthetic geosocial network, we combine a social graph with a collection of spatial objects. The two data components can be either generated or provided by the user. Therefore, we developed a *combiner* module for each network type in Sect. 2. Figure 7 shows the generation process. For simplicity, assume that both components are generated. The backend receives the number of vertices in the social network, the edges type (i.e., directed or undirected) and their labels, and the number of geometries (i.e., the number of spatial vertices to be created). The remaining parameters depend on the model used by the graph generator, the distribution and the geometry type for the spatial generator.

We briefly discuss the available combiners, which receive a .gr and .co file, as inputs. The G_s combiner randomly selects a subset of the input social vertices to become spatial, assigning them a geometry from the .co file. These vertex-to-geometry assignments are stored in the output .co file; the output .gr file is identical to the input. Unlike G_s, the G_c and G_p combiners extend the input network graph. They create a new spatial vertex for each geometry found in the input .co file, and new edges to connect these spatial vertices to the social ones found in the input .gr file. The difference is in how these new edges are created; G_c connects each spatial vertex to multiple social ones, while G_p to exactly one. Hence, the combiners also receive the parameters of a normal distribution to determine how many connections every spatial vertex will have (G_c), or how many connections to spatial vertices, every non-spatial will have (G_p).

4   Demonstration Scenarios

We will demonstrate both the generation and the visualization features. For the first, we will walk the attendees through the process considering two scenarios:

Scenario 1: Generating Networks from Scratch. We will demonstrate how to generate a geosocial network using only the network and spatial generators in GESONGEN. We will use different parameters to generate all three types of geosocial networks supported by our system and show the effect these parameters have on the final network. The attendees will also be able to explore the generation process and to create synthetic geosocial networks themselves.

Scenario 2: Generating Networks from Existing Datasets. We intend to also show how to generate a geosocial network from an existing network and/or a set of spatial objects; we will have various real-world social network and spatial datasets available. The attendees will have the opportunity to compare statistics of networks generated using real data and fully synthetic geosocial networks.

For the visualization, we will exemplify the three view options, i.e., social, spatial, and combined, explaining their advantages and limitations. Attendees will be able to interact with the interface, switching between different views, adding/removing vertices and edges from a generated network, and assigning labels. Moreover, they will be able to see the statistics for the generated networks and discuss how well they mimic real geosocial networks. Lastly, we will show the project management and export (to .gr/.co files) features of the system.

5   Conclusions

We developed the GESONGEN interactive interface to generate and visualize geosocial networks. The interface allows users to modify and extend generated networks by hand to meet their needs. GESONGEN is designed with extendability in mind. The generation process is decoupled from the network visualization which allows for incorporating additional graph and spatial generators. In the future, we plan to investigate this direction. Also, we will include new combiners to capture the correlation between socially interwoven and spatially close entities, and incorporate other generation approaches, e.g., from [10]. Another direction is to allow the visualization of larger graphs, by considering pre-rendering and interactive visualization techniques.

References

1. Albert, R., Barabási, A.L.: Emergence of scaling in random networks. Science **286**, 509–512 (1999)
2. Alizadeh, M., Cioffi-Revilla, C., Crooks, A.T.: Generating and analyzing spatial social networks. Comput. Math. Organ. Theory **23**(3), 362–390 (2017)
3. Armenatzoglou, N., Papadopoulos, S., Papadias, D.: A general framework for geo-social query processing. Proc. VLDB Endow. **6**(10), 913–924 (2013)
4. Beckmann, N., Seeger, B.: A benchmark for multidimensional index structures. Technical report, Philipps-Universität Marburg (2008). https://www.mathematik. uni-marburg.de/~rstar/benchmark/distributions.pdf
5. Bouros, P., Sacharidis, D., Bikakis, N.: Regionally influential users in location-aware social networks. In: SIGSPATIAL, pp. 501–504 (2014)
6. Chen, L., Liu, C., Zhou, R., Li, J., Yang, X., Wang, B.: Maximum co-located community search in large scale social networks. Proc. VLDB Endow. **11**(10), 1233–1246 (2018)
7. Doytsher, Y., Galon, B., Kanza, Y.: Querying socio-spatial networks on the world-wide web. In: WWW, pp. 329–332 (2012)
8. Erdos, P., Renyi, A.: On the evolution of random graphs. Publ. Math. Inst. Hungary. Acad. Sci. **5**, 17–61 (1960)
9. Fang, Y., Cheng, R., Li, X., Luo, S., Hu, J.: Effective community search over large spatial graphs. Proc. VLDB Endow. **10**(6), 709–720 (2017)
10. Gallagher, K., Anderson, T., Crooks, A.T., Züfle, A.: Synthetic geosocial network generation. In: LocalRec, pp. 15–24 (2023)
11. Gallagher, K., et al.: Human mobility-based synthetic social network generation. In: HANIMOB, pp. 23–26 (2022)
12. Katiyar, P., Vu, T., Eldawy, A., Migliorini, S., Belussi, A.: SpiderWeb: a spatial data generator on the web. In: SIGSPATIAL, pp. 465–468 (2020)
13. Kim, J., et al.: Location-based social network data generation based on patterns of life. In: MDM, pp. 158–167 (2020)
14. Kim, J., et al.: Simulating urban patterns of life: a geo-social data generation framework. In: SIGSPATIAL, pp. 576–579 (2019)
15. Levandoski, J.J., Sarwat, M., Eldawy, A., Mokbel, M.F.: LARS: a location-aware recommender system. In: ICDE, pp. 450–461 (2012)
16. Li, G., Chen, S., Feng, J., Tan, K., Li, W.: Efficient location-aware influence maximization. In: SIGMOD, pp. 87–98 (2014)
17. Mouratidis, K., Li, J., Tang, Y., Mamoulis, N.: Joint search by social and spatial proximity. TKDE **27**(3), 781–793 (2015)
18. Sarsour, A.A.R., Bouros, P., Chondrogiannis, T.: Towards generating realistic geosocial networks. In: LocalRec, pp. 25–28 (2023)
19. Sarwat, M., Levandoski, J.J., Eldawy, A., Mokbel, M.F.: LARS*: an efficient and scalable location-aware recommender system. TKDE **26**(6), 1384–1399 (2014)
20. Sun, Y., Sarwat, M.: A generic database indexing framework for large-scale geographic knowledge graphs. In: SIGSPATIAL, pp. 289–298 (2018)
21. Sun, Y., Sarwat, M.: Riso-Tree: an efficient and scalable index for spatial entities in graph database management systems. TSAS **7**(3), 12:1–12:39 (2021)

22. Vu, T., Migliorini, S., Eldawy, A., Bulussi, A.: Spatial data generators. In: Spatial-Gems @ ACM SIGSPATIAL (2019)
23. Watts, D.J., Strogatz, S.H.: Collective dynamics of 'small-world' networks. Nature **393**(6684), 440–442 (1998)
24. Winick, J., Jamin, S.: Inet-3.0: internet topology generator. Technical report, University of Michigan, Ann Arbor (2022). http://web.eecs.umich.edu/techreports/cse/02/CSE-TR-456-02.pdf

Entity Resolution and Integration

Accelerating Entity Resolution Through Vectorized Meta-blocking on GPUs

Nikolas Stamatopoulos[1]([✉]) [iD], Vassilis Stamatopoulos[1,2] [iD], Giorgos Alexiou[1] [iD],
Giorgos Giannopoulos[1] [iD], and George Papastefanatos[1] [iD]

[1] ATHENA Research Center, Athens, Greece
{nickstam,bstam,galexiou,giann,gpapas}@athenarc.gr
[2] Department of Computer Science and Engineering, University of Ioannina,
Ioannina, Greece

Abstract. This paper presents an approach to accelerate the Meta-blocking phase in entity resolution (ER) by leveraging GPU computational power. We enhance the performance of conventional meta-blocking algorithms by utilizing sparse matrix representations of block collections. Our proposed solution remains orthogonal to existing blocking and matching techniques, ensuring that their effectiveness is not compromised. By converting a standard block collection to a one-hot encoded sparse matrix and implementing block purging, block filtering, and edge pruning on GPUs, we achieve up to $40\times$ speedups compared to CPU-based implementations.

Keywords: Entity resolution · Data integration · Blocking · Meta-blocking

1 Introduction

Entity Resolution (ER) is an essential step in data preparation and integration, that identifies and merges records referring to the same real-world entities [4,5]. In traditional settings, it is part of data integration pipelines, where one or more datasets with duplicate entries are entirely cleaned before they become available for analysis. Other approaches address the challenges for data lake settings, differing the ER task for the analysis phase; i.e., ER is applied to subsets of the dataset(s) that the users are interested in [2,3]. Most recent advancements in Large Language Models (LLMs) have led to the adoption of pre-trained embeddings to improve the accuracy of entity matching [15]. These models, although highly effective on identifying duplicate pairs of entries, introduce significant computational overhead in the resolution process.

ER is inherently computationally intensive, and its scalability becomes a significant challenge when processing large datasets over LLMs. First, exhaustive pairwise comparisons between all entities in the input dataset(s) to identify duplicates, are impractical, as they have a quadratic time complexity [11]. Second, the inference time is also crucial to the efficiency of the entire ER process

© The Author(s), under exclusive license to Springer Nature Switzerland AG 2026
P. K. Chrysanthis et al. (Eds.): ADBIS 2025, CCIS 2676, pp. 141–150, 2026.
https://doi.org/10.1007/978-3-032-05727-3_14

and thus the number of redundant candidate pairs that is fed to the LLM for comparison should be kept small. To mitigate this, ER pipelines employ pre-processing steps: blocking, which groups potentially similar entities [6,7], and meta-blocking, which refines candidate pairs [13,14].

Despite meta-blocking's effectiveness on CPUs, its benefits diminish with GPU-based LLM matching. This is due to the CPU-centric implementations of state of the art meta-blocking algorithms, as well as the significant CPU-GPU data transfer overhead, creating a communication bottleneck. In this work, we propose a novel approach to significantly improve ER efficiency by implementing **GPU-accelerated Meta-blocking algorithms**, creating a cohesive, end-to-end GPU pipeline that complements LLM-based matching. We transform token-based block collections into a **one-hot encoded sparse matrix** using the **Coordinate List (COO) format** [1], then implement and optimize key Meta-blocking techniques (block purging, block filtering, and edge pruning) directly on the GPU. This achieves substantial speedups and efficiency gains without compromising ER effectiveness. Our key contributions include: GPU-accelerated implementation of core meta-blocking algorithms via a novel vectorized approach using COO sparse matrices; significant performance gains (accelerating meta-blocking by multiple orders of magnitude); analysis of trade-offs (GPU memory limitations, data transfer overhead); and an open-source implementation.

This paper is structured as follows: Sect. 2 covers related work. Section 3 details our methodology, while Sect. 4 presents the experimental evaluation of our approach. Finally, Sect. 5 concludes the paper.

2   Related Work

Entity Resolution (ER) identifies and merges records referring to the same real-world entities. To mitigate its quadratic complexity, ER employs pre-processing techniques: blocking and meta-blocking.

Blocking. Blocking strategies group potentially similar entities into blocks, limiting pairwise comparisons to entities within the same block [6,7] and significantly reducing the candidate pair space. Common approaches like token blocking use unique blocking keys. While crucial, blocking often generates redundant comparisons. Our GPU-accelerated meta-blocking framework is orthogonal and adaptable to any blocking method.

Meta-blocking. Meta-blocking refines the initial candidate pairs from blocking by eliminating redundant and non-matching comparisons while preserving true matches [12]. This involves *block refinement* (e.g., *Block Purging (BP)* and *Block Filtering (BF)*, which reduce block size or remove low-quality blocks) and *comparison refinement* (e.g., *Edge Pruning (EP)*, which prunes less promising edges from the comparison graph) [11]. Our work focuses on accelerating BP, BF, and Cardinality-based Edge Pruning (CEP) on GPUs.

CPU-based parallel meta-blocking techniques, based on MapReduce [8,9], incur overheads from task coordination and memory bottlenecks. In contrast, our approach leverages GPU parallelization, eliminating CPU-GPU data transfer and coordination overheads for end-to-end ER pipelines.

GPU-based Techniques in ER. GPU acceleration has gained traction in data processing, including ER. Recent efforts focus on GPU-based entity matching, particularly with Large Language Models (LLMs) [15]. While LLMs improve matching effectiveness, they are orthogonal to our work, which accelerates the *meta-blocking* stage. HyperBlocker [16] accelerates GPU-optimized rule-based *blocking*. In contrast, our contribution lies in GPU-accelerated *meta-blocking* algorithms that refine candidates from **any blocking method**. To our knowledge, ours is the first work exploring vectorizing meta-blocking operations for GPU-based acceleration, facilitating a comprehensive GPU-accelerated ER pipeline.

3   Methodology

Our methodology develops **GPU-accelerated meta-blocking algorithms** to enhance ER pipeline efficiency, crucial with rising computationally intensive matchers like LLMs [10,15]. Offloading meta-blocking to GPU drastically accelerates the process for large datasets and minimizes CPU-GPU data latency, streamlining end-to-end GPU workflows.

GPU integration is enabled by transforming blocking output into a one-hot encoded sparse matrix, stored directly in GPU memory using the **Coordinate List (COO) format**. COO efficiently stores only non-zero elements as $(row\_index, col\_index, value)$ triples, saving substantial space and adapting well for incremental updates in dynamic ER data.

Our GPU-accelerated meta-blocking pipeline sequentially applies **Block Purging (BP)**, **Block Filtering (BF)**, and **Edge Pruning (EP)**. This empirically proven order [2,3,14] maximizes efficiency and effectiveness. Our GPU implementation strictly increases processing speed for large datasets, producing *identical results* to CPU counterparts, thus maintaining overall ER accuracy.

In the following sections, we detail the sparse matrix representation and GPU-adapted meta-blocking techniques.

3.1   Blocking

To improve ER efficiency, blocking techniques group entities that are likely to match together, forming a block collection, which can be represented as a hashmap, referred to as the **Block Index (BI)** which maps blocks to entities.

Our method accepts any block collection, transforming it into a matrix for GPU processing (Fig. 1). Conceptually, a dense one-hot encoded matrix M ($|b| \times |e|$ blocks by entities) is formed on the CPU, but *crucially, it is never materialized*. Instead, M is immediately converted and stored as a **sparse matrix in GPU memory in COO format** [1] (Fig. 1, Step B). All storage and computations are performed using this sparse COO representation.

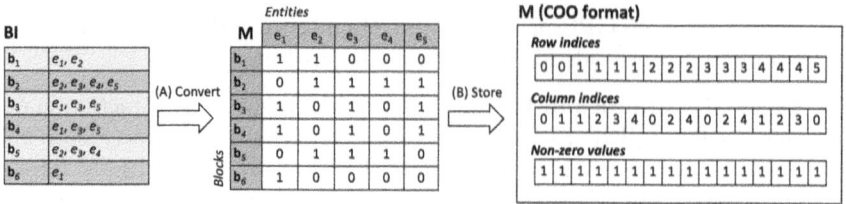

Fig. 1. Transformation of a Block Index (*BI*) to a conceptual dense matrix (Step A), then stored in GPU memory as sparse COO format (Step B). Dense matrix is not explicitly transferred.

3.2 Meta-blocking

Block Purging. The original Block Purging (BP) algorithm [13] removes oversized blocks based on a dynamically determined threshold t, identified by the first block whose size-to-total-comparisons ratio drops significantly compared to the next. In our GPU-accelerated approach, we transform the BP algorithm for vector operations on the sparse matrix M, as detailed in Algorithm 1. The process begins by computing row-wise sums of M to determine block sizes, forming vector S (lines 1-2), which is then sorted in descending order while maintaining original indices (line 3). The number of internal comparisons for each block is calculated as $\frac{S \circ (S-1)}{2}$, yielding vector C (line 4), where $\circ$ denotes the element-wise product. To identify the dynamic threshold, shifted versions of S and C, denoted S' and C', are computed via an upwards roll operation (lines 5-6). These are used to calculate the *criterion vector* with the formula $C \circ S' - sf \cdot (C' \circ S)$ (line 7). The cut-off point for purging is the position of the first negative value in this criterion vector (line 8). If such a point exists, all blocks (rows) in M up to this position are purged (lines 9–11).

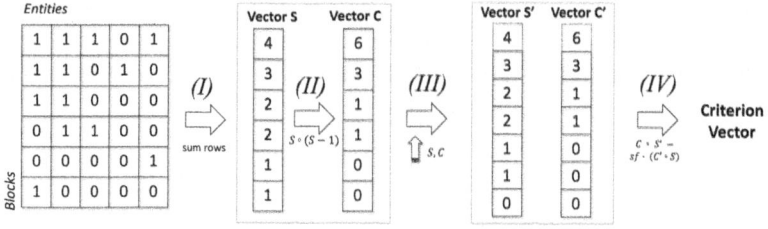

Fig. 2. Creation of the criterion vector for Block Purging performed on a 6×5 matrix.

Figure 2, illustrates the process of generating the criterion vector for Block Purging using a 6×5 matrix. We begin by calculating the reduced sum along the row axis, which produces vector S (step I). From there, vector C is computed using the formula $\frac{S \circ (S-1)}{2}$ (step II). We then generate vectors S' and C' (step III) by shifting S and C upwards. Finally (step IV), the criterion vector is determined using the expression $C \circ S' - sf \cdot (C' \circ S)$.

Algorithm 1: Block Purging

Input   : Sparse Matrix M (in COO format), percentage m, smoothing factor sf

Output: Purged Sparse Matrix M

1 $M \leftarrow \text{sortByRowSum}(M)$
2 $S \leftarrow \text{rowSums}(M)$
3 $S, \text{indexes} \leftarrow \text{sort\_desc}(S)$
4 $C \leftarrow \frac{S \circ (S-1)}{2}$
5 $S' \leftarrow \text{roll}(S)$
6 $C' \leftarrow \text{roll}(C)$
7 $criterion \leftarrow C \circ S' - sf \cdot (C' \circ S)$
8 $pos \leftarrow \text{findFirstNegative}(criterion)$
9 **if** $pos \neq None$ **then**
10 | $M_{purged} \leftarrow M[pos :]$
11 | **return** M_{purged}
12 **end**

Block Filtering. The original Block Filtering (BF) algorithm [13] improves computational efficiency by limiting the number of blocks each entity participates in. It leverages an inverse block index (IBI) where each entity e_i's blocks b_{e_i} are sorted by size, retaining e_i only in the smallest n blocks, where $n = p \times |b_{e_i}|$ and $p \in [0,1]$ is a predefined filtering parameter. In our GPU-based approach, detailed in Algorithm 2, a removal threshold T_j is calculated for each entity j by computing the reduced sum of its column in matrix M, multiplying by the filtering percentage k, and rounding down (lines 1–4). A "sparse" cumulative sum on M along its columns then creates matrix M', tracking for each element (i, j) the count of blocks larger than i that contain entity j (lines 5–10). With T and M' ready, a vectorized filtering is applied (lines 11–13) where an element (i, j) is retained if its value in M' is less than or equal to $T[j]$.

Figure 3 illustrates our Block Filtering implementation on matrix M using a filtering parameter $k = 50\%$, a value empirically proven to balance efficiency and effectiveness [3]. The process involves: (I) computing the reduced sum along the column axis of M; (II) multiplying this sum by k and rounding down to get the threshold vector T; (III) performing a sparse cumulative sum on M to obtain M' (where each row i is summed over following rows $j > i$, preserving zeros); and (IV) applying T as a step function over each row of M' to return the binary filtered matrix (V), where 0 signifies an entity filtered out in a block.

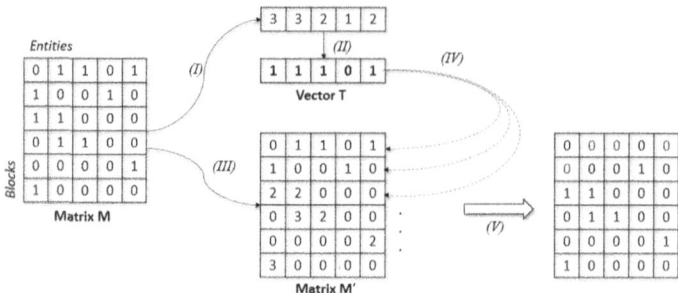

Fig. 3. Block Filtering performed on a 6×5 matrix with a filtering parameter $k = 50\%$.

Algorithm 2: Block Filtering

Input   : Sparse Matrix M (in COO format), percentage k
Output: Filtered Sparse Matrix M

1 $T \leftarrow$ zeros$(1, e)$;
2 **for** *each column j in M* **do**
3 $\quad$ $T[j] \leftarrow$ floor$(k \cdot$ sum$(M[:, j]))$;
4 **end**
5 $M' \leftarrow$ copy(M) ;
6 $C \leftarrow$ zeros(e) ;
7 **for** *each element (i, j) in M* **do**
8 $\quad$ $M'[i, j] \leftarrow M[i, j] + C[j]$;
9 $\quad$ $C[j] \leftarrow M'[i, j]$;
10 **end**
11 **for** *each element (i, j) in M'* **do**
12 $\quad$ **if** $M'[i, j] \leq T[j]$ **then**
13 $\quad\quad$ $M_{\text{filtered}}[i, j] \leftarrow 0$;
14 $\quad$ **else**
15 $\quad\quad$ $M_{\text{filtered}}[i, j] \leftarrow 1$;
16 $\quad$ **end**
17 **end**
18 **return** M_{filtered}

Edge Pruning. The Edge Pruning (EP) algorithm [11] improves efficiency by selectively reducing entity comparisons based on block membership similarity. It prunes comparisons if common blocks fall below a dynamic threshold (a fraction a $(0 < a \leq 1)$ of the fewer entity's total blocks), focusing on highly-overlapping pairs. Our GPU-accelerated *Cardinality-based Edge Pruning* uses sparse matrix multiplication: $M^T \times M$ yields Q, where $Q_{i,j}$ is the count of common blocks between entities i and j. Since Q can be dense, we process it in batches. Algorithm 3 outlines this vectorized approach: For each *batch_size* subset of M^T (line 2), it is multiplied by M to obtain Q_{batch} (common blocks for the current batch's entities) (line 3). Processed entities are then removed from M to free memory (line 4). Relevant connections in Q_{batch} are retained (line 5), and their cardinality weights are computed (lines 6–9). Finally, only the top $k_{\text{threshold}}$ values and corresponding entity pairs from the batch are kept (line 11). This batch-wise computation repeats until all entity connections are examined. After all batches, the top k lists are combined, and the overall top $k_{\text{threshold}}$ values are retained (line 13), ensuring graph-wide correctness.

Algorithm 3: Edge Pruning

Input   : Sparse Matrix M (in COO format), *batch_size*, $k_{\text{threshold}}$
Output: Pruned Edge List

1 **for** *each batch of size batch_size in M^T* **do**
2 $\quad$ $Q_{\text{batch}} \leftarrow M_{\text{batch}}^T \times M$
3 $\quad$ $M \leftarrow M.remove(Q_{\text{batch}})$
4 $\quad$ **for** *each (i,j) in Q_{batch}* **do**
5 $\quad\quad$ common_blocks $\leftarrow Q_{\text{batch}}[i,j]$
6 $\quad\quad$ $\log\_\text{term}_i \leftarrow \log_{10}\left(\frac{\text{no. of blocks}}{\text{column\_sum}[Q_{\text{batch}}.rows]}\right)$
7 $\quad\quad$ $\log\_\text{term}_j \leftarrow \log_{10}\left(\frac{\text{no. of blocks}}{\text{column\_sum}[Q_{\text{batch}}.columns]}\right)$
8 $\quad\quad$ $Q_{\text{batch}}[i,j] \leftarrow$ common_blocks $\times \log\_\text{term}_i \times \log\_\text{term}_j$
9 $\quad$ **end**
10 $\quad$ topK $\leftarrow$ getTopKValues($Q_{\text{batch}}, k_{\text{threshold}}$)
11 **end**
12 $finalTopK \leftarrow$ combineTopKLists(all topK lists, $k_{\text{threshold}}$)
13 **return** finalTopK

---

4   Experimental Evaluation

4.1   Experimental Setting

To thoroughly assess our method's performance and scalability, experiments were conducted on diverse datasets spanning multiple domains, as detailed in Table 1. These included: various sizes of Microsoft's Open Academic Graph Papers Schema (OAGP200k-5m)[1]; POI data for the US (GEOUS) from GeoNames[2]; and MusicBrainz datasets (MB2m, MB20m)[3], consisting of music metadata. Full details are presented in Table 1.

For CPU-based comparisons, we utilized the pyJedAI library[4], chosen for its established, comprehensive meta-blocking implementations and consistency with our Python setup. All experiments were conducted on commodity hardware: a 13th Gen Intel(R) Core(TM) i9 processor and an NVIDIA GeForce RTX 4090 with 24 GB of memory. Algorithms were implemented in Python v3.10, via the PyTorch library.

---

[1] https://www.microsoft.com/en-us/research/project/open-academic-graph/.
[2] https://www.geonames.org/postal-codes/postal-codes-us.html.
[3] https://musicbrainz.org/doc/MusicBrainz_Database.
[4] https://github.com/AI-team-UoA/pyJedAI.

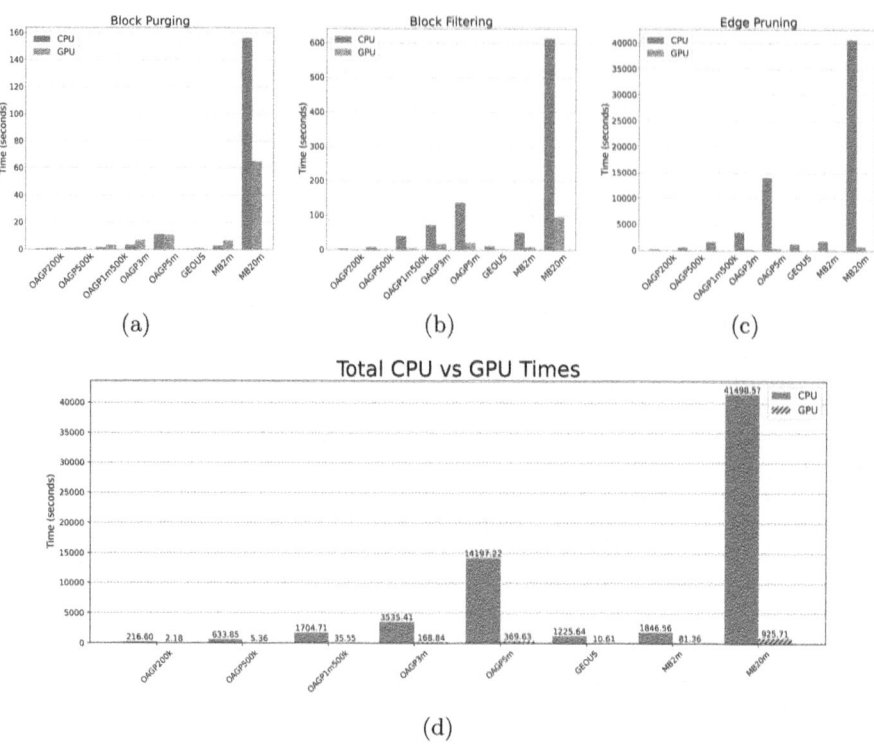

Fig. 4. Performance comparison of CPU and GPU across different tasks.

4.2   Evaluation of Our GPU-Based Implementation Against Standard CPU-Based Implementations

Benchmarking GPU-accelerated meta-blocking against CPU implementations across diverse datasets (Fig. 4) revealed significant speed enhancements and scaling insights.

Blocking: Performed on CPU using standard methods, taking *5.7 s for GEOUS and 282.9 s for MB20m*. Subsequent COO transformation and GPU transfer adds up to 1 s. These times are noted for reference, as our approach is orthogonal to the initial blocking method.

Block Purging (Fig. 4a): GPU-based BP was initially slower (few seconds) on smaller datasets but surpassed CPU on larger ones (e.g., OAGP5m, MB20m) due to superior scalability. This expected behavior for less complex tasks, combined with BP's inherent speed, means minor timing differences negligibly impact overall pipeline performance.

Block Filtering (Fig. 4b): GPU-accelerated BF significantly outperformed CPU across all datasets, with notable savings on *OAGP5m (19.75 s GPU vs.*

Table 1. Dataset statistics.

Name	#Entities	#Dups.	#Attrs.	#Blocks
OAGP200k	200k	1.5k	18	110k
OAGP500k	500k	9k	18	180k
OAGP1m500k	1.5m	35k	18	320k
OAGP3m	3m	135k	18	370k
OAGP5m	5m	340k	18	390k
GEOUS	1.5M	400k	1	90k
MB2m	2m	1.6m	12	104k
MB20m	20m	16m	12	435k

136.02 s CPU) and MB20m (94.04 s GPU vs. 612.64 s CPU). Crucially, CPU run-times grew exponentially, while GPU scaling remained nearly linear.

Edge Pruning (Fig. 4c): EP yielded the most substantial gains. GPU completed *OAGP5m in just 6 min (vs. approximately 4 h on CPU)* and *MB20m in 10 min (vs. approximately 16 h on CPU).* Like BF, GPU scales far more effectively with increasing dataset size.

Total CPU vs GPU Times (Fig. 4d): Aggregate comparison confirms GPU acceleration's substantial advantage across all meta-blocking tasks. GPU algorithms consistently completed faster than CPU, especially with increasing dataset sizes; *notably, 40 times faster for OAGP5m and MB20m.*

In summary, *while BP's GPU enhancements are marginal, significant gains in expensive BF and EP operations* strongly advocate GPU adoption for large, complex datasets. Specifically, *GPU is 7 times faster for Block Filtering and achieves up to 40 times reduction for Edge Pruning.*

5 Conclusions and Future Work

This work demonstrates significant ER performance improvements through GPU-accelerated Meta-blocking. By utilizing sparse COO matrices and GPU-optimized block purging, filtering, and edge pruning, we achieved substantial speedups (up to **40x faster**) and linear scalability over CPU methods. Our research highlights GPU's critical role in optimizing the entire ER pipeline. Future work involves integrating these algorithms into an end-to-end, GPU-accelerated ER system, incorporating state-of-the-art LLM-based blocking and entity matching.

Acknowledgments. This work was supported by the ExtremeXP project (EU Horizon program, GA 101093164).

References

1. Coordinate list (COO) format. https://documentation.sas.com/doc/en/ vdmmlcdc/8.1/casml/viyaml_textmine_details10.htm. Accessed 03 July 2024
2. Alexiou, G., Papastefanatos, G.: Query driven entity resolution in data lakes. In: Flouris, G., Laurent, D., Plexousakis, D., Spyratos, N., Tanaka, Y. (eds.) ISIP 2019. CCIS, vol. 1197, pp. 117–130. Springer, Cham (2020). https://doi.org/10. 1007/978-3-030-44900-1_8
3. Alexiou, G., Papastefanatos, G., Stamatopoulos, V., Koutrika, G., Koziris, N.: QueryER: a framework for fast analysis-aware deduplication over dirty data. In: 28th International Conference on Extending Database Technology (2025)
4. Altwaijry, H., Kalashnikov, D.V., Mehrotra, S.: QDA: a query-driven approach to entity resolution. IEEE Trans. Knowl. Data Eng. **29**(2), 402–417 (2017)
5. Altwaijry, H., Mehrotra, S., , Kalashnikov, D.V.: Query: a framework for integrating entity resolution with query processing. Proc. VLDB Endow. **9**(3), 120–131 (2015)
6. Baxter, R., Christen, P., Churches, T.: A comparison of fast blocking methods for record linkage. In: Workshop on Data Cleaning, Record Linkage and Object Consolidation, vol. 24, no. 9, pp. 25–27 (2003)
7. Christen, P.: A survey of indexing techniques for scalable record linkage and deduplication. IEEE Trans. Knowl. Data Eng. **24(9)**, 1537– 1555 (2012)
8. Efthymiou, V., Papadakis, G., Papastefanatos, G., Stefanidis, K., Palpanas, T.: Parallel meta-blocking: realizing scalable entity resolution over large, heterogeneous data. In: 2015 IEEE International Conference on Big Data (IEEE BigData 2015), Santa Clara, CA, USA, October 29–November 1 2015, pp. 411–420. IEEE Computer Society (2015)
9. Efthymiou, V., Papadakis, G., Papastefanatos, G., Stefanidis, K., Palpanas, T.: Parallel meta-blocking for scaling entity resolution over big heterogeneous data. Inf. Syst. **65**, 137–157 (2017)
10. Paganelli, M., Buono, F.D., Baraldi, A., Guerra, F.: Analyzing how BERT performs entity matching. Proc. VLDB Endow. (2022)
11. Papadakis, G., Ioannou, E., Palpanas, T., Niederee, C., Nejdl, W.: A blocking framework for entity resolution in highly heterogeneous information space. IEEE Trans. Knowl. Data Eng. **25**(2), 2665–268 (2013)
12. Papadakis, G., Koutrika, G., Palpanas, T., , Nejdl, W.: Meta-blocking: taking entity resolution to the next level. Trans. Knowl. Data Eng. **26**(8), 1946–1960 (2013)
13. Papadakis, G., Papastefanatos, G., Palpanas, T., Koubarakis, M.: Entity resolution for big data. In: EDBT, pp. 221–232 (2016)
14. Papadakis, G., Papastefanatos, G., Palpanas, T., Koubarakis, M.: Scaling entity resolution to large, heterogeneous data with enhanced meta-blocking. In: EDBT, pp. 221–232 (2016)
15. Zeakis, A., Papadakis, G., Skoutas, D., Koubarakis, M.: Pre-trained embeddings for entity resolution: an experimental analysis [experiment, analysis and benchmark]. In: arXiv (2023). https://doi.org/10.48550/arXiv.2304.12329
16. Zhu, X., Xie, M., Deng, T., Zhang, Q.: Hyperblocker: accelerating rule-based blocking in entity resolution using GPUs. Proc. VLDB Endow. (PVLDB) **18**(2), 308–321 (2024). https://doi.org/10.14778/3705829.3705847

Validating Data Provenance Polynomials

Paulo Pintor[1]([⊠])[iD], Rogério Luís de C. Costa[2][iD], and José Moreira[1][iD]

[1] IEETA, University of Aveiro, Aveiro, Portugal
{paulopintor,jose.moreira}@ua.pt
[2] CIIC, ESTG, Polytechnic of Leiria, Leiria, Portugal
rogerio.l.costa@ipleiria.pt

Abstract. This paper presents a validator of provenance polynomials that is compatible with semiring and semimodule-based formalisms proposed in the literature. The validations include checking whether all entries are valid and whether there are missing entries in annotations in Selection-Projection-Join-Union-Aggregation (SPJUA) operations and complex queries with subqueries. This approach is independent of any specific Database Management System (DBMS) and was used to validate provenance annotations generated by two frameworks over TPC-H benchmark queries. To the best of our knowledge, this is the first solution specifically designed to validate provenance polynomials.

Keywords: Data Provenance · Provenance Polynomials · Validation

1 Introduction

Data provenance is a key research area in the database community, focused on explaining how query results (tuples) are derived [7]. Provenance can be represented using algebraic structures such as semirings [5], monoids, and semimodules [1], and has been implemented in various systems [2,4,9]. However, as data volumes grow and queries become increasingly complex, verifying the correctness of the generated provenance annotations becomes more challenging. This highlights a significant gap, especially for developers creating or enhancing provenance-aware systems, and emphasises the need for automated methods to validate the correctness of provenance annotations.

This paper presents a method to validate whether the data provenance polynomials generated by a given system are correct. Building on the formalism proposed in [5] and extended to support aggregation in [1], the method supports monotone queries with SPJUA operations and subqueries, and is independent of the underlying DBMS. The proposed validation techniques can verify whether each term in a provenance annotation is valid, ensure that all terms contribute to the query result, and that the values of the tuples in the result can be derived from the corresponding provenance annotations. It is also possible to check for missing terms and validate the aggregation information as proposed in [1].

By addressing the lack of validation tools in provenance systems, our method adds a crucial layer of assurance to the provenance lifecycle by supporting the development of provenance annotation systems, facilitating correctness verification, and aiding in debugging - particularly when annotations grow to several megabytes.

2 Background and Related Work

Provenance semirings is a well-established formalism for capturing data provenance [5] that allows the representation of how database query results were obtained using mathematical expressions such as polynomials [6].

A provenance semiring is an algebraic structure $(K, 0, 1, +, \cdot)$ where K denotes a set of tokens, "$+$" is a commutative and associative operator with identity 0, "$\cdot$" is a commutative and associative operator with identity 1, and "$\cdot$" is distributive over "$+$" [3,5,8]. The constants 1 and 0 indicate whether a provenance token has contributed or not to including a tuple in the query result. This formalism supports Selection, Projection, Join and Union (SPJU) operations, where "$\cdot$" represents conjunctions (e.g., join and cartesian product) and "$+$" represents alternatives (e.g., union, projection and group by without functions). For instance, consider the relations R and S in Example 1, where $t1, \ldots, t7$ are tokens that identify each tuple (provenance tokens). The provenance of the result of query Q1 is listed in Example 1c. The polynomial in the first line shows that the name "A" can be obtained by joining $t1$ with $t5$, $t1$ with $t6$, $t2$ with $t5$, or $t2$ with $t6$. The entry $(t5 + t6)$ was generated by projecting the name from the relation B.

Example 1: Running example containing two tables and a query and its results with the provenance polynomials.

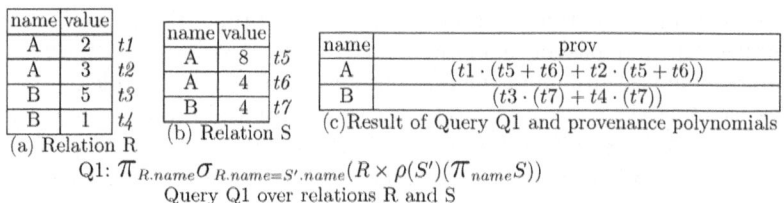

name	value	
A	2	$t1$
A	3	$t2$
B	5	$t3$
B	1	$t4$

(a) Relation R

name	value	
A	8	$t5$
A	4	$t6$
B	4	$t7$

(b) Relation S

name	prov
A	$(t1 \cdot (t5 + t6) + t2 \cdot (t5 + t6))$
B	$(t3 \cdot (t7) + t4 \cdot (t7))$

(c)Result of Query Q1 and provenance polynomials

Q1: $\pi_{R.name} \sigma_{R.name=S'.name}(R \times \rho(S')(\pi_{name} S))$

Query Q1 over relations R and S

This formalism is extended in [1] to capture the provenance of aggregation functions. Aggregations are represented as commutative monoids of the form $(M, +, 0)$, and the authors propose the use of K-semimodules to integrate the aggregation domain M (monoids) with provenance annotations rom a commutative semiring K. This integration is captured by the tensor product $K \otimes M$, which formally represents combinations of annotations and values, allowing provenance to be tracked through aggregation. Columns containing aggregated values can then be expressed as polynomials over these tensor expressions. To support grouping, the semiring is further extended to a commutative δ-semiring, denoted by $(K, 0_K, 1_K, +_K, \cdot_K, \delta_K)$, where the operator δ_K captures the presence of at least one contributing tuple in a group.

Since the aggregated columns are represented as polynomials, when there are comparisons or calculations on previously aggregated results, a symbolic expression is introduced to represent them with the $\delta - semirings$. This expression is denoted as $[x \alpha y]$, where both x and y are monoids, $\alpha \in \{=, <, \leq, >, \geq, \neq\}$. These variables can represent either simple values or polynomial expressions derived from other operations. When either x or y corresponds to a simple value m from the set M, they are mapped to the domain $K^M \otimes M$ using the function $\iota(m) = 1_K \otimes m$. For example, if y corresponds to m, the expression transforms into $[x \alpha 1_K \otimes m]$. Upon evaluation, once the values are embedded in the symbols, the expression is replaced by 1_K when the predicate is *True* or 0_K when *False*.

ProvSQL [9] and DataPROV[1] are systems that compute provenance polynomials based on the formulations in [1,5], yet the former does not adhere completely to the notation proposed in [1].

3   Validation of Provenance Semirings

In the following, it is considered that the result of an annotated query is a relation $R^k(a_1, \ldots, a_k, prov)$, such that a_i denotes an column in a query result. The attribute *prov* is a *k-semiring* [5] such as the annotations in Example 1.

The validation has two parts that can be performed independently: checking whether the result of the original query has remained unchanged (Validation 1) and checking the provenance polynomial itself. The latter is divided into two parts. The first (Validation 2) consists of checking whether all the terms in the provenance annotations are valid and correspond to the actual values in the query result. Each term represents an alternative derivation path for producing a tuple in the query result. This step ensures that there are no terms in the provenance annotation that do not contribute to the query result. The second (Validation 3) checks the completeness of the provenance annotations, i.e., verifying that all combinations of tokens (alternatives) capable of contributing to the query result are represented in the provenance. The techniques proposed in this section can validate SPJU queries, including subqueries, in which provenance is represented using *k-semiring* [5]. Checking aggregations (Validation 4) is covered in Sect. 4.

1 - Checking the Original Query Result. To verify that the result of the original query (R) is not changed, it is performed a projection of the attributes of the original query ($S = \pi_{a_1, \ldots, a_k}(R^k)$) and it is verified whether $R - S = S - R = \emptyset$.

2 - Checking Whether All Entries in an Annotation are Valid. Given that *prov* is a *k-semiring*, the symbols that can appear in the annotation are "+" for alternatives, "·" for joins, parenthesis when there are subqueries and the

---

[1] https://github.com/PauloPintor/DataPROV.

tokens $t_1, \ldots, t_n$. The first step consists of expanding the provenance polynomial using a symbolic solver, resulting in a new polynomial $prov'$ that contains only the symbols "+" and "·", the tokens and scalars.

Next, $prov'$ is split by the symbol "+" to get the individual terms of the polynomial ($terms = \{r_1, \ldots, r_m\}$), such that r_i is a monomial. The number of terms (monomials) is equal to the number of alternative ways to get the tuple in the query result. Each term corresponds to an individual token or a combination of tuples (join). To extract the tokens (s) of a term, r_i is split by the symbol "·" and the scalars are removed, yielding $tk_i = \{s_1, \ldots, s_p\}$, and a set of sets is built: $allTokens = \{\{tk_1\}, \ldots, \{tk_m\}\}$.

To complete the validation, it is necessary to verify whether each entry in $allTokens$ can retrieve the corresponding tuple in the query result. Two cases are considered. The first assumes that the tokens directly identify the base relations from which they originate, e.g., the tokens are in the format $relation.token$, $schema.relation.token$, or similar. In this case, assuming that $(a_1^j, \ldots, a_k^j, prov^j)$ denotes the j-th tuple in R^k and $g(tk_i)$ is a user-defined function that retrieves the base relation (and schema, if applicable) from tk_i. To validate the contribution of a token set $s_1, \ldots, s_p \in allTokens$, it suffices to check whether the projected tuple $(a_1^j, \ldots, a_k^j)$ can be obtained the query fragment in Eq. 1:

$$(a_1^j, \ldots, a_k^j) \in \pi_{a_1, \ldots, a_k}(\sigma_{condition}(R^k)) \tag{1}$$

such that $condition = g(s_1).prov^j = s_1 \wedge \cdots \wedge g(s_p).prov^j = s_p$. As tokens are unique identifiers, the right-hand expression always returns a single tuple. Example 2 shows the validation of the first provenance annotation in Example 1, step by step, where $g(t)$ is a function returning R or S.

Otherwise, if it is not possible to obtain the name of the relations from the tokens, then it is necessary to consider all possible mappings of tokens to base relations. For instance, if the base relations are R and S, then the condition of the selection in (1) is $(R.prov^j = s_1 \wedge S.prov^j = s_p) \vee (S.prov^j = s_1 \wedge R.prov^j = s_p)$.

However, when there are `left`, `right` or `outer` joins, it is needed to account for the presence of null values. For example, if there are three relations (R, S, V) while only having two tokens (e.g., $t1 \cdot t2$) it is required to ensure that all possible combinations of relations are considered. Specifically, it is necessary to first combine the relations in pairs $((R, S), (R, V), (S, V))$ and then associate all these combinations to the tokens. According to the semiring rules, if a joint relation contains more tokens than relations in the query, the polynomials are incorrect.

3 - Checking Whether There are Missing Entries in Annotations. To perform this validation it is necessary to find whether all tuples combined or aggregated to produce each tuple in the query result are represented in the provenance annotation. This is achieved by augmenting the query with a counter denoted $\bar{c}$, while preserving its original structure. If the number of such entries matches the number of terms (alternatives) in the provenance polynomial (Validation 2), it can be concluded that no terms are missing.

Example 2: Validation of a provenance polynomial

$$prov = (t1 \cdot (t5 + t6) + t2 \cdot (t5 + t6))$$
$$prov' = t1 \cdot t5 + t1 \cdot t6 + t2 \cdot t5 + t2 \cdot t6$$
$$terms = \{t1 \cdot t5, t1 \cdot t6, t2 \cdot t5, t2 \cdot t6\}$$
$$allTokens = \{\{t1, t5\}, \{t1, t6\}, \{t2, t5\}, \{t2, t6\}\}$$

...and testing condition (1) four times:

$$A = \pi_{name}(\sigma_{g(allTokens[1,1])=t1 \, \wedge \, g(allTokens[1,2])=t5}(R \times S))$$
$$A = \pi_{name}(\sigma_{g(allTokens[2,1])=t1 \, \wedge \, g(allTokens[2,2])=t6}(R \times S))$$
$$A = \pi_{name}(\sigma_{g(allTokens[3,1])=t2 \, \wedge \, g(allTokens[3,2])=t5}(R \times S))$$
$$A = \pi_{name}(\sigma_{g(allTokens[4,1])=t2 \, \wedge \, g(allTokens[4,2])=t6}(R \times S))$$

$$\pi_A(R) \rightarrow \gamma_{A; \, \text{count}(*) \rightarrow c}(R) \tag{2}$$

$$R \cup S \rightarrow \gamma_{U'.A; \, count(*) \rightarrow c} \, \rho(U')(R \overset{B}{\cup} S) \tag{3}$$

The selection, join and bag union/projection operations do not require validation, as the number of alternatives is 1. To evaluate the results, it is necessary to assess each row individually and compare if the obtained result in $\bar{c}$ within R is equivalent to the polynomial's solution when all tokens are replaced by 1.

To achieve this, it is necessary to propagate the column containing the number of tuples that have been aggregated or combined throughout the entire query. For the lowest-level query, the transformations (2) and (3) are applied. The propagation from this point depends on the operations at the higher levels. If a join is present between subqueries, the values from the columns must be multiplied. If an aggregation is applied over other aggregation, the values must be summed. In cases where both a join and an aggregation exist, the propagation requires a sum of the multiplications.

Using as an example Query Q1, presented in Example 1, and applying the rules to propagate $\bar{c}$, we obtain the query in Listing 4.

$$\gamma_{R.name; \, sum(\bar{c}) \rightarrow c} \sigma_{R.name=S'.name}(R \times \rho(S')(\gamma_{name; \, count(*) \rightarrow c}S)) \tag{4}$$

In this case, it is not necessary to perform a multiplication inside the sum since it is a join between a table and a subquery. The result of the query is $('A', 4), ('B', 2)$. Solving the polynomial for "A" we obtain: $1 \cdot (1+1) + 1 \cdot (1+1) = 4$ and for "B": $1 \cdot (1) + 1 \cdot (1) = 2$. Thus, there are no missing alternatives.

4   Validation of Aggregations

The validation of the aggregations proposed in this work adheres to the notation and definitions presented in [1]. It starts with aggregations with tuple grouping and progresses to selections over aggregations and nested aggregations. Aggregations without grouping are treated as set projections (see Sect. 3).

Aggregation with Grouping. Assuming that there is only one aggregation function to simplify the presentation, the result of a query with aggregations is a relation $R^{k \otimes M}(a_1, \ldots, a_{k-1}, m, prov)$, where the attribute a_k representing the aggregated values is now a monoid denoted as m and represented in polynomial form, and $prov$ is a δ-*semiring*. A δ-*semiring* is a k-*semiring* where aggregations are annotated with the symbol δ, to distinguish them from projections and unions. As the aggregation values in m are represented as formulas rather than numerical values, thus the validation must account for this representation.

For validating the monoid (m), the type of aggregated function (agg) is identified, its associated tags are removed and the symbols $\cdot$ and $\otimes$ are replaced by a multiplication $(\times)$ sign. This representation is named m' and the symbols it may hold are $+$, $\times$, parenthesis, scalars and tokens $(tk_1, \ldots, tk_p)$. If the aggregation function is `sum`, `count` or `avg`, the value 1 is assigned to all tokens and the polynomial is evaluated using a symbolic solver (Eq. 5). If the aggregation function is `min` or `max`, the scalars in m' are traversed and the value of the symbolic expression is the minimum or maximum of those values (Eq. 6). Finally, the result is compared to a_k (remains valid, even when subqueries exist).

$$a_k = polyval(m', tk_1 = 1, \ldots, tk_m = 1) \tag{5}$$

$$a_k = min/max(m') \tag{6}$$

To validate the δ-*semiring* $(prov)$, the δ symbols are removed and the validation method used for set projections is applied (Eq. 1). For instance, consider Query Q2 in Example 3 which consists of summing the values of $R \bowtie S$ grouped by name.

The column *total* is a monoid. Removing the tags, transforming the symbols into $+$ and $\times$ and solving the polynomial by replacing the tokens with the value 1, the result is $a_k = (1 \times 1) \times 2 + (1 \times 1) \times 2 + (1 \times 1) \times 3 + (1 \times 1) \times 3 = 10$ for the name A and $a_k = (1 \times 1) \times 5) + (1 \times 1) \times 1 = 6$ for B, as expected.

Example 3: Result of Query Q2 using the notation in [1]: Q2 : $\gamma_{R.name;\ sum(R.value) \to total}(R \bowtie_{R.name=S.name} S)$

name	total	prov
A	$(t1 \cdot t5) \otimes 2 + k \otimes sum(t1 \cdot t6) \otimes 2 + k \otimes sum(t2 \cdot t5) \otimes 3 + k \otimes sum(t2 \cdot t6) \otimes 3$	$\delta(t1 \cdot t5 + t1 \cdot t6 + t2 \cdot t5 + t2 \cdot t6)$
B	$(t1 \cdot t7) \otimes 5 + k \otimes sum(t4.t7) \otimes 1$	$\delta(t3 \cdot t7 + t4 \cdot t7)$

Selection over Aggregations. In the notation proposed in [1], the result of selections over aggregations, such as an `having` clause in SQL, is a relation $R^{k^M \otimes M}(a_1, \ldots, a_{k-1}, m, prov)$, where $prov$ has the structure $d._k mm$, d is a δ-semiring and mm is an annotation in the form $[m_1 \, \alpha \, m_2]$. The α denotes a comparison operator and m_1 and m_2 are monoids representing constants or polynomials. There is an entry for all possible alternatives, regardless the condition evaluates to *True* or *False*.

The δ-semiring d and the monoids $m1$ and $m2$ can be validated independently using the methods described above. The component in brackets represents a predicate. To evaluate its logical value, the symbolic expressions must be solved, using Eqs. 5 and 6.

Example 4: Result of Query Q3 using the notation in [1]: $Q3 \; : \; \sigma_{total=5}$ $\left(\gamma_{name; \, sum(value) \rightarrow total}(R) \right)$

name	total	prov
A	$t1 \otimes 2 +_{k \otimes SUM} t2 \otimes 3$	$\delta(t1 + t2)._k \; [t1 \otimes 2 +_{k \otimes SUM} t2 \otimes 3$ $= 1_k \otimes 5]$
B	$t3 \otimes 5 +_{k \otimes SUM} t4 \otimes 1$	$\delta(t3 + t4)._k$ $[t3 \otimes 5 +_{k \otimes SUM} t4 \otimes 1 = 1_k \otimes 5]$

For instance, the result in Example 4 lists the sum of the column *value* of the relation A in the running example, grouped by *name*, with a sum equal to 5 (Query Q3). The polynomials in the column *total* have the same structure as in the previous examples. The difference lies in the column *prov*. The δ-*semirings* $\delta(t1+t2)$ and $\delta(t3+t4)$ also keep the same structure as in the previous examples, but each of these is now combined with a predicate in brackets. By removing the tags and transforming the operators of the monoids as for the aggregation with grouping, we obtain $m' = t1 \times 2 + t2 \times 3$ in the first line and $m' = t3 \times 5 + t4 \times 1$. By assigning the value 1 to all tokens and calculating the values of the polynomials, we get $m' = 5$ and $m' = 6$, respectively. Thus, the predicate in the first line is *True* ($[5 = 5]$), while in the second line it is *False* ($5 = 6$). This explains why the second tuple will not appear in the query result.

Nested Aggregations. When performing operations on aggregations, the inputs can be of type $R^{k \otimes M}(a_1, \ldots, a_{k-1}, m, prov)$ or $R^{k^M \otimes M}(a_1, \ldots, a_{k-1}, m, prov)$. When aggregations are applied to other aggregations, the algebraic structures replacing the aggregated values (the monoid m) are similar to those of the *prov* column in Example 3.

The issue arises when there are selection over aggregations. In such cases, a tuple may contain multiple symbolic expressions enclosed in brackets, and it is excluded from the original query result only if all conditions evaluate to *False*. For instance, consider the result in Example 5 representing the sum of the values in *total* in Query Q3. The expression in blue is *True*, as previously mentioned, while the one in red is *False*. Since at least one expression is true, the tuple must appear in final query result. However, it is necessary to remove the information related to that expression (in the δ-semiring and the monoid in the aggregation column). This process yields the result shown in Example 6.

Example 5: Result of Query Q4: $\left(\gamma_{sum(total)\rightarrow sum\_total}(Q3)\right)$

2 sum_total	provenance
$(t1 \otimes 2 +_{k\otimes SUM} t2 \otimes 3) +_{k\otimes SUM} (t3 \otimes$ $5 +_{k\otimes SUM} t4 \otimes 1)$	$\delta(t1 + t2)._k[t1 \otimes 2 +_{k\otimes SUM} t2 \otimes 3 =$ $1_k \otimes 5] *_{k\otimes SUM} (t1 \otimes 2 +_{k\otimes SUM} t2 \otimes$ $3) +_{k\otimes SUM} \delta(t3 + t4)._k[t3 \otimes 5 +_{k\otimes SUM}$ $t4 \otimes 1 =$ $1_k \otimes 5] *_{k\otimes SUM} (t3 \otimes 5 +_{k\otimes SUM} t4 \otimes 1)$

Example 6: The result of Example 5, with only the symbolic expression where the result is 1_k

sum_total	provenance
$(t1 \otimes 2 +_{k\otimes SUM} t2 \otimes 3)$	$\delta(t1 + t2)._k[t1 \otimes 2 +_{k\otimes SUM} t2 \otimes 3 =$ $1_k \otimes 5] *_{k\otimes SUM} (t1 \otimes 2 +_{k\otimes SUM} t2 \otimes 3)$

The information represented by $*_{k\otimes SUM}$ captures the propagation of data within the aggregated column. It is used to compare against the aggregated values to verify their correctness. The aggregation column and the δ-semiring are then validated.

The approach presented in [1] to extend the semirings can be used in several ways depending on the evaluation and the values assigned to the tokens of the symbolic expressions and monoids. For example, assigning the value 0 to a token allows us to anticipate the result of the query if the corresponding tuple was removed. However, in this context, we are solely validating whether the representation correctly reproduces the original result.

5   Experimental Evaluation

The prototype is implemented in Python and is available on GitHub[2] and was tested with two data provenance systems. The validator uses an SQL parser[3] to extract query structure and rewrite queries according to Eqs. 2 and 3. It supports both $(K, 0, 1, +, \cdot)$ and $(K, 0, 1, \oplus, \otimes)$ notations, and assumes the provenance column is named *prov*. When inconsistencies are found, the validation stops and the user is informed which validation test failed.

We used ProvSQL and DataProv and the TPC-H benchmark with a dataset size of 100 MB. ProvSQL uses the notation proposed in [1], albeit with some variations, and creates a new column with the aggregation function. DataProv also allows a new column with the aggregation information. For testing purposes, a specific method was implemented to handle ProvSQL's notation for aggregated columns, assuming the new column name matches the original one with the suffix "_agg." To make the execution of validations more efficient, it was created and index for the provenance attribute in each of the TPC-H tables.

---

[2] https://github.com/PauloPintor/ProvenanceValidator.
[3] https://github.com/tobymao/sqlglot.

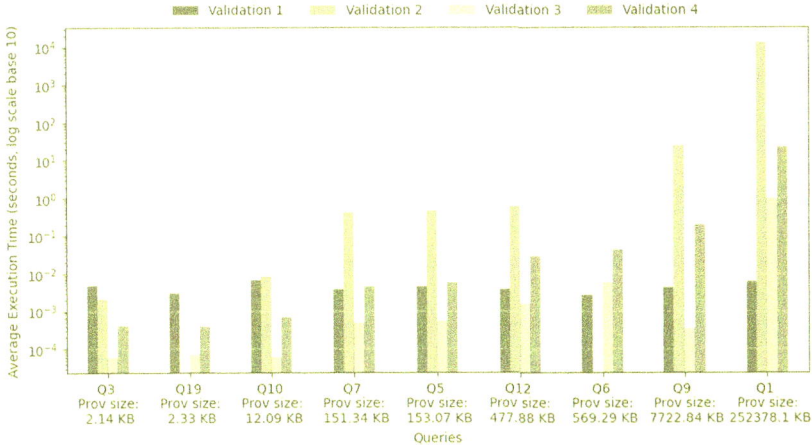

Fig. 1. The runtimes of the different methods in TPC-H Queries

Figure 1 shows the average execution times (in seconds) for validating queries supported by both ProvSQL and DataProv. Provenance size, shown below each query label, is computed as the difference between the result sizes with and without provenance. Validation types correspond to those in Sect. 3, with Validation 4 covering all aggregation-related checks from Sect. 4. Validation 1 refers to result validation and excludes query execution time.

Validation 1 checks whether the query result remains unchanged and is independent of provenance size. The execution time of validations 2 and 4 is directly correlated with the size of the provenance annotations, ranging from just 0.4 milliseconds to ≈ 215 min in the case of validation 2 for query Q1. The execution time of Validation 3 also tends to grow with the provenance size, except for Q9, whose structure requires fewer symbolic solver calls. Queries 6 and 19 lack results for Validation 2 due to containing only aggregation columns and were validated via Validation 4. Both frameworks are stable and passed all tests, except for DataProv, which failed on Query 16 due to missing support for `Count(Distinct ...)`.

6   Conclusions and Future Work

This paper presents a provenance polynomial validator built upon the formalisms and notations proposed in [1,5]. The validator allows for the following checks: (1) ensuring that the result of original query is not changed, (2) verifying of all the elements included in the provenance annotations are required, (3) confirming that the generated provenance polynomials can accurately retrieve all values in the query result, and (4) validates the aggregation information. The approach is DBMS- and framework-independent, making it a versatile tool.

The validator was tested using the queries from the TPC-H benchmark across two frameworks, ProvSQL and DataProv. As expected, the execution times of the validations are influenced by the number of token combinations and database accesses. However, calls to the symbolic solver for computing the values of aggregations and provenance polynomials have a major impact on overall execution times. This is reflected in the experiments performed in this work, as in the TPC-H benchmark, every query includes at least one aggregation function.

Future work includes improving usability through a user-friendly interface, greater flexibility in defining provenance and aggregation notation, and optimising validation performance—particularly for Validations 2 and 4.

Acknowledgments. This work is funded by National Funds through the FCT—Foundation for Science and Technology, in the context of the projects UIDB/00127/2020 and UIDB/04524/2020, by the PRR—Plano de Recuperação e Resiliência and by the Next Generation EU funds at University of Aveiro, through the scope of the Agenda for Business Innovation "NEXUS: Pacto de Inovação—Transição Verde e Digital para Transportes, Logística e Mobilidade" (Project no 53 with the application C645112083-00000059), and under the Scientific Employment Stimulus - Institutional Call - CEECINST/00051/2018.

References

1. Amsterdamer, Y., Deutch, D., Tannen, V.: Provenance for aggregate queries. In: Proceedings of the ACM SIGACT-SIGMOD-SIGART Symposium on Principles of Database Systems, pp. 153–164 (2011). https://doi.org/10.1145/1989284.1989302
2. Arab, B.S., Feng, S., Glavic, B., Lee, S., Niu, X., Zeng, Q.: GProM - a swiss army knife for your provenance needs. IEEE Data Eng. Bull. **41**(1), 51–62 (2018)
3. Buneman, P., Tan, W.C.: Data provenance: what next? SIGMOD Rec. **47**(3), 5–16 (2018). https://doi.org/10.1145/3316416.3316418
4. Glavic, B., Alonso, G.: Perm: processing provenance and data on the same data model through query rewriting. In: Proceedings of the International Conference on Data Engineering, pp. 174–185. IEEE (2009)
5. Green, T.J., Karvounarakis, G., Tannen, V.: Provenance semirings. In: Proceedings of the Twenty-Sixth ACM SIGMOD-SIGACT-SIGART Symposium on Principles of Database Systems, PODS 2007, pp. 31–40. ACM (2007)
6. Green, T.J., Tannen, V.: The semiring framework for database provenance. In: Proceedings of the 36th ACM SIGMOD-SIGACT-SIGAI Symposium on Principles of Database Systems, PODS 2017, pp. 93–99. ACM (2017)
7. Herschel, M., Diestelkämper, R., Lahmar, H.B.: A survey on provenance: What for? what form? What from? VLDB J. **26**, 881–906 (2017). https://api.semanticscholar.org/CorpusID:32007099
8. Senellart, P.: Provenance and probabilities in relational databases: from theory to practice. SIGMOD Rec. **46**, 5–15 (2017). https://doi.org/10.1145/3186549.3186551, 7, 5
9. Senellart, P., Jachiet, L., Maniu, S., Ramusat, Y.: ProvSQL: provenance and probability management in PostgreSQL. Proc. VLDB Endow. **11**(12), 2034–2037 (2018). https://doi.org/10.14778/3229863.3236253

Data Governance Maturity Models and Practices: A Systematic Literature Review

Luis Filipe Campos Cardoso[✉] and Edna Dias Canedo

Computer Science Department, University of Brasilia, Brasilia, DF 70910-900, Brazil
luisfilipe.tec@gmail.com, ednacanedo@unb.br

Abstract. Context: The exponential growth of data, coupled with regulatory requirements coming from the General Data Protection Regulation (GDPR) and Brazil's General Data Protection Law (LGPD), has elevated Data Governance (DG) to a strategic imperative for organizations. These regulations emphasize the critical need for structured approaches to ensure data compliance, quality, and security. **Objective**: This study aims to provide a comprehensive overview of DG maturity models and practical methodologies for assessing and enhancing organizational DG capabilities. **Method**: A Systematic Literature Review (SLR) was conducted, specifically focusing on DG maturity assessments, tools, and best practices from both academic and industrial perspectives. Following the SLR, 22 primary studies were selected, analyzed, and synthesized. This synthesis highlights the main features of existing models, tools, and practices employed in organizational DG. **Results**: The findings illuminate DG maturity models and a range of recurring practices. These practices, which include policy formalization, staff training, and iterative quality assessment, demonstrably support organizations in addressing challenges related to data integration, security, and strategic alignment. **Conclusions**: This work underscores the significance of DG maturity models and emerging recurring practices in structuring governance initiatives. The information gathered provides a comprehensive overview of this critical domain.

Keywords: Data Governance · Data Maturity · Systematic Literature Review · Data Governance Framework · Data Quality

1 Introduction

Data Governance (DG) is defined by [11] as the execution of authority, control, and shared decision-making over data asset management. According to [11], DG ensures compliance in data use and management while establishing clear principles, policies, procedures, and responsibilities. DG promotes collaboration across organizational areas, treating data as a strategic asset. It formalizes policies, standards, and procedures while ensuring compliance [1]. DG maximizes data

P. K. Chrysanthis et al. (Eds.): ADBIS 2025, CCIS 2676, pp. 161–170, 2026.
https://doi.org/10.1007/978-3-032-05727-3_16

value while minimizing costs and risks, which is increasingly relevant due to data growth and regulatory demands like the General Data Protection Regulation (GDPR) and Brazil's General Data Protection Law (LGPD). It also defines decision rights and responsibilities in data management [1].

Organizations that strategically manage data are more adaptable, as the quality of data hinges on strong governance practices [14]. Therefore, assessing DG maturity is important for enhancing data management practices and mitigating potential risks [10].

Despite its critical role in enforcing policies and standards [5], the implementation of DG remains challenging due to process complexities and limited organizational engagement in some cases, often resulting in ineffective practices [3]. To address these challenges, this study investigates existing maturity assessment models to help organizations evaluate their current DG maturity, identify weaknesses, and implement targeted improvements. We conducted a Systematic Literature Review (SLR) to systematically identify these assessments and recurring practices. The key findings of this study underscore the role of DG as a strategic enabler, important for ensuring data quality, security, and compliance with regulatory requirements.

2   Systematic Literature Review

According to [13], the Systematic Literature Review (SLR) was conducted in three phases: Planning, Conducting, and Reporting. The SLR stages were primarily led by the first author, with subsequent review and collaborative revisions by the second author to ensure accuracy. All generated artifacts are available on Zenodo for transparency and reproducibility.

The SLR was guided by the central objective: "To critically analyze maturity assessment methodologies and DG practices implemented in academia and industry, identifying effective tools, frameworks, and strategies that support the advancement of data maturity within organizations." This objective informed the development of the review protocol, including the formulation of research questions (RQs), selection criteria, and data analysis procedures.

Based on this objective, the following research questions (RQs) were defined to guide the review, focusing on the analysis of data maturity models and governance practices. RQ1: What maturity assessments in data governance are used in academia and industry, and what are their main characteristics? This question aims to identify and describe the maturity assessments applied across academic and industrial contexts. Understanding their main characteristics helps map existing tools and methodologies, as well as reveal patterns, strengths, and gaps. RQ2: What tools and practices are implemented in academia and industry to enable effective data governance? This question explores the tools and practices adopted to support and sustain efficient data governance. Examining their use across different environments contributes to a broader understanding of how organizations and academic institutions address governance challenges. RQ3: What are the main principles and best emerging practices in data governance and maturity assessments? Given the evolving nature of the field, this question seeks to consolidate established practices and identify emerging trends that may contribute to the advancement of data governance maturity.

The PICOC method (Population, Intervention, Comparison, Outcome, Context), as recommended by [13], structured the SLR and guided the definition of keywords and related terms, available on Zenodo. In this study, Population refers to the object of analysis, Intervention to the applied methods, Outcome to the expected impact, and Context to the study's scope. Comparison was not applicable.

To ensure study quality and relevance, specific inclusion and exclusion criteria were defined. Studies were included if they were peer-reviewed full-text articles (I01); addressed tools, practices, principles, or frameworks (I02); were published in English, Portuguese, or Spanish (I03); and focused on data governance or maturity assessment (I04). Only studies meeting all inclusion criteria were retained.

Conversely, studies were excluded if they were abstracts, posters, or reviews without full content (E01); papers under five pages unless highly relevant (E02); publications in other languages (E03); those unrelated to data governance or maturity (E04); or studies that lacked a clear methodology or practical application (E05). The full list of included and excluded studies is accessible on Zenodo.

For methodological rigor, a quality checklist comprising five key questions was applied. These questions assessed whether proposed practices were pragmatic and replicable (QA1); if the methodology was clear and appropriate (QA2); if validation methods (e.g., case studies, expert reviews) were present (QA3); if the results were clear and relevant to the research questions (QA4); and if limitations and threats to validity were addressed (QA5). Detailed results of this quality assessment are available on Zenodo.

A structured data extraction form was designed for consistent and comprehensive analysis, capturing article identification, objectives, keywords, and summaries. It also documented specific questions addressed and their answers, the type of data analysis (qualitative, quantitative, or mixed), results, and limitations. This ensured relevance, alignment with objectives, and consistency.

Figure 1 (available on Zenodo) illustrates the number of studies retained after each step. Initial filtering identified 23 duplicates, 13 studies not addressing data governance or maturity (E04), and 1 article in an unsupported language (E03). Consequently, 34 studies met all inclusion criteria (I01–I04), while 14 were excluded (E04, E03), and 23 were removed as duplicates. Subsequently, a quality assessment on the 34 eligible studies retained 22 studies that fully met all five quality criteria for data extraction.

The selected studies employed diverse research methods. Literature Review was most common (15 studies), followed by Case Studies (14 instances, often mixed-methods). Interviews were used in 11 studies, and Document Analysis in 5. Action Research and Focus Groups appeared less frequently. This methodological diversity highlights the breadth of approaches to investigating data maturity and governance. The extracted data is available at Zenodo.

3   Results

The analysis of the selected studies provides a comprehensive overview of data governance (DG) maturity assessments, tools, and best practices within both academic and industrial contexts. Findings are structured according to the defined research questions (RQs), covering: (1) prevalent DG maturity models, (2) key tools and practices supporting effective DG, and (3) main principles and emerging best practices driving improvements in DG maturity. This section synthesizes the methodologies, evaluation criteria, and strategic approaches identified. Table 1 (available on Zenodo) lists the primary studies selected for data extraction and outlines their contributions to each research question.

3.1   RQ1. Data Governance Maturity in Academia and Industry

Table 1 presents an overview of the data governance maturity models identified through the systematic literature review. These models serve as structured frameworks for organizations to assess their current state of data governance capabilities and define roadmaps for improvement.

A closer examination of Table 1 reveals that these models generally follow a progression from initial or ad hoc states to fully optimized and continuously improving processes. While they commonly emphasize core governance elements such as established policies, clearly defined roles and responsibilities, robust data quality mechanisms, and strong strategic alignment, each model is tailored to specific contexts and objectives. For example, the Stanford Data Governance Maturity Model and Loshin's Data Quality Maturity Model provide clear maturity levels but differ in their primary scope—the former offering a broader view of DG and the latter a focused assessment on data quality. In contrast, models like the Global Big Data Maturity Model (GBDMM) and the Consensual Big Data Maturity Assessment System (CBDAS) extend maturity assessments specifically to big data environments, addressing unique challenges associated with large-scale data processing. The Data Science Maturity Model (DSMM) caters to the manufacturing sector, focusing on the structuring of data science processes and their strategic alignment. Similarly, the Master Data Management Maturity Model (MD3M) is specialized for master data practices, emphasizing consistency and control across critical organizational data. For cloud-specific governance, the Cloud Data Governance Maturity Model (CDGM) integrates strategy, security, and quality in cloud environments, while the Big Data Maturity Model (BDMM) combines both technical and managerial dimensions. Beyond these, studies also identify specialized models [4, 22, 23] that tailor assessments for niche areas such as geographic data governance, open data initiatives, or specific audit frameworks. This diversity underscores the context-dependent nature of DG maturity assessment, where the choice of model is driven by an organization's specific operational landscape and strategic priorities.

Table 1. Data Governance Maturity Models Identified

Model	Description	Ref.
Stanford Data Governance Maturity Model	Evaluates six components (awareness, formalization, metadata, governance, data quality, master data) across three dimensions (people, policies, capabilities). Maturity ranges from ad hoc to continuous optimization.	[10,14,19]
Loshin's Data Quality Maturity Model	Covers eight components (quality expectations, data quality dimensions, policies, procedures, governance, standardization, technology, performance management) and five levels (Initial to Optimized).	[18]
Global Big Data Maturity Model (GBDMM)	Assesses company's readiness for big data projects by evaluating six domains (strategy alignment, data, people, governance, technology, and methodology).	[16,17]
Consensual Big Data Maturity Assessment System (CBDAS)	Assesses maturity in eight domains (e.g., data strategy, human interface), with application in consumer goods industries.	[15]
Data Science Maturity Model (DSMM)	Based on ISO/IEC 330xx, evaluates data science capabilities in manufacturing, focusing on process structure and strategic alignment.	[8,9]
Master Data Management Maturity Model (MD3M)	Defines five maturity levels and 13 focus areas such as data quality, ownership, and integration.	[12,20,21]
Cloud Data Governance Maturity Model (CDGM)	Addresses cloud-specific governance by integrating strategy, security, and quality aspects.	[6]
Big Data Maturity Model (BDMM)	Combines technical and managerial dimensions with emphasis on governance and strategic alignment.	[7]
Other Models	Cover specialized contexts such as geographic data governance, open data, and audit frameworks.	[4,22,23]

3.2   RQ2. Tools and Practices in Academia and Industry

The analyzed studies present a variety of tools and data governance practices implemented in both academia and industry that are crucial for promoting effective data governance. Table 2 summarizes the key examples identified in the literature, grouped by their primary function or focus area to provide a clearer understanding of their contributions to DG. These categories were derived iteratively during the data synthesis process, aiming to cluster similar tools and practices.

Table 2. Key Tools and Practices Identified in the Analyzed Studies

Tool or Practice	Description	Ref.
Policy Formalization and Staff Training	Establishment of documented policies and training for data stewards to address inconsistencies and enhance governance.	[10,19]
Evaluation Models and Frameworks	Use of maturity models such as GBDMM and CBDAS to evaluate and guide governance improvements. Models often integrate tools.	[15–17]
Data Quality Management Strategies	Application of frameworks such as DMBOK and ISO 8000 for continuous monitoring and validation using methods like the PDCA cycle.	[2,18]
Integration of Advanced Technologies	Adoption of machine learning, IoT, and scalable architectures to improve integration and manage data heterogeneity.	[2,9]
Metadata and Master Data Management Tools	Use of standardized metadata definitions and centralized master data models for consistency and control.	[6,12,21]
Geographic and Open Data Governance	Implementation of hierarchical models for geo-spatial data and legal frameworks to support coordination; policies for open data management.	[22,23]
Corporate and Strategic Practices	Alignment of data initiatives with corporate goals, leveraging infrastructure for real-time analytics and decision-making.	[7]
Reinforcement of Security and Privacy	Emphasis on risk management and compliance with privacy regulations to strengthen governance.	[4,6]

These findings, detailed in Table 2, illustrate a diverse set of tools and practices that strengthen data governance by aligning organizational strategies, leveraging emerging technologies, and applying structured frameworks. They collectively highlight the importance of policy formalization, continuous staff training, and iterative improvement cycles. The identified practices ensure that governance processes remain adaptable and effective across varied organizational contexts, encompassing both academic institutions and industrial enterprises.

3.3   RQ3. Main Principles and Best Emerging Practices

Based on the analyzed studies, Table 3 summarizes the identified key principles and emerging best practices in data maturity and governance evaluations. These categories were formulated to encapsulate common themes and recurring strategies observed across the primary studies, representing a synthesis of effective approaches for advancing DG.

Table 3. Key Principles and Best Practices in Data Governance

Principle/Practice	Description	Ref.
Definition of Roles and Automated Tools	Clear definition of organizational roles, use of automated tools for metadata collection, and implementation of quality standards to ensure interoperability and transparency.	[10]
Policy and Process Formalization	Adoption of frameworks such as DAMA DMBOK to formalize governance policies and improve data quality through stewardship.	[19]
Continuous Improvement Cycles and Supportive Organizational Culture	Iterative cycles based on domain priorities, supported by business intelligence systems and talent development.	[15]
Adoption of ISO Standards and Quantitative Metrics	Alignment with ISO/IEC 330xx and the use of quantitative indicators for evaluating data quality.	[8]
Lifecycle Management and Interoperability	Integrated management of the data lifecycle and promotion of platform interoperability.	[23]
Alignment with Strategic Objectives and Adaptive Governance	Use of agile governance methods and advanced technologies like machine learning to align data initiatives with organizational strategy.	[4, 17]
Open Data Improvement Practices	Iterative enhancement of open data policies with frameworks that promote quality and interoperability.	[22]
Iterative Quality Assessment Cycles	Implementation of regular quality evaluations and clear accountability structures.	[21]

These principles and practices collectively underscore the importance of clear roles, formalized policies, iterative improvement cycles, and adherence to recognized standards for advancing data governance maturity. They emphasize the value of aligning DG initiatives with overarching strategic objectives, leveraging agile methodologies, and incorporating advanced technologies to foster continuous improvement and seamless interoperability across data ecosystems. The recurring emergence of these themes across diverse studies reinforces their relevance as foundational elements for effective data governance.

3.4 Limitations and Future Work

This study's SLR presents limitations. Construct validity is influenced by search string definition and language restrictions (English, Portuguese, Spanish), potentially leading to omitted relevant studies. Internal validity threats, such as selection bias towards recent or highly cited works and manual data extraction, were mitigated by a multi-step screening process with documented criteria and a standardized extraction template. External validity is limited by the predominant academic focus, potentially underrepresenting real-world industry practices and practical DG implementation nuances.

Future work aims to address these limitations. To enhance external validity, research will include validation with data professionals and case studies from diverse organizational environments to bridge the theory-practice gap. Crucially, the development of a novel DG maturity model is proposed. Existing models [15–17] often lack conceptual clarity, contextual adaptability, and practical applicability, hindering their use, especially in early-stage organizations. This necessitates a more accessible and adaptable model. Finally, acknowledging the dynamic evolution of data technologies and regulations, future research will also explore how organizational cultural readiness, leadership engagement, and strategic alignment are an important enablers for DG initiative success, as their absence can diminish even comprehensive frameworks.

4 Conclusion

This study conducted a systematic analysis of DG maturity models, tools, and best practices, underscoring the fundamental role of DG in ensuring data quality, security, compliance, and strategic decision-making. The review identified prominent maturity models—such as the Stanford Maturity Model, Loshin's Data Quality Maturity Model, and the Global Big Data Maturity Model (GBDMM)—and highlighted critical practices, including policy formalization, iterative improvement cycles, ISO-aligned standards, and automation, as fundamental success factors.

This study contributes to the academic literature by synthesizing existing theoretical foundations and providing actionable guidance for organizations to systematically assess and enhance their data governance maturity.

References

1. Abraham, R., Schneider, J., vom Brocke, J.: Data governance: a conceptual framework, structured review, and research agenda. Int. J. Inf. Manage. **49**, 424–438 (2019). https://doi.org/10.1016/j.ijinfomgt.2019.07.008. https://www.sciencedirect.com/science/article/pii/S0268401219300787
2. Bena, Y.A., Ibrahim, R., Mahmood, J.: Current challenges of big data quality management in big data governance: a literature review. In: Saeed, F., Mohammed, F., Fazea, Y. (eds.) IRICT 2023. LNDECT, vol. 210, pp. 160–172. Springer, Cham (2024). https://doi.org/10.1007/978-3-031-59711-4_15

3. Bento, P., Neto, M., Côrte-Real, N.: How data governance frameworks can leverage data-driven decision making: a sustainable approach for data governance in organizations. In: 2022 17th Iberian Conference on Information Systems and Technologies (CISTI), pp. 1–5 (2022). https://doi.org/10.23919/CISTI54924.2022.9866895
4. Bernardo, B.M.V., Mamede, H.S., Barroso, J.M.P., dos Santos, V.M.P.D.: Data governance & quality management—innovation and breakthroughs across different fields. J. Innov. Knowl. **9**(4) (2024). https://doi.org/10.1016/j.jik.2024.100598
5. Boufassil, A., Bouhafer, F., Cherradi, M., El Haddadi, A.: Data catalog: approaches, trends, and future directions. In: 2023 17th International Conference on Signal-Image Technology & Internet-Based Systems (SITIS), pp. 369–376 (2023). https://doi.org/10.1109/SITIS61268.2023.00067
6. Cheng, G., Li, Y., Gao, Z., Liu, X.: Cloud data governance maturity model. In: 2017 8th IEEE International Conference on Software Engineering and Service Science (ICSESS), pp. 517–520. IEEE (2017). https://doi.org/10.1109/ICSESS.2017.8342968
7. Comuzzi, M., Patel, A.: How organisations leverage big data: a maturity model. Ind. Manag. Data Syst. **116**(8), 1468–1492 (2016). https://doi.org/10.1108/IMDS-12-2015-0495
8. Gökalp, M.O., Gökalp, E., Kayabay, K., Koçyiğit, A., Eren, P.E.: Data-driven manufacturing: an assessment model for data science maturity. J. Manuf. Syst. **60**, 527–546 (2021). https://doi.org/10.1016/j.jmsy.2021.07.011
9. Gökalp, M.O., Gökalp, E., Kayabay, K., Koçyiğit, A., Eren, P.E.: The development of the data science capability maturity model: a survey-based research. Online Inf. Rev. **46**(3), 547–567 (2022). https://doi.org/10.1108/OIR-10-2020-0469. cited by: 24
10. Harwanto, I.M., Hidayanto, A.N.: Data governance maturity assessment: a case study directorate general of corrections. In: 2022 International Conference on ICT for Smart Society (ICISS), pp. 01–06 (2022). https://doi.org/10.1109/ICISS55894.2022.9915243
11. International, D.: DAMA-DMBOK: Data Management Body of Knowledge, 2nd edn. Technics Publications, LLC, Denville (2017)
12. Iqbal, R., Yuda, P., Aditya, W., Hidayanto, A.N., Wuri Handayani, P., Harahap, N.C.: Master data management maturity assessment: case study of XYZ company. In: 2019 2nd International Conference on Applied Information Technology and Innovation (ICAITI), pp. 133–139. IEEE (2019). https://doi.org/10.1109/ICAITI48442.2019.8982123
13. Kitchenham, B., Charters, S.: Guidelines for performing systematic literature reviews in software engineering. Technical report. EBSE-2007-01, Technical report, EBSE Technical Report. EBSE-2007-01, UK (2007)
14. Kurniawan, D.H., Ruldeviyani, Y., Adrian, M.R., Handayani, S., Pohan, M.R., Rani Khairunnisa, T.: Data governance maturity assessment: A case study in it bureau of audit board. In: 2019 International Conference on Information Management and Technology (ICIMTech), vol. 1, pp. 629–634 (2019). https://doi.org/10.1109/ICIMTech.2019.8843742
15. Malacaria, S., Mauro, A.D., Greco, M., Grimaldi, M., Mignacca, B.: Toward the implementation of a consensual maturity model for big data in consumer goods companies. In: Knowledge Drivers for Resilience and Transformation, IFKAD 2022, pp. 2380–2401 (2022)
16. Mouhib, S., Anoun, H., Hassouni, L., Ridouani, M.: Analyzing the global big data maturity model domains for better adoption of big data projects. Int. J. Inf. Sci. Manage. **21**(4), 83–102 (2023). https://doi.org/10.22034/ijism.2023.1977940.0

17. Mouhib, S., Anoun, H., Ridouani, M., Hassouni, L.: Global big data maturity model and its corresponding assessment framework results. IAENG Int. J. Appl. Math. **53**(1) (2023). https://www.scopus.com/inward/record.uri?eid=2-s2. 0-85149629399. Cited by: 6

18. Nugraha, T.F., Wibowo, W.S., Genia, V., Fadhil, A., Ruldeviyani, Y.: A practical approach to enhance data quality management in government: case study of Indonesian customs and excise office. J. Inf. Syst. Eng. Bus. Intell. **10**(1), 51–69 (2024). https://doi.org/10.20473/jisebi.10.1.51-69

19. Permana, R.I., Suroso, J.S.: Data governance maturity assessment at PT. XYZ: case study of the data management division. In: 2018 International Conference on Information Management and Technology (ICIMTech), pp. 15–20 (2018). https://doi.org/10.1109/ICIMTech.2018.8528142

20. Qodarsih, N., Yudhoatmojo, S.B., Hidayanto, A.N.: Master data management maturity assessment: a case study in the supreme court of the republic of Indonesia. In: 2018 6th International Conference on Cyber and IT Service Management (CITSM), pp. 1–7. IEEE (2018). https://doi.org/10.1109/CITSM.2018.8674373

21. Spruit, M., Pietzka, K.: MD3M: the master data management maturity model. Comput. Hum. Behav. **51**, 1068–1076 (2015). https://doi.org/10.1016/j.chb.2014. 09.030

22. Visintin, L., Todesco, J.L., Álvaro Ostuni Gauthier, F.: Open data maturity model: a reference matrix for organizations. In: Proceedings of the 14th Seminar on Ontology Research in Brazil (ONTOBRAS 2021) and 5th Doctoral and Masters Consortium on Ontologies (WTDO 2021), vol. 3050, pp. 291–296 (2021). https://www. scopus.com/inward/record.uri?eid=2-s2.0-85123288169. Conference: ONTOBRAS 2021; Florianópolis, Brazil

23. Xue, C., Zhang, Y., Jia, D.: History, current situation and future challenges of china's geographic information data governance. In: 2022 29th International Conference on Geoinformatics, pp. 1–5 (2022). https://doi.org/10.1109/ Geoinformatics57846.2022.9963876

An Evaluation of Energy Consumption for Deep Learning-Based Privacy Preserving Record Linkage

Emmanouil Sokorelis[1], Alexandros Karakasidis[1(✉)] [iD],
Eftychios Protopapadakis[1] [iD], and Chairi Kiourt[2] [iD]

[1] Department of Applied Informatics, University of Macedonia, Thessaloniki, Greece
{ics21087,a.karakasidis,eftprot}@uom.edu.gr
[2] Archimedes/Athena Research Center, Marousi, Greece
chairiq@athenarc.gr

Abstract. Deep Neural Network models require significant energy resources, increasing the need to balance performance with sustainability, a field known as Green AI. This paper investigates the intersection of accuracy, privacy, and energy efficiency within Deep Learning-based, Privacy-Preserving Record Linkage. Through a series of experiments, the effects of key parameters on matching efficiency and energy consumption are explored, outlining the impact of encoding, noise addition, and deep learning model configurations. The findings indicate clear trade-offs between energy consumption, privacy, and matching performance.

Keywords: Green AI · Entity Resolution · Differential Privacy · Bloom Filters · Energy Efficiency

1 Introduction

Record linkage is a key task in domains like healthcare, e-commerce, and data management systems, often requiring large-scale data processing. It involves identifying the same real-world entity among distinct record-based datasets. As data is dirty, approximate matching techniques should be employed, making this problem non-trivial to address.

This task becomes even more complex when data privacy must be preserved in cases where the real-world entities that have to be identified are humans, so their personal information should be safeguarded. This leads to the Privacy-Preserving Record Linkage problem (PPRL) where, on top of the complexity of the classical record linkage problem, no other information should be leaked regarding the entities to be linked.

With the ubiquitous use of Artificial Intelligence in the last years, it has been more than expected that PPRL would have seen benefits from such techniques [14]. However, the benefits of applying deep learning to PPRL come with trade-offs. Artificial Intelligence-based techniques, especially those relying on

P. K. Chrysanthis et al. (Eds.): ADBIS 2025, CCIS 2676, pp. 171–180, 2026.
https://doi.org/10.1007/978-3-032-05727-3_17

Deep Learning, require significant computational resources, which in turn usually result in high energy consumption and environmental impact. For example, training large neural networks can emit carbon dioxide equivalent to the amount produced by five cars in their lifetime [19]. This has led to increasing interest in Green AI, which aims to balance environmental sustainability with AI performance.

To this end, we consider the latest development in PPRL [14] and study the convergence of matching performance, privacy, memory, and energy efficiency. Our contributions are the following. (1) We investigate the matching behavior of the said method in a variety of setups, considering different neural network architectures and Bloom filter encoding methods. (2) We examine the energy consumption incurred, the power required, the memory occupied, and the runtime elapsed for each of these setups. (3) We identify a sweet spot that provides a clear trade-off between matching performance, privacy, and energy consumption.

2   Related Work

A recent survey on PPRL may be found in [7]. Bloom filters combined with *n-grams* [14,17] is one of the most popular approaches to PPRL. The resulting bit vectors are ANDed for assessing matching status via a separate server referred to as the *Linkage Unit*. However, Bloom filter-based solutions have certain vulnerabilities, requiring additional hardening measures [6].

Beyond Bloom Filters, bit vectors and Locality Sensitive Hashing have been combined [18], at an increased computational cost [3]. A two-step hash method with quasi-identifiers converted into *n-grams* is proposed in [13], requiring, however, a Linkage Unit and a threshold to perform approximate matching. Differential privacy has also been used for PPRL, with current solutions focusing on categorical and numerical attributes [15], while this work focuses on strings.

Proposed cryptographic methods exhibit high computational costs. Homomorphic encryption is also susceptible to certain types of attacks [5]. Garbled circuits [2] need further investigation regarding size and reusability [16]. Fuzzy Vaults [12] relying on polynomial reconstruction through interpolation should also be further investigated in terms of time and matching performance.

There have been several recent developments in AI applied to Entity Resolution, which is a general form of the Record Linkage problem. Deep Learning techniques have been applied to Entity Resolution tasks [10,11]. More recently, the growing capabilities of Large Language Models (LLMs) have been leveraged to enhance performance [9]. However, to the best of our knowledge, no existing approach utilizes LLMs in a privacy-preserving manner. Furthermore, there has been no prior effort to evaluate the energy efficiency of deep learning methods specifically within the context of Privacy-Preserving Record Linkage (PPRL).

3   Methodology

In this section, we present the method we have evaluated and the approach we have followed to measure energy consumption.

3.1 Deep Learning-Based Privacy Preserving Entity Resolution

Let us begin by presenting the Deep Learning-based PPRL method [14] implemented[1] in our evaluation. The presence of two dataholders is assumed, while a Linkage Unit (LU) is used to facilitate the process, which begins with records from both datasets being encoded into Bloom filters. To enhance privacy, differential privacy is employed to introduce controlled noise into the filters. Next, all possible encoded record pairs are generated, and their similarity and difference features are computed. These extracted features serve as inputs for a deep learning model, which is trained and evaluated to optimize matching accuracy while preserving privacy. Next, we describe these steps in more detail.

First, each record is tokenized into bigrams and added to its own Bloom filter, each of which relies on two parameters: capacity and error rate. Capacity refers to the maximum number of elements the Bloom filter can store. The error rate, or false positive rate, represents the probability that an element will be incorrectly identified as part of the set. These two parameters directly impact the size of the Bloom filter. Higher capacities and lower error rates lead to increased Bloom filter sizes while maintaining lower error rates. Afterwards, noise is introduced into the filters, using Randomized Response, a fundamental method in differential privacy. This is implemented through bit flipping, a process in which certain bits within the Bloom filter are randomly altered to obscure the original data. However, excessive noise may degrade Bloom filter utility due to excessive distortion of the stored representation.

Next, pairs of records are generated, and features are computed for comparison. This process involves calculating the Cartesian product of the encoded datasets, producing all possible pairs of records. While the Cartesian product is a straightforward approach, it is computationally expensive, especially for large datasets. However, it remains necessary in the absence of blocking techniques that could reduce the number of candidate pairs. Once candidate pairs are generated, the next task is to calculate features that quantify similarities and differences between the records. This process uses three key metrics: Jaccard similarity, Dice similarity, and Hamming distance. The training phase involves feeding these computed features into a neural network, enabling the model to assess similar and dissimilar record pairs.

3.2 Energy Consumption Evaluation

A device's energy consumption can be assessed in two ways. The first one is externally, by using a specialized plug with power and energy measuring capabilities, which measures the overall energy consumption of a device. In the case of a computer, it will measure everything, also including all running processes. This is not particularly useful in this case, since we need to isolate the consumption of the operations related to the PPRL process we investigate.

The second approach, which we adopted, is using internal system interfaces and software-based estimations. We relied on Intel's Running Average Power

---

[1] Available at: https://github.com/manossokorelis/Greener-AI-PPRL.

Limit (RAPL) interface which provides a mechanism for estimating and controlling the power consumption of various processor components, such as the CPU package, DRAM, and uncore domains. RAPL does not measure energy through physical sensors but relies on model-based estimations, which are fast, low-overhead, and accessible via software using Model Specific Registers (MSRs). Despite its indirect nature, RAPL has been shown to offer reasonably accurate energy estimations for profiling and power-aware computing research [4].

To access energy and power metrics, we used Perun [8], a Python package that has access to the RAPL interface. We used Perun as a command-line tool, performing 5 executions for each test case we considered, letting it calculate and average the measurements it retrieved. As this has been a CPU-only assessment, we measured CPU time, power, and energy. Also, we considered memory occupation.

3.3   Experimental Setup

For this evaluation, we used two samples from the North Carolina's voters dataset[2] consisting of 10,000 records with 25% overlap, meaning that 2,500 records between these two datasets are identical. In order to simulate real-world conditions where data contain errors, the second sample was corrupted, using the German Record Linkage Center's data corrupter [1], with one edit distance operation at a random field per row, so that a join operation using these quasi-identifiers yields an empty result set. We have used five attributes for matching: 'last name', 'first name', 'middle name', 'address', and 'suburb'. We assume that the attributes chosen comprise a candidate key.

Matching performance is evaluated through Precision and Recall. Precision is defined as the fraction of the relevant elements among the retrieved elements, while Recall is defined as the fraction of the retrieved relevant elements divided by the total relevant elements: $Precision = \frac{TP}{TP+FP}$ and $Recall = \frac{TP}{TP+FN}$.

Regarding the Bloom filter, two capacities have been considered, of 100 and 200 elements. The error rate, representing the false positive rate of the Bloom filter, was either 0.1 or 0.01, with a higher error rate increasing the likelihood of false positives, while a lower error rate increases the filter size. The flip probability, which introduces privacy-preserving noise into the Bloom filter, had two values: 0, and 0.01. A flip probability of 0 means no noise is introduced. Higher probabilities increase privacy but may potentially affect matching performance.

The Neural Network architecture followed a sequential design varying between 1, 2 and 3 hidden layers, tested with 4, 8, and 16 neurons per layer, followed by a batch normalization layer and a ReLU activation function. The final layer did not include a sigmoid activation function, as the loss function used was Binary Cross Entropy with Logits. This loss function operates directly on the raw logits, eliminating the need for a separate sigmoid layer and improving computational efficiency. Before training the model, the features were split into training, validation, and test sets with a ratio of 60:20:20. The results of these

---

[2] Available at: https://dl.ncsbe.gov/?prefix=data.

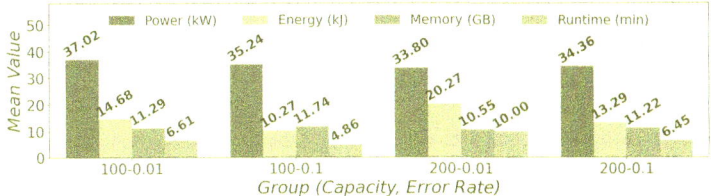

Fig. 1. Data Operations: Power, Energy, Memory and Runtime.

experiments helped examine the trade-offs between matching performance, privacy preservation, execution times, and energy efficiency. Training epochs varied between 5, 10, and 20. The batch size was 8192. To address class imbalance in the dataset, class weights were incorporated into the loss function, giving more importance to the minority class (matching pairs) during training. The model was trained using the Adam optimizer with a learning rate of 0.001, selected for its adaptability and efficient gradient computation.

4   Experimental Results

In this section, we present the results of our empirical evaluation, focusing on two phases. First, the energy consumption during feature generation, and second, the training of the Neural Network and its validation. In this second phase, we discuss results for matching performance and energy consumption.

4.1   Feature Generation Energy Consumption

Evaluation results of power, energy, memory and execution time required for creating the dataset are illustrated in Fig. 1. The X-axis stands for the encoding parameters of the specific dataset, while the Y-axis stands for the mean of the measured values. Pairs in each group represent the Bloom filter capacity and error rate, respectively. It is easy to discern that the average power required does not fluctuate significantly. Furthermore, increasing the size of the BF (capacity, error rate) affects energy consumption, especially when increasing capacity. Execution time is increased as well when increasing Bloom filter size, as more hash functions are used and more computations are performed. Memory required for the feature generation is not affected.

4.2   Matching Performance

For matching performance, Recall in all experiments was equal to 1. As such, we only discuss Precision results, which are illustrated in Figs. 2(a)–2(f). In all figures, the X-axis represents the number of training epochs, while the Y-axis represents average Precision. Each line corresponds to a Bloom filter size and neuron count per hidden layer (all layers have the same number of neurons). Each

---

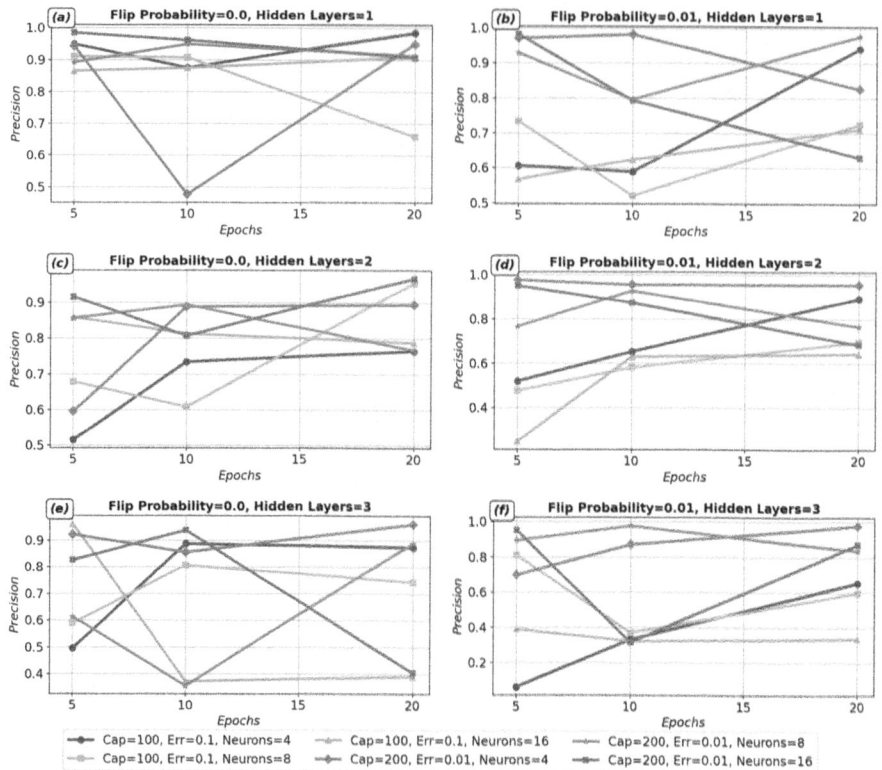

Fig. 2. Precision results across different flip probabilities and hidden layer setups.

row of figures shows how Precision varies across Neural Network architectures with 1, 2, and 3 hidden layers, respectively. Figures in the left column illustrate results without any noise added to the Bloom filters, while the ones on the right illustrate results with noise, with a flip probability equal to 0.01.

Let us begin with the case of using one hidden layer in the Neural Network. Figure 2(a) illustrates the case where no noise is added to the Bloom filter before training. Figure 2(b) shows the situation when noise has been added. Large Bloom filters (Capacity = 200, Error rate = 0.01) seem to be more resilient to noise addition than the small ones (Capacity = 100, Error rate = 0.1). When no noise is added, using 16 neurons offers good behavior for all cases, especially with a small number of training epochs. When noise is added, considering the large Bloom filters, using a small number of training epochs is a safe choice. Increasing the number of training epochs does not necessarily improve matching performance.

Next, there is the case of using Neural Networks with two hidden layers, considering again Bloom filters without (Fig. 2(c)) and with noise (Fig. 2(d)). For the first case, in most setups, increasing the number of epochs leads to an increase in Precision, with the same happening as the number of neurons rises for the small Bloom filters. For large Bloom filters, the number of neurons does not seem to affect Precision. Moving to the set of experiments where noise has been added to the Bloom filters, (Fig. 2(d)) it is easy to discern two groups of lines: those representing the small Bloom filters at the bottom and those for the large ones at the top. Again, large Bloom filters behave better when noise is added. Also, increasing the number of training epochs for the small Bloom filters leads to increased precision in general, something that does not hold in the case of the large Bloom filters. For large Bloom filters, smaller counts of hidden neurons per layer seem to favor performance.

Finally, there is the configuration of three hidden layers. Here, in the case of no noise (Fig. 2(e)), models with fewer neurons behave better, especially as the number of training epochs increases. More complex models have worse performance, especially when the number of neurons per hidden layer increases. Adding noise now (Fig. 2(f)), models using large Bloom filters again outperform those using small ones. Increasing the number of training epochs is beneficial for models with 4 neurons per hidden layer.

Overall, we may conclude that using large BFs and many training epochs yields good results for 4 neurons per hidden layer and 5 or ten epochs for one or two layers, or 20 training epochs for two and three hidden layers. Another alternative is to use 8 neurons in two or three hidden layers and a medium number of epochs.

4.3   Power, Energy, Time and Memory

We now turn to the energy consumption analysis. Runtime power usage, energy consumption, execution time, and memory occupation results for 4 neurons per hidden layer, as they exhibited overall superior performance in the previous set of experiments, are illustrated in Fig. 3. Each row corresponds to a different number of hidden layers, while the first column shows results without noise in the Bloom filter, and the second column shows results with noise.

First, it is easy to discern that flip probability does not have an impact on the power required, energy consumption, runtime, or memory of Neural Network training and evaluation (Figs. 3(a), 3(c), 3(e)). Thus we will mainly focus on the case of using noise (Figs. 3(b), 3(d), 3(f)).

Energy consumption increases with the complexity of the Neural Network, and grows in a linear manner with the number of training epochs across all figures. This reflects the cumulative computational workload from repeated training iterations. Required power increases when increasing the number of epochs and the number of hidden layers. Memory occupation is not significantly affected, while execution time increases with the complexity of the Neural Network.

An interesting discussion topic comes up when comparing energy-wise the addition of a hidden layer with 4 neurons against the use of more epochs for training. In this case, the following occurs. Doubling the number of epochs from 5 to 10 increases energy consumption by 26–27% for 1 hidden layer in all cases, while it reaches 37% for three hidden layers. Increasing the number of training epochs from 10 to 20 leads to an increase of 36–37% for 1 hidden layer and at

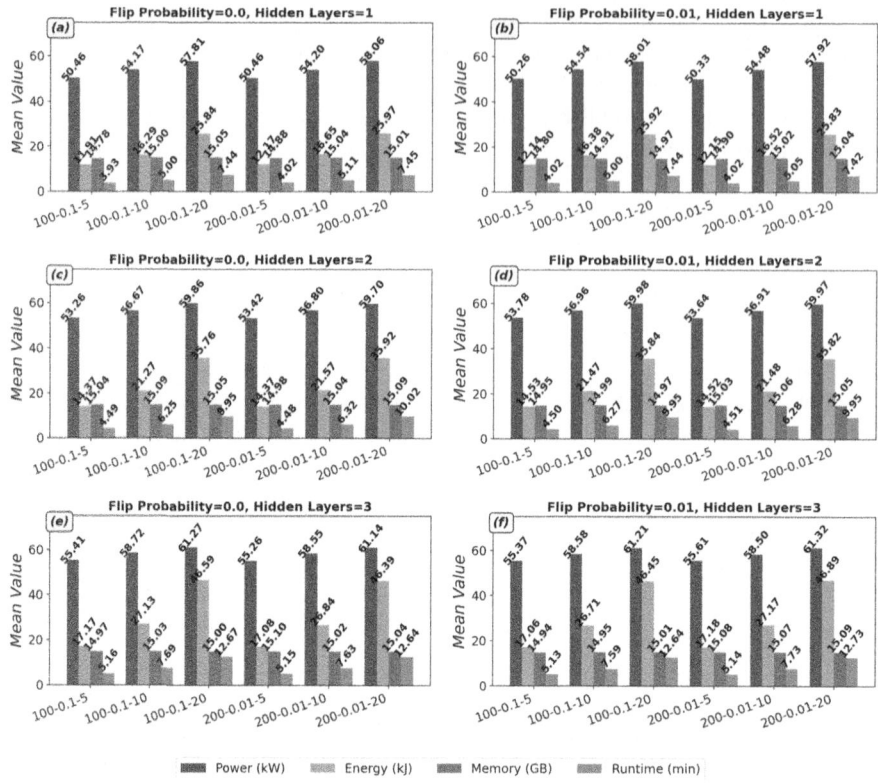

Fig. 3. NN Power, Energy, Time and Memory for training and evaluation (4 neurons/ layer).

42% for 3 hidden layers. On the other hand, keeping the number of epochs steady and increasing the number of layers does not incur this cost. In particular, going from 1 hidden layer to 2 for 10 training epochs increases energy consumption by 23%, while moving to 3 layers increases energy by an additional 21% on average. As such, it appears to be more energy-efficient to add an additional hidden layer of neurons than to double training epochs.

A clear trade-off emerges between maximizing precision and minimizing energy consumption. This balance is summarized in Table 1. This occurs when large Bloom filters with small error rates are used. The highest precision was achieved by a model with 1 hidden layer trained for 10 epochs and a capacity of 200, reaching a precision of 0.9819 with an energy cost of 16.52 kJ. However, the best trade-off between precision and energy use was provided by a 2-hidden-layer model trained for only 5 epochs, which attained 0.9762 precision while consuming just 14.52 kJ. The most energy-efficient configuration, maintaining, at the same time, a strong overall precision of 0.9709, was a simpler model with 1 hidden layer and 5 training epochs, which consumed only 12.15 kJ. These

Table 1. Precision and Energy Tradeoffs for NN Models with 4 Neurons/Layer.

Precision	Energy (kJ)	Hidden Layers	Epochs	Capacity
0.9819	16.52	1	10	200
0.9762	14.52	2	5	200
0.9709	12.15	1	5	200

findings suggest that moderate model depth and shorter training can achieve performance-to-efficiency balance without substantially sacrificing accuracy.

5   Conclusions

In this paper, we evaluated the performance of a Deep Learning-based method for Privacy Preserving Record Linkage with respect to matching performance and energy consumption. Our findings indicate that there is a sweet spot that comprises a good tradeoff between energy consumption and matching performance. Our next steps are directed towards exploring the behavior both energy-wise and performance-wise with more datasets employing additional Machine Learning-based methods.

Acknowledgment. This work has been partially supported by project MIS 5154714 of the National Recovery and Resilience Plan Greece 2.0 funded by the European Union under the NextGenerationEU Program.

Disclosure of Interests. The authors have no competing interests to declare that are relevant to the content of this article.

References

1. Bachteler, T., Reiher, J.: A test data generator for evaluating record linkage methods. Technical report, German RLC Work. Paper No. wp-grlc-2012-01 (2012)
2. Chen, F., et al.: Perfectly secure and efficient two-party electronic-health-record linkage. IEEE Internet Comput. **22**(2), 32–41 (2018)
3. Christen, P., Ranbaduge, T., Schnell, R.: Linking Sensitive Data - Methods and Techniques for Practical Privacy-Preserving Information Sharing. Springer, Cham (2020)
4. David, H., Gorbatov, E., Hanebutte, U., Khanna, R., Le, C.: RAPL: memory power estimation and capping. In: Proceedings of the 16th ACM/IEEE International Symposium on Low Power Electronics and Design (ISLPED), pp. 189–194. ACM (2010)
5. Essex, A.: Secure approximate string matching for privacy-preserving record linkage. IEEE Trans. Inf. Forensics Secur. **14**(10), 2623–2632 (2019)
6. Franke, M., Sehili, Z., Rohde, F., Rahm, E.: Evaluation of hardening techniques for privacy-preserving record linkage. In: 24th International Conference on Extending Database Technology, pp. 289–300. OpenProceedings.org (2021)

7. Gkoulalas-Divanis, A., Vatsalan, D., Karapiperis, D., Kantarcioglu, M.: Modern privacy-preserving record linkage techniques: an overview. IEEE Trans. Inf. Forensics Secur. **16**, 4966–4987 (2021)
8. Gutiérrez Hermosillo Muriedas, J.P., Flügel, K., Debus, C., Obermaier, H., Streit, A., Götz, M.: perun: benchmarking energy consumption of high-performance computing applications. In: Cano, J., Dikaiakos, M.D., Papadopoulos, G.A., Pericàs, M., Sakellariou, R. (eds.) Euro-Par 2023: Parallel Processing, pp. 17–31. Springer, Cham (2023)
9. Li, H., et al.: On leveraging large language models for enhancing entity resolution. arXiv preprint arXiv:2401.03426 (2024)
10. Li, X., Talburt, J.R., Li, T., Liu, X.: When entity resolution meets deep learning, is similarity measure necessary? In: Arabnia, H.R., Ferens, K., de la Fuente, D., Kozerenko, E.B., Olivas Varela, J.A., Tinetti, F.G. (eds.) Advances in Artificial Intelligence and Applied Cognitive Computing. TCSCI, pp. 127–140. Springer, Cham (2021). https://doi.org/10.1007/978-3-030-70296-0_10
11. Lv, Y., Qi, L., Huo, J., Wang, H., Gao, Y.: Joint multi-field Siamese recurrent neural network for entity resolution. In: Geng, X., Kang, B.-H. (eds.) PRICAI 2018. LNCS (LNAI), vol. 11013, pp. 482–490. Springer, Cham (2018). https://doi.org/10.1007/978-3-319-97310-4_55
12. Mullaymeri, X., Karakasidis, A.: Using fuzzy vaults for privacy preserving record linkage. In: The 23rd International Workshop on Design, Optimization, Languages and Analytical Processing of Big Data. CEUR Workshop Proceedings, vol. 2840, pp. 101–110. CEUR-WS.org (2021)
13. Ranbaduge, T., Christen, P., Schnell, R.: Secure and accurate two-step hash encoding for privacy-preserving record linkage. In: Lauw, H.W., Wong, R.C.-W., Ntoulas, A., Lim, E.-P., Ng, S.-K., Pan, S.J. (eds.) PAKDD 2020. LNCS (LNAI), vol. 12085, pp. 139–151. Springer, Cham (2020). https://doi.org/10.1007/978-3-030-47436-2_11
14. Ranbaduge, T., Vatsalan, D., Ding, M.: Privacy-preserving deep learning based record linkage. IEEE Trans. Knowl. Data Eng. **36**(11), 6839–6850 (2023)
15. Rao, F., Cao, J., Bertino, E., Kantarcioglu, M.: Hybrid private record linkage: separating differentially private synopses from matching records. ACM Trans. Priv. Secur. **22**(3), 15:1–15:36 (2019)
16. Saleem, A., Khan, A., Shahid, F., Alam, M., Khan, M.K.: Recent advancements in garbled computing: how far have we come towards achieving secure, efficient and reusable garbled circuits. J. Netw. Comput. Appl. **108**, 1–19 (2018)
17. Schnell, R., Bachteler, T., Reiher, J.: Privacy-preserving record linkage using bloom filters. BMC Med. Inform. Decis. Mak. **9**, 41 (2009)
18. Smith, D.: Secure pseudonymisation for privacy-preserving probabilistic record linkage. J. Inf. Secur. Appl. **34**, 271–279 (2017)
19. Strubell, E., Ganesh, A., McCallum, A.: Energy and policy considerations for deep learning in NLP (2019). https://arxiv.org/abs/1906.02243

Doctoral Consortium School Invited Talks

Machine Learning for Query Optimization in Knowledge Graphs

Maribel Acosta[(✉)] [ID]

Technical University of Munich, Munich, Germany
`maribel.acosta@tum.de`

Abstract. Query optimization is a main component of graph databases and triple stores that host knowledge graphs (KGs). Traditional symbolic optimizers rely on heuristics and cost models that are prone to inaccuracies, leading to suboptimal execution plans. Recent advances in machine learning (ML) provide promising solutions to address these limitations by learning from data and queries to enhance cardinality estimation, cost prediction, and plan enumeration. This work surveys the emerging landscape of ML-based query optimization over KGs, including learned, neuro-symbolic, and (fully) neural approaches. We discuss the architecture and trade-offs of these systems, present preliminary results, and highlight open challenges for future research.

Keywords: query optimization · neural networks · neuro-symbolic AI

1 Introduction

Knowledge Graphs (KGs) have become a core component of modern AI systems, enabling applications such as semantic search, recommender systems, and retrieval-augmented generation (RAG). Querying KGs efficiently is critical for these applications, yet query optimization over KGs remains an open problem. Query engines must evaluate complex graph pattern queries, typically expressed in SPARQL (for RDF) or Cypher (for property graphs), but traditional optimization techniques often fail to cope with the structural properties of KGs.

Conventional query optimizers rely on symbolic methods, including rule-based rewriting, heuristic strategies, and cost-based plan selection. These methods assume the availability of accurate dataset statistics and predictable query behavior, which seldom hold in large-scale, heterogeneous graph data. Consequently, suboptimal plans frequently arise, leading to execution times that vary by orders of magnitude. To address these shortcomings, recent works explore the application of machine learning (ML) to augment or replace components of the query optimizer. ML models can generalize from past queries or observed data patterns to improve critical tasks such as cardinality estimation, cost modeling, and plan enumeration. Initial results suggest that learned components can outperform their symbolic counterparts in challenging scenarios.

© The Author(s), under exclusive license to Springer Nature Switzerland AG 2026
P. K. Chrysanthis et al. (Eds.): ADBIS 2025, CCIS 2676, pp. 183–191, 2026.
https://doi.org/10.1007/978-3-032-05727-3_18

In this work, we explore the landscape of ML-enhanced query optimization over KGs. We provide a systematized view of symbolic, neural, and neuro-symbolic approaches, drawing on recent developments from both the database and AI communities. We also provide a roadmap of advantages and challenges for deploying these approaches in real-world query engines. The main points of this work are:

- An overview of learned components for query optimization over KGs.
- An introduction to neuro-symbolic and neural query optimizers.
- Preliminary results for neuro-symbolic and neural optimizers over KGs.
- A discussion of open challenges, including generalization across datasets, adaptability for evolving knowledge, and interpretability.

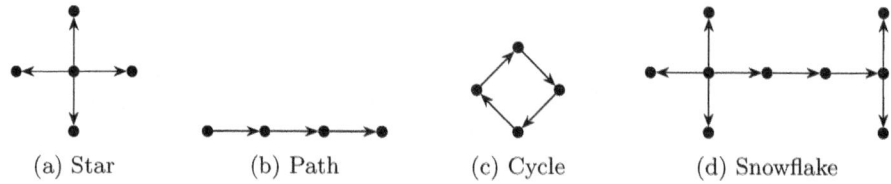

(a) Star   (b) Path   (c) Cycle   (d) Snowflake

Fig. 1. Examples of different graph query shapes. Complex queries (e.g., snowflakes) are created by combining several query shapes (e.g., stars and paths).

2   Preliminaries: Query Optimization over KGs

Query optimization is the problem of selecting an efficient execution strategy from a set of logically equivalent query plans. A *query plan* is a tree [7], where the leaves correspond to the access methods – in the case of KGs, these are graph patterns that can match nodes or edges over the graph – and the intermediate nodes correspond to the query operators—e.g., joins, unions, optionals, filters.

In traditional systems, query optimizers are built from three main symbolic components. (i) *Cardinality estimators* predict the number of results produced by subqueries. (ii) *Cost models* estimate the computational cost of each plan, using the cardinality estimations, i.e., the larger the number of results produced by a subplan, the higher its cost. (iii) *Planners* traverse the space of equivalent plans to select the most efficient one. These components depend on statistical summaries and assumptions about data independence and uniformity.

While such assumptions are often effective in relational databases with well-defined schemas, they frequently break down in graph databases and triple stores. Graph data is typically characterized by schema heterogeneity, where entities vary widely in their attributes, leading to semi-structured or even schema-less datasets. Furthermore, the high connectedness of graph data means that queries often involve complex joins traversing intricate topologies, and query performance is strongly influenced by the specific query shape (Fig. 1). Additionally, real-world graphs exhibit skewed correlations, following power-law distributions that invalidate the assumption of attribute independence commonly used in traditional optimization. These factors contribute to inaccurate estimations, poor plan selection, and long query execution times.

Therefore, optimizing queries over KG requires approaches that are tailored to the graph-specific characteristics and challenges. Recent approaches model query optimization as a learning problem, where components of the optimizer are replaced with ML models trained on historical query workloads or sampled data. This paradigm shift has led to a spectrum of hybrid systems, from symbolic optimizers augmented with learned estimators to fully neural architectures. In the following sections, we examine each of these directions in detail.

3 Query Optimization with Machine Learning

3.1 Learned Cardinality Estimation

In cost-based optimization, the cardinality estimator is a fundamental component and has been the primary focus of ML research in this domain. In KGs, symbolic estimators often fail due to data sparsity, irregularity, and skewed data distributions. Machine learning offers an alternative by learning mappings from query structures to estimated result sizes. For this, two main paradigms have emerged: query-driven and data-driven approaches.

Query-driven approaches treat cardinality estimation as a supervised regression task. Models such as Multi-Layer Perceptrons (MLP) and Graph Neural Networks (GNNs) are trained on queries paired with observed cardinalities. These methods rely on encodings of query graphs (e.g., adjacency tensors) and node embeddings to capture structural and semantic information. Examples of these approaches include SPACE [3], the supervised LMKG variant [4], GNCE [11], and LSS [16]. The main advantage of query-driven models is their ability to capture complex correlations between query structure and data distribution without needing access to the full dataset. However, their performance typically depends on the diversity and representativeness of the training query workload. These models may struggle with generalization to unseen query shapes or data updates, requiring retraining or fine-tuning to adapt to evolving KGs.

Data-driven approaches model the underlying data distribution directly, using techniques like autoregressive models. These approaches sample subgraphs (e.g., stars, paths) to learn joint probability distributions over entities and relations, which are then used to estimate cardinalities. This paradigm has been mainly applied in relational databases using sum-product networks (SPNs) [8]. For knowledge graphs and graphs, a few approaches have been proposed under this paradigm, such as the unsupervised LMKG variant [4], and the learned sketches for subgraph counting?, which combines GNNs with sketching techniques for approximate frequency estimation. An advantage of data-driven models is their ability to capture deep correlations inherent in the data without relying on query workloads, making them potentially more generalizable across query patterns. However, these methods face scalability challenges when applied to large and highly connected Kgs, as modeling the complete data distribution may require large computational resources. Additionally, most existing data-driven models are tailored to simple query structures and may not readily extend to arbitrary graph queries.

3.2   Learned Cost Models

Beyond estimating cardinalities, some learned systems aim to predict query plan latency directly. These learned cost models bypass symbolic cost formulas by training regression models on encodings of query plans paired with observed runtimes [9,12]. This approach eliminates the need for manually calibrated cost functions, which are often inaccurate across diverse data and hardware settings.

Feature-based models such as those proposed by Hasan and Gandon [5] and Zhang et al. [15] rely on handcrafted features that characterize the query's algebraic structure—such as the number of triple patterns, the types and counts of joins (e.g., star, chain, snowflake), the depth and width of the query plan tree, and selectivity estimates from available statistics. These features are input into traditional machine learning algorithms like decision trees, support vector machines, or ensemble models to regress the predicted execution time.

More recently, PlanRGCN [10] introduced a *structure-aware* and representation learning-based approach that models SPARQL query plans as directed acyclic graphs, where nodes represent query operators and edges capture data dependencies. PlanRGCN employs Relational Graph Convolutional Networks (R-GCNs) to encode both the structural topology of the query plan and semantic information derived from the underlying RDF graph, such as predicate types and entity features. By embedding query plans into latent spaces that preserve structural and relational characteristics, PlanRGCN can more accurately predict query runtimes, outperforming feature-based baselines.

In general, feature-based models depend heavily on the choice and engineering of input features, which may not capture deeper interactions in complex queries. In contrast, structure-aware models mitigate this by learning representations directly from the plan topology, but they require significant computational resources for training and may struggle with generalization when faced with previously unseen query patterns or dataset characteristics.

3.3   Learned Plan Enumeration

Plan enumeration is traditionally guided by heuristics or cost-based strategies. ML can enhance this step by learning to navigate the plan space efficiently.

Reinforcement learning (RL) has been used to train agents that incrementally construct query plans by choosing the next join operation. States encode partial plans; actions correspond to join candidates; rewards reflect cost or latency. To date, ReJOOSp [13] is one of the few efforts applying RL to SPARQL join ordering. It encodes SPARQL queries using matrix representations and trains an RL agent to decide the join order that minimizes runtime.

Neural planners often rely on plan representations such as adjacency matrices, graph-based encodings, or tree structures, which are then processed using MLPs, Tree-LSTMs, or GNNs. Yet, most advanced neural planner techniques like the one proposed by Zu et al. [14] are predominantly studied in the relational setting and have not been fully adapted to the particular characteristics of KGs.

Overall, these advances suggest that KG neural planners are an open research problem, particularly in designing models to handle KG-specific challenges such as heterogeneous schemas, different query shapes, and skewed distributions.

Table 1. Comparison of neuro-symbolic and neural optimizers

Aspect	Neuro-Symbolic	Neural
Architecture	Symbolic+learned components	End-to-end neural architecture
Components	Learned cardinality, cost model, or planner (in isolation)	Entire pipeline: plan representation, cost, plan traversal
Plan Representation	Symbolic (trees, algebraic rules)	Differentiable (e.g., soft adjacency matrices, diff. cost model)
Learning Objective	Component-wise accuracy (e.g., q-error, cost prediction)	Global plan quality (e.g., cost surrogate loss)
Training	Component-specific labeled data	End-to-end with feedback
Interpretability	Moderate to high	Low
Fallback to Symbolic	Easy to integrate and revert	Difficult (no symbolic interface)

4   Building Neuro-symbolic and Neural Query Optimizers

Neuro-symbolic and neural optimizers represent two distinct approaches to integrating machine learning into query optimization [1]. Table 1 summarizes the main differences between these approaches. In the following, we provide further details on these optimizer architectures and present preliminary results of their application to query optimization over KGs.

4.1   Neuro-symbolic Optimizers

Neuro-symbolic query optimizers integrate one or more learned components (cf. Sect. 3) into symbolic architectures. The goal here is to leverage the strengths of both paradigms: the reliability and explainability of symbolic logic with the flexibility and predictive power of neural models. These hybrid systems maintain symbolic control logic, allowing them to preserve plan correctness.

An Exemplary Neuro-symbolic Optimizer. We implemented a neurosymbolic optimizer by extending the nLDE engine [2], an adaptive query processor designed for querying RDF KGs. In its symbolic form, nLDE relies on cost-based optimization techniques that use traditional cardinality estimators to guide the planner.

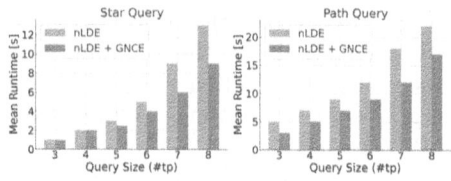

Fig. 2. Performance of a neuro-symbolic and a symbolic optimizer on SWDF [11].

To build a neuro-symbolic optimizer within nLDE, we replace the symbolic cardinality estimator with GNCE [11]. GNCE leverages GNNs and KG embeddings to predict the cardinalities of subqueries, capturing both the structural properties of the query graph and the semantics of entities and predicates. The predicted cardinalities from GNCE feed into the cost model to guide the planner. Preliminary results are shown in Fig. 2. The neuro-symbolic optimizer (nLDE+GNCE) outperforms the symbolic optimizer (nLDE) in large starshaped queries (more than 5 triple patterns) and produces better plans in all path queries.

4.2   Neural Optimizers

Neural optimizers represent a more radical departure. They aim to learn the entire optimization pipeline without explicit symbolic reasoning. Instead, they treat plan selection as a continuous optimization problem, where: (i) plans are represented using differentiable structures, such as *soft adjacency matrices,* (ii) neural networks predict plan quality using surrogate loss functions (e.g., estimated cost), and (iii) gradient-based methods traverse the plan space by updating the plan representation iteratively.

Neural optimizers have several advantages. First, they enable a unified, end-to-end learning process that does not rely on rules or heuristics, potentially reducing the need for manual system tuning. Second, by operating directly in a continuous or differentiable plan space, neural optimizers can explore novel plan forms that are not easily reachable through algebraic transformations. Finally, because the learning signal can incorporate direct feedback from query performance, such as cost, neural optimizers can implicitly capture latent data and plan characteristics that are difficult to model symbolically.

A Preliminary Neural Optimizer. We are currently developing a neural optimizer and have encountered several challenges in the process. One key difficulty is designing representations that can effectively encode query plans in a continuous space, as illustrated in Fig. 3a. Another challenge involves ensuring that the soft representations produced by the neural model correspond to valid execution plans, since unconstrained outputs may violate logical or structural requirements (Fig. 3b). Additionally, crafting effective and stable training objectives has proven complex, as the loss functions must reliably guide the optimizer toward high-quality plans while navigating noisy and often sparse feedback from

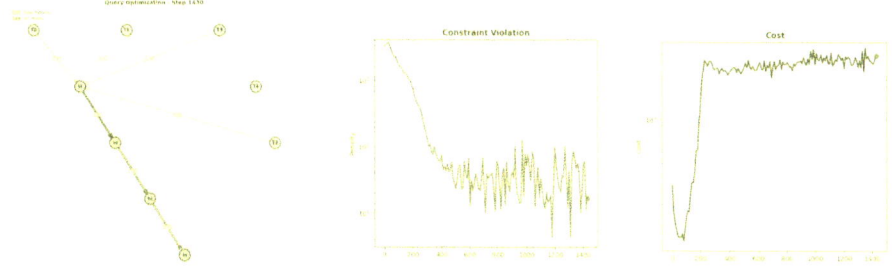

(a) Soft plan representation   (b) Ensuring valid plans   (c) Optimization objective

Fig. 3. Training a neural optimizer. (a) Access methods (triple patterns) and query operators (joins) are connected in a soft adjacency matrix with real numbers. (b) Minimization of violations over epochs to ensure valid plans. (c) Convergence of the learning objective (plan log cost) over epochs to find good plans.

execution metrics (Fig. 3c). Our preliminary results over star and path queries over the LUBM KG suggest that neural optimizers can find better plans for large queries in comparison to optimizers that traverse the space of plans using greedy strategies or Dynamic Programming.

5   Challenges and Future Directions

Despite promising advances, the integration of ML into query optimization for (graph) databases introduces several open challenges:

Generalization and Robustness: Learned components often perform well on training distributions but may struggle to generalize to new query shapes, unseen entities, or dynamic datasets. Therefore, robustness techniques [6,7] can be extended to cope with the unreliable predictions from learned components.

Training Data and Scalability: Acquiring representative training data, especially for large KGs, remains resource-intensive. Efficient sampling, synthetic query generation, and transfer learning could mitigate this issue.

Interpretability: Neural models are often opaque boxes, making it difficult for developers and administrators to debug the optimizer's decisions. Explainability techniques and hybrid neuro-symbolic approaches may improve interpretability.

Adaptive and Online Learning: Static models fail to capture changes in data distributions, hardware environments, or workload patterns. Adaptive mechanisms like eddies [2] that change the plan during execution can address this, though they introduce runtime overhead.

Uncertainty Quantification: Techniques that quantify prediction uncertainty or confidence can be used to trigger fallbacks to symbolic components or adjust plan enumeration strategies.

6 Conclusion

Machine learning is reshaping the landscape of query optimization in graph databases and triple stores. Current experimental results over diverse KGs show that ML provides effective solutions to overcome the limitations of traditional symbolic techniques. While neuro-symbolic architectures offer a practical compromise between performance and reliability, fully neural approaches suggest a future in which entire query plans are synthesized through end-to-end learning.

However, realizing this vision requires addressing open challenges related to generalization, interpretability, adaptivity, and scalability. Continued research at the intersection of databases and machine learning will be critical to advance these next-generation optimizers and ensure their practical applicability in complex, real-world queries and knowledge graphs.

Acknowledgments. The author thanks Tim Schwabe and Roman Mishchuk for the preparation of the experimental results presented in this paper.

References

1. Acosta, M., Qin, C., Schwabe, T.: Neuro-symbolic query optimization in knowledge graphs. In: Handbook on Neurosymbolic AI and Knowledge Graphs, Frontiers in Artificial Intelligence and Applications, vol. 400, pp. 624–643. IOS Press (2025)
2. Acosta, M., Vidal, M.-E.: Networks of linked data eddies: an adaptive web query processing engine for rdf data. In: Arenas, M., et al. (eds.) ISWC 2015. LNCS, vol. 9366, pp. 111–127. Springer, Cham (2015). https://doi.org/10.1007/978-3-319-25007-6_7
3. Aytimur, M., Chondrogiannis, T., Grossniklaus, M.: Space: cardinality estimation for path queries using cardinality-aware sequence-based learning. Proc. ACM Manage. Data **3**(3), 1–26 (2025)
4. Davitkova, A., Gjurovski, D., Michel, S.: LMKG: Learned Models for Cardinality Estimation in Knowledge Graphs (2022)
5. Hasan, R., Gandon, F.: A machine learning approach to SPARQL query performance prediction. In: 2014 IEEE/WIC/ACM International Joint Conferences on Web Intelligence (WI) and Intelligent Agent Technologies (IAT), vol. 1, pp. 266–273. IEEE (2014)
6. Heling, L., Acosta, M.: Cost-and robustness-based query optimization for linked data fragments. In: International Semantic Web Conference, pp. 238–257 (2020)
7. Heling, L., Acosta, M.: Robust query processing for linked data fragments. Semant. Web **13**(4), 623–657 (2022)
8. Hilprecht, B., Schmidt, A., Kulessa, M., Molina, A., Kersting, K., Binnig, C.: Deepdb: learn from data, not from queries! Proc. VLDB Endow. **13**(7), 992–1005 (2020)
9. Marcus, R., et al.: Neo: a learned query optimizer. Proc. VLDB Endow. (PVLDB) **12**(11), 1705–1718 (2019)
10. Mohanaraj, A., Lissandrini, M., Hose, K.: Planrgcn: predicting sparql query performance. Proc. VLDB Endow. **18**(6), 1621–1634 (2025)

11. Schwabe, T., Acosta, M.: Cardinality estimation over knowledge graphs with embeddings and graph neural networks. Proc. ACM Manage. Data **2**(1), 1–26 (2024)
12. Sun, J., Li, G.: An end-to-end learning-based cost estimator. Proc. VLDB Endow. (PVLDB) **13**(3), 307–319 (2019)
13. Warnke, B., Martens, K., Winker, T., Groppe, S., Groppe, J., Adhiyaman, P.: Rejoosp: reinforcement learning for join order optimization in sparql. Big Data Cognit. Comput. **8**(7), 71 (2024)
14. Yu, X., Li, G., Chai, C., Tang, N.: Reinforcement learning with Tree-LSTM for join order selection. In: 36th IEEE International Conference on Data Engineering (ICDE), pp. 1297–1308. IEEE (2020)
15. Zhang, W.E., Sheng, Q.Z., Qin, Y., Taylor, K., Yao, L.: Learning-based sparql query performance modeling and prediction. World Wide Web **21**(4), 1015–1035 (2018)
16. Zhao, K., Yu, J.X., Zhang, H., Li, Q., Rong, Y.: A learned sketch for subgraph counting, pp. 2142–2155. Association for Computing Machinery (2021)

Vector Databases and Language Models: Synergies and Challenges

Toni Taipalus[✉][iD]

Tampere University, Tampere, Finland
`toni.taipalus@tuni.fi`

Abstract. Vector databases are a critical component in modern system infrastructures. In this study, we discuss the principles behind vector database management systems, with a focus on their features, the concept of vector embeddings, and similarity search mechanisms. Furthermore, we examine the synergies between vector databases and language models, which rely on vector embeddings for semantic search and retrieval-augmented generation. We also discuss the challenges arising from the integration of language models with vector databases. Through this discussion, we aim to provide early-stage researchers with an overview of the integration of vector databases and language models.

Keywords: vector databases · vectorization · database · data management · language model · retrieval-augmented generation

1 Introduction

Language models, despite revolutionizing natural language processing, face well-known limitations: they can hallucinate false facts [5], struggle with out-of-date knowledge, and incur escalating costs as their parameters grow [8]. Integrating language models with vector databases offers a compelling solution to these issues [8].

In the retrieval-augmented generation (RAG) paradigm, a language model's parametric knowledge is supplemented with non-parametric memory from an external vector database. This synergy enables the model to fetch relevant information on the fly, which grounds the responses in up-to-date data and potentially reduces hallucinations. Early work on *k-nearest-neighbor* language models demonstrated that augmenting a neural language model with a vector-indexed datastore of examples dramatically improved the model's ability to recall rare factual patterns, consequently reducing perplexity on long-tail content without model re-training [10]. Modern RAG systems build on this idea by using neural dense embeddings to retrieve semantically relevant documents which the language model then uses to produce more informed outputs [11]. In specific knowledge-intensive tasks, retrieval augmentation has enabled smaller language models to approach the performance of substantially larger models. For example, the RETRO model has been shown to achieve comparable performance to

P. K. Chrysanthis et al. (Eds.): ADBIS 2025, CCIS 2676, pp. 192–202, 2026.
https://doi.org/10.1007/978-3-032-05727-3_19

GPT-3 on certain knowledge-intensive benchmarks, while using 25 times fewer parameters [2].

In this study, we discuss system-level underpinnings of synergies similar to the one above, focusing on how vector databases support neural retrieval for language models. We examine dense embedding generation, indexing and storage in vector databases, query latency considerations, integration pipelines for language models and vector databases, the current challenges, and emerging research opportunities. The discussion strives for domain-agnosticism, and targets data management and retrieval aspects relevant to many applications.

2   Background Concepts

2.1   Vector Embeddings

Vector embeddings are numerical representations of data elements. Data objects such as words, images, or nodes in a graph can be mapped into continuous vector spaces. These embeddings capture semantic or structural relationships, and allow for the search of vectors that are similar to (but not the same as) another vector. In this study, we use two-dimensional vectors for illustration purposes and simplicity, but such vector may have dimensions (i.e., elements) in the hundreds or thousands. As an example, Fig. 1a shows the vectors of two plays, PA and PB, in a two-dimensional vector space. The elements of the vectors correspond to the amount of comedy and tragedy in the plays. Based on the positions of the vectors in the vector space, we can see that play PA is relatively comic and not very tragic, and that play PB is the opposite.

Using text data as an example of creating vector embeddings from data objects, the creation of vector embeddings is grounded in the distributional hypothesis, which posits that linguistic items with similar distributions have similar meanings. This is the basic principle of models like *Word2Vec*, which learns embeddings by predicting a word based on its context, or predicting surrounding words given a target word [15]. Such models utilize large corpora to capture co-occurrence statistics, resulting in vector spaces where semantic relationships are reflected in geometric proximity. As an example, when searching for "vitamin deficiency symptoms", a suitable embedding model might retrieve documents about iron deficiency and B12 anemia. A poor embedding model might return blog posts about "vitamin shopping" or "best supplements in 2025". All these are technically close, but not all are semantically useful.

Different embedding techniques are also better suited for different geometric proximity calculations (cf. Figure 1b). Mismatch between vector geometry and similarity function can lead to reduced effectiveness. Furthermore, Embeddings trained on in-domain data (e.g., biomedical text for a health app) drastically improve recall (i.e., finding all relevant items) and precision (i.e., avoiding irrelevant ones).

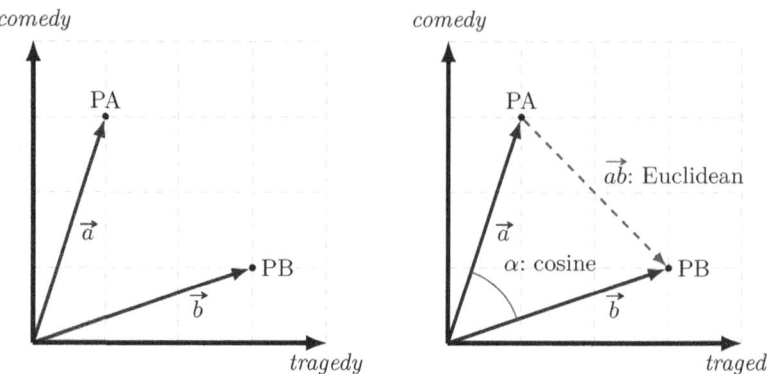

(a) The elements of the vectors measure the amount of tragedy and comedy

(b) Two methods for measuring vector similarity

Fig. 1. Two-dimensional vector space with two vectors and two methods of measuring similarity: Euclidean distance and cosine similarity; adapted [19]

2.2 Vector Databases

Vector databases, or rather vector database management systems (VDBMS), can provide the means of efficiently storing vectors and metadata associated with them in forms of different vector indices. As with data objects other than vectors, indices form the basis of efficient data retrieval. Additionally, VDBMSs can provide different ways of measuring vector similarity. Commonly used methods are Euclidean distance, cosine similarity, and inner product. The two former are illustrated in Fig. 1b. One can likely see the implications of different similarity search methods in the figure.

In addition to vector-specific indices and searching, different vector databases provide different functionality, ranging from software libraries (e.g., FAISS [9]), to fully-fletched DBMSs with role-based access control, concurrency, and replication and sharding (e.g., Milvus [22]). Several DBMSs following some other database paradigm have also adopted features for vector data management, for example PostgreSQL, Redis, and MongoDB. Currently, the maximum number of vector dimensions in these systems are measured in thousands, while dedicated systems such as Milvus, Pinecone, Weaviate, and Manu [4] can manage dimensions in the tens of thousands.

2.3 Language Models

Simplified, language models function by converting textual input into high-dimensional vector representations (e.g., $\mathbb{R}^{768}$), which enables them to capture semantic relationships and contextual nuances. These models process input within a fixed-size context window, which defines the maximum number of tokens the model can consider at once. Expanding this context window enhances the model's ability to understand longer inputs. However, computational complexity of standard self-attention in transformers is $O(n^2)$ with respect to the sequence length n. Newer architectures such as Longformer [1] are specifically designed to reduce this cost for long-context scenarios.

Training and fine-tuning language models are typically resource-intensive processes that require substantial computational power and time. For example, training large-scale models can incur costs running into millions of dollars, making it impractical for many applications. To mitigate these challenges, vector databases are employed to provide external context to language models without the need for retraining. By storing precomputed vector embeddings of relevant data, these databases enable efficient retrieval of relevant data objects based on semantic similarity. When a prompt is issued, the system retrieves the most relevant vectors from the database and incorporates them into the model's context window, enhancing the model's responses with up-to-date and domain-specific information.

3   Supplementing Retrieval with Vector Databases

RAG refers to techniques that combine an language model with a retrieval mechanism to incorporate external knowledge. A typical RAG system first encodes a user query into a vector representation, then performs a similarity search in a vector database of background documents, and finally feeds the top-ranked retrieved documents (or their content) into the language model's context before generating the answer [11] (cf. Figure 2). This pipeline effectively gives the language model access to an external knowledge base in real-time. The approach was introduced in knowledge-intensive NLP tasks [11], showing that a language model (BART in their case) augmented with a learned retriever outperformed fully-parametric models on open-domain QA benchmarks.

Subsequent systems have strengthened this paradigm. For example, Atlas [6], a pretrained retrieval-augmented model achieved higher accuracies on QA with 50 times fewer parameters than a 540 billion-parameter PaLM model by carefully training the retriever and generator together on knowledge tasks. Another line of work from the database community treats the vector store as a reliable long-term memory for language models [24]: rather than packing all world knowledge into model weights, one can store factual data in a vector database and let the model retrieve it as needed [8]. This design not only improves factual accuracy, but also allows updating the knowledge base without retraining the language model, which is a major advantage for keeping up with evolving information.

In practice, retrieval augmentation has been shown to mitigate hallucinations in several benchmark tasks, provided the retrieved information is relevant and accurate [8]. Even at inference time, an language model can be queried in a semi-open-book manner. That is, in experimental setups, frozen language models have demonstrated improved rare-token prediction accuracy by retrieving nearest neighbor tokens from a vectorized representation of the training corpus [10].

Across the studies mentioned, a clear picture emerges: dense neural retrieval is a key enabler for making language models more knowledgeable, accurate, and efficient by offloading memory to an external vector database.

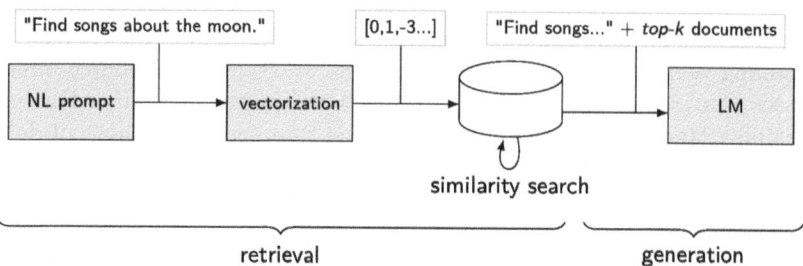

Fig. 2. The general principle of retrieval-augmented generation: the vectorized natural language (NL) prompt is used for vector similarity search to find n closest documents, which are then used in tandem with the natural language prompt for the language model (LM), enhancing the prompt with context

4   Vector Storage and Retrieval

Central to neural retrieval is the use of dense embeddings. Instead of sparse keyword matches, RAG pipelines represent text as high-dimensional vectors (hundreds or even thousands of dimensions) generated by transformer encoders [21] or language model embeddings. These vectors capture semantic similarity, meaning that a query vector will lie close (e.g., in cosine or Euclidean distance) to vectors of documents on the same topic to retrieve relevant information even when there are no shared keywords [2]. Storing and searching through millions or billions of such vectors is the core function of a vector database. This task is non-trivial: the vectors live in a continuous space lacking obvious structure, and brute-force search would be prohibitively slow, as each distance computation involves hundreds of multiplications.

Instead, approximate nearest neighbor (ANN) algorithms are used to index the embeddings and accelerate queries. One popular approach is building a small-world graph index (HNSW, or hierarchical navigable small world), where each vector is a node connected to its neighbors in such a way that a greedy graph traversal yields a near-optimal set of results [14] (cf. Figure 3). Malkov and Yashunin's HNSW method exemplifies this, allowing sub-millisecond retrieval on million-scale datasets with high recall by navigating a hierarchical graph structure instead of exhaustive scanning [14].

Another class of methods uses vector quantization to compress and partition the space. Product quantization (PQ) compresses each vector into a short code by splitting the space into subspaces âĂŞ dramatically reducing memory usage and enabling fast coarse search at some cost to accuracy [7]. Modern vector

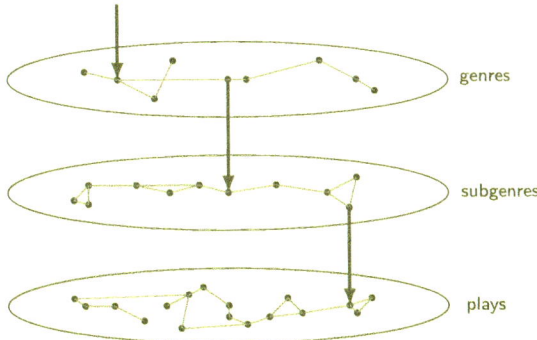

Fig. 3. Basic operating principle of an HNSW index: an arbitrary number of layers of indices are used to limit the search space; vector search starts at the top layer (blue arrow) from a random vector; each vector stores a list of pointers to a small set of neighbor vectors (red lines); ANN is used to find the most suitable match in a relatively small vector space, which then points to the next layer; again ANN is used in a subset of vectors to find the most suitable match, which is again followed to find the actual search results (Color figure online)

databases often combine different techniques: for example, the FAISS library [9] provides inverted file indexes with product quantization and HNSW graphs, supporting billion-scale similarity search on GPUs [9]. SPANN exemplifies a two-tier index architecture optimized for billion-scale corpora, although such systems remain more common in research contexts than production use today [3]. The common goal of these structures is to balance recall (retrieving true nearest neighbors) with efficiency (time and computer memory).

In practice, an ANN index can retrieve top-k neighbors from a million-vector set in just a few milliseconds with 95% recall, a sweet spot for many language model augmentation scenarios. Vector databases incorporate these algorithms under the hood, managing the indexing, compression, and search operations transparently. For instance, Milvus [22] is an open-source vector DBMS that offers multiple index types and automatically chooses an index depending on data scale and latency requirements. By leveraging such indices in optimized settings, vector databases can support similarity queries with latencies approaching those of traditional keyword search engines.

Finally, many VDBMSs support hybrid search or hybrid operators, meaning that in addition to the query vector, a set of more traditional conditions are imposed on the metadata, such as `price > 50` [18]. This limits the vector search to a subset of vectors in the vector space. Vector indices can also be constructed in subsets of vectors (i.e., shards), and metadata used in assigning certain vectors to certain shards, e.g., electronics to one shard, utilities to another.

5   Performance Considerations

Building a high-performance vector database-backed language model pipeline requires careful system-level design. One consideration is the latency budget for retrieval. In interactive applications (e.g. a question-answering chatbot), the vector lookup must be relatively fast not to bottleneck the language model's response. A typical vector search on tens of thousands of embeddings can execute in a few milliseconds with an optimized ANN index in memory [3]. However, if the knowledge corpus is very large (e.g., hundreds of millions of entries), keeping the entire index in RAM may be unfeasible. In such cases, systems turn to disk-based or distributed solutions: SPANN, for example, demonstrates that a hybrid disk/RAM index was shown to serve a billion-scale vector corpus with only 64GB of memory, achieving query latencies on the order of tens of milliseconds with high recall [3].

Another approach is to exploit hardware parallelism. FAISS and other libraries can batch distance computations to utilize GPUs efficiently [9], yielding significant speedups for large query loads. The vector database acts as the orchestrator to route similarity computations to the appropriate hardware and to manage caching of vector data. Caching is indeed a valuable optimization in cases when certain queries or documents are frequently accessed, as the vector database can cache their embeddings or results in faster storage. In an end-to-end RAG pipeline, there is also a trade-off between retrieval time and generation time. Language model inference is typically the slower component (especially for large language models), so one might tolerate, say, 50 ms of retrieval delay from the vector database, if it substantially improves the quality of a generation that takes 2 s.

System designers often overlap retrieval with other stages (e.g. start the language model on the user query while concurrently fetching documents, then concatenate results when ready) to hide latency. The integration between language model and vector database can be tight or loose. A tight integration might use the language model's internal representation to continually fetch new information mid-generation (iterative retrieval), whereas a loose integration uses a fixed retrieved set obtained before generation. Many practical implementations use an orchestration layer (such as a middleware or libraries like LangChain) to manage the sequence. That is, embed the query, query the vector store, retrieve top-N texts, and finally construct an augmented prompt for the language model. Each interface crossing between the language model and the database incurs overhead, so some research explores training the retriever and generator jointly so that the two components hand off information more fluidly [2].

Ensuring scalability requires monitoring how retrieval performance scales with data size. Vector databases often support sharding or partitioning of the index across nodes to handle very large corpora, with a slight loss in recall due to partition boundaries [3] (cf. Figure 4). Another issue is consistency in distributed settings: when multiple language model instances share access to a vector database, inserts to the database may need to be immediately visible to all models, which calls for transactional or eventually-consistent replication pro-

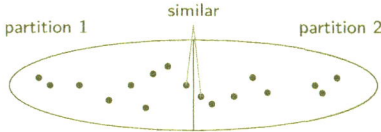

Fig. 4. Similar vectors near the partition boundaries can be missed if only a part of the partitions are searched

tocols in the vector database. Such requirements pose limitations to the selection of the vector database.

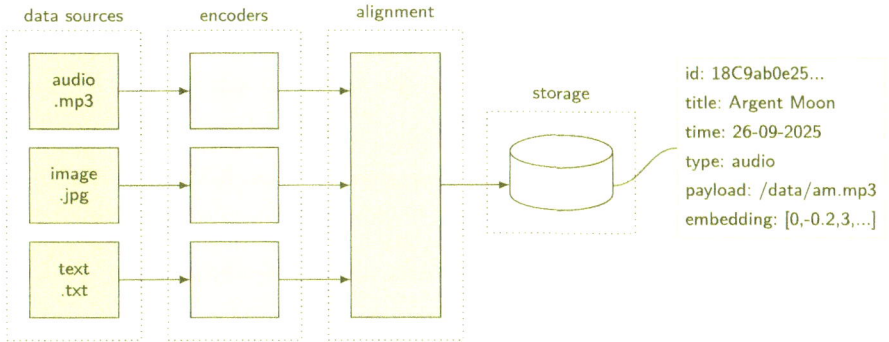

Fig. 5. Vectorization allows the storage and querying of multi-modality data in a single format: vectors; multi-modality data such as images and text are encoded with respective encoders into vector embeddings, these embeddings are then aligned into a single vector space, ensuring that vector embeddings of similar data objects are close to each other; the vectors are then stored into a vector database; some techniques such as CLIP and CLAP already provide the alignment without need for additional processing though other software libraries or models

6   Future Research

Beyond text, vector databases can store embeddings of images, audio, code, graphs, etc. An exciting current research direction is enabling language models to retrieve not just text documents but other modalities (images or knowledge graph substructures) to ground their understanding (Fig. 5). Multi-modality poses additional challenges to already recognized limitations in embedding models, data quality, and retrieval accuracy [23]. Early research explored unified systems that support hybrid queries combining vector similarity with structured constraints (e.g. temporal or graph-based conditions) [3]. In this regard, vector databases contribute to addressing challenges reminiscent of those seen in multi-model databases, such as the absence of a unified query language, which have been discussed in prior work [13].

When an language model's internal knowledge conflicts with the retrieved data, the result can be an inconsistent or confusing answer. Mechanisms to detect and resolve these conflicts are needed. One idea is to train language models to defer to retrieved evidence in critical domains, i.e., effectively learning a form of truth alignment with the external database. Another aspect is keeping the database in sync with the world: unlike static model weights, a vector store can be updated continuously. Techniques for incremental indexing (i.e., updating indexes without full rebuilds) and for handling concept drift (i.e., when new data shifts the embedding space) will be crucial for systems where real-time consistency is required [3].

From hardware perspective, GPUs already excel at vector calculations due to parallelism and floating-point performance, yet there is interest in quantized or compressive transformers that integrate vector search natively, as well as hardware like dedicated ANN accelerators [17]. In our opinion, there is a need for more comprehensive benchmarks that evaluate the combined system of a language model and a vector database holistically. While components are individually benchmarked (e.g. ANN benchmarks for vector search [9], and natural language processing benchmarks for language model accuracy), the field lacks standardized tasks that measure end-to-end performance, including response quality and latency under varying loads, similarly to relational DBMS benchmarking. Developing such benchmarks would drive research into more efficient and effective integration.

Privacy and security pose additional research avenues. Embedding textual data could leak private information, and external retrieval might introduce adversarial inputs to the language model. Emerging work has begun to explore privacy-preserving vector retrieval mechanisms, such as secure ANN and filtered retrieval, which are likely to become increasingly important as such systems are adopted in sensitive domains. It remains an open questions whether this is a training-related challenge, whether such privacy issues should be ensured with guardrails outside model training, or perhaps with hybrid queries, which account not only similarity search but also search conditions on metadata (cf. right-hand side of Fig. 5).

Finally, even though vectors are not a novel way of storing data, the field has advanced rapidly in terms of popularization and usability of different tools and systems. In our opinion, vector databases and adjacent technologies are more than ripe for more *applied research* in different and exciting domains. Although we have seen concepts are opportunities in fields such as healthcare [16], cyber security [20], and finance [12], current vector data management techniques provide a vast frontier of opportunities that now only lacks more imagination.

7   Conclusion

Vector databases and large language models together form a powerful architecture for generative artificial intelligence. The dense neural retrieval capabilities of vector databases complement the generative features of language models and enable systems that not only produce coherent language, but reason over timely, model-external knowledge. In this study, we reviewed how dense embeddings serve as the lingua franca between language models and vector database, and how efficient indexing and storage make real-time retrieval feasible at scale. Key system challenges such as reducing query latency, building robust integration pipelines, and ensuring scalability have seen rapid progress, yet continue to face challenges. As future research addresses the outlined directions which span from better indexes and multi-modal retrieval to alignment and security, we can expect language models to become more reliable and versatile by leveraging the right database systems behind the scenes. This interdisciplinary synergy connects the advances in data management and language models.

References

1. Beltagy, I., Peters, M.E., Cohan, A.: LongFormer: the long-document transformer. arXiv preprint arXiv:2004.05150 (2020)
2. Borgeaud, S., Sifre, L., et al.: Improving language models by retrieving from trillions of tokens. In: Proceedings of the 39th International Conference on Machine Learning (ICML), pp. 2206–2240 (2022)
3. Chen, Q., et al.: SPANN: highly-efficient billion-scale approximate nearest neighbor search. In: Advances in Neural Information Processing Systems (NeurIPS 2021), vol. 34, pp. 5199–5212 (2021)
4. Guo, R., et al.: Manu: a cloud native vector database management system. arXiv preprint arXiv:2206.13843 (2022)
5. Huang, L., et al.: A survey on hallucination in large language models: principles, taxonomy, challenges, and open questions. ACM Trans. Inf. Syst. **43**(2), 1–55 (2025)
6. Izacard, G., et al.: Atlas: few-shot learning with retrieval augmented language models. J. Mach. Learn. Res. **24**(251), 1–43 (2023)
7. Jégou, H., Douze, M., Schmid, C.: Product quantization for nearest neighbor search. IEEE Trans. Pattern Anal. Mach. Intell. **33**(1), 117–128 (2011)
8. Jing, Z., et al.: When large language models meet vector databases: a survey (2024)
9. Johnson, J., Douze, M., Jégou, H.: Billion-scale similarity search with GPUs. IEEE Trans. Big Data **7**(3), 535–547 (2019)
10. Khandelwal, U., Levy, O., Jurafsky, D., Zettlemoyer, L., Lewis, M.: Generalization through memorization: nearest neighbor language models. In: Proceedings of the International Conference on Learning Representations (ICLR) (2020)
11. Lewis, P.S.H., et al.: Retrieval-augmented generation for knowledge-intensive NLP tasks. In: Advances in Neural Information Processing Systems (NeurIPS 2020), vol. 33, pp. 9459–9474 (2020)
12. Liu, X., Zhu, J.: FinanceQA-Agent: a high-precision comprehensive financial technology question-answering intelligent system based on vector databases. In: 2024 International Conference on Advances in Electrical Engineering and Computer Applications (AEECA), pp. 576–582. IEEE (2024)
13. Lu, J., Holubová, I.: Multi-model databases: a new journey to handle the variety of data. ACM Comput. Surv. **52**(3), 1–18 (2019). https://doi.org/10.1145/3323214
14. Malkov, Y.A., Yashunin, D.A.: Efficient and robust approximate nearest neighbor search using hierarchical navigable small world graphs. IEEE Trans. Pattern Anal. Mach. Intell. **42**(4), 824–836 (2020)

15. Mikolov, T., Chen, K., Corrado, G., Dean, J.: Efficient estimation of word representations in vector space. arXiv preprint (2013)
16. Ng, K.K.Y., Matsuba, I., Zhang, P.C.: RAG in health care: a novel framework for improving communication and decision-making by addressing LLM limitations. NEJM AI **2**(1), AIra2400380 (2025)
17. Pan, J.J., Wang, J., Li, G.: Survey of vector database management systems. VLDB J. **33**(5), 1591–1615 (2024)
18. Pan, J.J., Wang, J., Li, G.: Vector database management techniques and systems. In: Companion of the 2024 International Conference on Management of Data, pp. 597–604. ACM (2024)
19. Taipalus, T.: Vector database management systems: fundamental concepts, use-cases, and current challenges. Cogn. Syst. Res. **85**, 101216 (2024). https://doi.org/10.1016/j.cogsys.2024.101216
20. Taipalus, T., Grahn, H., Turtiainen, H., Costin, A.: Utilizing vector database management systems in cyber security. In: Proceedings of the 23th European Conference on Cyber Warfare and Security. Academic Conferences International Ltd (2024)
21. Vaswani, A., et al.: Attention is all you need. Adv. Neural Inf. Process. Syst. **30** (2017)
22. Wang, J., et al.: Milvus: a purpose-built vector data management system. In: Proceedings of the 2021 ACM SIGMOD International Conference on Management of Data, pp. 2614–2627 (2021)
23. Wang, M., et al.: Must: an effective and scalable framework for multimodal search of target modality. In: 2024 IEEE 40th International Conference on Data Engineering (ICDE), pp. 4747–4759. IEEE (2024)
24. Zhang, Y., Yu, Z., Jiang, W., Shen, Y., Li, J.: Long-term memory for large language models through topic-based vector database. In: 2023 International Conference on Asian Language Processing (IALP), pp. 258–264. IEEE (2023)

Ethical and Equitable Data Science: Bridging Social Justice and Technical Innovation *ADBIS 2025 Doctoral Consortium Lecture*

Genoveva Vargas-Solar[✉] ⓘ

CNRS, Univ Lyon, INSA Lyon, UCBL, LIRIS, UMR5205, 69221 Écully, France
genoveva.vargas-solar@cnrs.fr

Abstract. This paper summarises the fundamental background of the Doctoral Consortium titled *"Ethical and Equitable Data Science"*. It introduces FREDA, a methodology for designing ethical, frugal, and equitable data and algorithm-driven science. It bridges technical innovation with social justice by integrating data sovereignty, fairness-aware analytics, and community-in-the-loop infrastructure.

Rooted in decolonial and feminist perspectives, FREDA addresses transparency, accountability, and epistemic diversity through policy-aware Spark pipelines, federated learning, and negotiated resource dispatching. A case study illustrates how sovereignty-aware pipelines enable community control, minimize extractivism, and embed plural, justice-centered values into AI systems.

Keywords: Responsible Data Science · frugal algorithm design · data sovereignty · decolonial data science · equity-aware AI

1 Introduction

The Fourth Paradigm, introduced by Jim Gray [19], defines a data-intensive model of scientific discovery highlighting the integration of large, diverse datasets to uncover patterns and drive cross-disciplinary insight. This paradigm shifted science by placing data at the heart of research. It promoted open data, scalable infrastructure, and analytics to tackle complex problems. Data becomes a core knowledge source, requiring new tools, skills, and ethical safeguards. The 2025 "Stanford HAI AI Index Report" echoes these principles by stressing benchmark datasets, scalable systems, and interdisciplinary collaboration. As Artificial Intelligence (AI) impacts fields like healthcare and climate, the report underscores rising concerns around transparency, provenance, and social-environmental harms—issues central to the Fourth Paradigm's call for ethical, accountable data science.

Supported by the AAP program of the CNRS and the project FRIENDLY of the transversal projects program of the LIRIS.

P. K. Chrysanthis et al. (Eds.): ADBIS 2025, CCIS 2676, pp. 203–215, 2026.
https://doi.org/10.1007/978-3-032-05727-3_20

This paper examines the ethical, social, and fairness implications of data-driven decision-making introduced in the ADBIS 2025 Doctoral Consortium (DC) titled *"Ethical and Equitable Data Science"*. The paper introduces a novel methodology, FREDA, for addressing scientific problems through data, algorithms and computing infrastructure guided by ethics, responsibility, equity and inclusion. FREDA frames transparency, accountability, and bias mitigation as essential principles, illustrating through case studies how misuse of data can perpetuate harm and inequality. It explores broader impacts—socio-economic, political, and environmental—with emphasis on data sovereignty, equity, and marginalized voices. Fairness metrics and decolonial methods are introduced to address epistemic and cultural diversity. The paper outlines techniques introduced in the ADBIS 2025 DC course for detecting and reducing bias, including audits, inclusive data practices, and equity-aware design. Regulatory and ethical frameworks provide guidance for navigating legal and moral complexities. The paper closes with best practices for building fair, trustworthy models, advocating sustainable, non-extractive data practices rooted in plural and community-driven knowledge.

The remainder of the paper is organized as follows: Sect. 2 reviews background and related work on ethical and equitable data science. Section 3 introduces a methodology, FREDA, that integrates ethics and equity into data- and algorithm-driven research. Section 4 describes a use case demonstrating this approach. Section 5 concludes with perspectives and future directions.

2   Related Work

Ethical and equitable data science [41] rests on three interconnected pillars: data, algorithms, and infrastructures. These are not merely technical domains but fundamental to solving multidisciplinary problems where human, social, and environmental dimensions intersect. Notably, the concept of *body territory*[1] highlights the situated nature of data and its implications.

Ethics, Diversity, Equity, and Inclusion in Data. A major challenge in responsible and equitable data science is the profound absence of diverse datasets that meaningfully represent underrepresented communities—such as women, people with non-binary gender identities, indigenous populations, and individuals from marginalized socio-economic and educational backgrounds. Most existing datasets are shaped by Western-centric paradigms, overwhelmingly curated in dominant languages, and reflective of majority world norms, thereby erasing plural ways of knowing, seeing, and experiencing the world. The lack of inclusive data is not merely a gap but a form of epistemic exclusion that perpetuates algorithmic biases and misrecognition. Collecting data from marginalized groups

---

[1] The body territory refers to the inseparability between the human body and the land or territory one inhabits [8].

involves navigating complex ethical terrain: issues of trust, consent, community-defined relevance, and the risk of extractivism must be addressed through participatory and culturally grounded methods. In contexts where real-world data is absent or inaccessible due to historical invisibilisation, there is a growing conversation around the cautious use of synthetic data to introduce alternative narratives and correct imbalances. However, the generation and integration of such data must be carefully negotiated to avoid reifying stereotypes or reproducing dominant epistemologies under the guise of inclusion. Ultimately, addressing these absences demands not only technical interventions but also political and methodological shifts that centre the voices, values, and sovereignties of those historically excluded from digital knowledge infrastructures.

The field has evolved from initial concerns about privacy and fairness [3,14] to examining how sociotechnical systems perpetuate inequality. Benjamin [5], Noble [29], and others argue that algorithmic harms are embedded in structural racism and historical erasure. For example, Buolamwini and Gebru [7] revealed significant racial and gender bias in facial recognition systems. Efforts to address these challenges emphasize community-centred governance, accountability, and inclusive data practices [12,26]. Initiatives like *Data Feminism* [15] and *Design Justice* [10] advocate for infrastructures shaped by lived experiences and structural analysis of power.

Considering that human-related data are often the backbone of many experiments performed in computer and data science, the objective is to exhibiting the implications of using biased human-related datasets during the preprocessing stage of data analytics pipelines. On the technical side, frameworks such as IBM's AI Fairness 360 [4], fairness-aware sampling [21,24], and fairness-aware SQL rewriting [22,35] offer computational mechanisms to embed fairness in data processing and decision-making. Institutions like NYU's Center for Responsible AI [31,36] also promote transparency tools and co-design methodologies. Recent contributions extend fairness frameworks to structured queries, exploratory interfaces, and graph analytics. Stoyanovich [39] introduced responsibility by design, while Amer et al. [1,2] explored fairness-aware data exploration. Pitoura [44] examined algorithmic bias in graph-based rankings. Srivastava [38] focused on fairness in SQL and entity resolution. These developments shift fairness from a constraint to a foundational design principle. Their integration into decolonial and relational data environments fosters more inclusive and accountable systems.

Responsible AI. Refers to the development, deployment, and governance of artificial intelligence systems in ways that prioritize ethical values, societal well-being, human rights, and long-term sustainability. Responsible AI research bridges algorithmic transparency and systemic equity. While early efforts focused on fairness, accountability, and transparency (FAT) [13,14], feminist and decolonial critiques argue for deeper transformation beyond compliance.

Technical approaches include fairness definitions [3], interpretability [13], and transparency tools like model cards [27]. NYU's Responsible AI Initiative [31] adds documentation and impact assessments. Critical perspectives from Birhane [6], Mohamed et al. [28], and others argue that AI must interrogate whose knowledge is encoded and whose interests are served. Participatory co-design [36] and Indigenous data sovereignty [23] advance ethical concerns around data ownership and governance. Best practices now combine differential privacy, fairness audits, and pluralistic epistemologies. Ethical AI must embed these within structures that reflect justice, care, and social accountability.

Accessibility to Infrastructure. Equitable access to infrastructure is essential for inclusive AI and data science, especially in fields like health, agriculture, and climate research. Initiatives like EOSC [9], the African Open Science Platform [30], and Google Earth Engine [18] seek to democratize access to data and computation. Scalable frameworks such as Spark, Dask, and Kubernetes have enabled AI workflows in hybrid and cloud-native environments [45], yet disparities persist in low-connectivity regions [32]. Emerging approaches such as federated learning [20], edge computing [37], and mesh networks offer alternatives better suited to resource-constrained contexts.

Beyond technical concerns, critical work by U. Mejias and N. Couldry [11,25] and Y.E. Aguilar Gil[2] theorize *digital colonialism*—how cloud infrastructures replicate extractive logics, concentrating control among a few global actors. These critiques challenge assumptions of neutrality and universality in digital infrastructure. Activist and research collectives like Data Género [17], FA+IR [16], and Via Libre propose decentralized, feminist infrastructures grounded in transparency, care, and autonomy. Their work emphasizes community ownership, epistemic plurality, and democratic negotiation of infrastructure. Infrastructures must thus be reimagined not just as technical enablers but as contested sites of power. Relational, context-aware systems rooted in territorial sovereignty can resist extractive models and foster durable, inclusive innovation.

Digital independence in Europe reflects a growing effort to secure technological sovereignty through control over infrastructure, data, and standards—reducing reliance on dominant global platforms and fostering open, ethical, and regionally grounded alternatives. This ambition resonates with demands from decolonial movements, which similarly call for self-determined infrastructures that respect local knowledge systems, cultural values, and community governance. In both contexts, the push for digital autonomy is not only technical but deeply political: it seeks to reclaim agency over how technology is built, accessed, and used, challenging asymmetries of power embedded in global digital ecosystems [40].

Discussion. Feminist and decolonial approaches offer deep critiques of AI systems, linking ethical issues to systemic inequities rather than technical flaws. Positionality is key—knowledge is relational and situated. Networks like FA+IR [16] promote care-based, participatory AI centred on lived experience and intersectionality. These approaches inform inclusive documentation, participatory

---

[2] https://restofworld.org/2020/saving-the-world-through-tequiology/.

audits, and value-driven pipeline design [10,15]. Decolonial computing calls for resisting epistemic extractivism and promoting local sovereignty [28,33,34]. Challenges persist: integrating local knowledge, scaling consent, and reducing infrastructural dependency. Ethical AI requires hybrid methods that combine technical precision with structural critique to build plural, redistributive, and just technologies. The DC course introduces a methodology that proposes guidelines for designing end-to-end frugal, fair and responsible Data and Algorithm-Driven Science solutions.

3 FREDA Methodology: Frugal, Responsible, Equitable Data Algorithm Driven Science

The *ADBIS 2025 DC lecture* introduces FREDA, a methodology that proposes a technical and epistemically grounded framework for designing data and algorithm-driven scientific systems that are equitable, sustainable, and locally accountable. It integrates three pillars: (1) data sovereignty[3], (2) responsible, frugal algorithm training, and (3) community-in-the-loop infrastructure dispatching, in alignment with decolonial values and fairness metrics (Fig. 1).

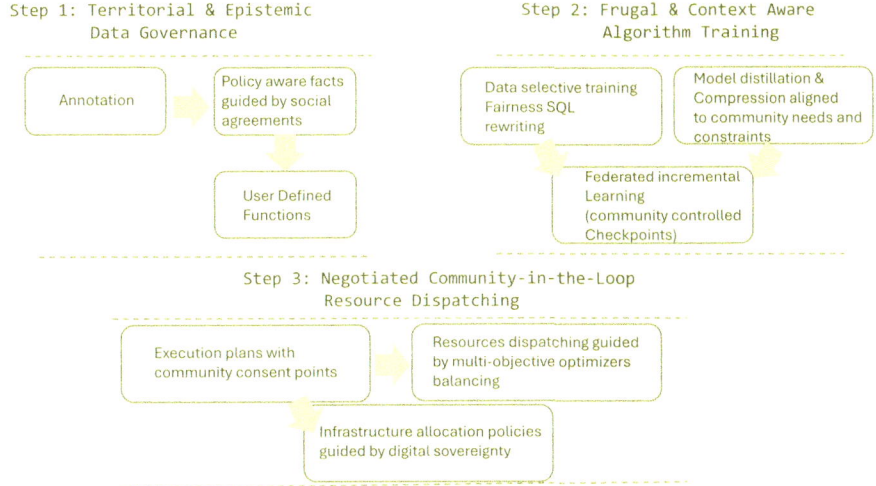

Fig. 1. Frugal, Fair, and Responsible Data and Algorithm-Driven Science Methodology

---

[3] Data sovereignty refers to the principle that data is subject to the laws and governance structures within the nation, community, or territory in which it is collected or belongs. More broadly, it embodies the right of individuals, groups, or nations—especially Indigenous and historically marginalized communities—to control the access, use, sharing, and storage of their data according to their values, knowledge systems, and governance principles.

Step 1: Territorial and Epistemic Data Governance. The DC course begins by discussing data sovereignty and epistemic data governance. Input datasets are annotated using culturally grounded vocabularies and metadata extensions (e.g., tribal license IDs, consent levels, symbolic representations) stored within Delta or Hudi lakehouses[4]. These metadata layers inform both access and computation:

– Spark SQL is extended with policy-aware operators that integrate tags for filtering, masking, or aggregating data in accordance with sovereignty agreements. Using Databricks Unity Catalog, one can implement row-level filters and column masks. For example:

Listing 1.1. Tribal-Aware SQL: UDFs, Row Filters, and Column Masks

```
1  -- Define a SQL UDF to enforce tribal authorization
2  CREATE FUNCTION tribal_filter(tribal_auth_tag STRING)
3    RETURN IF(
4      is_account_group_member(current_user()), --
         group-based logic
5      TRUE,
6      tribal_auth_tag = get_user_tribal_tag(current_user())
7    );
8
9  -- Attach row-level filter to the table
10 ALTER TABLE tribal.events
11   SET ROW FILTER tribal_filter ON (tribal_auth_tag);
12
13 -- Define and attach a column mask on 'location'
14 CREATE FUNCTION mask_location(loc STRING)
15   RETURN CASE
16     WHEN is_account_group_member('sovereign_tribe')
17       THEN loc ELSE 'REDACTED' END;
18
19 ALTER TABLE tribal.events
20   ALTER COLUMN location
21   SET MASK mask_location;
```

This ensures that Spark automatically filters or redacts data based on sovereignty tags :contentReferenceindex=1.
– User Defined Functions (UDFs) enforce symbolic transformations (e.g., place names, gender markers) that reflect situated vocabularies, ensuring data harmonization is epistemically respectful. To harmonize place names into local terms:

---

[4] Delta Lake by Databricks and Apache Hudi of the Hadoop ecosystem are open-source technologies designed to bring ACID transactions, schema enforcement, and data versioning to data lakes—transforming them into lakehouses (i.e., unified systems that combine the flexibility of data lakes with the reliability and performance of data warehouses).

Listing 1.2. Registering a UDF for Situated Place Names

```scala
// Scala map for situated place names
val placeMap = Map(
  "mountainX" -> "Popocatepetl",   // example Nahuatl name
  "riverY"    -> "Kukulkan"        // example Maya name
)

spark.udf.register("to_local_place", (stdName: String) =>
  placeMap.getOrElse(stdName, stdName)
)

spark.sql("""
  CREATE OR REPLACE VIEW local_places AS
  SELECT
    id,
    to_local_place(place_std) AS place_name,
    ...
  FROM tribal.base_data
""")
```

This UDF replaces standardized toponyms with community-validated names, respecting epistemic localization.

– The catalogue layer integrates community-curated constraints through Ranger or Amundsen policies to produce traceable, queryable, and enforceable access. With tools like Apache Ranger or Amundsen, you enforce sovereignty metadata in the catalogue. Use Ranger UI to set row-filter policies:

```
Service: hive_sql
Resource: tribal.events
Policy: tribal_row_policy
Condition: tribal_auth_tag = '${USER.tribal_tag}'
```

Listing 1.3. Row-Level Access Control via Apache Ranger

```
Mask column 'gender_marker'
WHEN NOT group_member('sovereign_tribe')
APPLY REDACTION 'X';
```

Listing 1.4. Column Masking Policy for Sensitive Attributes

This design ensures not only legal compliance but also community-centred governance through reversibility, localization, and data-as-relational-practice principles [23,43].

Step 2: Frugal and Context-Aware Algorithm Training. To mitigate energy-intensive and extractivist AI development practices, model training must be constrained by both technical and ethical limits. Our framework introduces:

– Data-selective training via fairness-aware SQL rewriting [22,35], enabling selection of balanced, contextually relevant training subsets to avoid overfitting and group harm.

- Model distillation and compression, where large pre-trained models are distilled into smaller domain- and context-specific models suitable for edge devices, aligned with community needs and constraints.
- Federated and incremental learning, reducing central processing costs by training models on distributed, locally owned datasets, with community-controlled checkpoints [42].

Frugality here is both environmental (reducing carbon impact) and epistemic (preventing overfitting to dominant knowledge regimes). Community participation defines acceptable accuracy-efficiency trade-offs and performance fairness thresholds [41].

Step 3: Negotiated, Community-In-The-Loop Resource Dispatching.
AI pipelines are operationalized using resource dispatching strategies that reflect social and environmental constraints [42]. We propose a negotiated DAG execution model for Spark and lakehouse workflows:

- Execution plans include community consent points—intervention moments where stakeholders may pause, validate, or reject a transformation based on ethical, epistemic, or practical criteria.
- Resource dispatching is guided by multi-objective optimizers balancing latency, energy cost, data locality, and fairness objectives. Community-defined fairness weights (e.g., priority for underserved groups) are encoded as runtime parameters in Spark DAG optimizers.
- Infrastructure allocation policies (cloud/edge/on-premise) are made adaptive: systems prioritize local or low-power devices when possible, respecting digital sovereignty and minimizing cloud dependency.

We extend the notion of dispatching from resource allocation to epistemic negotiation, wherein infrastructure decisions (e.g., parallelism level, model retraining frequency) are co-designed with communities and reflect socio-environmental impacts.

Integration and Feedback Loops. A feedback loop ensures continuous alignment with justice-centred metrics:

- Human-in-the-loop explainability modules provide voice and text interfaces for questioning model outputs.
- Logging and provenance are used not only for debugging but for auditable accountability, exposing the power dynamics embedded in data transformations and model behaviour.
- Stakeholders iteratively correct or annotate outputs, shaping model evolution via participatory mechanisms.

Rationale. This methodology foregrounds relational infrastructure, frugal computation, and territorial knowledge governance as the foundation of ethical and responsible science. It blends Spark-based pipelines, federated learning, and policy-aware analytics with feminist, indigenous, and decolonial commitments to care, sustainability, and sovereignty. The goal is not only fairer science but science that is accountable to the communities, environments, and knowledge systems it seeks to serve.

4   Use Case: Federated, Sovereignty-Aware Analytics of Feminicide Clusters in Mexico

This use case applies the proposed methodology to detect *geographical clusters of feminicides* potentially associated with *drug cartel activity*, while ensuring *data sovereignty, privacy*, and *decolonial accountability* for families and communities affected by gender violence and disappearance in Mexico.

Context and Stakeholders

- **Data owners**: *Madres buscadoras*, NGOs, family collectives, and local networks documenting feminicides and disappearances.
- **Data types**: Geo-tagged incident reports, victim profiles, community testimonies, forensic clues, temporal trends.
- **Constraints**: Extreme sensitivity of the data; risks to informants and survivors; need for anonymity, non-traceability, and contextual control over computation.

1. Territorial and Epistemic Data Governance. All datasets are stored *locally* on secure community-run servers—often in rural or semi-urban areas with low bandwidth. Data are *annotated with epistemic metadata*, including:

- `type_of_control`: "familial", "NGO-restricted", "not-to-be-shared"
- `dissemination_level`: "public", "protected_view", "for_legal_action_only"
- `territorial_scope`: "Sonora", "Chiapas", etc.

Access is governed by a *Delta Lake-based sovereignty catalogue*. Spark reads are subject to real-time permission checks, and schema-on-read masking is applied to sensitive fields.

Decision Point 1 (Pre-processing): Families may pause ingestion, request masking, or retract contributions via a secure interface.

2. Frugal, Federated Learning. Federated Spark ML pipelines train clustering models (e.g., DBSCAN) to detect spatial-temporal clusters. Local community nodes train models and share encrypted, differentially private updates.

- No central data pooling
- No location backpropagation
- Granularity and radius of clusters are negotiated

Decision Point 2 (Model Training): Communities approve cluster resolutions, scope, or may halt training for specific territories.

3. Negotiated Dispatching and Infrastructure Use. The system prioritizes edge computing and mesh networks. Workflows are adjusted in real-time based on energy, consent, and latency.

Decision Point 3 (Infrastructure): Community members approve use of devices, time schedules, and dispatching strategies.

4. Analysis and Interpretation. Outputs are anonymized, multi-scalar heatmaps correlating feminicide clusters with cartel zones. Families validate outputs before release.

Decision Point 4 (Dissemination): Families control map storage, public access, and expiration policies.

Table 1. Phases and Community Control Points in Sovereignty-Aware Analytics

Phase	Control Point	Who Decides	Technical Means
Pre-processing	Inclusion, masking	Families, NGOs	Sovereignty catalog, Spark UDFs
Model Training	Cluster granularity	Communities	Federated ML controllers
Dispatching	Time, location, power	Technicians, NGOs	DAG constraints, edge scheduling
Dissemination	Map access, duration	Families, Data stewards	Filters, token-based storage

Table 1 summarises the control points during the phases of the methodology, decision maker and technical means. This use case shows the viability of justice-aware, frugal analytics pipelines that:

- Preserve community control at each stage;
- Minimize cloud dependency;
- Localize sovereignty within data, algorithms, and infrastructure;
- Embed participatory ethics and privacy-by-design in Spark.

5   Conclusion and Future Work

This paper proposed a sovereignty-aware, ethical methodology, FREDA, for data- and algorithm-driven science that centres fairness, decolonial values, and community control, introduced in the *ADBIS 2024 DC*. Through the integration of Spark-based analytics, federated learning, and territorial data governance, we demonstrated how technical systems can support plural epistemologies and minimize extractive computation. The feminicide use case in Mexico illustrates how analytics pipelines can embed consent, interruptibility, and local negotiation, shifting power toward affected communities and away from centralized infrastructures.

The DC lecture discusses future work will focus on developing participatory optimization methods, relational fairness metrics, and revocable consent layers across distributed systems. We also envision extending FREDA across territories with diverse legal and epistemic regimes while supporting interpretability and trust through localized, community-facing tools. More broadly, the approach affirms that responsible data science must go beyond procedural ethics and toward infrastructural justice—grounding AI in care, autonomy, and social accountability.

References

1. Amer-Yahia, S.: Exploratory data science: from fairness to multi-party analytics. In: Proceedings of the 2022 International Conference on Management of Data (SIGMOD), pp. 2950–2954. ACM (2022). https://doi.org/10.1145/3514221.3536107
2. Amer-Yahia, S., Chakroun, I., Mahabadi, S., Montassier, A.: Fairness in data exploration. IEEE Data Eng. Bull. **43**(2), 55–66 (2020)
3. Barocas, S., Selbst, A.D.: Big data's disparate impact. Calif. Law Rev. **104**(3), 671–732 (2016)
4. Bellamy, R.K., et al.: Ai fairness 360: an extensible toolkit for detecting, understanding, and mitigating unwanted algorithmic bias. IBM J. Res. Dev. **63**(4/5), 4-1 (2019)
5. Benjamin, R.: Race After Technology: Abolitionist Tools for the New Jim Code. Polity (2019)
6. Birhane, A.: Algorithmic injustice: a relational ethics approach. Patterns **2**(2), 100205 (2021)
7. Buolamwini, J., Gebru, T.: Gender shades: intersectional accuracy disparities in commercial gender classification. In: Proceedings of the Conference on Fairness, Accountability, and Transparency (FAT*), pp. 77–91 (2018)
8. Cabnal, L.: El relato de las violencias desde mi territorio cuerpo-tierra. En tiempos de muerte: cuerpos, rebeldías, resistencias **4**, 113–126 (2019)
9. Commission, E.: European open science cloud strategic implementation plan (2020). https://ec.europa.eu/research/openscience/pdf/eosc-strategic-implementation-roadmap.pdf
10. Costanza-Chock, S.: Design Justice: Community-Led Practices to Build the Worlds We Need. MIT Press, Cambridge (2020)
11. Couldry, N., Mejias, U.A.: The Costs of Connection: How Data is Colonizing Human Life and Appropriating it for Capitalism. Stanford University Press, Redwood City (2019)
12. Crawford, K., Paglen, T.: Atlas of AI: Power, Politics, and the Planetary Costs of Artificial Intelligence. Yale University Press, Yale (2021)
13. Doshi-Velez, F., Kim, B.: Towards a rigorous science of interpretable machine learning. arXiv preprint arXiv:1702.08608 (2017)

14. Dwork, C., Hardt, M., Pitassi, T., Reingold, O., Zemel, R.: Fairness through awareness. In: Proceedings of the 3rd Innovations in Theoretical Computer Science Conference (ITCS), pp. 214–226 (2012)
15. D'Ignazio, C., Klein, L.F.: Data Feminism. MIT Press, Cambridge (2020)
16. FA+IR Network: Statement on feminist ai research in latin america (2023). https://feministdata.ai/latinamerica
17. Faur, E., Moreno, L.: Algoritmos para la igualdad: Herramientas para la equidad de género en la sociedad digital (2022). https://datagenero.org.ar/algoritmos-para-la-igualdad/
18. Gorelick, N., Hancher, M., Dixon, M., et al.: Google earth engine: planetary-scale geospatial analysis for everyone. In: Remote Sensing of Environment, vol. 202, pp. 18–27 (2017). https://doi.org/10.1016/j.rse.2017.06.031
19. Hey, T., Tansley, S., Tolle, K.M., et al.: The Fourth Paradigm: Data-Intensive Scientific Discovery, vol. 1. Microsoft Research, Redmond (2009)
20. Kairouz, P., McMahan, B., et al.: Advances and open problems in federated learning. Found. Trends ® Mach. Learn. **14**(1–2), 1–210 (2021). https://doi.org/10.1561/2200000083
21. Kearns, M., Neel, S., Roth, A., Wu, Z.S.: Preventing fairness gerrymandering: auditing and learning for subgroup fairness. In: Proceedings of the 35th International Conference on Machine Learning (ICML), pp. 2569–2577 (2018)
22. Koutras, C., Nikolakopoulos, K., Vazirgiannis, M.: Fair data generation via query rewriting. In: Proceedings of the 44th International ACM SIGIR Conference on Research and Development in Information Retrieval, pp. 1099–1108 (2021)
23. Kukutai, T., Taylor, J.: Indigenous Data Sovereignty: Toward an Agenda. ANU Press, Canberra (2016)
24. Lakkaraju, H., Leskovec, J., Kleinberg, J., Leskovec, J.: Identifying unknown unknowns in the open world: representations and policies for guided exploration. In: Advances in Neural Information Processing Systems (NeurIPS), pp. 3198–3206 (2017)
25. Mejias, U.A., Couldry, N.: Colonialism of the digital kind: data as a resource and the data colonialism framework. Telev. New Media **20**(4), 336–349 (2019). https://doi.org/10.1177/1527476419831640
26. Milan, S., Treré, E.: Latin American visions for a feminist internet: reflections on the movement and the challenges ahead. Internet Policy Rev. **9**(4) (2020)
27. Mitchell, M., Wu, S., Zaldivar, A., et al.: Model cards for model reporting. In: Proceedings of the Conference on Fairness, Accountability, and Transparency (FAT* 2019), pp. 220–229 (2019)
28. Mohamed, S., Png, M.T., Isaac, W.: Decolonial AI: decolonial theory as sociotechnical foresight in artificial intelligence. Phil. Technol. **33**, 659–684 (2020)
29. Noble, S.U.: Algorithms of Oppression: How Search Engines Reinforce Racism. NYU Press, New York (2018)
30. Ochu, E., Wafula, J.: African open science platform: advancing open science in Africa (2020). https://africanopenscience.org.za
31. Passi, S., Barocas, S., et al.: Managing bias in algorithmic decision making: an interdisciplinary review. Ann. Rev. Sociol. **47** (2021)
32. Rest of World: Saving the world through tequiology (2020). https://restofworld.org/2020/saving-the-world-through-tequiology/. Accessed 14 July 2025
33. Ricaurte, P.: Data epistemologies, the coloniality of power, and resistance. In: Milan, S., Treré, E. (eds.) Data Justice and the Right to the City, pp. 201–216. University of Westminster Press (2019). https://doi.org/10.16997/book33.n. https://www.westminsterpapers.org/article/id/840/

34. Roio, D.: Decolonizing algorithms: reading computer science through the lens of indigenous epistemologies. In: Proceedings of the Decolonizing the Digital Conference. University of Bologna, Bologna, Italy (2021). https://decolonizingthedigital.org/roio-decolonizing-algorithms

35. Salimi, B., Koch, C., Weikum, G.: Capuchin: causal database repair for algorithmic fairness. In: Proceedings of the 2019 International Conference on Management of Data (SIGMOD), pp. 653–668. ACM (2019)

36. Sambasivan, N., Doshi, A., Basu, S., Cutrell, E., Toyama, K.: Re-imagining algorithmic fairness in india and beyond. In: Proceedings of the 2021 CHI Conference on Human Factors in Computing Systems, pp. 1–18 (2021). https://doi.org/10.1145/3411764.3445512

37. Shi, W., Cao, J., Zhang, Q., Li, Y., Xu, L.: Edge computing: vision and challenges. IEEE Internet Things J. **3**(5), 637–646 (2016). https://doi.org/10.1109/JIOT.2016.2579198

38. Srivastava, D., Karimi, M., Salimi, B.: Querying provenance for fairness. In: Proceedings of the 2021 International Conference on Management of Data (SIGMOD), pp. 573–578. ACM (2021). https://doi.org/10.1145/3448016.3457562

39. Stoyanovich, J.: Responsibility by design. In: Proceedings of the 2020 ACM SIGMOD International Conference on Management of Data, pp. 2833–2838. ACM (2020). https://doi.org/10.1145/3318464.3380591

40. Vargas-Solar, G.: The hidden voice of artificial intelligence: navigating the unseen barriers to inclusion in scientific discourse. Inclus-Elab **Special Issue**(2), 15–32 (2025). https://site.unibo.it/inclus-elab/en/documenti/inclus-elab-unibo-special-issue-2.pdf

41. Vargas-Solar, G.: Towards responsible and fair data science: resource allocation for inclusive and sustainable analytics. arXiv preprint arXiv:2502.11459 (2025)

42. Vargas-Solar, G., et al.: Techno/ecofeminism in action: fair and responsible resource allocation for sustainable data science pipelines. In: CEUR Workshop Proceedings, vol. 3946, pp. 1–9. CEUR-WS (2025)

43. Walter, M., Suina, M.: Indigenous data, indigenous methodologies and indigenous data sovereignty. Int. J. Soc. Res. Methodol. **22**(3), 233–243 (2019)

44. Zehlike, M., Pitoura, E., Castillo, C.: Fairness in rankings and recommendations: an overview. ACM SIGMOD Rec. **50**(2), 16–27 (2021). https://doi.org/10.1145/3476963.3476966

45. Zhu, Y., Hou, Y., Fu, G.: Understanding and improving the representativeness of climate models through high-performance cloud computing. Environ. Model. Softw. **96**, 75–89 (2017). https://doi.org/10.1016/j.envsoft.2017.06.017

Data Integration for Data Science: Solutions and Still Open Problems

Robert Wrembel$^{(\boxtimes)}$ ⓘ

Poznan University of Technology, Poznań, Poland
robert.wrembel@cs.put.poznan.pl

Abstract. This paper, accompanying a talk at the *ADBIS 2025 Doctoral Consortium School*, provides an overview of various data integration (DI) architectures - from virtual through physical to hybrid. It also presents an architecture for integrating stream data from robotic devices and introduces a novel concept for managing data source (DS) connectors. Additionally, the paper discusses selected current trends in applying machine learning to DI problems and outlines a few open research challenges. The insights presented in the talk and paper are drawn from our practical experience in data integration projects across the financial, IT, and intelligent farming sectors.

Keywords: data integration · data science · data integration process · machine learning

1 Introduction

Data science, as a research and technological area focuses on architectures, techniques, and algorithms for extracting knowledge from vast, distributed, and heterogeneous datasets. It applies various research and technological solutions, like integration architectures; data homogenization, cleaning, imputation, (a.k.a. data wrangling); analytics - from on-line analytical processing to analyze trends to advanced machine learning (ML) algorithms to build prediction models. Each of these solutions represent a distinct field of research itself. A fundamental task in data science is data integration (DI), i.e., before being analyzed, data must be made available in a unified format suitable for a particular type of analysis. For example, data are organized into the multidimensional data model for standard on-line analytical processing, whereas for building prediction models by means of neural networks data are transformed into vectors of values.

Therefore, the first fundamental task in data science is to build a DI architecture. The architectures must provide data not only in a common format, suitable for analytics, but also clean (curated data), so that analytical techniques produce reliable results.

In this paper, which accompanies my talk at the *ADBIS 2025 Doctoral Consortium School*[1], I will cover core solutions in DI. With an extensive (but not

---

[1] https://adbis2025.github.io/Calls/doctoral-consortium-school.html.

© The Author(s), under exclusive license to Springer Nature Switzerland AG 2026
P. K. Chrysanthis et al. (Eds.): ADBIS 2025, CCIS 2676, pp. 216–229, 2026.
https://doi.org/10.1007/978-3-032-05727-3_21

exhaustive) yet manageable bibliography, PhD students can easily find additional materials on the topics discussed, laying the groundwork for more comprehensive study. Section 2 overviews multiple DI architectures - from virtual through physical to hybrid. Section 3 outlines an architecture for handling robotic devices and their dynamicity. Section 4 presents a novel concept of managing DS connectors, currently promoted by IBM.

Recent advancements in artificial intelligence (AI) are also reflected in their application to data integration; some trends and examples are listed in Sect. 5. Even though data integration has been an active field of research for decades, there are still open research and technological problems, which will also be outlined in the talk (Sect. 5). The content of this paper and the talk are based on the experience gained from a few data integration projects that we have done in the financial, IT, and intelligent farming industries. Our previous publications serve as a foundation for the content presented in this talk and paper, e.g., [7–10, 15, 16, 85–88].

2   Data Integration Architectures

Over time, a few fundamental data integration architectures have been proposed and developed, falling into three main categories: virtual, materialized, and hybrid. All these architectures aim at making available data coming from heterogeneous, distributed, and independent data sources (DSs). Typically, DSs are connected by a *data integration layer*. In general, this layer is responsible for the following tasks: (T1) ingesting data from DSs, (T2) unifying data into the same data model, data structures, and formats suitable for analytical and ML/AI applications, (T3) cleaning (e.g., removing errors, inconsistencies) and homogenizing values, (T4) discovering duplicated data (a.k.a. data deduplication, entity resolution, entity matching), and (T5) making integrated data available for applications. These tasks are implemented by means of *DI processes*.

A generalized DI architecture is shown in Fig. 1. *Integrated data* delivered by the DI layer can be made available either as virtual, as materialized, or as hybrid. In a virtual DI architecture data reside in DSs and are integrated on demand. In a materialized architecture, integrated data are persistently store in a data repository. In a hybrid architecture, some data are integrated on demand, whereas other data are stored persistently.

Virtual DI architectures include **federated databases** [19, 31, 66] and **mediator**-based systems [20, 84]. In these architectures, data are made available on demand by a software that is responsible not only for tasks T1-T5 but also for: decomposing user queries into sub-queries and routing them into appropriate DSs for execution, transforming the routed sub-queries into programs understandable and executable in the DSs, and transforming and integrating results returned by the queried DSs. The main difference between the federated and mediated architecture is that the first one is used to integrate databases built on the same data model (relational). The mediated architecture is applied to integrate not only databases but also other types of DSs.

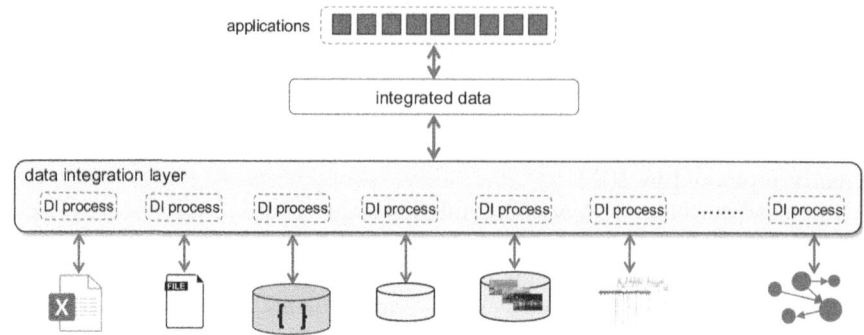

Fig. 1. A generalized data integration architecture

The first representative of a physical DI architecture is a **data warehouse** (DW) [81]. The integration is implemented by means of extract-transform-load (ETL) processes [74,79] (or their ELT variant). These processes run the afore-mentioned tasks T1-T5. This architecture is foremost in application domains like insurance, finances, trading, sales, which process large volumes of simple data, e.g., strings, numbers, and dates.

The standard DW architecture extended with capabilities of collecting data that arrive constantly to a system as data streams is called **lambda** [38,55]. It includes two data processing lanes - the standard, which stores data in a DW and the real-time one, which accesses data stream. The architecture was developed in order to be able to merge and analyze static data from a DW and stream data at the same time. Both lanes are integrated in a dedicated integration layer.

For integrating big data, a **data lake** (DL) was proposed. It is a repository that stores heterogeneous data ingested from DSs in their original formats [41, 56]. Such data have to be further homogenized by DI processes, to produce data available for applications, e.g., [45]. In a pure DL architecture, data are unified on demand, like in the virtual architectures. This class of DI architectures is **hybrid**, as it combines physical and virtual integration.

In a **data lakehouse** [33,43,72,93] data coming from a DL are first unified by DI processes and then physically stored in one or more DWs, which are part of the whole architecture. Each DW provides data prepared for specific analytical applications.

Two technological concepts called (1) a **data mesh** [14,28,39,62] and (2) **data spaces** [11,34,52,80] have recently gained popularity in research and business, e.g., [48,71]. Both of them offer offer on demand lightweight integration. Both concepts can be regarded as a data architecture and data governance approach, where data ownership is decentralized. Each component in a mesh or data space is a DS having a dedicated owner responsible for maintaining its data clean and up-to-date. Such a DS can be made accessible via a standardized interface.

A possible implementation of a data mesh and data spaces is by means of micro-services [2], i.e., software components that offer specialized services, e.g., accessing a given DS, error detection, data wrangling. These components are independent on each other and communicate over a network. The modularity of services allows for independent development, deployment, management, adequate resource allocation, implementation in a language the most suitable for a given task, extensions with new functionalities (e.g., by deploying a new service), and scaling of individual services. Architectures based on micro-services are frequently adopted to facilitate the integration and movement of large data volumes across heterogeneous systems [3, 23, 68].

An orthogonal classification of DI architectures distinguishes: polyglot, multi-store, and polystore systems (see [77] for an overview). **Polyglot** allows to access multiple DSs built on the same data model by means of multiple access interfaces, e.g., SQL-like, procedural. **Multi-store** allows to integrate DSs built on various data models and to access data via a single interface, e.g., a query language. **Polystore** allows to integrate DSs built on multiple data models by means of multiple access interfaces.

3   Data Integration Architecture for IoRT

In recent years, traditional agriculture undergoes dramatic changes towards the so-called smart, precision, or sustainable agriculture. These changes aim at making food production more efficient, sustainable, and harmless to the environment. This goal can only be achieved if farm management is data driven. To this end, the Internet of Robotic Things (IoRT) devices (e.g., sensors, cameras, and scanners deployed on both ground and aerial robots), data engineering technologies, and data analytics technologies are applied to farm management. Multiple R&D projects have already started; see for example a few projects in the intelligent/eco farming industry: GIS4IoRT [24], GreenFieldData [53], SPARKLE [6], Boost [1], Epi-Agri [4].

Data driven farm management can be achieved only if data produced by IoRT devices are made available for analytics in an integrated and timely manner. This requirement transforms to a DI architecture for IoRT.

IoRT devices produce highly heterogeneous data that differ in types, formats, and structures. For example, simple sensors (e.g., for measuring humidity, temperature) produce numbers, timestamps, and texts; imaginary devices produce 2-dimensional static images, video sequences (e.g., cameras), and 3-dimensional images (e.g., LiDaR scanners). All these data are stamped with time, to form time-series and trajectories [13].

The IoRT devices produce data that are available as streams. Such data have to be accessible in real time and at the same time they have to be stored in a persistent repository (like a DW or DL). Real-time access to data (real-time querying) is needed to monitor robots, their work, and crops (or animals). Data from the repository are used for analyzing trends, building optimization models, and building prediction models.

Moreover, the IoRT infrastructure is highly dynamic. It results from: (1) new robotic devices that can be dynamically deployed in fields and (2) unstable, limited, or unavailable WiFi in fields, causing that devices moving into areas without a network coverage disappear temporarily from the system. As a consequence, accessing data produced by them is limited or impossible.

Querying such a dynamic system requires additional functionality. First, the system must be able to dynamically include an IoRT device into the architecture, once the device is operational and within a WiFi range. Second, the system must be able to dynamically estimate a query execution time and be able to route a query to the appropriate data source (device), e.g., offering the fastest response or the highest data quality. Third, the results of queries must be equipped with metadata describing the quality of the result, like: an image resolution decrease or the completeness of a query result, which allows to estimate how much data is missing due to temporarily unavailable devices.

In order to address the aforementioned challenges we proposed a specialized data integration architecture [89]. Its main components are visualized in Fig. 2. Components marked as *DI process* represent wrappers [20] to DSs that are: (1) robotic devices including sensors, 2D cameras, and 3D scanners (like LiDaR) and (2) a repository of historical data produced by these devices. The robotic devices produce streams of data that are ingested by a *queuing system*. Then queues of interest are accessed by DI processes. At the same time, the streaming data are uploaded into the *data repository*. Its content is queried by applications that search for correlation patterns and trends.

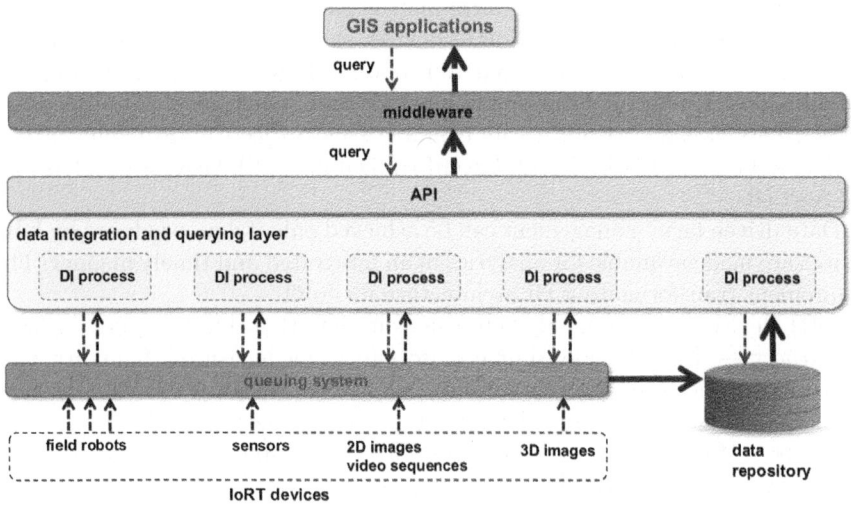

Fig. 2. The architecture of the *GIS4IoRT* system

Data provided by these DSs are pre-processed, integrated (as much as possible), and correlated by the *data integration and querying layer*, accessible via *API*. The correlation applies to data of different modalities that are related to the same real-world phenomenon. For example, text data describing a field (geographical coordinates and dimensions, the type of a crop cultivated there, the type of soil) can be correlated with images of this field. This layer is also responsible for translating queries arriving from GIS applications via *middleware*.

Typically, IoRT architectures process data at *edge*, i.e., on devices and on *fog*, i.e., at an intermediary layer. Such an architecture requires particular **resource management**, which is provided by the *middleware*.

4   Connectors as a Service Architecture

DS connectors are integral components of every DI architecture. They offer standardized interfaces for accessing DSs, which use different data access protocols and data formats. Connectors are typically available as software - libraries of connectors (LCs). Notice that the number of different data sources (and connectors) reaches more than 700 [5]. Big DI integration projects connect at least dozens of different DSs. For such architectures the standard management of connectors by means of LCs expose problems in [17,18], which are briefly listed in this section.

– Maintainability of connectors - they cannot be easily deployed and migrated between platforms, taking into account their multiple versions and dependencies between these versions and versions of DSs. Moreover, connectors have to be upgraded regularly (e.g., security patches, new capabilities, licensing). As a consequence, all services that embed these connectors are impacted.
– Security measures, like encryption, authentication, access control, data integrity, and auditing mechanisms must be implemented in DSs and in an integration system, on top of connectors.
– Scalability - the throughput provided by connectors for variable data volumes and large number of parallel ingesting processes may be low. Connectors in an LC may not be optimized for every possible use case, and may not provide the best performance or scalability for a particular application. As a consequence, multiple different implementations (often in different programming languages) of the same connector may be needed, which again increases the complexity of software dependencies.

For a flexible integration of multiple DSs, the *Library of Connectors as a Service* (LCaaS) may offer some advantages [17]. Its basic components are shown in Fig. 3. Connectors to various DSs located in the *LCaaS* layer map various native interfaces of DSs into a common access interface. This interface is used in *applications* willing to access the DSs by means of the *connection server*. The *dispatcher* is responsible for instantiating a connector for a given DS and forwarding request to this connector, similarly as the task of a mediator in the mediated DI architecture.

Such an approach frees developers from the complexities of managing connectors and therefore enhances his/her productivity by embedding crucial services directly within the LCaaS layer. This means the LCaaS layer takes the responsibility of: connector management (including upgrades, versioning, and maintaining software dependencies), metadata management, data governance,

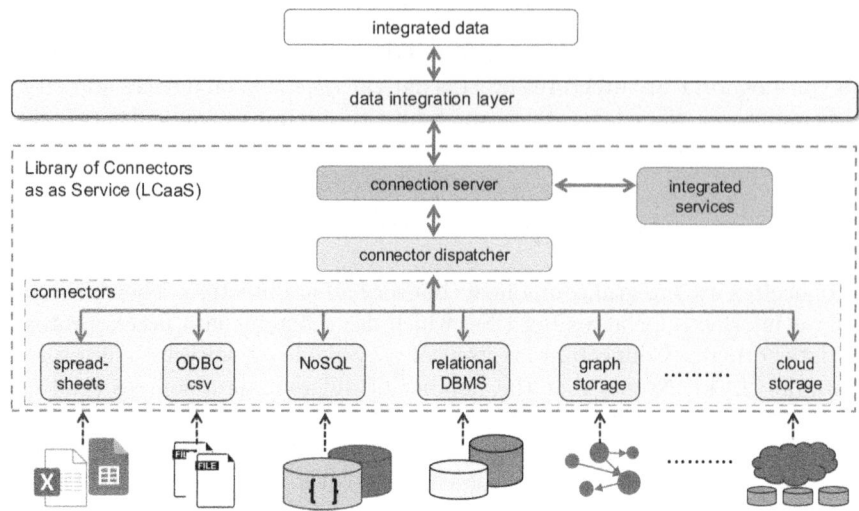

Fig. 3. The architecture of the *Library of Connectors as a Service*

and the secure handling of data vaults/credentials. It also handles data access policy, provides vital data access monitoring, and collects runtime statistics.

5   Trends and Open Problems

This paper summarizes selected topics in data integration, i.e., core architectures for integrating: (1) table-like data and complex data, (2) robotic devices, and (3) for offering a pool of DS connectors as a service (which is promoted and developed by IBM).

New trends show the application of ML/AI in various tasks of data integration.

- Detecting errors and outliers: popular methods include clustering (e.g., [35, 70,76,83]) and neural networks, mostly for time series (e.g., [25,37,91]).
- Imputing missing data: recent works focus on applying neural networks that learn from past interactions attribute values that are probable for imputing (e.g., [40,50,51,90,92]).
- Data deduplication has been very intensively researched (see for example state-of-the-art reports [9,26,59]). Recent trends focus on classification models (e.g., [30,67,73]), learning schemes for record comparisons (e.g., [12,36,47]), learning record matching rules (e.g., [27]), and active/self learning techniques (e.g., [22,44,46,82]). Recent ML methods apply complex neural networks (e.g., [21,29,54,78,94,95]) as the main component of pre-trained language models or (recently) large language models.
- Schema matching identifies semantically equivalent schema objects across diverse DSs. ML models are applied for learning patterns and relationships

from existing (typically human annotated) data and mappings. The learned patterns and matching rules are then used for matching unknown schemas [64]. Based on the same idea, recent solutions apply neural networks [57], pre-trained language models [58,96,97], and LLMs for schema matching [63,69].

– Designing DI pipelines can be supported by a recommender engine that learns from historical designs of pipelines, like in [65].

Even though the aforementioned novel solutions are able to solve some problems in DI, these solutions still need to be researched further. The below list of open problems reflects my personal view on the topic.

– Methods for data wrangling proved to work well for simple data. Industry domains like agriculture, smart cities, and healthcare produce large volumes of highly heterogeneous data that also have to be cleaned. Data complexity makes the process of data wrangling much more complex and time costly.

– Deduplication is a complex and well-addressed problem in research and industry, with numerous solutions available. It can generally be addressed in two ways: either as a four-stage pipeline (detailed for example in [26,32,49,60,61]) or as a black-box machine learning/AI model. In the first case, each task is supported by multiple algorithms and the problem is which algorithm to choose for a given dataset. An end-to-end recommender approach is still missing. In the second case, the deduplication model suffers from explainability and it needs a substantial set of training data, which may often be unavailable.

– DI pipelines process large volumes of data and therefore their execution must be efficient. A DI pipeline is typically build by a human from pre-defined components available in a design environment. Some software assists in designing a pipeline by proposing the right component to choose. However, there is no guarantee that a design pipeline is optimal. For this reason, cost-based optimization methods (using similar techniques as in SQL query optimization) are tempting. But the problem here is much complex than in query optimization. First, cost functions of building blocks of a pipeline may be difficult to formulate. Second, the performance of the pipeline depends on the sequence of tasks and re-ordering them is a computationally complex problem [42,75]. This difficulty is further increased when user defined functions (UDFs) are included in a pipeline. Since UDFs are often implemented as black-boxes neither their internals nor cost models are available.

To sum up, even though DI research has been active for decades and even though AI solutions help in developing some DI tasks, there are still hard open problems to solve. What is more, the application of AI solutions generate new interesting problems.

Acknowledgements. this research is partially funded from the EU project *Chist-Era call 2023*, entitled *Development of a Plug-and-Play Middleware for Integrating Robot Sensor Data with GIS Tools in a Cloud Environment*, managed by the National Science Centre (NCN), Poland, grant no. 2024/06/Y/ST6/00136.

References

1. Boost. ERASMUS-EDU-2021-PI-ALL-INNO. https://www.project-boost.eu/
2. Build software with fine-grained, loosely coupled services. https://developer.ibm. com/depmodels/microservices/. IBM Developer
3. Cloud Native Computing Foundation annual survey. https://www.cncf.io/reports/ cncf-annual-survey-2022/
4. EIP-AGRI Network. https://ec.europa.eu/eip/agriculture/
5. Fivetrain platform. Accessed June 2025
6. SPARKLE: Sustainable Precision Agriculture: Research and Knowledge for Learning). Erasmus+ project. https://sparkle-project.eu/
7. Ali, S.M.F., Wrembel, R.: From conceptual design to performance optimization of ETL workflows: current state of research and open problems. VLDB J. **26**(6), 777–801 (2017). https://doi.org/10.1007/s00778-017-0477-2
8. Ali, S.M.F., Wrembel, R.: Framework to optimize data processing pipelines using performance metrics. In: International Conference on Big Data Analytics and Knowledge Discovery (DAWAK). LNCS 12393 (2020)
9. Andrzejewski, W., Bębel, B., Boiński, P., Wrembel, R.: On customer data deduplication - research vs. industrial perspective: - lessons learned from a r&d project in the financial sector. In: European Conference on Advances in Databases and Information Systems, volume 2186 of *CCIS*. Springer, (2024)
10. Andrzejewski, W., Bębel, B., Boiński, P., Wrembel, R.: On tuning parameters guiding similarity computations in a data deduplication pipeline for customers records: experience from a R&D project. Inform. Syst. **121** (2024)
11. Ayala, C.P., Bilalli, B., Gómez, C., Mazón, J.N., Romero, O.: Challenges to enforce data quality in data spaces. In: International Workshop on Design, Optimization, Languages and Analytical Processing of Big Data (DOLAP) @EDBT/ICDT, volume 3931. CEUR-WS.org (2025)
12. Bianco, G.D., Gonçalves, M.A., Duarte, D.: BLOSS: effective meta-blocking with almost no effort. Inform. Syst. **75** (2018)
13. Bimonte, S., et al.. Technological and research challenges in data engineering for sustainable agriculture. In: International Workshop on Big Data in Emergent Distributed Environments (BiDEDE) @SIGMOD. ACM (2024)
14. Bode, J., Kühl, N., Kreuzberger, D., Hirschl, S., Holtmann, C.: Data mesh: best practices to avoid the data mess. CoRR, abs/2302.01713 (2023)
15. Bodziony, M., Ciesielski, B., Lehnhardt, A., Wrembel, R.: On reasoning about black-box UDFs by classifying their performance characteristics. In: International Conference on Information Systems Development (ISD) (2024)
16. Bodziony, M., Morawski, R., Wrembel, R.:. Evaluating push-down on nosql data sources: experiments and analysis paper. In: International Workshop on Big Data in Emergent Distributed Environments(BiDEDE) @ ACM SIGMOD/PODS Conference. ACM (2022)
17. Bodziony, M., Wrembel, R.: Data source connectors layer as a service - design patterns. In: Intereanional Workshop on Design, Optimization, Languages and Analytical Processing of Big Data (DOLAP), vol. 3369 of CEUR Workshop Proceedings. CEUR-WS.org (2023)
18. Bodziony, M., Wrembel, R.: On developing data connectivity services for industrial applications. In: International Conference on Information Systems Development (ISD) (2024)

19. Bouguettaya, A., Benatallah, B., Elmargamid, A.: Interconnecting Heterogeneous Information Systems. Kluwer Academic Publishers, ISBN 0792382161 (1998)
20. Brezany, P., Tjoa, A.M., Wanek, H., Wöhrer, A.: Mediators in the architecture of grid information systems. In: International Conference Parallel Processing and Applied Mathematics (PPAM), vol. 3019 of LNCS. Springer (2003)
21. Chen, R., Shen, Y., Zhang, D.: Gnem: a generic one-to-set neural entity matching framework. In: The Web Conference (WWW). ACM (2021)
22. Chen, X., Xu, Y., Broneske, D., Durand, G.C., Zoun, R., Saake, G.: Heterogeneous committee-based active learning for entity resolution (healer). In: European Conference on Advances in Databases and Information Systems (ADBIS), volume 11695 of LNCS. Springer (2019)
23. Chia, D.: Scaling data pipelines on kubernetes. https://airbyte.com/blog/scaling-data-pipelines-kubernetes (2020)
24. Chist-Era. Development of a plug-and-play middleware for integrating robot sensor data with gis tools in a cloud environment (GIS4IoRT). Chist-Era Project Call 2023. https://www.chistera.eu/projects-call-2023, https://ncn.gov.pl/sites/default/files/pliki/chistera-wrembel-en.pdf
25. Cho, Y., Lee, J., Ham, G., Jang, D., Kim, D.: Generality-aware self-supervised transformer for multivariate time series anomaly detection. Appl. Intell. **55**(7) (2025)
26. Christophides, V., Efthymiou, V., Palpanas, T., Papadakis, G., Stefanidis, K.: An overview of end-to-end entity resolution for big data. ACM Comput. Surv. **53**(6), 127:1–127:42 (2021)
27. Cohen, W.W., Richman, J.: Learning to match and cluster large high-dimensional data sets for data integration. In: ACM SIGKDD International Conference on Knowledge Discovery and Data Mining (KDD). ACM (2002)
28. Dehghani, Z., et al.: Data Mesh: Delivering Data-Driven Value at Scale. O'Reilly, ISBN 1492092398 (2022)
29. Ebraheem, M., Thirumuruganathan, S., Joty, S., Ouzzani, M., Tang, N.: Distributed representations of tuples for entity resolution. VLDB Endowment **11**(11) (2018)
30. Elfeky, M.G., Verykios, V.S., Elmagarmid, A.K.: Tailor: A record linkage tool box. In: International Conference on Data Engineering (ICDE). IEEE Computer Society (2002)
31. Elmagarmid, A., Rusinkiewicz, M., Sheth, A.: Management of Heterogeneous and Autonomous Database Systems. Morgan Kaufmann Publishers, ISBN 1-55860-216-X, (1999)
32. Elmagarmid, A.K., Ipeirotis, P.G., Verykios, V.S.: Duplicate record detection: a survey. IEEE Transactions on Knowledge and Data Engineering **19**(1) (2007)
33. Errami, S.A., Hajji, H., Kadi, K.A.E., Badir, S.: Spatial big data architecture: From data warehouses and data lakes to the lakehouse. J. Parall. Distrib. Comput. **176** (2023)
34. Franklin, M.J.: Dataspaces: progress and prospects. In: Sexton, A.P. (ed.) BNCOD 2009. LNCS, vol. 5588, pp. 1–3. Springer, Heidelberg (2009). https://doi.org/10.1007/978-3-642-02843-4_1
35. Fumanal-Idocin, J., Rodríguez-Martínez, I., Indurain, A., Minárová, M., Bustince, H.: Almost aggregations in the gravitational clustering to perform anomaly detection. Inform. Sci. **612** (2022)
36. Gagliardelli, L., Papadakis, G., Simonini, G., Bergamaschi, S., Palpanas, T.: Generalized supervised meta-blocking. VLDB Endowment **15**(9) (2022)

37. Gao, C., Ma, H., Pei, Q., Chen, Y.: Dynamic graph-based graph attention network for anomaly detection in industrial multivariate time series data. Appl. Intell. **55**(6) (2025)
38. Gillet, A., Leclercq, É., Cullot, N.: Lambda+, the renewal of the lambda architecture: category theory to the rescue. In: La Rosa, M., Sadiq, S., Teniente, E. (eds.) CAiSE 2021. LNCS, vol. 12751, pp. 381–396. Springer, Cham (2021). https://doi.org/10.1007/978-3-030-79382-1_23
39. Goedegebuure, A., et al.: Data mesh: a systematic gray literature review. ACM Comput. Surv. **57**(1) (2025)
40. Gondara, L., Wang, K.: MIDA: multiple imputation using denoising autoencoders. In: Pacific-Asia Conference Advances in Knowledge Discovery and Data Mining (PAKDD), volume 10939 of LNCS (2018)
41. Hai, R., Koutras, C., Quix, C., Jarke., M.: Data lakes: a survey of functions and systems (extended abstract). In: International Conference on Data Engineering (ICDE) (2024)
42. Halasipuram, R., Deshpande, P.M., Padmanabhan, S.: Determining essential statistics for cost based optimization of an ETL workflow. In: EDBT (2014)
43. Harby, A.A., Zulkernine, F.H.: From data warehouse to lakehouse: a comparative review. In: IEEE Big Data (2022)
44. Jain, A., Sarawagi, S., Sen, P.: Deep indexed active learning for matching heterogeneous entity representations. VLDB Endowment **15**(1) (2021)
45. Jemmali, R., Abdelhédi, F., Zurfluh, G.: DLToDW: transferring relational and noSQL databases from a data lake. SN Comput. Sci. **3**(5) (2022)
46. Jurek, A., Hong, J., Chi, Y., Liu, W.: A novel ensemble learning approach to unsupervised record linkage. Inform. Syst. **71** (2017)
47. Kejriwal, M., Miranker, D.P.: A two-step blocking scheme learner for scalable link discovery. In: International Workshop on Ontology Matching @ISWC, vol. 1317. CEUR-WS.org (2014)
48. Kernstock, P.: Building industrial data platforms - a case study on introducing data mesh. In: European Conf. on Information Systems (ECIS) (2024)
49. Köpcke, H., Rahm, E.: Frameworks for entity matching: a comparison. Data Knowl. Eng. **69**(2) (2010)
50. Krishnan, S., Wang, J., Wu, E., Franklin, M.J., Goldberg, K.: Activeclean: interactive data cleaning for statistical modeling. VLDB Endowment **9**(12) (2016)
51. Mecca, G., Papotti, P., Santoro, D., Veltri, E.: BUNNI: learning repair actions in rule-driven data cleaning. ACM J. Data Inform. Qual. **16**(2) (2024)
52. Morejón, A., Berenguer, A., de Espona, L., Tomás, D., Mazón, J:. Exploring content-based catalogs for enhanced discovery services in data spaces. In: International Workshop on Design, Optimization, Languages and Analytical Processing of Big Data (DOLAP) @EDBT/ICDT, vol. 3931. CEUR-WS.org (2025)
53. MSCA. IoRT data management and analysis for sustainable agriculture (GreenFieldData). HORIZON-MSCA-2024-DN-01 (Marie Skłodowska-Curie Actions Doctoral Networks 2024. https://maiage.inrae.fr/node/3242
54. Mudgal, S., et al.: Deep learning for entity matching: a design space exploration. ACM, In SIGMOD Int. Conf. on Management of Data (2018)
55. Munshi, A.A., Mohamed, Y.A.I.: Data lake lambda architecture for smart grids big data analytics. IEEE Access **6**, 40463–40471 (2018)
56. Nargesian, F., Zhu, E., Miller, R.J., Pu, K.Q., Arocena, P.C.: Data lake management: challenges and opportunities. VLDB Endowment **12**(12) (2019)d

57. Oh, H., Kulvatunyou, B.S., Jones, A.T., Finin, T.: Employing word-embedding for schema matching in standard lifecycle management. J. Industri. Inform. Integr. **38** (2024)
58. Pan, Z., Yang, M., Monti, A.: Schema matching based on energy domain pretrained language model. Energy Inform. **6**(1) (2023)
59. Papadakis, G., Ioannou, E., Palpanas, T.: Entity Resolution: Past, Present And Yet-to-come. In Int, Conf on Extending Database Technology (EDBT) (2020)
60. Papadakis, G., Skoutas, D., Thanos, E., Palpanas, T.: Blocking and filtering techniques for entity resolution: a survey. ACM Comput. Surv. **53**(2) (2020)
61. Papadakis, G., Tsekouras, L., Thanos, E., Giannakopoulos, G., Palpanas, T., Koubarakis, M.: Domain- and structure-agnostic end-to-end entity resolution with jedai. SIGMOD Record **48**(4) (2019)
62. Papp, A., Bub, U., Lähteenoja, V., Kuikkaniemi, K., Turpeinen, M., Jokela, S.: Data mesh and data space: a comparative analysis with a focus on governance. In: International Conference Innovations for Community Services, vol. 2513 of CCIS. Springer (2025)
63. Parciak, M., Vandevoort, B., Neven, F., Peeters, L.M., Vansummeren, S.: Schema matching with large language models: an experimental study. In: Workshops @VLDB (2024)
64. Rahm, E., Bernstein, P.A.: A survey of approaches to automatic schema matching. VLDB J. **10**(4) (2001)
65. Redyuk, S., Kaoudi, Z., Schelter, S., Markl, V.: Assisted design of data science pipelines. VLDB J. **33**(4) (2024)
66. Rusinkiewicz, M., Czejdo, B.D., Embley, D.W.: An implementation model for muldidatabase queries. In: International Conference on Database and Expert Systems Applications (DEXA). Springer (1991)
67. Sarawagi, S., Bhamidipaty, A.: Interactive deduplication using active learning. In: ACM SIGKDD International Conference on Knowledge Discovery and Data Mining (KDD). ACM (2002)
68. Saucedo, A.C.M., Rodríguez, G., Rocha, F.G., dos Santos, R.P.: Migration of monolithic systems to microservices: a systematic mapping study. Inform. Softw. Technol. **177** (2025)
69. Seedat, N., van der Schaar, M.: Matchmaker: self-improving large language model programs for schema matching. CoRR abs/2410.24105 (2024)
70. Shi, P., Zhao, Z., Zhong, H., Shen, H., Ding, L.: An improved agglomerative hierarchical clustering anomaly detection method for scientific data. Concurr. Comput.: Pract. Exper. **33**(6) (2021)
71. Sienkiewicz, M.: From data silos to data mesh: a case study in financial data architecture. In: International Conference on Database and Expert Systems Applications (DEXA), volume appear of LNCS (2025)
72. Sienkiewicz, M., Wrembel, R.: Managing data in a big financial institution: conclusions from a r&d project. In: Workshops of the EDBT/ICDT Joint Conference
73. Silva, J.A., Pereira, D.A.: A multiclass classification approach for incremental entity resolution on short textual data. Int. J. Business Intell. Data Mining **18**(2) (2021)
74. Simitsis, A., Skiadopoulos, S., Vassiliadis, P.: The history, present, and future of ETL technology (invited). In: International Workshop on Design, Optimization, Languages and Analytical Processing of Big Data (DOLAP) @EDBT/ICDT, vol. 3369 CEUR Workshop Proceedings (2023)
75. Simitsis, A., Vassiliadis, P., Sellis, T.K.: State-space optimization of ETL workflows. IEEE Trans. Knowl. Data Eng. **17**(10), 1404–1419 (2005)

76. Song, S., Li, C., Zhang, X.: Turn Waste Into Wealth: On Simultaneous Clustering And Cleaning Over Dirty Data. ACM, In ACM SIGKDD Int. Conf. on Knowledge Discovery and Data Mining (2015)

77. Tan, R., Chirkova, R., Gadepally, V., Mattson, T.G.: Enabling query processing across heterogeneous data models: a survey. In: IEEE Big Data (2017)

78. Thirumuruganathan, S., et al.: Deep learning for blocking in entity matching: a design space exploration. VLDB Endowment **14**(11) (2021)

79. Thomsen, C.: ETL. Springer, In Encyclopedia of Big Data Technologies (2019)

80. Trujillo, J., Candela, G., Reina-Reina, A.: Data analytics and artificial intelligence in the new scenario of data spaces. In: International Workshop on Design, Optimization, Languages and Analytical Processing of Big Data (DOLAP) @EDBT/ICDT, vol. 3931. CEUR-WS.org (2025)

81. Vaisman, A.A., Zimányi, E.: Data Warehouse Systems - Design and Implementation, 2nd edn. Springer, Data-Centric Systems and Applications (2022)

82. Wang, Q., Vatsalan, D., Christen, P.: Efficient interactive training selection for large-scale entity resolution. In: Pacific-Asia Conference Advances in Knowledge Discovery and Data Mining (PAKDD), volume 9078 of LNCS. Springer (2015)

83. Wenz, V., Kesper, A., Taentzer, G.: Clustering heterogeneous data values for data quality analysis. ACM J. Data Inform. Qual. **15**(3) (2023)

84. Wiederhold, G.: Mediators in the architecture of future information systems. Computer **25**(3) (1992)

85. Wrembel, R.: Data integration, cleaning, and deduplication: Research versus industrial projects. In: International Conference on Information Integration and Web Intelligence (iiWAS), vol. 13635 LNCS (2022)

86. Wrembel, R.: Data integration revitalized: From data warehouse through data lake to data mesh. In: International Conference on Database and Expert Systems Applications (DEXA), vol. 14146 LNCS (2023)

87. Wrembel, R.: Data integration in the AI era: research trends and still open issues. In: International Conference on Big Data Analytics and Knowledge Discovery (DAWAK), volume to appear of LNCS (2025)

88. Wrembel, R.: On three missing pieces in the data integration puzzle. In: Workshops of the EDBT/ICDT Joint Conference volume 3946. CEUR-WS.org (2025)

89. Wrembel, R.: On integrating robotic data with GIS tools in a cloud environment. In: Workshops of the EDBT/ICDT Joint Conference, vol. 3946. CEUR-WS.org (2025)

90. Wu, R., Zhang, A., Ilyas, I.F., Rekatsinas, T.: Attention-based learning for missing data imputation in HoloClean. In: Conference on Machine Learning and Systems (MLSys) (2020)

91. Xu, D., Xia, T., Hou, J., Xiang, Y., Xuan, Q.: MSR-GAN: multi-scales decomposition representations for unsupervised anomaly detection. Appl. Intelli. **55**(8) (2025)

92. Yoon, J., Jordon, J., van der Schaar, M.: GAIN: missing data imputation using generative adversarial nets. In: International Conference on Machine Learning (ICML), volume 80 of Machine Learning Research (2018)

93. Zaharia, M., Ghodsi, A., Xin, R., Armbrust, M.: Lakehouse: a new generation of open platforms that unify data warehousing and advanced analytics. In: Conference on Innovative Data Systems Research (CIDR),(2021)

94. Zeakis, A., Papadakis, G., Skoutas, D., Koubarakis, M.: Pre-trained embeddings for entity resolution: an experimental analysis. VLDB Endowment **16**(9) (2023)

95. Zhang, W., Wei, H., Sisman, B., Dong, X.L., Faloutsos, C., Page, D.: Autoblock: a hands-off blocking framework for entity matching. In: International Conference on Web Search and Data Mining (WSDM). ACM (2020)
96. Zhang, Y., Di, M., Luo, H., Xu, C., Tsai, R.T.: SMUTF: schema matching using generative tags and hybrid features. Inform. Syst. **133** (2025)
97. Zhang, Y., et al.: Schema matching using pre-trained language models. In: International Conference on Data Engineering (ICDE) (2023)

MADEISD 2025: 7th Workshop on Modern Approaches in Data Engineering and Information System Design

Enhancing Data Interoperability in Multi-platform Lakehouses with Apache Iceberg

Muhammad Hassan Shafiq$^{(\boxtimes)}$, Zheying Zhang(iD), and Kostas Stefanidis(iD)

Tampere University, Tampere, Finland
hassanshafiq17@outlook.com,
{zheying.zhang,konstantinos.stefanidis}@tuni.fi

Abstract. Managing data across diverse platforms poses significant challenges, including data duplication, vendor lock-in, and inconsistent governance. Lack of a unified table format often leads to complex pipelines, increased storage costs, and hindered interoperability. Apache Iceberg, with its platform-agnostic design, presents a solution by providing a consistent table format for large-scale analytical workloads while addressing cross-platform data accessibility. In this paper, we study the use of Apache Iceberg as a unified table format to enable interoperability between Snowflake and Databricks, with data stored on Amazon S3. Experimental setups include accessing Snowflake-managed Iceberg tables in Databricks and vice versa. Key focus areas include examining query performance, metadata synchronization, and the challenges of managing consistent data across platforms. Optimization strategies, specifically data reordering, were applied to test improvements in query performance for various workloads. The results show that Iceberg reduces the complexity of data management by automating metadata handling and synchronization, ensuring real-time data consistency. Query performance showed improvement in medium-complexity queries with optimized Iceberg tables, while highlighting potential areas for further optimization in full-table scans. These findings underscore Iceberg's potential as a scalable, efficient solution for modern data lake architectures.

Keywords: Data Interoperability · Metadata Synchronization · Data Lakes

1 Introduction

The rapid expansion of data in enterprises has driven organizations to adopt cloud-based solutions due to their flexibility, ease of resource provisioning, and cost-effective pay-as-you-go models. Traditional data warehouses have been widely used for structured data processing, offering strong consistency and query optimization. However, they are expensive and often lack the scalability needed for handling semi-structured and unstructured data [1,7,11], leading to the rise

P. K. Chrysanthis et al. (Eds.): ADBIS 2025, CCIS 2676, pp. 233–247, 2026.
https://doi.org/10.1007/978-3-032-05727-3_22

of data lakes. While data lakes provide scalable and cost-efficient storage, they suffer from governance issues, data inconsistency, and slow query performance, making them inefficient for enterprise analytics [6]. To address these challenges, the data lakehouse architecture has emerged, combining the transactional support of data warehouses with the flexibility and scalability of data lakes [2].

According to a report from Virtana [12], over 80% of enterprises have adopted a multi-cloud strategy, with 78% running workloads across more than three public cloud platforms. Enterprises use these multiple cloud platforms, including Databricks[1], Snowflake[2], and others to address diverse needs: some excel in big data and machine learning, while others specialize in SQL analytics and business reporting. This multi-platform approach introduces significant challenges, including data duplication, manual synchronization, and increased storage costs. Moving large volumes of data between platforms requires custom integration pipelines, leading to high operational overhead and governance complexities. Furthermore, metadata management varies across platforms, making schema evolution and access control enforcement inconsistent. Query performance also differs due to variations in storage optimization and indexing strategies. As Mone has pointed out, "Metadata management has emerged as a significant bottleneck in big data systems, where different platforms often have disparate ways of accessing and managing data" [3]. Without a standardized approach, ensuring real-time data synchronization across platforms remains a major challenge.

To mitigate these issues, organizations require a unified table format that facilitates efficient metadata handling, transactional consistency, and interoperability across cloud platforms. Furthermore, a unified table format would enhance cross-platform data sharing and accessibility. By having a common metadata structure, organizations could streamline the integration of data pipelines, reducing the risk of data mismatches and versioning errors. This consistency would support more accurate analytics and reporting, enabling them to make data-driven decisions with greater confidence. As multi-platform data ecosystems expand, a unified table format offers a clear path to building flexible, efficient architectures that leverage each platform's strengths without the complexity of managing separate systems. It enables a single source of truth, reduces operational overhead, and supports diverse analytics without redundant data copies. Several open table formats, including Delta Lake[3] Apache Hudi[4] and Apache Iceberg[5] have been developed to address these needs. While Delta Lake is tightly integrated with the Databricks ecosystem and Apache Hudi is optimized for real-time ingestion, Apache Iceberg has gained significant traction due to its vendor-neutral architecture, broad compatibility, and advanced data management capabilities. Iceberg provides ACID (Atomicity, Consistency, Isolation, Durability) transactions, hidden partitioning, schema evolution without

---

[1] https://www.databricks.com/.
[2] https://www.snowflake.com/en/emea/.
[3] https://delta.io/.
[4] https://hudi.apache.org/.
[5] https://iceberg.apache.org.

full table rewrites, and time travel for historical analysis. Its ability to support multiple query engines, including Spark[6] Trino[7] Flink[8], Hive[9], Databricks[10], Snowflake[11], and Redshift[12], makes it a strong candidate for enabling multi-platform data interoperability.

This work aims to understand the structure of Apache Iceberg, its features, how it fits in the modern data architecture for cross-platform integrations, and the reasons behind its growing adoption across cloud providers. This study also explores the use of Iceberg as a unified table format for cross-platform data access between Snowflake and Databricks, leveraging S3[13] as the underlying storage layer. It aims to assess query performance across different workloads, analyzing execution latency for simple, medium, and complex queries. Additionally, it investigates Iceberg's ability to maintain real-time data synchronization and schema evolution across platforms, ensuring consistency without requiring extensive manual intervention. By identifying potential limitations and areas for improvement, this research contributes to a deeper understanding of Iceberg's role in modern multi-platform data architectures and its potential to drive seamless, efficient, and interoperable data management.

The rest of the paper is structured as follows: Sect. 2 covers related work; Sect. 3 uncovers the current integration option and structure of Iceberg Table; Sect. 4 presents a bi-directional integration between Snowflake and Databricks; Sect. 5 outlines the experimental design; Sect. 6 presents results; and Sect. 7 concludes with key contributions.

2   Background and Related Work

In the early days of data management, data was primarily stored in flat files, which provided a simple storage mechanism but suffered from redundancy, inconsistency, and lack of efficient query capabilities. As data volumes increased, Relational Database Management Systems (RDBMS) emerged, inspired by the Relational Model [4], to offer structured storage with ACID transactions ensuring data integrity. While RDBMS provided efficient indexing and query optimization, they were limited by proprietary storage formats that were tightly coupled with their specific implementations. The increasing demand for large-scale distributed data processing led to the development of Hadoop [4], which introduced the Hadoop Distributed File System (HDFS) for scalable storage and MapReduce as a parallel processing framework. However, writing complex MapReduce jobs required significant expertise, limiting its usability. To address this, Apache

---

[6]   https://iceberg.apache.org/spark-quickstart/.
[7]   https://trino.io/docs/current/connector/iceberg.html.
[8]   https://iceberg.apache.org/docs/1.4.3/flink-connector/.
[9]   https://iceberg.apache.org/docs/latest/hive/.
[10] https://docs.databricks.com/en/delta/uniform.html.
[11] https://docs.snowflake.com/en/user-guide/tables-iceberg.
[12] https://docs.aws.amazon.com/redshift/latest/dg/querying-iceberg.html.
[13] https://aws.amazon.com/s3/.

Hive[14] introduced an SQL-like language, i.e. HiveQL, making it easier for analysts to query big data without extensive programming knowledge.

As data processing evolved, columnar file formats such as Apache Parquet[15], ORC[16] were introduced to improve storage efficiency and query performance. These formats enabled faster analytical processing by optimizing data compression and retrieval patterns. As data needs grew beyond traditional systems, the concept of data lakes emerged, offering a more flexible, scalable way to store vast amounts of raw and processed data. Modern data lakes are cloud-based storage systems designed to accommodate data in its original format, whether it is structured, semi-structured, or unstructured [10]. Unlike traditional file systems or databases, data lakes decouple storage from computing, making it possible to scale them independently based on demand. Building on the foundation of HDFS, data lakes take the benefits of distributed storage to the next level. While HDFS was designed for on-premises environments, data lakes leverage cloud-native architecture to provide infinite scalability, high availability, and cost-effective storage. Unlike files in directories in HDFS, Data is stored as objects in data lakes, which are durable and can be accessed from anywhere. However, data lakes suffered from a lack of transactional consistency, governance challenges, and poor query performance. This makes it difficult to maintain data integrity across multiple analytical workloads [9].

To address the limitations of traditional data lakes, modern open table formats were developed to introduce transactional consistency, schema evolution, and efficient metadata management and support the lakehouse architecture [8]. The open table architecture (see Fig. 1) extends traditional data lake storage by implementing a metadata layer that tracks schema changes, partitions, and transactional states. By decoupling the table structure from the underlying storage, these formats offer enhanced data consistency, reliability, and easy management of large datasets in data lakes. These formats separate logical and physical data organization by abstracting the physical file structure. They track the table's state, including partitions, within a metadata layer at the file level.

Existing research has largely focused on comparing data warehouse and data lake architectures, examining how data lakes enable the storage and processing of both structured and unstructured data [1,10]. Additionally, studies have explored the emergence of the lakehouse architecture, which integrates the strengths of both approaches, offering enhanced performance and efficiency compared to querying raw files directly in data lakes [2].

Further research has extensively compared the most widely used lakehouse table formats, such as Delta Lake, Apache Hudi, and Apache Iceberg. These studies focus on performance benchmarking, transaction guarantees, schema evolution capabilities, and metadata handling efficiency [9]. Comparisons have analyzed read and write performance, assessing each format's ability to handle

---

[14] https://hive.apache.org/.
[15] https://parquet.apache.org/.
[16] https://orc.apache.org/.

real-world workloads involving frequent updates, deletes, and large-scale batch processing.

Data integration remains a fundamental challenge since the early days of database systems, particularly in merging data from diverse sources into a unified, coherent view. Research has explored the issues associated with cross-platform data integration, including schema heterogeneity, data quality, and query rewriting—especially in autonomous, distributed systems [13,14]. These challenges are further amplified in modern architectures, where data is spread across different cloud platforms, technologies, and formats.

Research has also explored various data integration techniques and identifies recurring challenges such as data heterogeneity, scalability, and performance across platforms [14]. These limitations often stem from a lack of standardized data representation and tight coupling between processing engines and storage formats.

This study tackles the challenges outlined above by proposing the use of the Apache Iceberg open table format to standardize and automate data integration across platforms. It specifically examines Iceberg's effectiveness in a multi-platform environment, focusing on its query performance, metadata synchronization, schema evolution, and scalability when used with Snowflake, Databricks, and data stored in Amazon S3. Unlike previous implementations that rely on copying terabytes of data across platforms—a process that is both tedious to maintain and prone to scalability and performance issues—this research aims to provide a comprehensive understanding of Iceberg's potential as a unified table format for seamless cross-platform data sharing, all while maintaining a single copy of the data.

3   Enabling Cross-Platform Data Interoperability

3.1   Introduction to Medallion Architecture in Modern Lakehouse Systems

As organizations increasingly adopt cloud-native architectures and leverage multiple data platforms, the need for structured, scalable data processing pipelines has grown substantially. One widely adopted paradigm that addresses this need is the Medallion Architecture, which provides a layered approach to data organization within lakehouse environments [15]. The architecture is composed of three layers:

- **Bronze Layer**: This foundational layer captures raw, unfiltered data from diverse sources, such as transactional systems, IoT devices, logs, and APIs, and stores it in a cost-effective, schema-flexible format within a data lake or lakehouse. This layer ensures that all incoming data is preserved in its original form for future processing or reprocessing.
- **Silver Layer**: The intermediate layer applies data transformation workflows including cleansing, deduplication, normalization, and the enforcement of referential integrity. These transformations are commonly performed using ETL

or ELT tools and can occur within the same platform or be transferred to another engine optimized for data processing.

- **Gold Layer**: At the top of the medallion stack, this layer contains curated datasets tailored for business intelligence, advanced analytics, and machine learning workflows. Data in this layer is typically aggregated, joined, and enriched to support domain-specific use cases, such as real-time dashboards, forecasting models, and operational analytics.

To fully leverage each layer's capabilities, different tools and platforms optimized for specific tasks, such as cloud warehouses for BI, lakehouses for transformation, and separate engines for machine learning, are deployed in different layers based on the architecture and use case. This heterogeneous tooling introduces the need for robust data integration pipelines that ensure consistency and accessibility across layers and systems. Traditionally, several different integration approaches have been developed to enable cross-platform data integration [14]:

1. **ETL Pipelines**: These extract data from one system, transform it, and load it into another. While widely used, ETL workflows are often rigid, complex to scale, and prone to latency and data duplication.
2. **Data Virtualization**: Offers real-time access to data across platforms without physical movement. However, it introduces performance bottlenecks and lacks robust support for transactional guarantees, particularly in analytical workloads.
3. **API-Based Integration**: Exposes data or services through APIs to facilitate communication between platforms. While flexible, APIs require custom development, introduce security and versioning challenges, and may not be optimized for large-scale, high-throughput data transfers.
4. **Centralized Data Lakes**: Use cloud object storage to aggregate raw and processed data from multiple systems. Though flexible, data lakes typically lack standardized support for table-level operations, making schema evolution, version control, and concurrent access management difficult.

3.2 Standardizing Data Integration Across Platforms Using Apache Iceberg

To address the fragmentation caused by legacy methods, we will adopt the Iceberg format to enhance cross-platform integrations. Iceberg tables are an open-source, high-performance table format designed to enable reliable and efficient data lake analytics. Developed initially by Netflix and now governed by the Apache Software Foundation, Iceberg addresses many limitations of older table formats. The official definition describes Iceberg as "an open table format for huge analytic datasets" that brings SQL-like reliability and performance to the realm of distributed data lakes [8]. Iceberg's architecture is optimized for cloud environments, allowing efficient integration with modern data processing frameworks while eliminating issues related to file listing operations in object stores.

Layered Architecture of Iceberg Tables. Iceberg employs a structured, layered architecture that enhances scalability, query performance, and data consistency. The architecture consists of three key layers as seen in Fig. 2.

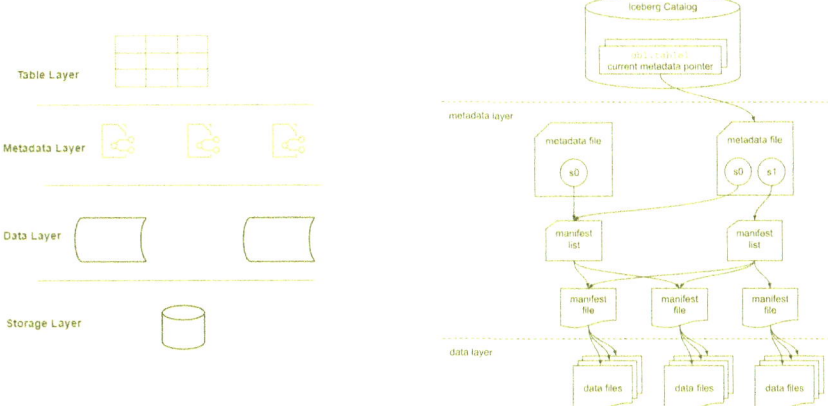

Fig. 1. Open Table Format Architecture

Fig. 2. Iceberg Table Structure.

The Metadata Layer in Iceberg plays a crucial role in managing table state through structured metadata files. The Snapshot Metadata File captures schema, partition details, and links to the Manifest List, enabling time travel and rollback. The Manifest List maps snapshots to Manifest Files, which store data file paths and column-level statistics for optimized query performance. The Data Files, stored in formats like Parquet, ORC, or Avro, ensure compatibility with various tools while supporting partition evolution without rewriting data. The Data Storage Layer supports diverse storage systems like Amazon S3 and Google Cloud Storage, preserving native formats without conversion overhead. The iceberg catalog integrates with execution engines like Spark, Flink, Trino, and Presto, leveraging metadata to optimize queries, ensure consistency, and prune unnecessary files. This structured approach enhances data lake visibility, allowing efficient data tracking via metadata pointers instead of costly file scanning. Iceberg's layered architecture improves performance, scalability, and interoperability across analytics platforms.

Why Choose Iceberg? Iceberg's architecture is purpose-built for efficient cross-platform integration. Its metadata layer sits atop the data lake, allowing query engines to quickly access the latest snapshot without scanning the full dataset. The pointer-based structure adds clarity and control, bringing true visibility to the data lake. As a result, Iceberg is gaining widespread adoption and native support across diverse processing engines. With extensive support from both commercial and open-source engines, and the recent announcement that S3[17] supports storing tables in Iceberg format, Iceberg ensures flexibility, interoperability, and a vendor-neutral approach for modern data management. By offering sophisticated metadata handling, schema evolution, and partition pruning, Iceberg enables precise data management, unlocking the true potential of data lakes for high-performance analytics. Its vendor-agnostic design ensures compatibility with a wide array of query engines and storage systems, empowering organizations to maintain flexibility and avoid vendor lock-in. This open and scalable architecture positions Iceberg as a key enabler for modern data engineering, seamlessly integrating with both existing data lakes and cutting-edge analytics platforms. For these reasons, companies are increasingly focusing their attention on the iceberg format, leading to its growing adoption.

4   Bi-Directional Data Integration Between Snowflake and Databricks

To address the challenges of data interoperability and consistency across platforms, in this section we will investigate the practical implementation of Apache Iceberg tables as a unified table format between Snowflake and Databricks, with data stored in Amazon S3. The experiments are designed to explore the research question regarding the challenges and limitations of using Iceberg to enable cross-platform data interoperability. This approach aims to simplify the complexities of manual pipeline management while ensuring compatibility and interoperability across diverse systems.

4.1   Architecture

The diagram 3 illustrates a generalized architecture for integrating Apache Iceberg tables across platforms (Snowflake and Databricks). The architecture comprises four main components. The source engine manages the Iceberg table, ingesting incoming data, typically representing the silver or gold layers in the Medallion Architecture, and writing it to a centralized data lake, such as Amazon S3. It also interacts with a catalog that maintains metadata files and pointers to the latest snapshot, ensuring transactional consistency. Both data and metadata are stored in the data lake, which serves as the central repository. The source engine requires write access to S3, typically controlled through IAM roles. A query engine, used for analytics, BI, or ML workloads, connects to the catalog to fetch the latest snapshot metadata and then reads the corresponding data files from the lake. Access to the catalog can be established via JDBC/ODBC, service principals, or native connectors, depending on the system. The query engine also requires appropriate read permissions to access the data in S3. This architecture ensures consistent, scalable, and efficient cross-platform data access using a single, authoritative Iceberg table. The study uses the publicly available

---

[17] https://aws.amazon.com/about-aws/whats-new/2024/12/amazon-s3-tables-apache-iceberg-tables-analytics-workloads/.

Yellow Taxi Trip Records[18] from NYC TLC, covering the first quarter of 2024. With over 9.5 million rows and 19 columns, the dataset offers rich, real-world attributes like timestamps, geolocations, and payment info—ideal for data lake and analytical workload scenarios.

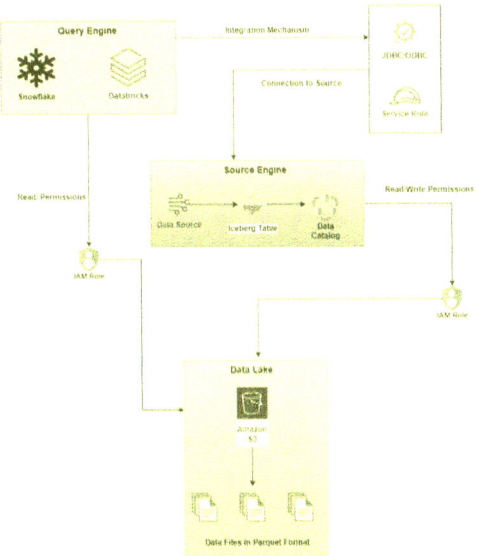

Fig. 3. Querying Iceberg Tables Across Platforms

5 Experimental Design

This section presents the methodology for evaluating Iceberg table integration across platforms, emphasizing real-world scenarios encountered by data teams. The experiment aimed to assess how efficiently Snowflake and Databricks can access and process Iceberg-managed tables while ensuring real-time synchronization and query optimization.

5.1 Integration Feasibility

Several detailed steps were followed[19],[20] to allow bidirectional integration of Iceberg tables between Snowflake and Databricks.

---

[18] https://www.nyc.gov/site/tlc/about/tlc-trip-record-data.page.

[19] https://medium.com/snowflake/how-to-integrate-databricks-with-snowflake-managed-iceberg-tables-7a8895c2c724.

[20] https://www.databricks.com/blog/read-unity-catalog-tables-in-snowflake.

Snowflake Managed Iceberg Tables in Databricks
Step 1: Create an Amazon S3 bucket in the `eu-north-1` region.
Step 2: Configure IAM role with read/write permissions for Snowflake to access S3.
Step 3: Create external volume to manage data transfers between Snowflake and S3.
Step 4: Create Snowflake-managed Iceberg table using the external volume and taxi dataset.
Step 5: Apply security restrictions by creating a dedicated read-only user for Databricks.
Step 6: On Databricks side, configure IAM role with read-only permissions.
Step 7: Set up Databricks cluster with 10.4 LTS runtime and required libraries (Iceberg, AWS bundle, JDBC) and the Snowflake user.

Databricks Managed Iceberg Tables in Snowflake
Step 1: Enable UniForm on the Delta table in Databricks (during creation or via `ALTER TABLE`).
Step 2: Register Unity Catalog in Snowflake via Databricks service principal and OAuth credentials.
Step 3: Grant Snowflake access to S3 with IAM role, trust policies, and define external volume.
Step 4: Create Iceberg table in Snowflake referencing Unity Catalog metadata.
Step 5: Test integration by updating data in Databricks and verifying reflection in Snowflake (via auto-refresh or `ALTER ICEBERG TABLE REFRESH`).

5.2   Data Synchronization with Varying Dataset Sizes

To evaluate performance and scalability, datasets of 3 million, 6 million, and 9 million rows were loaded into Iceberg tables, and synchronization performance was observed. The test also included simultaneous access to multiple Iceberg tables to assess concurrent query execution.

5.3   Schema Evolution Across Platforms

Schema evolution refers to structural changes in a dataset, such as adding new columns, removing existing ones, or modifying column definitions (e.g., changing a column's data type or renaming it). It often represents a major challenge in data pipelines, as changes to the structure of data can disrupt workflows and lead to pipeline failures during data movement or integration. Schema evolution was tested by adding a new `RATING column` (VARCHAR) to an Iceberg table and updating it with default values. Additionally, type conversion from VARCHAR to INT, which required a workaround—creating a new column `RATING_INT`, applying `TRY_CAST`, and renaming the column after validation.

5.4   Latency and Performance Testing

Latency and performance form a critical part of this experiment, as they demonstrate how queries perform on both internally managed and externally managed Iceberg tables. The dataset used for this evaluation contains over 9.5 million rows. To enhance the analysis, external tables were also included in the evaluation. External tables refer to datasets that reside in an external data lake, such as AWS S3 in this case, where the query engine directly accesses the data files without a dedicated metadata management layer. While this setup is similar to Iceberg tables in terms of data location, the key difference lies in the presence of Iceberg's metadata layer. Furthermore, an optimization was applied to the Iceberg table to enhance query performance. The data was reordered based on a TIMESTAMP column, which should allow the query engine to read data more efficiently by leveraging the sorted structure. Queries were designed in three categories:

- **Simple Queries** – Basic operations like COUNT(*) to measure row counts.
- **Medium Queries** – Arithmetic or conditional computations based on column grouping with conditional logic.
- **Complex Queries** – Aggregations, with full join.

6   Results and Discussion

In this section, we will discuss the results to evaluate the performance and efficiency of the integrations.

6.1   Integration Feasibility

The results showed that Iceberg table integrations between Snowflake and Databricks were efficient and effective. When accessing Snowflake-managed Iceberg tables from Databricks, Snowflake handles only lightweight metadata retrieval, while Databricks performs the actual computation. Testing showed metadata retrieval times within milliseconds. When Snowflake queried Databricks-managed Iceberg tables, metadata retrieval worked as expected, but additional setup was required to establish connectivity with Databricks' Iceberg catalog.

6.2   Data Synchronization with Varying Dataset Sizes

Scaling the dataset size had no impact on synchronization. Netflix, managing petabytes of data with Iceberg[21], further proves its scalability. The results were the same for multiple datasets as they could be accessed concurrently without issues, demonstrating the reliability of the integration. This capability ensures

---

[21] https://netflixtechblog.com/optimizing-data-warehouse-storage-7b94a48fdcbe.

that both current and future datasets can be easily managed and queried across platforms without requiring significant reconfiguration or manual intervention. If this process were implemented without Iceberg tables, it would require significant overhead. For instance, syncing data between platforms would involve setting up multiple services on both ends to ensure data transfer. Additionally, scheduled jobs would be needed to refresh tables, taking considerable time and resources. Data quality checks would also be necessary at each step to verify the accuracy and consistency of the data. By contrast, Iceberg simplifies this process considerably. Since the data remains in the data lake and is accessed via metadata updates, there is no need for duplicative jobs or complex synchronization processes. As data resides solely in the data lake, any platform that supports Iceberg can query it, giving organizations the flexibility to use different platforms for varying use cases.

6.3   Schema Evolution Across Platforms

Schema evolution tests showed that Iceberg handles changes like adding, removing, or updating columns efficiently, with updates reflected instantly on querying platforms—thanks to its metadata layer as they read the schema definition directly from the latest metadata files. This eliminates the need for manual schema updates when the schema isn't explicitly defined. However, direct type conversion (e.g., VARCHAR to INT) isn't supported and requires workarounds. Despite this, features like column renaming and dropping worked seamlessly. We also observed differences in data type handling across platforms (see Table 1), and noted that Iceberg only supports microsecond precision for TIME and TIMESTAMP types in both v1 and v2. Overall, Iceberg's native support for schema evolution reduces pipeline breakage and simplifies data operations by keeping platforms in sync automatically. This capability significantly simplifies data operations, allowing teams to focus on their analytical and operational goals without worrying about the complexities of the manual schema management.

Table 1. Differences between column data types

Snowflake Datatypes	Databricks Datatypes
Number	Decimal
Float	Double
Varchar	String
Timestamp Ntz(6)	Timestamp

6.4   Latency and Performance

Figure 4 shows Databricks-managed Iceberg tables were tested on both Databricks and Snowflake, showing notable performance differences. In

Databricks, simple queries showed minimal difference between standard Iceberg and external tables, while medium queries executed 2 s faster than external tables and 4 s faster than standard Iceberg tables after optimization. For complex queries, external tables outperformed others. In Snowflake, Iceberg tables outperformed external tables by 4.5 times for medium queries and 9 times for hard queries. Optimized Iceberg tables showed a slight performance improvement over standard tables, highlighting the benefits of optimizations.

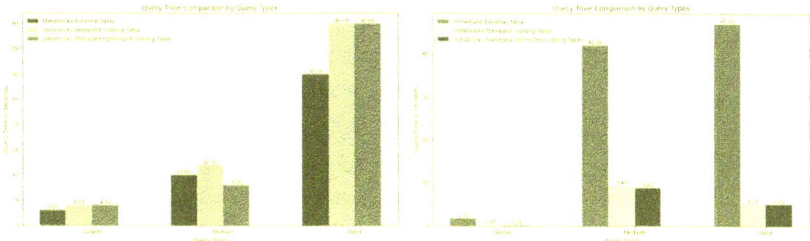

Fig. 4. Databricks managed Iceberg Table in Databricks(left), Snowflake(right)

Next Snowflake-managed Iceberg tables were tested on both Databricks and Snowflake, showing notable performance differences. In Snowflake, Iceberg tables outperformed external tables, being 9 times faster for medium queries and 12 times faster for hard queries, likely due to Snowflake's use of its own catalog. In Databricks, external tables generally performed better, except for medium-sized queries as seen in Fig. 5. Queries leveraging Iceberg features and scanning fewer partitions were notably faster. Performance in Databricks varied based on query type and operations.

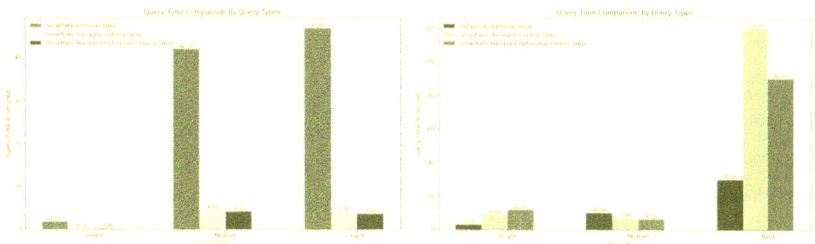

Fig. 5. Snowflake managed Iceberg Table in Snowflake(left), Databricks(right)

This reveals several interesting observations. Iceberg tables outperformed external tables for queries that did not require a full table scan, such as conditional calculations. This was consistent for both self- managed and external-managed Iceberg tables in Databricks. However, in cases of full table scans, external tables demonstrated faster performance. In Snowflake, Iceberg tables outperformed external tables across all query types, particularly when managed by Snowflake. Iceberg tables also offer the advantage of automated synchronization, ensuring near real-time data access and consistency, unlike external tables that require periodic refreshes, leading to delays, extra costs, and the risk of querying outdated data.

6.5   Challenges and Limitations

Integrating external Iceberg tables in Databricks with Unity Catalog[22] enabled presents challenges, as Unity Catalog uses Delta Lake as the primary metadata format, limiting access to external Iceberg tables. While Databricks can read Iceberg tables from external platforms when Unity Catalog is disabled, compatibility issues remain. Another important consideration is that as data size increases, it's crucial to implement policies for expiring outdated metadata and data files. This ensures that the data doesn't grow unnecessarily, helping to keep storage usage in check according to specific requirements. While Iceberg tables offer a promising solution for automating data synchronization across platforms, their query performance for full table scans on certain platforms still falls short and requires further optimization.

7   Conclusions

In this paper, we have explored the evolution of data management, the challenges posed by different offerings, the emergence of the lakehouse architecture, the advent of modern open table formats, and the complexities that arise when using multiple table formats across different platforms. The challenges organizations face in ensuring interoperability among these formats underscore the pressing need for a unified table format. This need becomes even more critical in a data-driven world where data sharing and efficient analytics are pivotal for success.

Apache Iceberg has emerged as a strong solution for modern data challenges, offering performance, scalability, and cross-platform consistency. Its growing adoption enables seamless interoperability between platforms like Snowflake and Databricks, reducing latency, improving query efficiency, and simplifying data management. Our comparative study showed that Iceberg enables near real-time synchronization, handles schema evolution effectively, and minimizes data duplication by maintaining a single data copy. While query performance was strong—especially when avoiding full table scans—some integration aspects still need refinement. Adopting Iceberg can streamline workflows, reduce operational burdens, and unlock the full potential of data assets, empowering organizations to make faster, more effective data-driven decisions.

Future research can explore several promising avenues to extend the findings of this paper. A key focus could be evaluating Databricks functionality when Unity Catalog begins supporting external Iceberg tables, with an emphasis on

---

[22] https://www.databricks.com/product/unity-catalog.

performance analysis. Additionally, expanding Iceberg's integration with platforms beyond Snowflake and Databricks could highlight its potential as a truly universal table format. Investigating the adoption of advanced catalog systems, such as Apache Polaris[23] for managing metadata instead of relying solely on storage engines, could enhance management and query performance. Furthermore, advance optimization techniques could be tested for improving query performance. Iceberg's time travel feature presents an exciting opportunity for assessing its applications in historical data analysis and auditing. Exploring these areas will provide valuable insights into Iceberg's role in modernizing data ecosystems and driving broader adoption across diverse platforms.

References

1. Ravat, F., Zhao, Y.: Data lakes: trends and perspectives. In: International Conference on Database and Expert Systems Applications, pp. 304–313 (2019)
2. Armbrust, M., et al.: Lakehouse: a new generation of open platforms that unify data warehousing and advanced analytics. CIDR. Vol. 8 (2021)
3. Mone, A.: The metadata challenge in big data systems. J. Data Manag. **23**(4), 15–27 (2020)
4. Codd, E.F.: A relational model of data for large shared data banks. Commun. ACM **13**(6), 377–387 (1970)
5. Jain, P., Kraft, P., Power, C., Das, T., Stoica, I., Zaharia, M.: Analyzing and Comparing Lakehouse Storage Systems. In: CIDR (2023)
6. Nargesian, F., Zhu, E., Miller, R.J., Pu, K.Q., Arocena, P.C.: Data lake management: challenges and opportunities, PVLDB, vol. 12, no. 12 (2019)
7. Christophides, V., Efthymiou, V., Stefanidis, K.: Entity resolution in the web of data. In: Synthesis Lectures on the Semantic Web: Theory and Technology, Morgan & Claypool Publishers (2015). ISBN 978-3-031-79467-4
8. Errami, S.A., et al.: Spatial big data architecture: from data warehouses and data lakes to the Lakehouse. J. Parallel Distrib. Comput. **176**, 70–79 ((2023))
9. Armbrust, M., et al.: Delta lake: high-performance ACID table storage over cloud object stores. PVLDB **13**(12), 3411–3424 (2020)
10. Singh, J., Singh, G., Bhati, B.S.: The Implication of Data Lake in Enterprises: A Deeper Analytics. ICACCS (2022)
11. Brasileiro Araújo, T., Efthymiou, V., Christophides, V., Pitoura, E., Stefanidis, K.: TREATS: fairness-aware entity resolution over streaming data. Inf. Syst. **129**, 102506 (2025)
12. https://www.virtana.com/press-release/virtana-research-finds-more-than-80-of-enterprises-have-a-multi-cloud-strategy-and-78-are-using-more-than-three-public-clouds/
13. Doan, A., Halevy, A., Ives, Z.: Principles of data integration. Elsevier (2012)
14. Bagam, N.: Data integration across platforms: a comprehensive analysis of techniques, challenges, and future directions. Int. J. Intell. Syst. Appl. Eng. **12**, 902–919 (2024)
15. Bhatt, S., Sekar, D.: Data Warehousing Modeling Techniques and Their Implementation on the Databricks Lakehouse Platform (2022). https://www.databricks.com/blog/2022/06/24/data-warehousing-modeling-techniques-and-their-implementation-on-the-databricks-lakehouse-platform.html

---

[23] https://polaris.apache.org/.

Executable Semantics for Teaching Concatenative Stack-Based DSLs: The Case of StackLang

William Steingartner[1]([✉])[iD] and Wolfgang Schreiner[2][iD]

[1] Technical University of Košice, Košice, Slovakia
william.steingartner@tuke.sk
[2] Johannes Kepler University Linz, Linz, Austria
Wolfgang.Schreiner@risc.jku.at

Abstract. In the context of teaching computer science, many domain-specific languages (DSLs) used for data manipulation and transformation follow imperative paradigms, yet their semantics remain informal or tool-dependent. This paper proposes a pedagogical framework based on executable formal semantics to improve conceptual understanding and practical competence in such DSLs. Using a minimal imperative DSL developed as a teaching tool to illustrate arithmetic and data transformations, we define its syntax and semantics using denotational semantics and develop an executable interpreter directly derived from the formal rules. The framework enables students to explore and visualize the effects of each language construct, reason about program behavior, and verify correctness properties. We present a case study in which we focus on the gradual use of cross-curricular relationships and gradually build a comprehensive package for students that draws on knowledge from several courses focused on formal methods in software engineering. The paper concludes with a discussion of the potential of this methodology to bridge the gap between formal methods and practical education in the field of computer science.

Keywords: DSL · executable semantics · formal semantics · SLANG tool · StackLang · university didactics

1 Introduction

Interactive education is becoming increasingly popular. Perhaps the reason was the previous period when teaching was forced to be done only remotely and teachers lost direct contact with students and thus the possibility of direct feedback or discussion. Distance education has its limitations and lacks many elements of the classic form of teaching, e.g. the loss of feedback and practical interaction in remote settings. In the education of young IT experts, the possibility of practical testing of concepts is key and thus helps to achieve a deeper understanding of the subject matter, or to acquire knowledge and skills more quickly and acquire new competencies. Another important aspect is the development of abstraction in understanding problems and connecting multiple perspectives in solving them, the application of necessary procedures in solving problems, as well as the ability to easily reorient to other technologies depending on the requirements and input specification of the problem. It is not unusual that inter-subject relationships are often weak, or are not applied as mandatory prerequisites or co-requisites. This results in isolated subjects or content coverage of subjects, which, although outwardly exhibit certain relationships with each other, are actually applied to an insignificant extent.

The subjects such as Formal Semantics and Program Verification, Formal Languages, Compiler Construction, Formal Methods in Computer Science, or Data Structures and Algorithms are part of the curriculum of many universities when teaching computer science and similar fields, while the emphasis in teaching is placed depending on the orientation of the study program (often the main criterion is the technical or natural science direction), on the overall content of the curriculum, and on the resulting profile of graduates.

In this article, we present the connection of several useful results together with positive experiences from teaching, research and development. We focus on a pedagogical case study, which in the pedagogical area is based on the connection of the principles of data structures (working with a stack), formal languages (design of a language grammar), formal semantics of programs (individual elements of the language are assigned a specific meaning by a selected semantic method), while from the perspective of paradigms and languages, the concatenative/compositional style of the resulting language is be supported. From the perspective of research and development, it will be based on the achieved results of fruitful cooperation in the design and definition of the denotational semantics of a specific experimental concatenative/compositional language. The properties of this language inspired us and served as the basis for building a language, which we gave the simple name StackLang. It is a domain-specific language – a language designed for evaluating expressions on a stack (in its sense as a language of some abstract machine). For its semantic modeling and implementation of executable semantics, we used a tool that is the result of our fruitful bilateral collaboration SLANG.

The project designed in this way is not a complete language, but allows its addition and expansion according to requirements or assignments. It can be successfully integrated into teaching within the mentioned subjects, since by its nature and properties it draws precisely from interdisciplinary relationships. It is a successful example of how theoretical knowledge can be linked with practical skills in an academic environment during teaching and as an inspiration for further research.

Imperative DSLs are common in data engineering (e.g., scripting in ETL, streaming rules, transformation pipelines). Such DSLs include SQL for relational data processing, Pig Latin for data transformations, and streaming DSLs like Apache Flink's DataStream API. Lack of formal semantics leads to difficulties in reasoning, teaching, and verifying these DSLs. As a problem statement we see that students often use DSLs as "black boxes" with unclear behavior and no way to reason about correctness.

Executable semantics provides a crucial link between theoretical models and practical understanding, offering an effective way to teach the internal mechanisms of DSLs. Our goal is to provide an educational framework that enables students to understand and experiment with formal definitions, interpreter design, and semantic analysis. The presented approach is equally beneficial for broader computer science curricula, where students can gain valuable insights into language design and semantics.

The paper is organized as follows. Section 2 covers the view on similar technologies and approaches which led to executable semantic specifications and their role in teaching. In Sect. 3, we present a comprehensive review of the proposed DSL named StackLang with focus to its syntax and semantics. Next, in Sect. 4, the procedure for implementing the specification and executable semantics of StackLang using the SLANG tool is presented. Finaly, the Sect. 5 concludes our paper.

2   Related Work and Focus Areas

The formalization and development of our StackLang DSL using SLANG is in line with several ongoing research topics in the area of executable semantics and tool-supported language definitions. In this section, we review current work in the context of the existing literature, focusing on its contributions to the semantics of executable programs, stack-based languages, and learning frameworks.

Some well-known executable semantic frameworks such as the K Framework and PLT Redex have proven useful in defining and implementing operational semantics of programming languages. The work of G. Roşu et al. on the K Framework [10] provides a robust basis for the systematic definition of language semantics and also serves as a reference point for the SLANG approach. PLT Redex [3] demonstrates how operational rules can be expressed as executable specifications, paralleling the SLANG methodology in its focus on rule definition and testing.

Formal verification approaches and techniques used in stack-based languages emphasize systematic specification and support for proof generation. The implementation of SLANG as a tool for generating executable semantic frameworks, such as StackLang, reflects methods used in works such as the formal verification of a stack-based virtual machine in Isabelle/HOL [4] and the formalization of simple stack machines in Rocq [1]. These approaches demonstrate the potential of SLANG to incorporate verification functions, thereby supporting the correctness and reliability of programs in StackLang.

Domain-specific languages (DSLs) often use executable semantics to bridge the gap between high-level specifications and low-level implementations. Existing research shows how executable semantics can formalize the behavior of a DSL and offer predictable and reusable sets of rules. Developing with a DSL using SLANG thus builds StackLang into this tradition, providing a structured framework for defining and implementing the semantics of a language while maintaining modularity and extensibility.

Finally, executable semantic frameworks have proven to be very useful as educational tools that allow students to simplify the study of semantic methods. Our intention was for SLANG (at the local level) to support such educational goals by allowing students to explore semantic definitions through direct experimentation and visualization.

This placement confirms the relevance of the current work within the broader landscape of executable semantics, formal verification, DSL design, and educational frameworks.

Compared to the presented related works such as K Framework or PLT Redex, our approach focuses specifically on simplicity for pedagogical purposes. StackLang and its implementation in SLANG are designed as minimal examples that can be fully understood and experimented within an educational context.

3   The StackLang DSL

In this section, an overview about the project StackLang is provided. We present the motivation and its origins in our teaching. Then we focus on the syntax and semantica and the supporting (prototyping) tool.

3.1   Overview of StackLang

StackLang is a very simple domain-specific language inspired by language KKJ published in [5]. A similar stack-oriented machine language, though with a more general machine model and a small-step operational semantics, is described in [12]. StackLang currently supports the following instructions which enable arithmetic computations and basic stack manipulations:

- LOAD n – pushes integer n onto the stack,
- ADD – pops two values, pushes their sum,
- SUB – pops two values, pushes their difference,
- MUL – pops two values, pushes their product,
- DUP – duplicates the top of the stack.

The instruction set is not limited in any way, so it is quite flexible and allows us to easily and conveniently extend the syntax of the language. The basic specification was inspired by the simple task of visualizing a calculation and the associated exercises of a basic automaton, whose language contained only four instructions: ADD, SUB, MUL, LOAD a. The exercise is available in document [18], the only change is the order of stack display, where in our specification the top of the stack is always on the left in the line notation (in accordance with the approach e.g. in the work [6]). This simple language, after a successful project with the KKJ language, became an inspiration for further work on the subjects Data Structures and Algorithms, as well as on the subject Formal Languages, which are included in the curriculum of most universities. In the first mentioned, it served to simulate calculations (as a theoretical model for working with a stack), in the second mentioned as a template for creating a simple interpreter.

At the beginning, the creation of the interpreter was left to the definition of rules and the use of some language processor generators (usually ANTLR [7]). However, since the *rôle* of abstract machines in computer science (and especially in branches of theoretical computer science) is quite broad (see [2]), it was a logical step to include this concept in the content plan of the course Semantics of Programming Languages. Here, the goal was to formulate denotational and operational semantics for this DSL. Of course, different DSLs require different modifications or adaptations of the definition of the relevant semantic methods, e.g. for the Robot language (originally published here: [9]), which is inspired by the Robot Karel language [8], several works dealing with the definition of semantics have been published. For example, in the work [17], we formulated the denotational and natural semantics of one of its variants and subsequently proved the equivalence of these semantic methods. It has also been shown that the concepts acquired in this course can be transferred to the related subject Development of Domain-Specific Languages, which is more practically oriented, but without paradigmatic knowledge of DSLs and the principles of their semantic modeling, it would be difficult to properly understand their design and implementation in real applications. We note that paradigmatic knowledge of DSLs and the principles of their semantic modeling have a fundamental impact on the practical teaching of the subject, because they allow students not only to understand the theoretical foundations, but also to effectively apply these principles in the design and implementation of their own domain-specific languages. Thanks to this symbiosis, students will not only learn the theoretical foundations, but also gain practical skills that will allow them to develop high-quality and effective domain-specific languages.

3.2   Syntax of StackLang

As we mentioned earlier, StackLang is a concatenative language. In this language, simple arithmetic calculations are formulated, which are implemented using memory. As is usual for most automata and theoretical computational models (for example, an abstract machine for operational semantics [6]), this memory is a stack and StackLang is in principle a stack language. The stack can be empty or contain only integer values (in a simplified model, in advanced projects this can be modified and expanded as needed).

The basic project of the StackLang language is prepared for students with only the above arithmetic instructions and an instruction for inserting a value into the stack (we have abandoned the traditional notation PUSH, since the opposite operation POP is not directly defined here – it is not needed). The syntax is therefore expressed by production rules:

- for instructions

$$instr ::= \text{ADD} \mid \text{SUB} \mid \text{MUL} \mid \text{LOAD } a \mid \text{DUP}, \tag{1}$$

- for code

$$c ::= \varepsilon \mid instr\ c,$$

where c stands for the sequence of instructions (or code) to be executed and formally it is an element of the syntactic domain **Code**, $instr$ is an element of the syntactic domain **Instruction**, argument a is a numeral, $a \in$ **Num** (a string of digits, terminal), and element ε is the empty sequence. As a delimiter, a single space is used.

3.3   Formal Semantics of StackLang

Since StackLang is purely stack-based language and it has no memory or variables, the state of execution consists only of the stack. In a similar sense, we define a semantic domain whose elements are stacks. We use the evaluation stack to evaluate arithmetic expressions. Formally, it is a list of values **Stack** $= \mathbb{Z}^*$. A stack is thus a sequence of integers with the topmost element depicted on the left (Fig. 1). Depicted is the stack after input sequence

LOAD 16 LOAD 12 LOAD 14 LOAD 10.

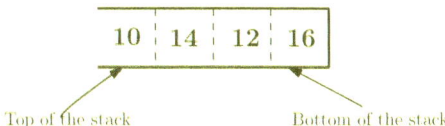

Fig. 1. Visual representation of the stack.

The next step is to formulate the semantic function. The semantic function maps an instruction and a stack state to a new stack state:

$$\mathscr{S} : \textbf{Instruction} \times \textbf{Stack} \rightarrow \textbf{Stack}.$$

For a given instruction i and stack s, we write the semantic function as

$$\mathscr{S}[\![i]\!]s = s'$$

where s' is the resulting stack, and $s, s' \in$ **Stack**.

Denotational semantics for specific instructions is straightforward and is defined as follows.

- Instruction LOAD a. For an instruction LOAD a, where $a \in \mathbb{Z}$:

$$\mathscr{S}[\![\text{LOAD}\ a]\!]s = s \cdot [a].$$

The stack s is concatenated with the new value a at the top.

- Addition instruction ADD. For an instruction ADD, the top two elements are removed (popped), and their sum is pushed:

$$\mathscr{S}[\![\texttt{ADD}]\!]s = \begin{cases} s' \cdot [a+b]\,, & \text{if } s = s' \cdot [a,b]\,, \\ \bot, & \text{if } len(s) < 2. \end{cases}$$

  The stack s' is the stack without the top two elements. The sum $a+b$ is pushed onto the resulting stack.
- Instruction DUP duplicates the top of the stack – it simply replicates the topmost value and pushes it onto the stack:

$$\mathscr{S}[\![\texttt{DUP}]\!]s = \begin{cases} s' \cdot [a,a]\,, & \text{if } s = s' \cdot [a]\,, \\ \bot, & \text{if } len(s) < 1. \end{cases}$$

- Because the denotational semantics is fully compositional, the semantics of the sequence of codes is defined as follows:

$$\mathscr{S}[\![c_1\ c_2]\!]s = \begin{cases} s'', \text{ if there exists } s' \text{ such that } \mathscr{S}[\![c_1]\!]s = s' \\ \quad\quad \text{and } \mathscr{S}[\![c_2]\!]s' = s'', \\ \\ \bot, \text{ if } \mathscr{S}[\![c_1]\!]s = \bot \\ \quad\quad \text{or if there exists } s' \text{ such that } \mathscr{S}[\![c_1]\!]s = s' \\ \quad\quad \text{but } \mathscr{S}[\![c_2]\!]s' = \bot. \end{cases}$$

Other arithmetic operations are defined in an analogous manner.

As an example, we consider the following StackLang program:

LOAD 2 LOAD 3 ADD

and we will trace its execution from the initial state to the final state. The initial state is an empty stack: $s_0 = [\,]$. Applying the semantic rule for the instruction LOAD 2:

$$\mathscr{S}[\![\texttt{LOAD 2}]\!]s_0 = s_0 \cdot [2] = [2].$$

Next, we apply the rule for LOAD 3:

$$\mathscr{S}[\![\texttt{LOAD 3}]\!][2] = [3,2].$$

Now, we apply the rule for ADD. The stack has at least two elements, so we proceed with the addition:

$$\mathscr{S}[\![\texttt{ADD}]\!][3,2] = [3+2] = [5].$$

Thus, the final state of the stack after executing the entire program is $s_f = [5]$. The denotation of the whole program is given by a composition as:

$$\mathscr{S}[\![\texttt{LOAD 2 LOAD 3 ADD}]\!]s = (\mathscr{S}[\![\texttt{ADD}]\!] \circ \mathscr{S}[\![\texttt{LOAD 3}]\!] \circ \mathscr{S}[\![\texttt{LOAD 2}]\!])\,s$$

and the whole process of calculation is depicted in Fig. 2.

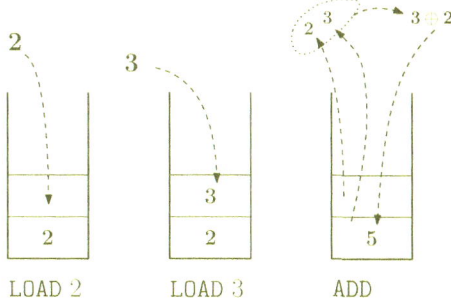

Fig. 2. The calculation on the stack (standard top to bottom stack view).

The definition of the semantics of other instructions (including user-defined ones) is similar. For a well-defined syntax, the semantics are formulated straight-forwardly, analogously to the work [5] for the KKJ language.

The possible extension of syntax and the definition of semantics form the basis for understanding the structure of the StackLang language and the abstraction of defining the meaning of its syntactic elements. In this case, students must understand how new instructions are defined: that an instruction has its own syntax, semantics, and thus extends the properties and computational capabilities of the language.

The design estimates simplicity and clarity with minimal syntax while incorporating imperative features such as arithmetic operations, control structures, and stack operations. The syntax consists solely of instructions defined by specific grammar rules.

4   Implementing Semantics and Generating Tools

For pedagogical purposes, it is advantageous to use tools that support formal design and prototyping of programming languages, thereby allowing students to apply the knowledge, skills and competencies acquired in previous courses, such as Data Structures and Algorithms, Formal Languages and Semantics of Programming Languages. SLANG represents such a tool, which combines theoretical foundations with the ability to generate executable code based on formal specifications.

4.1   SLANG Tool

SLANG [13, 16] is a language engineering tool that supports the definition and execution of formal language semantics, enabling the formal design, implementation, and prototyping of programming languages from their specifications. It allows defining abstract syntax, semantics, and type systems using logical rules and algebraic structures. Compared to existing tools that focus on small-step operational semantics (e.g. Maude, K interpreter), SLANG emphasizes denotational semantics and big-step operational semantics. This straightforward implementation is based on structural recursion, which allows generating directly executable Java code.

Based on the formal language specification, SLANG automatically generates:

- Parser – converts the text representation of the program into an abstract syntax tree (AST) using the ANTLR4 tool,
- Printer – creates a text representation of the program from the AST,
- Type checker – implements static type checking based on inference rules,
- Interpreter – implements the language semantics defined via big-step rules or denotational semantics.

Unlike tools like Maude or K, which require specific meta-languages to define semantics and use rewriting engines or external proof assistants, SLANG uses built-in Java code to express functions and relations. This approach minimizes the need to formalize mathematical entities in the meta-language and at the same time allows the use of standard or custom Java libraries, which simplifies the development of semantics and increases the efficiency of the execution of the generated code. Figure 3 shows the architecture of SLANG, highlighting how specifications are transformed into executable interpreters.

Fig. 3. Compact architecture of the SLANG tool (horizontal layout).

As an example of a successful application of the SLANG tool, we refer to the implementation of a prototype of a subset of the SQL language, where a parser, printer, type checker and interpreter were automatically generated based on a formal specification of syntax, type system and denotational semantics [15]. Another application is the design of the EFSM language for machine control based on extended finite state machines, where SLANG was used to implement a control system for a model of an idealized robotic vacuum cleaner, including visualization of its behavior [14].

4.2 Executable Semantics via SLANG

The formal semantics of StackLang is defined using executable semantics via the SLANG tool. By translating formal rules into interpreter definitions, SLANG enables a faithful and precise implementation of the semantics. This approach ensures consistency in rule application, facilitates testing through executable specifications, and promotes the reusability of formal definitions across different contexts.

In this work, SLANG is used for formal specification of the syntactic domains of the StackLang language, definition of semantics using denotational equations, and automatic generation of an interpreter including visualization of the program execution process.

The SLANG specification of this language looks as follows:

```
language StackLang { // file StackLang.txt
    target java {...}
    domains {...}
    printer {...}
    parser antlr4 {...}
    function ... {...} ...
}
```

The next step in language definition is the introduction of syntactic domains, which represent abstract syntactic trees for individual language constructs. For each syntactic domain, grammatical rules are defined that determine the possible constructions of abstract syntactic trees:

```
domains {
    Number = Unsigned[NUM] + Negative[NUM];
    Command = Add + Sub + Mul + Dup
                    + Seq[Command,Command] + Load[Number];
    Program = Prog[Command];
}
```

This definition of numerals ensures that StackLang will be able to work with integers, i.e. with both positive and negative values.

SLANG uses the ANTLR4 tool [7] to automatically generate a parser that converts a text representation of a program into an abstract syntax tree (AST). The parser analyzes the input program text based on a formal grammar and creates a hierarchical structure that represents individual syntactic elements and their mutual relationships. The definition of specification rules for the parser requires separate rules for each domain. Here we present a shorthand notation for selected instructions.

```
parser antlr4 {
    domain Number {
        case # n=dNUM # -> Unsigned[n];
        case # '-' n=dNUM # -> Negative[n];
    }

    domain Command {
        case # c1=dCommand c2=dCommand #
                    -> Seq[c1,c2];
        case # 'ADD' # -> Add;
        case # 'DUP' # -> Dup;
        case # 'LOAD' n=dNumber # -> Load[n];
```

```
        . . .
    }

    domain Program {
        case # c=dCommand EOF # -> Prog[c];
    }
}
```

In addition to the parser, SLANG also generates a printer that performs the opposite transformation – from the abstract syntax tree back to the text representation of the program. This process is useful for verifying the correctness of the syntax tree, as well as for visualizing and analyzing the structure of the program. In this way, SLANG ensures consistency between the text and abstract representations of the program. Following is the shortened code for printer.

```
printer {
    domain Number {
        case Unsigned[n]  -> # _result = n; #;
        case Negative[n]  -> # _result = '-' + n; #;
    }

    domain Command {
        case Add    -> # _result = "ADD"; #;
        case Dup    -> # _result = "DUP"; #;
        case Load[n] -> # _result = "LOAD " + n;#;
        case Seq[c1,c2] ->
                    # _result = c1 + "; " + c2; #;
        . . .
    }

    domain Program {
        case Prog[c] -> # _result = c.toString(); #;
    }
}
```

Denotational semantics in SLANG defines the meaning of a language by mapping abstract syntactic domains to values from the corresponding semantic domains. For each syntactic domain, a mathematical function is defined that maps well-formed syntactic trees of that domain to values from the corresponding semantic domain. This approach allows the semantics of a language to be expressed using mathematical equations that determine the meaning of individual syntactic constructs. Semantic functions thus provide a formal basis for the denotational semantics of a language, ensuring a consistent mapping between syntax and meaning [11,19]. We only provide an shortened version for specific instructions.

```
function S[[Command]]: #Stack<Integer>#
-> #Stack<Integer># {
```

```
equation S⟦Add⟧(s0) = s
{
    a: #Integer# = #s0.pop();#;
    b: #Integer# = #s0.pop();#;
    #s0.push(a+b);#
    #s= (Stack<Integer>) s0.clone();#
}
equation S⟦Dup⟧(s0) = s
{
    a: #Integer# = #s0.pop();#;
    #s0.push(a);#
    #s0.push(a);#
    #s= (Stack<Integer>) s0.clone();#
}
equation S⟦Load[n]⟧(s0) = s
{
    i = N⟦n⟧;
    #s0.push(i);#
    #s = (Stack<Integer>) s0.clone();#
}
equation S⟦Seq[c1,c2]⟧(s) = s2
{
    s1 = S⟦c1⟧(s);
    s2 = S⟦c2⟧(s1);
}
...
}
```

4.3 SLANG's Role in Language Processor Generation

The main goal of using the SLANG tool is to generate a set of Java classes that
form the basic components of a language processor – a parser, a type checker, an
interpreter, and classes representing abstract syntax nodes. The parser, created
using ANTLR4, converts text input into an abstract syntax tree, whose nodes are
represented by concrete Java classes with methods for accessing subnodes and
performing semantic operations. The type checker implements logical inference
rules as methods for verifying the type correctness of individual AST nodes and
allows them to be extended with additional rules, however, due to the simplicity
of StackLang, we did not explicitly implement inference rules in our project.
The interpreter contains methods implementing either denotational semantics
via functional equations or operational semantics via big-step transitions; in
this project, we only introduce denotational semantics for our language. All of
these classes are designed to be self-contained, reusable, and easily integrated
into larger software systems, allowing them to be used directly as modules in
applications, further extended, tested, and connected to other libraries or services
within the Java ecosystem.

Finally, we present a short example of functionality, where after running the generated interpreter, the input is a syntactically correct program in the StackLang language and the result is an enumerated expression. We evaluate the simple expression $5001^2 - 4999^2$ (which results to 20000) in two ways. The first program (the expression written in StackLang instructions) is as follows:

$$\text{LOAD 4999 DUP MUL LOAD 5001 DUP MUL SUB} \qquad (2)$$

The result of the calculation is shown in the top part of Fig. 4. The expression can also be evaluated by rewriting it to the form $(5001 - 4999) \times (5001 + 4999)$. The second program is written as follows:

$$\text{LOAD 4999 5001 SUB LOAD 4999 LOAD 5001 ADD MUL} \qquad (3)$$

The result of the calculation is shown in the bottom part of Fig. 4.

```
d:\Libs\SLANG\languages>StackLang
LOAD 4999 DUP MUL LOAD 5001 DUP MUL SUB
^Z
LOAD 4999; DUP; MUL; LOAD 5001; DUP; MUL; SUB
[20000]


d:\Libs\SLANG\languages>StackLang
LOAD 4999 LOAD 5001 SUB LOAD 4999 LOAD 5001 ADD MUL
^Z
LOAD 4999; LOAD 5001; SUB; LOAD 4999; LOAD 5001; ADD; MUL
[20000]
```

Fig. 4. Evaluation of the program (2) – top and (3) – bottom.

As can be seen, both instruction sequences yield the same result; hence, they are semantically equivalent.

4.4   How StackLang can be Used in Teaching

This approach is interesting in that it allows easy integration into the pedagogical process. For example, at the beginning of the lectures, it is possible to show a live demo – from the construction of the language to the functional interpreter, with the possibility of direct modification/extension of the given language. This also includes changes in semantics. This will help students understand even more complex concepts much easier. This approach will also allow progress in self-assessment exercises because students can write their own programs in the created language and verify the results of their execution themselves, thus supporting active involvement in the language and its semantics. At the next stage, tasks related to the extension and modification of the language, its syntax and subsequently semantics are possible (comprehensive from formal specification to implementation).

In the current version of the StackLang language interpreter, it only provides the final calculation result for a syntactically correct input program. Here, it is possible to encourage students to expand the resulting statement to a step-by-step visualization of individual calculation steps and thus allow them to follow the entire calculation process (displaying the stack after each step), which is another prerequisite for easy understanding of calculations, the relationship between input and output, and possible debugging. Step-by-step calculation is also possible at the SLANG level, but the generated Java classes can be included in user's project and thus add animation or visualization on the user application side.

5   Conclusion

In this article, we presented one approach to university teaching with the aim of connecting knowledge and skills from multiple subjects so that the potential of the acquired knowledge gradually increases. In future research, we want to focus on expanding the concepts acquired in this development and research step and focus on supporting the expansion of formal methods in conjunction with modern technologies, such as large language models, and explore whether they can also be indirectly used to support teaching. For example, we can explore how the LLMs can support teaching, formalization, and even semi-automatic reasoning in the context of our StackLang and similar domain-specific languages. Potential tasks for students include assisting with code generation, verifying examples, or supporting student tasks with natural language explanations. In this context, we plan to integrate solutions with examples of domain-specific languages (DSLs) from real applications, which will allow for a more effective demonstration of concepts on specific cases. Here, we can focus on using a StackLang-like DSL for more realistic or industry-inspired case studies. This could include embedded computing in edge devices, data stream processing, or educational DSLs tailored to teach data engineering patterns. The next step is the development of automated tools for analysis and verification of the created implementations. These tools could leverage existing executable semantics to verify student-written programs or detect inefficiencies. The created semantic methods and tools can more widely support educational activities, contributing to the connection of academic training with current trends in software engineering.

Acknowledgments. This work was supported by the Aktion Österreich Slowakei project grant Nr. 2024-05-15-001 "Formalizing and Generating Executable Implementations of Domain-Specific Languages", and by KEGA project 030TUKE-4/2023 "Application of new principles in the education of IT specialists in the field of formal languages and compilers", granted by the Cultural and Education Grant Agency of the Slovak Ministry of Education.

References

1. Barras, B., et al.: The Coq Proof Assistant: Reference Manual. Coq Project, INRIA, version 6.3.1 edn. (2000). https://rocq-prover.org
2. Diehl, S., Hartel, P., Sestoft, P.: Abstract machines for programming language implementation. Futur. Gener. Comput. Syst. **16**(6), 739–751 (2000)
3. Felleisen, M., Findler, R.B., Flatt, M.: Semantics Engineering with PLT Redex, 1st edn. The MIT Press, Cambridge (2009)
4. Marmsoler, D., Brucker, A.D.: Isabelle/solidity: a deep embedding of solidity in isabelle/HOL. Form. Asp. Comput. **37**(2) (2025). https://doi.org/10.1145/3700601
5. Mihelič, J., Steingartner, W., Novitzká, V.: A denotational semantics of a concatenative/compositional programming language. Acta Polytechnica Hungarica **18**(4), 13–28 (2021). https://doi.org/10.12700/APH.18.4.2021.4.13
6. Nielson, H.R., Nielson, F.: Semantics with Applications: An Appetizer. Undergraduate Topics in Computer Science. Springer, Heidelberg (2007). https://doi.org/10.1007/978-1-84628-692-6
7. Parr, T.: The Definitive ANTLR 4 Reference. Pragmatic Bookshelf, 2nd edn. (2013). https://pragprog.com/titles/tpantlr2/the-definitive-antlr-4-reference
8. Pattis, R.E.: Karel the Robot: A Gentle Introduction to the Art of Programming, 2nd edn. John Wiley & Sons Inc., Hoboken (1994)
9. Pereira, M.J.V., Mernik, M., da Cruz, D.C., Henriques, P.R.: Program comprehension for domain-specific languages. Comput. Sci. Inf. Syst. **5**(2), 1–17 (2008). https://doi.org/10.2298/CSIS0802001P
10. Roşu, G., Şerbănută, T.F.: An overview of the K semantic framework. J. Logic Algebraic Program. **79**(6), 397–434 (2010). https://doi.org/10.1016/j.jlap.2010.03.012
11. Schmidt, D.A.: The Structure of Typed Programming Languages. MIT Press, Cambridge (1994). https://mitpress.mit.edu/books/structure-typed-programming-languages
12. Schreiner, W.: Thinking Programs. Texts & Monographs in Symbolic Computation, 1st edn. Springer, Cham (2021). https://doi.org/10.1007/978-3-030-80507-4
13. Schreiner, W., Steingartner, W.: The SLANG Semantics-Based Language-Generator — Tutorial and Reference Manual (Version 1.0.*). Technical report 23-13, Research Institute for Symbolic Computation (RISC), Johannes Kepler University Linz, Austria (2023). https://doi.org/10.35011/risc.23-13
14. Schreiner, W., Steingartner, W.: Semantics-based rapid prototyping of a machine controller language. In: Novitzká, V., Szakál, A. (eds.) Informatics 2024, 2024 IEEE 17th International Scientific Conference on Informatics, pp. 348–353. IEEE, Poprad (2024). https://doi.org/10.1109/Informatics62280.2024.10900792
15. Schreiner, W., Steingartner, W.: Semantics-Based Rapid Prototyping of a Subset of SQL. Technical Report 25-02, Research Institute for Symbolic Computation, Johannes Kepler University (2025). https://doi.org/10.35011/risc.25-02
16. The SLANG Semantics-Based Language Generator (2023). https://www.risc.jku.at/research/formal/software/SLANG
17. Steingartner, W., Novitzká, V., Schreiner, W.: Proof of equivalence of semantic methods for a selected domain-specific language. J. Appl. Math. Comput. Mech. **23**(2), 79–92 (2024). https://doi.org/10.17512/jamcm.2024.2.07

18. University of Glasgow: Algorithms & Data Structures: MSc in Information Technology and MSc in Software Development – Examination Paper (2013). https://www.dcs.gla.ac.uk/%7Edaw/teaching/ADS/Exams/paper.2013.pdf. Accessed 12 May 2025
19. Winskel, G.: The Formal Semantics of Programming Languages—An Introduction. MIT Press, Cambridge (1994). https://mitpress.mit.edu/books/formal-semantics-programming-languages

Exploring Big Data Maturity Models: Findings from a Systematic Literature Review

Marija Đukić$^{(\boxtimes)}$ ⓘ and Ivan Luković ⓘ

Faculty of Organizational Sciences, University of Belgrade, Belgrade, Serbia
{marija.djukic,ivan.lukovic}@fon.bg.ac.rs

Abstract. The potential Big Data offers to organizations is widely recognized; however, translating this potential into business value requires a certain level of maturity. Maturity models have emerged as valuable tools to help organizations assess their current state and guide their progress toward higher maturity levels. We conducted this systematic literature review to explore and evaluate existing Big Data Maturity Models (BDMMs) and draw insights from related domains that could be transferred to the Big Data context. The final selection included 72 papers, 18 presenting BDMMs. The findings reveal consistent patterns across models in terms of design approach, architecture, purpose, and typology, with some distinctions in orientation and domain applicability. Notably, most models are primarily descriptive and only a few have the capacity for prescriptive use. Key limitations include a lack of detailed documentation and the absence of support tools, such as assessment software and visualization reports to aid decision-making.

Keywords: Big Data · Big Data Analytics · Maturity Model · Literature Review

1 Introduction

Over the last decade and a half, Big Data has become a fundamental component supporting a wide range of scientific disciplines [1]. Domains that benefit from it include education, engineering, and healthcare, to name a few. The number of academic publications addressing the topic of Big Data has significantly increased during this time [1]. Big Data can be defined as "a term that describes high volumes of high velocity and high-variety information assets that require new forms of processing to enable enhanced decision making, insight discovery, and process optimization" [2]. Although there is no unified definition between industry and academia, there is general acknowledgement that Big Data includes vast quantities of data from various sources and comprises various data types that exceed the processing capabilities of traditional databases and techniques. Therefore, organizations must invest in infrastructure and technologies designed to capture, store, and analyze this data.

The importance of Big Data comes from its potential to add a competitive advantage to organizations operating in highly competitive environments. Therefore, they must develop mechanisms to support decision-making that simultaneously satisfy customer expectations and foster business excellence through the alignment of processes with

P. K. Chrysanthis et al. (Eds.): ADBIS 2025, CCIS 2676, pp. 264–279, 2026.
https://doi.org/10.1007/978-3-032-05727-3_24

their business objectives and strategies. However, many organizations still struggle to effectively integrate Big Data into their operations or value chains [3]. This gap can be attributed to organizations that have not yet reached the maturity level required to take full advantage of Big Data's potential. Maturity refers to "a state of being complete, perfect or ready" [4]. To achieve the desired state of maturity, a progressive, evolutionary path from the initial to the target stage must be followed. The concept of "maturity models" emerged to help organizations identify their position along this path [5]. Maturity models (MMs) are recognized as instruments designed to evaluate the maturity of processes or organizations across a sequence of maturity levels. Each level describes specific criteria and characteristics that must be fulfilled to claim that position [5]. Furthermore, the descriptive elements of subsequent levels often serve as best-practice guidelines for attaining higher stages of maturity. During maturity assessment, a snapshot of the organization regarding the given criteria is made [5]. This enables stakeholders to understand necessary changes and implement appropriate measures to facilitate the transformation process. The Capability Maturity Model (CMM), introduced in 1986, pioneered this approach by evaluating software process maturity in organizations. Since then, numerous maturity models have emerged across various domains.

In the context of Big Data maturity models (BDMMs), considering the rapid evolution and velocity of data, there is a clear need to measure Big Data maturity [6]. Maturity evaluation enables organizations to establish their current position relative to selected assessment criteria. The BDMM serves as a guide for assessing an organization's current state concerning its data assets, technology infrastructure, and human resources required to reach desired maturity levels [6]. The BDMM is a tool for evaluating an organization's performance, achievements, and competencies concerning Big Data, along with important aspects that might improve the organization's standing. Specific metrics can be retrieved and prioritized based on the as-is analysis, which can also be used to make recommendations for progress toward higher maturity levels.

While the potential of Big Data for organizations is well recognized, in practice many still struggle to translate this potential into business value [7]. To fully leverage Big Data, organizations must reach a certain maturity level, which is not straightforward due to various obstacles. The technologies required to process Big Data are either relatively new or have only recently gained widespread adoption in an organization. Moreover, data is generated at all levels and by all processes of an organization, making Big Data an organization-wide concern that needs to be based on appropriate data collection and extraction to provide useful business insights. Lack of clear strategic alignment and resistance to change further complicates the effective integration of Big Data initiatives. Considering all of this, assessing and understanding organization's Big Data maturity is required to bridge gaps, align efforts, and ultimately realize the full potential of data-driven decision-making.

The objective of this research is to explore and evaluate existing approaches for developing Big Data maturity models, with particular emphasis on models' capacity to provide actionable, context-specific recommendations. For this purpose, a systematic literature review (SLR) was conducted to retrieve the largest possible range of relevant articles. Although the SLR provides a detailed overview of the field, several aspects need further analysis. To structure this research, we formulated the following hypotheses regarding the current state of Big Data maturity models:

H1: Most Big Data maturity models emphasize technical components (e.g., data, infrastructure) over organizational factors.

H2: Existing Big Data maturity models differ in their dimensions, structure, and assessment methods.

H3: Existing Big Data maturity models lack mechanisms for generating context-specific recommendations.

The purpose of our SLR is to summarize evidence of the state-of-the-art BDMMs, identify gaps in the current research, and uncover opportunities for future work.

This paper is organized as follows: Sect. 2 provides a review of related work in the field. Section 3 outlines the objective and purpose of this SLR, while Sect. 4 details the research methodology. Section 5 presents the results from related categories, followed by Sect. 6, which focuses on the findings specific to the Big Data category. Section 7 provides an in-depth discussion with recommendations for future research, and Sect. 8 concludes the paper.

2   Related Work

Big Data has emerged as a critical component for decision-making and maintaining competitiveness in an unpredictable marketplace. An organization's success with Big Data initiatives is linked to its maturity in this domain. Despite the growing significance, BDMMs represent a relatively recent trend in literature, with a limited number of published articles, and consequently, there are not many published review articles in this area. To address this limitation and gain a broader understanding of the area, the articles beyond the specific domain of Big Data are also included. Selected review articles are presented in Table 1. The selection of these articles was based on their rigorous analysis of maturity models, specifically of their design, development, assessment methods, and identified limitations. Each article includes extensive coverage of digital libraries and presents detailed findings with clear and transparent reporting.

Examining the published literature reviews highlights the importance of maturity models, while also revealing opportunities for additional insights by further reviewing the published research. With an interdisciplinary perspective, this study employs a thorough search that extends beyond the domain of BDMMs, where the insights are considered transferable and relevant to the Big Data context. Also, the extensive set of digital libraries is included to increase the likelihood of covering all relevant articles, supplemented by backward citation searches. All articles included in the review were assessed using established quality criteria to ensure their academic rigor.

Table 1. Summary of selected reviews on maturity models

Reference	Scope	Sources	Restrictions	Quality Assessment
[6]	assessment models for Big Data implementation; strategic assessment in focus (12 articles)	ACM DL, IEEE Explore, Scopus, Springer Link, Science Direct, Google Scholar + backward citation search	Time frame: Jan 2010–Apr 2016, publications: industrial papers, experimental research papers	Yes
[8]	Big Data assessment models; dimensions and assessment tools (15 articles)	ACM DL, IEEE Xplore, Springer Link, Scopus, EBSCOhost, Google, Google Scholar + backward citation search	Time frame: 2007–2022, included white papers, technology reports, developers' websites	Yes
[9]	maturity models in the AI domain; design approach, architecture and purpose of use (15 articles)	ACM DL, IEEE Xplore, SpringerLink, Scopus, Web of Science, Pro-Quest, Google Scholar, Dimensions.ai	Time frame: 2015–2020, included theses and dissertations	No
[10]	data and analytics maturity models; structures and content (38 articles)	Scopus, ScienceDirect, Web of Science, EBSCOhost + backward citation search	Peer reviewed articles only	No
[11]	data analytics maturity models for SMEs, BDMMs included (18 articles)	ScienceDirect, Web of Science, PubMed + backward citation search	Time frame: 2000–2024, included non-peer reviewed articles	No

3 Purpose of the Review

Big Data offers businesses several advantages, such as better decision-making, and the capacity to extract valuable insights from large and complex datasets. Organizations must, however, build certain infrastructure, governance, and skill-related capacities to realize these advantages. The organization's current capabilities and its target state can be represented through maturity levels. In response, maturity models have been introduced to help organizations evaluate their preparedness and navigate the path toward more effective use of Big Data. These models are intended not only to reflect an organization's current standing but also to provide practical guidance for advancing to higher levels of maturity.

This research aims to thoroughly examine and evaluate existing approaches for developing Big Data maturity models, considering aspects such as development methods, architecture, design approaches, and intended purpose. The models' capacity to offer practical recommendations is given particular attention, as is the extent to which these recommendations can be customized for organizational factors. A set of research questions has been defined to guide the research:

RQ1: Which design approaches, components and assessment methods are commonly used when developing maturity models, particularly in the Big Data domain?

RQ2: How, and to what extent, do existing maturity models for Big Data integrate recommendation mechanisms?

4   Research Methodology

This study follows the systematic literature review methodology outlined by [12]. The first step was to develop a review protocol that guided subsequent steps. The protocol established research questions that directed the article selection, along with a search strategy, inclusion and exclusion criteria for identifying relevant publications, quality assessment parameters, and procedures for data extraction and synthesis. Additionally, the Preferred Reporting Items for Systematic Reviews and Meta-Analyses (PRISMA-S) guidelines [13] were integrated into research to improve the transparency and rigor of review reporting.

4.1   Data Sources and Search Strategy

The search approach began with a selection of relevant keywords, which were then combined to form search strings. Given the focus of the review, the term "big data maturity model" is acknowledged as an essential notion. Combinations of the three keyword categories "big data," "maturity," and "model" are utilized to retrieve relevant articles. Table 2 lists all the synonyms used in each category. The search scope was expanded outside the Big Data domain to encompass related domains to obtain a deeper understanding and additional insights. To reduce the number of search strings needed, wildcard symbols were included. An example of a search string used is as follows:

TITLE ((* data* OR business* OR information system OR artificial intelligence OR "BDA" OR "IS" OR "AI" OR "ERP") AND (maturit* OR capabilit* OR capacit* OR competenc* OR readiness) AND (model OR framework OR level OR assessment OR evaluation OR analysis OR measure*)).

The search strings are adapted to meet each database's specific requirements while maintaining research validity by ensuring no relevant references are overlooked. The titles of publications were searched, as applying these terms to abstracts or full-text searches led to many irrelevant results.

Table 2. Keywords used to form search strings

Keyword	Synonym
big data	big data analytics, big data analysis, data analytics, data analysis, business analytics, business analysis, business intelligence, data science, information system, BDA, AI, IS, ERP
maturity	capability, capacity, competence, readiness
model	framework, level, assessment, evaluation, analysis, measure, measurement

The search strategy included multiple electronic databases. ACM Digital Library, IEEE Xplore, Scopus, ScienceDirect, and Web of Science were chosen as the main databases following the recommendations by [14], while Springer Link, AIS eLibrary, and Semantic Scholar are used as supplementary resources (Fig. 1). Following the same guidelines, certain databases, like Google Scholar, were excluded because they contain non-peer-reviewed material, such as theses and reports, which might not have undergone proper academic review.

The search was initiated on February 3, 2025, and concluded on March 25, 2025. Initially, 5.262 articles were retrieved and imported into the Mendeley reference management system. During stage 1, titles were screened to assess relevance to the review's focus. Articles unrelated to maturity were excluded at this point, resulting in 546 remaining articles. After removing duplicates, 426 unique articles remained. In stage 2, the abstracts of these articles were analyzed to determine their relevance to the review, leading to the selection of 87 articles for further analysis. The backward snowballing method is then used to find additional relevant documents not captured in the first search. This process produces 13 additional articles, bringing the total to 100 articles subjected to full-text review in stage 3. Additional 16 papers were excluded following the full-text review, resulting in a final selection of 84 articles for inclusion in the study.

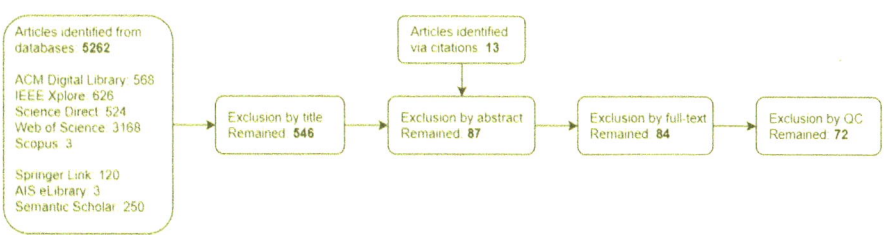

Fig. 1. Stages of study selection process

4.2 Inclusion and Exclusion Criteria

Recognizing the significance of the selection phase in maintaining the validity of the literature review, a set of inclusion (IC) and exclusion criteria (EC) was defined and applied throughout the screening process. The review included articles that proposed a model or framework for assessing maturity in the context of Big Data or related domains (IC1), and only those published in academic outlets were considered, specifically peer-reviewed journals and conference proceedings (IC2). Accordingly, articles were excluded if they did not address maturity in Big Data or related areas (EC1), lacked peer review, such as research-in-progress papers or earlier versions of published studies (EC2), or were not written in English (EC3). No restrictions were placed on the publication date, allowing for a comprehensive temporal scope.

4.3   Quality Assessment

Each of the 84 articles that remained after stage 4 was evaluated separately based on a set of quality criteria (Table 3). These criteria, collectively, served to assess the potential contribution of each article to the review. The criteria responses are scored as Yes (1), Partially (0.5), or No (0). Based on this, 12 articles were excluded (score below 2.5), resulting in a final set of 72 papers selected for data extraction and synthesis.

Table 3. Quality assessment criteria

Number	Question
QC1	Is the objective of the article clearly stated and consistent with the development of a maturity model?
QC2	Is the domain for which the maturity model was developed described in the article?
QC3	Does the article outline the maturity model's structure, including levels, dimensions, and constructs?
QC4	Does the article explain the assessment method, including data collection, measurement, and validation?

4.4   Data Extraction and Synthesis

The final set of selected articles was analyzed to extract key information relevant to the review. The classification schema (Table 4) was adopted with a set of categories adapted from [9] to aid analysis. When a study did not provide sufficient information for a given category, it was marked as NA.

The articles were coded based on the core research domains they sought to contribute to, enabling the development of a meaningful categorization. The categories were established to group domains that share common characteristics. These include Information System Infrastructure (ISI), Business-Oriented Analytics (BOA), Advanced Computational Intelligence (ACI), and Big Data, which is the primary focus of this review. The distribution of articles across these categories is illustrated in Fig. 2.

Table 4. Classification schema

Class	Value	Description
Research Method	Analytical	Conceptual, mathematical, or statistical methodologies
	Empirical	Experimental or observational approaches (i.e. case studies, content analysis)
	Theoretical	Grounded in established theories or existing maturity models (i.e. CMM)
Design approach	Top-down	During the design phase, the maturity levels were established first, and then the items of assessment
	Bottom-up	The assessment items were first identified during the design phase and then the maturity levels
Purpose of use	Descriptive	Explains the current state of an organization's processes or capabilities
	Prescriptive	Provides guidance on how to progress to higher maturity levels
	Comparative	Enables benchmarking by comparing maturity levels across organizations or sectors in the organization
Architecture	Quantitative	Relies on metrics to provide maturity score
	Qualitative	Descriptive assessments or categorical ratings to evaluate maturity
Typology	Maturity grid	Matrix-based model aligning maturity levels with capability stages
	Structured model	Hierarchical model with formalized stages (e.g. CMM)
	Likert-like	Questionnaire-based assessment using ordinal scales
Orientation [10]	Technology	Focus on digital technologies and their integration into organization's operations
	Data	Focus on organization's ability to manage and utilize data assets across entire analytical process
	Organization	Focus on internal structure, processes and people
Domain	Generic	Broad, cross-industry applicability
	Specific	Designed for a particular industry or organizational context
Number of levels		The number of maturity levels as the degree of maturity

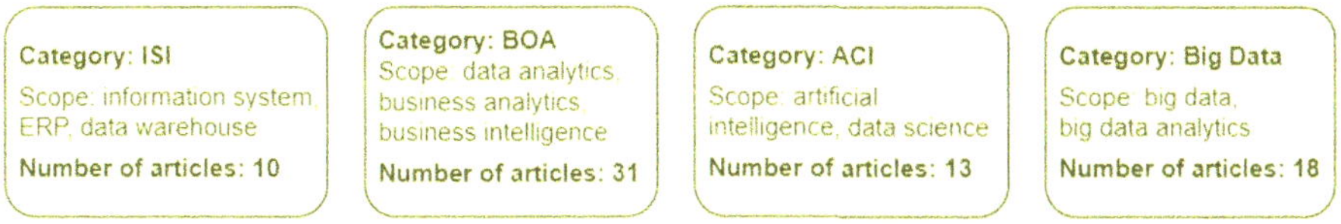

Category: ISI
Scope: information system, ERP, data warehouse
Number of articles: 10

Category: BOA
Scope: data analytics, business analytics, business intelligence
Number of articles: 31

Category: ACI
Scope: artificial intelligence, data science
Number of articles: 13

Category: Big Data
Scope: big data, big data analytics
Number of articles: 18

Fig. 2. Categories of articles

5 Results

This section presents the findings from the categories Information System Infrastructure, Business-Oriented Analytics, and Advanced Computational Intelligence.

5.1 Category ISI

The Information System Infrastructure category comprises 10 articles, each of which was analyzed using the established classification scheme. More recent studies explored maturity in domains such as ETL processes and cloud databases, while earlier works primarily focused on ERP systems. Five out of ten articles proposed maturity models, whereas the other five addressed readiness assessments. The aim of these two approaches is different: generally, readiness assessment is conducted prior to the start of the maturity process, while maturity assessment evaluates the status within a progressive maturity journey [15]. Three of the articles focus on readiness proposed assessment models. In comparison with maturity models, assessment models have a guiding framework and bound their analysis to an evaluative reference model [16].

Regarding the research methods employed in the development of proposed solutions, four articles adopted an analytical approach, relying on literature reviews of existing models. The other four utilized a mixed-methods approach, combining findings from literature with insights gathered from industry experts. Concerning design approaches, four articles adhered to a bottom-up approach, while the rest took a top-down approach. As noted by [17], maturity models can have descriptive, prescriptive, or comparative purposes. Most of the models in this category are descriptive (7), and a smaller number serves a comparative function (3). Five models are organization-oriented, while the most frequent number of maturity levels is five. Additionally, the review discovered that seven solutions were developed for generic applications, while three were designed for specific purposes, such as cloud-native databases or data warehouse capabilities.

A positive aspect is that eight of the proposed solutions were empirically validated, with seven of them tested in real-world case study scenarios. However, only two of the solutions included a visual component to communicate maturity levels to users. A clear limitation is the lack of supporting documentation, as only one article provided detailed documentation for its proposed model. Furthermore, no recommendation mechanism was included in any of the articles to help the pursuit of a higher maturity level.

5.2 Category BOA

The Business-Oriented Analytics category includes 31 articles, addressing topics such as data governance and data quality in addition to exploring business intelligence maturity. Two articles targeted readiness assessments, while six proposed models were based on the CMM. Most articles proposed maturity models, other than two assessment models and five maturity frameworks. The most common research methods were mixed-method (9 articles) and analytical approach (8 articles). Additionally, three articles presented solely theoretical models. Nearly half of the articles (15) followed a bottom-up design approach, while only two adopted a top-down approach. Most models are descriptive in nature (24), though five were also described for prescriptive application. Nine articles focused on organizational analysis, while 13 were data-oriented. The most common number of defined maturity levels remained five. Furthermore, the review showed that most solutions (18) were tailored to specific domains, such as higher education or software-intensive companies, whereas 11 were developed for general applications.

The positive side is that 19 of the proposed solutions were validated empirically, with 16 of them tested in real-world case studies. Five models included some form of recommendation to facilitate progression to higher maturity levels. However, only two solutions offered supporting documentation.

5.3 Category ACI

The Advanced Computational Intelligence category includes 13 articles, all of which were published after 2019, indicating that the topic of maturity in the field of artificial intelligence is relatively new. Eight of these articles proposed maturity models, while three focused on readiness assessment and presented frameworks tailored to that purpose. In total, seven of the identified solutions were designed for specific applications, five targeting particular domains such as the public sector or SMEs, and two tailored to specific purposes, such as trustworthy AI systems and AI ethics. In terms of design methodology, nine articles followed a bottom-up approach, whereas none used a top-down one. Six models were developed using an analytical method based on literature reviews, while another six used a mixed-methods approach. Most models in this category are descriptive in nature (7 models), with a smaller subset serving a comparative purpose (4 models). The majority adopted a qualitative architectural design, and once again, the most defined number of maturity levels across the models was five.

A significant limitation across the articles is the absence of supporting documentation, as none provided comprehensive documentation for their proposed models. Similarly, none of the studies incorporated a recommendation mechanism to guide advancement to higher maturity levels. The validation aspect shows slightly more promise; five of the proposed solutions underwent empirical validation, although only one was tested in a real-world case study setting.

6 Results of Category Big Data

The Big Data category comprises 18 articles (Table 5), six of which focus on readiness assessment. Of the nine maturity models proposed, two were applied to evaluate organization readiness, instead of maturity. The predominant research methods were mixed-method (8 articles) and analytical approach (5 articles), with one article proposing solely theoretical model. Most articles (11) applied a bottom-up design approach, and only four followed a top-down manner. In terms of architecture, a nearly equal distribution was observed between qualitative (7 articles) and quantitative (6 articles) assessment approaches. Most models are descriptive in nature (12), although three were also described for prescriptive use. Seven articles were organization-oriented, while another seven covered all three predefined orientational aspects. The largest share of models (9) was structured using hierarchical, formalized stages. Some articles provided insufficient information to determine the typology, while for [23] and [27], the typology classification was deemed not applicable. Five maturity levels were most often defined in this category as well. Regarding domains, nine models were developed for general application, and the remaining nine for specific domains, with the public sector being the most common.

Table 5. Classification schema for BDMMs

MM	Research method	Design approach	Purpose of use	Architecture	Typology	Orientation	Domain	Number of levels
[15]	empirical	bottom-up	descriptive	qualitative	structured model	organization	specific	3
[18]	theoretical	bottom-up	comparative	quantitative	maturity grid	all 3	generic	5
[19]	mixed	top-down	descriptive, prescriptive	qualitative	structured model	all 3	generic	5
[20]	analytical	bottom-up	descriptive	both	maturity grid	all 3	specific	4
[21]	mixed	bottom-up	descriptive	both	structured model	NA	generic	NA
[22]	NA	top-down	descriptive	both	maturity grid	data	generic	5
[23]	analytical	bottom-up	descriptive	quantitative	NA	data	generic	NA
[24]	mixed	bottom-up	comparative	qualitative	structured model	all 3	specific	4
[25]	mixed	bottom-up	NA	quantitative	NA	all 3	generic	NA
[26]	mixed	bottom-up	descriptive	quantitative	structured model	all 3	specific	6
[27]	empirical	top-down	comparative	quantitative	NA	technology	specific	NA
[28]	mixed	bottom-up	descriptive	qualitative	hybrid	organization	specific	5
[29]	empirical	NA	comparative	quantitative	NA	organization	specific	4
[30]	analytical	bottom-up	descriptive, prescriptive	both	structured model	all 3	generic	5
[31]	mixed	bottom-up	descriptive	quantitative	NA	organization	generic	NA
[32]	analytical	NA	descriptive	qualitative	structured model	organization	specific	5
[33]	mixed	top-down	descriptive, prescriptive	both	structured model	organization	generic	5
[34]	analytical	NA	NA	qualitative	structured model	organization	specific	5

6.1 Limitations

The findings of this SLR reveal several limitations in existing BDMMs (Table 6). [17] mentioned that maturity models are frequently introduced as theoretical concepts and lack practical applicability. This review showed that 13 models incorporated a self-assessment, where both the model and its dimensions were tested for validity and usefulness. Validation through real-world case studies was conducted for 12 models, four of which were only partially validated. On the other hand, only three solutions were compared with other existing maturity models and most of them do not consider organizational heterogeneity, such as differences in company size. This implies there is still a moderate gap between theoretical development and practical implementation. Furthermore, most of the existing models lack comprehensive guidance and support tools such as software tools as assessment instrument and visualization reports for decision making. Many models rely on text-based questionnaires, often using Likert scales for assessment. Assessment methods were identified only for six models, and just two included a dedicated software tool. Seven models offered visual reports (three only partially), while three models provided comprehensive documentation. This presents a significant barrier to practical implementation, since poor documentation impacts models usability and adoption. The development procedure was fully outlined for three and only partially for another seven models, indicating a lack of methodological transparency in most solutions, which limits the ability to assess the rigor of model development. Although not designed as prescriptive, three models did include recommendations for advancing to higher maturity levels, indicating a limited capacity of existing models to offer actionable, context-specific guidance.

Table 6. Limitations of existing BDMMs (F – full limitation, P – partial limitation)

Limitations	Existing Big Data maturity models																	
	[15]	[18]	[19]	[20]	[21]	[22]	[23]	[24]	[25]	[26]	[27]	[28]	[29]	[30]	[31]	[32]	[33]	[34]
Development method not specified	F	P	P	F		F	P	P	P			F	F	F	P	F	P	F
Assessment method not specified	F		P		P	P	F	F		F	F	P			P		P	F
No software assessment tool	F	F	F	F	F	F	F	F	F	F	F	F			F	F	F	F
Poor documentation about the model	F	F	F		F	F	F	F	F			F	F	F	F	F	F	F
Self-assessment of maturity model	F	F				F								F				F
Limited validation		P		F	F	F		P			P				F	P	F	F
No comparison with existing MMs	F	F		F		F	F	F	F	F		F	F	F	F	F	F	F
No evaluation in real case study		P			F	F	F		P		P				F	P	F	F
No visualization reports	F	F	F	P	F	P	F		F		P	F			F	F	F	F
No recommendations	F	F		F	F	F	F	F	F	F	F	F	F		F	F		F

7   Discussion

The continuous advancement of Big Data necessitates approaches to measure and improve organizational maturity. Contrary to the first hypothesis, which stated that BDMMs emphasize technical components over organizational factors, the findings reveal different picture. The SLR showed that while maturity models in other domains are predominantly organization-oriented, BDMMs frequently adopt a more holistic perspective, addressing technical, data, and organizational aspects. The complexity of Big Data demands simultaneous progress in all three aspects. The most frequently identified dimensions in BDMMs include data governance, infrastructure, organization, people, information technology, strategic alignment, and culture. Additionally, most BDMMs derive their structure from mixed-method research. This combination pushes models beyond just technical focus by capturing both theoretical rigor and real-world demands. Also, iterative refinement is commonly used in BDMMs, enabling model designers to reconsider initial biases and incorporate feedback. Some implications for future models can be drawn. Interdependencies between dimensions should be tested, e.g. do technical advancements require parallel organizational changes. Moreover, model adaptability can allow organizations to weight dimensions based on their priorities (e.g. in sectors with strict regulatory requirements, data governance may take precedence over infrastructure). Most models are empirically validated, however mixed-method validation should become standard practice. While organizational aspects can be evaluated qualitatively, technical aspects often require quantitative indicators. This approach also bridges the gap between academia and practice, where researchers require quantified reliability, while practitioners need evidence that models work in real settings.

As SLR revealed consistent patterns across maturity models, the results do not support the second hypothesis, which relates to model structure. Most models follow a bottom-up design approach and often rely on qualitative architectures for maturity assessment. Regarding typology, most solutions are structured models with formalized, hierarchical stages. Bottom-up models prioritize identifying assessment items before defining maturity levels, favoring solutions grounded in practical challenges rather than solely theoretical constructs. Moreover, in complex and rapidly evolving domains like Big Data, initiating model development from granular assessment enables greater adaptation to organizational diversity. Conversely, top-down approaches often use predefined maturity levels and carry risks of oversimplification. The dominance of qualitative architecture can be attributed to the fact that organizational maturity often involves intangible factors that are difficult to quantify. Qualitative scales are adaptable to diverse industries; however, they can be interpreted inconsistently as subjective assessments are harder to validate. Also, they often lack clarity on the degree of improvement required. Several models inherited levels from the CMM, which contributes to the dominance of structured models in BDMMs typology. The hierarchical stages simplify strategic planning for organizations by providing a clear, intuitive roadmap to monitor progress and serve as concrete milestones. However, such models carry the assumption of linear progression, while in practice progress is often non-linear with uneven advancements across dimensions. The five-level is the most common maturity structure, likely due to its balance between simplicity and granularity. Further research could explore whether customization influences methodological choices, e.g. whether industry-specific MMs are more likely to adopt a bottom-up approach, as it allows collecting context-specific dimensions. Assessment methods determine how organizations evaluate their maturity. According to our analysis, existing solutions dominantly use questionnaires assessed through Likert scales. Despite widespread adoption due to its simplicity and ease of implementation, this approach presents several limitations. Likert scales depend on self-reported or evaluator-based judgments, making them subject to bias. Moreover, they tend to reduce multidimensional maturity into single numeric scores.

The lack of recommendation mechanisms in existing BDMMs confirms the third hypothesis. Notably, only three models offer limited guidance on how organizations can advance to higher maturity levels. This gap diminishes the practical value of these models, as organizations are left without clear pathways for improvement despite having their maturity assessed. As a result, the absence of a prescriptive approach indicates that reviewed MMs do not provide a complete image to both assess and increase maturity in organization. To address the current lack of actionable guidance, we aim to develop a prescriptive Big Data maturity model. The development process will be grounded in a clearly stated methodology, such as Design Science [35], and will incorporate real-case validation to ensure both academic rigor and practical usability. Moreover, the proposed model must include an integrated assessment toolkit featuring software-supported evaluation and visualization dashboards to facilitate decision-making.

8 Conclusion

Big Data holds significant potential for organizations but translating that potential into business value requires a certain level of maturity. The importance of Big Data maturity models is to evaluate the current state and guide progress toward higher maturity levels. Therefore, we conducted this SLR to gain knowledge from publications on this topic and draw insights from related domains that can be transferred to the Big Data context. The final selection included 72 papers, of which 18 presented BDMMs.

This SLR revealed consistent patterns across maturity models within the defined article categories. In response to the first research question, most models adopt a bottom-up design approach and primarily serve a descriptive purpose, often employing qualitative architectures for maturity assessment. Regarding typology, most solutions are structured models with formalized, hierarchical stages. However, some distinctions do emerge. While maturity models from other categories are predominantly organization-oriented, BDMMs are evenly split between organization-oriented and those covering all three aspects. Regarding domains, MMs covered with this review are mostly developed for specific applications, while BDMMs are evenly split between generic applicability and specific domains. Assessment methods of the maturity concept are often undocumented. Many models rely on text-based questionnaires, typically using Likert scales for assessment. The absence of standardized assessment methods, even for well-designed models, presents a risk of inconsistent implementation in practice.

In response to the second research question, most of the maturity models included in this SLR serve descriptive purpose, with a smaller subset serving a comparative function. Only eight models were described as having prescriptive potential, offering some form of recommendation, three of which are BDMMs. This indicates that most models aim to determine the current state, without providing concrete guidance on how organization can advance to higher levels of maturity. The limited presence of prescriptive approaches in existing maturity models can impact on their practical implementation and broader adoption, as organizations are left without actionable improvement plans.

This SLR highlights the common characteristics and persistent gaps in the development of Big Data maturity models. Although existing solutions are effective in diagnosing the current maturity level, most fall short in providing practical improvement strategies, which is an area that requires primary attention moving forward.

Acknowledgments. This research is supported by the Faculty of Organizational Sciences, University of Belgrade.

Disclosure of Interests. The authors have no competing interests to declare that are relevant to the content of this article.

Materials. Supplementary material is available from the corresponding author upon request.

References

1. Tosi, D., Kokaj, R., Roccetti, M.: 15 years of Big Data: a systematic literature review. J. Big Data **11**(1), 73 (2024)
2. Beyer, M.A., Laney, D.: The importance of 'big data': A definition. Stamford, CT: Gartner (2012). https://www.gartner.com/en/documents/2057415
3. Maroufkhani, P., Wagner, R., Wan Ismail, W.K., Baroto, M.B., Nourani, M.: Big data analytics and firm performance: a systematic review. Information **10**(7), 226 (2019)
4. Simpson, J.A., Weiner, E.S.C.: The Oxford English Dictionary. Oxford University Press, Oxford. UK (1989)
5. Becker, J., Knackstedt, R., Pöppelbuß, J.: Developing maturity models for IT management: a procedure model and its application. Bus. Inf. Syst. Eng. **1**, 213–222 (2009)
6. Adrian, C., Abdullah, R., Atan, R., Jusoh, Y.Y.: Towards developing strategic assessment model for big data implementation: a systematic literature review. Int. J. Adv. Soft Comput. Appl **8**(3), 173–192 (2016)
7. Mikalef, P., Boura, M., Lekakos, G., Krogstie, J.: Big data analytics and firm performance: findings from a mixed-method approach. J. Bus. Res. **98**, 261–276 (2019)
8. Al-Sai, Z.A., et al.: Big data maturity assessment models: a systematic literature review. Big Data Cogn. Comput. **7**(1), 2 (2022)
9. Sadiq, R.B., Safie, N., Abd Rahman, A.H., Goudarzi, S.: Artificial intelligence maturity model: a systematic literature review. PeerJ. Comput. Sci. **7**, e661 (2021)
10. Langer, B.: Understanding data & analytics maturity: a systematic review of maturity model composition. Schmalenbach J. Bus. Res. 1–23 (2025)
11. Marohn, R., Li, Y.: Data Analytics Capability Maturity Models for Small and Medium Enterprises–A Systematic Literature Review (2024)
12. Kitchenham, B.A.: Guidelines for performing systematic literature reviews in software engineering. Version 2.3. Keele University and University of Durham. EBSE technical report (2007)
13. Rethlefsen, M.L., et al.: PRISMA-S: an extension to the PRISMA statement for reporting literature searches in systematic reviews. Syst. Rev. **10**, 1–19 (2021)
14. Gusenbauer, M., Haddaway, N.R.: Which academic search systems are suitable for systematic reviews or meta-analyses? evaluating retrieval qualities of google scholar, PubMed, and 26 other resources. Res. Synth. Meth. **11**(2), 181–217 (2020)
15. Ariansyah, K., Setiawan, A.B., Hikmaturokhman, A., Ardison, A., Walujo, D.: Big data readiness in the public sector: an assessment model and insights from Indonesian local governments. J. Sci. Technol. Policy Manage. **16**(2), 252–278 (2025)

16. Tarhan, A., Turetken, O., Reijers, H.A.: Business process maturity models: a systematic literature review. Inf. Softw. Technol. **75**, 122–134 (2016)
17. Pöppelbuß, J., Röglinger, M.: What makes a useful maturity model? A framework of general design principles for maturity models and its demonstration in business process management (2011)
18. Areerakulkan, N., Pongpech, W.A.: A dempster-shafer big data readiness assessment model. In: ICEIS, vol. 2, pp. 581–585 (2021)
19. Comuzzi, M., Patel, A.: How organisations leverage big data: a maturity model. Ind. Manage. Data Syst. **116**(8), 1468–1492 (2016)
20. Corallo, A., Crespino, A.M., Del Vecchio, V., Gervasi, M., Lazoi, M., Marra, M.: Evaluating maturity level of big data management and analytics in industrial companies. Technol. Forecast. Soc. Chang. **196**, 122826 (2023)
21. Dremel, C., Overhage, S., Schlauderer, S., Wulf, J.: Towards a capability model for big data analytics. Wirtschaftsinformatik **2017**, 1141–1155 (2017)
22. Farah, B.: A value based big data maturity model. J. Manage. Policy Pract. **18**(1), 11–18 (2017)
23. Farzaneh, M., Mozaffari, F., Ameli, S.P., Karami, M., Mohamadian, A., Arianyan, E.: Designing an organizational readiness framework for big data adoption. In: 2018 9th International Symposium on Telecommunications (IST), pp. 387–391. IEEE (2018)
24. Fornasiero, R., Kiebler, L., Falsafi, M., Sardesai, S.: Proposing a maturity model for assessing artificial intelligence and big data in the process industry. Int. J. Prod. Res. **63**(4), 1235–1255 (2025)
25. Gupta, M., George, J.F.: Toward the development of a big data analytics capability. Inf. Manage. **53**(8), 1049–1064 (2016)
26. Hausladen, I., Schosser, M.: Towards a maturity model for big data analytics in airline network planning. J. Air Transp. Manag. **82**, 101721 (2020)
27. Joubert, A., Murawski, M., Bick, M.: Measuring the big data readiness of developing countries–index development and its application to Africa. Inf. Syst. Front. **25**(1), 327–350 (2023)
28. Klievink, B., Romijn, B.J., Cunningham, S., de Bruijn, H.: Big data in the public sector: uncertainties and readiness. Inf. Syst. Front. **19**(2), 267–283 (2017)
29. Limpeeticharoenchot, S., Cooharojananone, N., Chavarnakul, T., Charoenruk, N., Atchariyachanvanich, K.: Adaptive big data maturity model using latent class analysis for small and medium businesses in Thailand. Expert Syst. Appl. **206**, 117965 (2022)
30. Mouhib, S., Anoun, H., Ridouani, M., Hassouni, L.: Global big data maturity model and its corresponding assessment framework results. IAENG Int. J. Appl. Math. **53**(1) (2023)
31. Nasrollahi, M., Ramezani, J.: A model to evaluate the organizational readiness for big data adoption. Int. J. Comput. Commun. Control. **15**(3) (2020)
32. Olszak, C.M., Mach-Król, M.: A conceptual framework for assessing an organization's readiness to adopt big data. Sustainability **10**(10), 3734 (2018)
33. Pour, M.J., Abbasi, F., Sohrabi, B.: Toward a maturity model for big data analytics: a roadmap for complex data processing. Int. J. Inf. Technol. Decis. Mak. **22**(01), 377–419 (2023)
34. Sulaiman, H., Cob, Z.C., Ali, N.A.: Big data maturity model for Malaysian zakat institutions to embark on big data initiatives. In: 2015 4th International Conference on Software Engineering and Computer Systems (ICSECS), pp. 61–66. IEEE (2015)
35. Wieringa, R.: Design Science Methodology for Information Systems and Software Engineering. Springer-Verlag, Berlin Heidelberg (2014)

Improving Data Discovery Effectiveness: Experimental Evaluation of Content-Based Catalogs in Data Spaces

Adriana Morejón[(✉)][iD], Alberto Berenguer[iD], Lucía de Espona[iD], David Tomás[iD], and Jose-Norberto Mazón[iD]

Department of Software and Computing Systems, University of Alicante, Carretera San Vicente del Raspeig s/n, 03690 San Vicente del Raspeig, Spain
adriana.morejon@ua.es

Abstract. Effective data discovery is crucial for collaboration and innovation in data sharing contexts. Data spaces combine the secure sharing features of data ecosystems with economic aspects of data markets in a federated environment. Traditional data catalogs, which primarily rely on high-level metadata (e.g., dataset name, license, keywords), often fail to adequately convey dataset utility to potential consumers. Our solution proposes content-based catalogs to enhance data discovery within data spaces through three key components: (i) high-quality descriptive metadata, (ii) representative data samples, and (iii) advanced discovery services. These components enable consumers to effectively find datasets that align with their requirements and evaluate their relevance prior to access. In this paper, we demonstrate through extensive experimentation across multiple contexts and data quality levels that our sampling technique significantly enhance dataset discoverability while preserving data provider sovereignty.

Keywords: data space · data discovery · data catalog · data sample · metadata

1 Introduction

In today's data-driven landscape, data ecosystems facilitate secure data sharing with emphasis on interoperability and privacy, while data markets focus on monetization and economic incentives. Data spaces emerge from this confluence [30], enabling seamless sharing while preserving data sovereignty [24] and supporting discovery for various applications [17]. Data discovery is critical as consumers must identify appropriate datasets, driving innovation in efficient discovery mechanisms [6]. Traditionally, discovery relies on metadata catalogs [14,20,29], which provide inventories describing data content, structure, and quality without exposing the actual data [21,28]. However, metadata alone often lacks detailed content information, hindering precise dataset identification for specific requirements [5]. While content-based discovery could address this limitation, it creates an Arrow information paradox [1] where exposing data to demonstrate value risks diminishing sovereignty.

Our previous work [31] tackled that problem by proposing content-based catalogs with selected metadata and data samples to reduce exposure risk while demonstrating data value, enabling consumers to assess relevance without full exposure until formal sharing agreements are established. Specifically, our approach of catalog includes: (i) content-based metadata providing relevant insights on the data format and structure, (ii) data samples by determining which is the subset of data that behaves in the same way that the original dataset for data discovery, as well as (iii) a discovery service [3] built upon a metadata repository harvesting the above information across the multiple data providers of the data space.

Therefore, while previous work laid the theoretical foundations, the motivation of the present paper is to experimentally explore how word embeddings and clustering techniques can be used to extract representative subsets from larger datasets. To do so, we utilized tabular data to compare our sampling approach against simpler table reduction techniques, such as selecting random rows or extracting the first k rows from tables. We conducted three experiments to assess the integration of discovery services with content-based descriptive metadata and data sampling techniques: first, an intrinsic evaluation comparing original and reduced tables via descriptive variables; second, an extrinsic assessment measuring the effectiveness of reduced tables in information retrieval tasks, demonstrating our method excels particularly with lower-quality data; and third, a performance assessment comparing indexing times between full and reduced tables.

The remainder of this paper is structured as follows: Sect. 2 presents related work to the field of data discovery and sample generation in the context of data spaces; Sect. 3 describes our architecture for data spaces that considers a content-based catalog and data samples for data discovery services; the experimental evaluation of the system is described in Sect. 4; finally, Sect. 5 sketches out conclusions and future work.

2 Related Work

A data space is a federated infrastructure enabling trustworthy data sharing while ensuring interoperability and provider sovereignty [33], creating value through simplified data exchange and innovation [19]. Building on this concept, multiple research initiatives have produced reference architectures to accelerate data space development [2], with the International Data Spaces Association (IDSA)[1] establishing influential standards through its Reference Architecture Model (IDS-RAM) [27], which aligns with GAIA-X principles [26].

Dataset discovery has the task of identifying and retrieving relevant datasets from data repositories, data lakes, or as we focus on this case, data spaces

---

[1] https://internationaldataspaces.org/.

[8,15,16]. One of the key elements of reference data space architectures for data discovery is the data catalog: a structured inventory of available datasets including their associated metadata. Key elements typically include semantic annotations for understanding the datasets' content, technical specifications like data formats and access endpoints, and governance information that describes access policies, usage constraints, licensing, and data quality metrics. Despite their importance in data spaces, current catalog implementations exhibit critical limitations that impair data discovery effectiveness: *metadata quality deficiencies* prevent consumers from accurately assessing dataset relevance; *excessive dependence on providers* who often lack incentives or expertise results in superficial catalog entries; and *content opacity* forces consumers to access full datasets for relevance evaluation, contradicting fundamental sovereignty principles.

Researchers have proposed several methods to discover relevant datasets. Text-based search enables users to express information needs through keyword queries that match against metadata, data content, or both [37,38,40]. In contrast, dataset-content-oriented approaches allow users to provide a query table to search for relevant or related datasets [7,11,13,23,36]. These discovery techniques leverage various technologies, including graph-based methods, statistical models, and large language models, all aimed at optimizing the identification of relevant tables.

Despite these advances in dataset discovery methods, several significant challenges persist. Most existing approaches assume direct access to dataset content for indexing and matching, an assumption incompatible with the sovereignty principles fundamental to data spaces. Furthermore, current techniques often struggle with semantic heterogeneity across domains, where the same concept may be expressed differently in various datasets. Scalability remains problematic in federated environments with thousands of distributed datasets, as content-based matching becomes computationally expensive. Additionally, most methods lack mechanisms to assess the trustworthiness and quality of discovered datasets, critical factors in a productive data space.

While using data samples for discovery rather than complete datasets improves computational efficiency, it introduces the critical challenge of generating representative samples that maintain both utility and privacy guarantees. A variety of sample generation techniques have emerged to address this challenge. Random representative sampling remains a foundational method, providing broad coverage with minimal bias, though it may underrepresent rare patterns [25]. More advanced approaches include Virtual Sample Generation (VSG), which synthesizes data points through interpolation or neural methods to enrich sparse datasets [9]. Bootstrapping methods, popular in machine learning, generate multiple resampled datasets with replacement to support robust training and uncertainty estimation [22]. Meanwhile, synthetic data generation techniques using models such as GANs and variational autoencoders simulate realistic sample while preserving privacy, facilitating discovery in restricted-access environments [10]. Other strategies include active sampling, where data is selectively retrieved based on query relevance or informativeness [35], and clustering-based sampling, which ensures semantic diversity by extracting representative elements from identified clusters [12]. Dimensionality reduction methods, such as PCA or autoencoder-based embeddings, also guide sampling by preserving the global structure of high-dimensional data [32]. Lastly, metadata-driven techniques leverage high-quality semantic annotations and quality metrics to inform targeted sample extraction [39].

However, these existing approaches exhibit critical limitations taking into account the context in which we want to use the information, i.e. a data discovery context. Methods like random, bootstrap, or synthetic sample generation, lack semantic awareness, often overlooking the underlying meaning or contextual relationships between data attributes. Moreover, methods based solely on metadata quality are ineffective in cases where metadata is sparse, inconsistent, or superficial.

To address these limitations, our solution leverages semantic embeddings and unsupervised clustering techniques to drive the sampling process. The experimentation performed with our proposed solution shows its effectivity regarding the challenges of data discovery, by integrating content-based metadata and relevant data samples into the reference architecture. Furthermore, the data discovery mechanisms balance effective content-based relevance with strict sovereignty preservation and computational efficiency.

3   Proposed Solution

We have extended the reference architecture proposed by the IDSA [27], with demonstrated compatibility with other frameworks. We made use of the IDSA's open source ready-to-use Minimally Viable Data Space (IDS-MVDS)[2].

In the IDS ecosystem, metadata and data are hierarchically structured according to the IDS information model[3]. At its core, the *resource* entity contains essential metadata (title, description, license) of data objects. These resources are organized into *catalogs*, while each resource contains *representations* describing data formats and *artifacts* with technical details like checksums. Artifacts reference *contract agreements* that govern usage between providers and consumers, with contracts potentially containing multiple *rules* representing IDS Usage Control Patterns.

Our solution [31], shown in Fig. 1, enhances the IDS-MVS architecture with a discovery service that automatically extracts data schemas and generates representative samples stored locally on the provider side. These samples, linked to original resources via IDS metadata standards, are freely accessible to data space participants without negotiation, enabling consumers to evaluate dataset relevance before initiating negotiations for complete datasets, thus significantly improving the discovery process.

---

[2] https://github.com/International-Data-Spaces-Association.

[3] https://international-data-spaces-association.github.io/DataspaceConnector/
Documentation/v5/DataModel.

In order to facilitate data discovery, we integrate InferIA [3], a search engine software provider with the IDS data space ecosystem through a suite of components depicted in Fig. 1:

- **Data consumer**: allows the browser to collect metadata available in the data space.
- **Browser**: queries the Metadata Broker periodically to retrieve data providers and their catalogs, then requests resource listings and metadata for each catalog.
- **Metadata storage**: structures and optimizes data collected by the browser and stores word embeddings generated by the embedding microservice.
- **Embedding microservice**: generates embeddings from given tables using a large language model (LLM).
- **Search microservice**: provides the data retrieval algorithm according to our previous work [4,5,34].
- **API**: serves as middleware between the Search microservice and the Web interface.

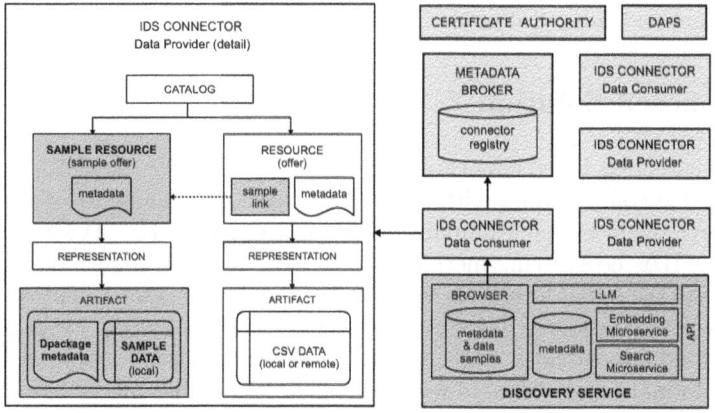

Fig. 1. Overview of the dataspace from IDSA (yellow) and our contribution (blue). (Color figure online)

The following subsections detail our primary architectural contributions: first introducing our content-based catalog for data spaces, then examining data sample generation techniques and their seamless integration into the catalog infrastructure.

3.1 Content-Based Metadata

Our solution integrates content-based metadata at both resource and data content levels. At the resource level, metadata follows the IDS standard with fields like title, description, and license added by data providers. This is automatically enhanced with content-based metadata derived from the data itself, formatted using the Datapackage standard[4] from the Frictionless Data project [18], and stored locally at the artifact level alongside a representative data sample. When consumers request the sample resource, they simultaneously receive this enriched metadata.

3.2   Generating Data Samples

Considering tabular data, various table reduction approaches to create data samples have been applied: (i) randomly sampling rows from the original table, (ii) sequential sampling by selecting the first ten rows of the original table, and (iii) a novel reduction approach we have developed based on clustering techniques to obtain representative rows from the original table. This cluster reduction approach converts table data into numerical vectors using word embeddings, which are then grouped into clusters based on their similarities.

With the data transformed into a numerical vector space, clustering techniques are applied to group table rows based on their similarity across multiple dimensions. We selected K-means as our clustering algorithm for its efficiency and scalability with large datasets, particularly when working with high-dimensional numerical vectors. K-means divides data into a fixed number of clusters by minimizing within-cluster variance, creating groups that are internally cohesive yet distinct from one another.

The K-means algorithm partitions the data into k clusters to maximize similarity with a reduced table of target rows. Once these clusters are calculated, we select the row closest to each cluster's center, resulting in a reduced table of k rows. These centroids effectively represent the central tendencies of the data distribution, preserving general patterns and variability.

4   Experimental Evaluation

In this section, we describe the comprehensive evaluation of our proposed cluster through intrinsic, extrinsic, and temporal assessments. The intrinsic assessment compared original tables with their reduced counterparts to measure information preservation. The extrinsic assessment evaluated practical utility by employing reduced tables in information retrieval tasks within our proposed data space architecture, serving as a data discovery process. We benchmarked our cluster reduction technique against simpler methods such as random row selection and first rows selection. Additionally, we analyzed indexing performance differences between original and reduced datasets.

---

[4] https://datapackage.org/standard/data-package/.

4.1   Evaluation Dataset

We selected a real-world dataset provided by a data space provider in the experiments conducted to evaluate the effectiveness of our table reduction methodology. The dataset comprises 100 tables extracted from a tourism open data portal from the Valencia Region in Spain, TData[5]. This collection was carefully selected to encompass a wide array of data structures and column types, providing a robust testing ground for assessing the adaptability and performance of the reduction approach. Table 1 offers an overview of the statistics of the evaluation dataset.

Table 1. Statistics of the collected dataset.

Characteristic	
Number of tables	100
Total number of rows	748,088
Total number of columns	1,926
Total number of text columns	1637
Avg. number of rows	7,480.88
Avg. number of columns	19.26
Avg. number of text columns	16.37
Max. number of rows	90,792
Max. number of columns	44
Max. number of text columns	43

4.2   Intrinsic Assessment

To evaluate the effectiveness of the proposed table reduction approach, a comparison was conducted between the original tables and their reduced versions, using descriptive statistics as the basis for similarity assessment. This was an indicator of how well the reduced tables retained the statistical properties of the original data.

Seven descriptive statistics were calculated along each column: the mean, median, standard deviation, minimum, maximum, as well as 25th and 75th percentiles. These seven metrics were calculated for each of the columns, and the resulting values are stored in a vector representing each table, thus allowing us to quantitatively compare tables. The similarity between the original and reduced tables was determined by calculating the cosine similarity between these vectors, providing a metric of alignment between the two tables.

---

[5] https://tdata.dlsi.ua.es/.

We evaluated the table reduction approach across three distinct methods to benchmark its performance. In the first method, we applied the cluster reduction methodology described in Sect. 3.2. In the second method, we created a reduced table by randomly sampling rows from the original dataset. In the third method, we simply selected the first ten rows from the initial table, representing a sequential selection without methodological design.

For all three methods, we generated reduced tables with a standardized 10-row count. Cosine similarity analysis yielded matching scores of 0.933 (random reduction), 0.935 (first ten rows), and 0.934 (cluster reduction), enabling direct effectiveness comparison of dimensional reduction techniques. These nearly identical scores suggest our proposed reduction approach offers no significant advantage over simpler random or sequential selection methods when preserving statistical features from the original dataset. The minimal variation in similarity scores (0.002 maximum difference) indicates that further research is needed to develop methods that better preserve distinctive characteristics across original tables.

4.3   Extrinsic Assessment

In order to evaluate the table reduction utility for data discovery services, we conducted experiments comparing the performance of the tables against the original datasets using embeddings generated by the JinaAI[6] Spanish model, specifically the *jina-embeddings-v2-base-es* variant. This systematic comparison allowed us to thoroughly assess how well reduced tables perform relative to original datasets in information retrieval tasks.

The experimentation was carried out by first selecting randomly 10 rows from each original table to serve as a query in the experiments. Next, embeddings were created for both the original dataset and its three reduced versions as described in Sect. 3.2. Finally, search metrics were calculated using the queries against the original and sample tables.

We established three search contexts by generating indexes from distinct metadata components in our proposed content-based catalog. In *Context 1*, indexes incorporated only the table content. *Context 2* expanded this approach by including both table content and column names. *Context 3*, applicable to tables with available column descriptions, generated indexes using a comprehensive combination of column names, column descriptions and table content.

To evaluate retrieval quality, we employed two metrics: Mean Reciprocal Rank (MRR), which measures the position of the first relevant retrieved result, and Precision at k (P@k metric), which calculates the proportion of relevant results among the top k retrievals. These metrics provide an objective assessment of each approach's performance. In our results, we highlight statistically significant differences between the random and clustering reduction methods with asterisks, as these two approaches emerged as the most competitive. A

---

[6] https://jina.ai/embeddings/.

single asterisk (*) indicates statistical significance with $p < 0.05$, while double asterisks (**) denote stronger significance with $p < 0.01$.

Table 2 shows the search metrics results using reduction techniques in the three different contexts. The results reveal interesting trends and variations across contexts and methods:

- *Context 1*: Both Cluster reduction and Random reduction yield similar results, slightly below the original dataset but still noticeably better than First rows reduction. Among all methods, First rows reduction performs the weakest, with its MRR dropping to 0.6392 and P@1 falling to 0.5426, indicating its limitations in this context.
- *Context 2*: The Cluster approach provides the better reduction scores. However, Random reduction and First rows reduction perform significantly worse in this context, with First rows reduction being the weakest. Its MRR of 0.6311 and P@1 of 0.5362 reflect a clear gap in performance compared to Cluster reduction.
- *Context 3*: Cluster reduction achieves the highest scores in MRR and P@1 metrics. With an MRR of 0.7856 and a P@1 of 0.7043, Cluster reduction demonstrates good effectiveness in this context. Both Random reduction and First rows reduction also show substantial improvement in Context 3 compared to the other contexts. Although their performance is slightly lower than that of Cluster reduction, they achieve high results, with MRR scores of 0.7508 and 0.6720, respectively.

Overall, Cluster reduction offers the best performance in reduction metrics, particularly excelling in Context 3. On the other hand, Random reduction and First rows reduction generally fall behind in most scenarios, with First rows reduction frequently yielding the weakest results. These findings also highlight the variability in method effectiveness depending on the specific context.

To ensure a realistic evaluation of the data discovery service in our data space architecture, we considered varying data quality levels, particularly regarding sparsity. Therefore, next experiments introduced different percentages of null values into the indexed tables, reflecting the common scenario where tables often contain incomplete rows or missing data. This approach allowed us to assess the service's performance under conditions that mirror the imperfect data typically encountered in real-world scenarios.

Table 3 presents different levels of null values that were introduced into the original tables for Context 1, Context 2, and Context 3, respectively. In these three contexts, four levels of null value percentages were established: 10%, 25%, 50%, and 75% of missing values. In all these experiments, it is evident that the method based on the First rows consistently yields the worst results, just as it did when working without null values. Therefore, we will focus on discussing the results of comparing the Cluster reduction with the Random reduction.

Table 3 presents the impact of introducing varying levels of null values (10%, 25%, 50%, and 75%) into the original tables across all three contexts. Consistent with our previous findings, First rows reduction consistently produces the

Table 2. Search metrics combining different contexts (Ctx) and reduction techniques.

	Reduction	MRR	P@1	P@5	P@10
Context 1	Original data	0.8875	0.8340	0.9919	0.9979
	Cluster	0.7327*	0.6521*	0.8415*	0.9511**
	Random	0.7100	0.6351	0.8043	0.9191
	First rows	0.6392	0.5426	0.7723	0.8766
Context 2	Original data	0.7304	0.6479	0.8777	0.9447
	Cluster	0.6668*	0.5713*	0.7979*	0.9266*
	Random	0.6537	0.5606	0.7840	0.9160
	First rows	0.6311	0.5362	0.7755	0.8745
Context 3	Original data	0.8974	0.8473	0.9839	0.9979
	Cluster	0.7856**	0.7043*	0.9161**	0.9613**
	Random	0.7508	0.6720	0.8624	0.9613
	First rows	0.7279	0.6376	0.8710	0.9290

poorest results, mirroring its performance in datasets without null values. Given this consistent underperformance, our analysis will focus primarily on comparing Cluster reduction with Random reduction methods.

In *Context 1*, the comparison between Cluster and Random reduction shows similar performance trends across all levels of null values. At 10% null values, Cluster reduction slightly outperforms Random reduction in all the metrics. At 50% and 75% null values, Cluster reduction maintains a narrow advantage over Random reduction. The differences are more clear at 25% null values, where Cluster reduction surpasses Random reduction more clearly in MRR (0.7807 vs. 0.7288) and P@1 (0.6840 vs. 0.6351). This suggests that Cluster reduction has a good behavior in medium sparse datasets. This trend is consistent across the other two contexts. In *Context 2*, it is evident that in situations with a medium number of null values (25%), the Cluster reduction method clearly outperforms the Random reduction across all metrics. The same pattern is observed in *Context 3*, where the cluster-based method shows a significant advantage over the random method. This suggests that in real-world scenarios, where tables often contain a medium proportion of null values, the proposed Cluster reduction can be significantly more effective than the Random reduction (Fig. 2).

4.4 Indexing Performance

Indexing speed is a critical performance metric for evaluating search system efficiency, particularly in dynamic environments managing large datasets like data spaces. This study compares the indexing times of two data catalogs employing distinct approaches: Catalog A indexes complete data tables, calculating embeddings for each row to ensure comprehensive dataset representation, while Catalog B utilizes our proposed Cluster reduction method (detailed in Sect. 3.2)

Table 3. Search metrics for different null percentages (N%) with Context 1, 2, and 3.

	N%	Reduction	MRR	P@1	P@5	P@10
Context 1	10%	Original data	0.8803	0.8198	0.9896	1.0000
		Cluster	0.7172*	0.6250	0.8510*	0.9615**
		Random	0.6990	0.6063	0.8323	0.9302
		First rows	0.6380	0.5333	0.7792	0.8927
	25%	Original data	0.8656	0.7968	0.9684	0.9979
		Cluster	0.7329**	0.6347*	0.8663**	0.9716**
		Random	0.6847	0.5916	0.8095	0.9368
		First rows	0.6264	0.5200	0.7758	0.8979
	50%	Original data	0.8054	0.7198	0.9344	0.9958
		Cluster	0.7043**	0.6010**	0.8521**	0.9688**
		Random	0.6543	0.5427	0.8031	0.9115
		First rows	0.6131	0.5146	0.7510	0.8656
	75%	Original data	0.6900	0.5688	0.8646	0.9667
		Cluster	0.5919**	0.4677**	0.7750**	0.9052**
		Random	0.5506	0.4146	0.7490	0.8677
		First rows	0.4907	0.3740	0.6479	0.8063
Context 2	10%	Original data	0.7403	0.6438	0.8865	0.9865
		Cluster	0.6633**	0.5594*	0.8073	0.9458**
		Random	0.6414	0.5427	0.7885	0.9104
		First rows	0.6047	0.4833	0.7646	0.8938
	25%	Original data	0.7052	0.5874	0.8716	0.9937
		Cluster	0.6542**	0.5337*	0.8232**	0.9463**
		Random	0.6158	0.5032	0.7726	0.9032
		First rows	0.6031	0.4853	0.7674	0.9053
	50%	Original data	0.6410	0.5125	0.8156	0.9531
		Cluster	0.6033**	0.4688**	0.7938**	0.9281**
		Random	0.5540	0.4292	0.7208	0.8542*
		First rows	0.5461	0.4302	0.6979	0.8583
	75%	Original data	0.5123	0.3698	0.7052	0.9000
		Cluster	0.4932**	0.3594**	0.6625**	0.8333**
		Random	0.4238	0.2854	0.6021	0.7729
		First rows	0.4229	0.2896	0.6042	0.7438
Context 3	10%	Original data	0.8932	0.8368	0.9905	1.0000
		Cluster	0.7688*	0.6726*	0.9137*	0.9842**
		Random	0.7429	0.6537	0.8811	0.9695
		First rows	0.7286	0.6379	0.8463	0.9526
	25%	Original data	0.8811	0.8180	0.9723	0.9979
		Cluster	0.7807**	0.6840*	0.9298**	0.9851*
		Random	0.7288	0.6351	0.8745	0.9691
		First rows	0.7257	0.6287	0.8713	0.9532
	50%	Original data	0.8282	0.7526	0.9421	0.9958
		Cluster	0.7527**	0.6484*	0.9032**	0.9853**
		Random	0.7083	0.6116	0.8379	0.9526*
		First rows	0.7203	0.6211	0.8495	0.9495
	75%	Original data	0.7319	0.6211	0.8916	0.9800
		Cluster	0.6668*	0.5568*	0.8168*	0.9495*
		Random	0.6523	0.5389	0.8116	0.9316
		First rows	0.6594	0.5432	0.8211	0.9442

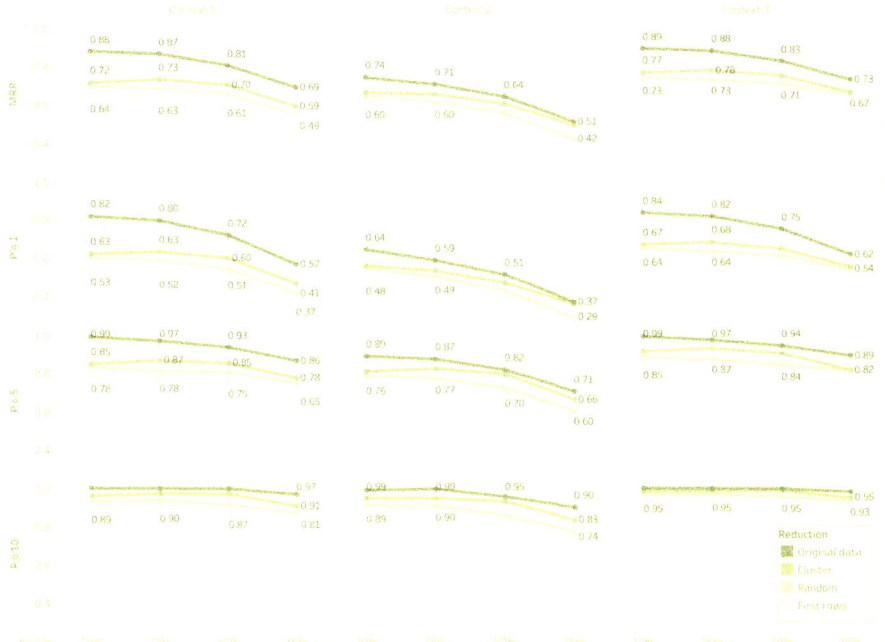

Fig. 2. Search metrics for different null percentages (N%) with Context 1, 2, and 3. A consistent colour scheme has been used to facilitate the interpretation of the results: black represents Original data, red corresponds to reduction by Clustering, blue indicates Random reduction and yellow represents First rows of the dataset. (Color figure online)

to generate embeddings from reduced table versions. This latter approach was specifically developed to address the performance limitations observed when processing full datasets.

In our experimental setup, Catalog A processed datasets averaging 7,480 rows and 19 columns, while Catalog B utilized the same tables reduced to just 10 rows per table through our clustering approach. The results demonstrate that Catalog B significantly improved indexing efficiency, achieving approximately 184× faster processing without compromising system performance. Specifically, Catalog B completed indexing in 32 s total (12 s for indexing plus 20 s for the reduction process) compared to Catalog A's 5,891 s. Moreover, this performance gap widened as dataset size increased, indicating superior scalability of the Catalog B approach for larger datasets.

These findings highlight the important trade-off between comprehensive data representation and processing efficiency, emphasizing the need for optimized indexing algorithms that balance performance requirements with data fidelity, particularly in data-intensive environments with high throughput demands.

5   Conclusions and Future Work

In this paper, we evaluated our proposed architecture for data spaces that integrates content-based metadata generation and discovery mechanisms, balancing data sovereignty principles with effective dataset identification. Our approach addresses data discovery challenges by combining descriptive metadata with content-based sampling, overcoming limitations of traditional metadata catalogs.

Our evaluation revealed key insights into table reduction methods. Using both intrinsic and extrinsic metrics, we demonstrated that our Cluster reduction approach outperforms simpler techniques, particularly with inconsistent data quality. While statistical analysis showed minimal differences in preserving intrinsic dataset characteristics (variation under 0.002), the extrinsic evaluation showed Cluster reduction consistently outperformed Random and First rows approaches across different contexts. This advantage was more pronounced with sparse datasets containing high proportions of null values—common in real-world applications.

Our Cluster reduction approach delivered substantial efficiency gains, enabling approximately 184× faster indexing speeds without compromising search quality. This scalability advantage increased with dataset size, establishing a foundation for effective data discovery in federated environments while protecting provider sovereignty. These findings highlight the important trade-off between comprehensive data representation and processing efficiency, emphasizing the need for optimized indexing algorithms in data-intensive environments.

Future work should focus on developing more sophisticated reduction techniques that better preserve distinctive statistical characteristics while maintaining search performance advantages, and exploring adaptive sampling methodologies that adjust to dataset characteristics. Additionally, validating these approaches across various sectoral data spaces would provide insights into their practical applicability across different domains.

Acknowledgements. This work is part of CLIO project (TSI-100130-2024-69), funded by Spanish Ministry of Digital Processing and by NextGeneration EU; and CIAICO/2022/019 project, funded by Generalitat Valenciana (Spain).

References

1. Azcoitia, S.A., Laoutaris, N.: A survey of data marketplaces and their business models. ACM SIGMOD Rec. **51**(3), 18–29 (2022)
2. Bacco, M., Kocian, A., Chessa, S., Crivello, A., Barsocchi, P.: What are data spaces? Systematic survey and future outlook. Data Brief **57**, 110969 (2024)
3. Berenguer, A., Alcaraz, O., Tomás, D., Mazón, J.: From research on data-intensive software to innovation in data spaces: a search service for tabular data. IEEE Softw. **41**(3), 59–66 (2024). https://doi.org/10.1109/MS.2024.3359333

4. Berenguer, A., Mazón, J., Tomás, D.: Word embeddings for retrieving tabular data from research publications. Mach. Learn. **113**(4), 2227–2248 (2024). https://doi.org/10.1007/s10994-023-06472-0
5. Berenguer, A., Tomás, D., Mazón, J.: Tabular open government data search for data spaces based on word embeddings. In: Gallinucci, E., Golab, L. (eds.) Proceedings of the 25th International Workshop on Design, Optimization, Languages and Analytical Processing of Big Data (DOLAP 2023). CEUR Workshop Proceedings, vol. 3369, pp. 61–70. CEUR-WS.org (2023). https://ceur-ws.org/Vol-3369/paper6.pdf
6. Bhandari, A., et al.: Examples are all you need: iterative data discovery by example in data lakes. In: CIDR (2022)
7. Bogatu, A., Fernandes, A.A.A., Paton, N.W., Konstantinou, N.: Dataset discovery in data lakes. In: 2020 IEEE 36th International Conference on Data Engineering (ICDE), pp. 709–720 (2020). https://doi.org/10.1109/ICDE48307.2020.00067
8. Castelo, S., Rampin, R., Santos, A., Bessa, A., Chirigati, F., Freire, J.: Auctus: a dataset search engine for data discovery and augmentation. Proc. VLDB Endow. **14**(12), 2791–2794 (2021). https://doi.org/10.14778/3476311.3476346
9. Chu, M.T., Khosla, R.: Index evaluations and business strategies on communities of practice. Expert Syst. Appl. **36**(2, Part 1), 1549–1558 (2009). https://doi.org/10.1016/j.eswa.2007.11.053
10. Cinquini, M., Giannotti, F., Guidotti, R.: Boosting synthetic data generation with effective nonlinear causal discovery. In: 2021 IEEE Third International Conference on Cognitive Machine Intelligence (CogMI), pp. 54–63 (2021). https://doi.org/10.1109/CogMI52975.2021.00016
11. Cong, T., Nargesian, F., Jagadish, H.V.: Pylon: Semantic table union search in data lakes (2023). https://arxiv.org/abs/2301.04901
12. Cui, Y., Jin, Z., Jiang, J.: A novel supervised feature extraction and classification fusion algorithm for land cover recognition of the off-land scenario. Neurocomputing **140**, 77–83 (2014). https://doi.org/10.1016/j.neucom.2014.03.034
13. Dong, Y., Xiao, C., Nozawa, T., Enomoto, M., Oyamada, M.: Deepjoin: Joinable table discovery with pre-trained language models (2023). https://arxiv.org/abs/2212.07588
14. Eichler, R., Gröger, C., Hoos, E., Schwarz, H., Mitschang, B.: Data shopping—how an enterprise data marketplace supports data democratization in companies. In: International Conference on Advanced Information Systems Engineering, pp. 19–26. Springer (2022)
15. Esmailoghli, M., Schnell, C., Miller, R.J., Abedjan, Z.: Blend: A unified data discovery system (2024). https://arxiv.org/abs/2310.02656
16. Fan, G., Wang, J., Li, Y., Zhang, D., Miller, R.: Semantics-aware dataset discovery from data lakes with contextualized column-based representation learning (2023). https://arxiv.org/abs/2210.01922
17. Fernandez, R.C., Subramaniam, P., Franklin, M.J.: Data market platforms: trading data assets to solve data problems. Proc. VLDB Endowment **13**(12), 1933–1947 (2020)
18. Fowler, D., Barratt, J., Walsh, P.: Frictionless data: making research data quality visible. Int. J. Digit. Curation **12**(2), 274–285 (2017)
19. Gieß, A., Möller, F., Schoormann, T., Otto, B.: Design options for data spaces. In: Thirty-First European Conference on Information Systems (ECIS 2023) (2023)
20. Gröger, C.: There is no AI without data. Commun. ACM **64**(11), 98–108 (2021)

21. Hauff, M., Comet, L.M., Moosmann, P., Lange, C., Chrysakis, I., Theissen-Lipp, J.: FAIRness in dataspaces: the role of semantics for data management. In: The Second International Workshop on Semantics in Dataspaces, co-located with the Extended Semantic Web Conference (2024)
22. Hewagamage, P., Mihiranga, A., Perera, D., Fernando, R., Thilakarathna, T., Kasthurirathna, D.: Computer-vision enabled waste management system for green environment. In: 2021 3rd International Conference on Advancements in Computing (ICAC), pp. 276–281 (2021). https://doi.org/10.1109/ICAC54203.2021.9671222
23. Hu, X., et al.: Automatic table union search with tabular representation learning. In: Rogers, A., Boyd-Graber, J., Okazaki, N. (eds.) Findings of the Association for Computational Linguistics: ACL 2023, pp. 3786–3800. Association for Computational Linguistics, Toronto, Canada (2023). https://doi.org/10.18653/v1/2023.findings-acl.233
24. Hummel, P., Braun, M., Tretter, M., Dabrock, P.: Data sovereignty: a review. Big Data Soc. **8**(1), 2053951720982012 (2021)
25. Imielinski, T., Mannila, H.: A database perspective on knowledge discovery. Commun. ACM **39**(11), 58–64 (1996). https://doi.org/10.1145/240455.240472
26. International Data Spaces Association: Position Paper - GAIA-X and IDS (2021). https://internationaldataspaces.org/wp-content/uploads/dlm_uploads/IDSA-Position-Paper-GAIA-X-and-IDS.pdf
27. International Data Spaces Association: IDS Reference Architecture Model, Version 4.0 (2022). https://docs.internationaldataspaces.org/ids-knowledgebase/ids-ram-4/introduction/1_1_goals_of_the_international_data_spaces
28. Jahnke, N., Otto, B.: Data catalogs in the enterprise: applications and integration. Datenbank-Spektrum **23**(2), 89–96 (2023)
29. Labadie, C., Legner, C., Eurich, M., Fadler, M.: Fair enough? Enhancing the usage of enterprise data with data catalogs. In: 2020 IEEE 22nd Conference on Business Informatics (CBI). vol. 1, pp. 201–210. IEEE (2020)
30. Möller, F., et al.: Industrial data ecosystems and data spaces. Electron. Mark. **34**(1), 1–17 (2024)
31. Morejón, A., et al.: Exploring content-based catalogs for enhanced discovery services in data spaces. In: Maté', A., Lissandrini, M. (eds.) Proceedings of the 27th International Workshop on Design, Optimization, Languages and Analytical Processing of Big Data (DOLAP 2025). CEUR Workshop Proceedings, vol. 3931, pp. 79–84. CEUR (2025). https://ceur-ws.org/Vol-3931/short7.pdf
32. Nargesian, F., Asudeh, A., Jagadish, H.V.: Responsible data integration: next-generation challenges. In: Proceedings of the 2022 International Conference on Management of Data, pp. 2458–2464. SIGMOD '22, Association for Computing Machinery, New York, NY, USA (2022). https://doi.org/10.1145/3514221.3522567
33. Otto, B.: A federated infrastructure for European data spaces. Commun. ACM **65**(4), 44–45 (2022). https://doi.org/10.1145/3512341
34. Pilaluisa, J., Tomás, D., Navarro-Colorado, B., Mazón, J.: Contextual word embeddings for tabular data search and integration. Neural Comput. Appl. **35**(13), 9319–9333 (2023). https://doi.org/10.1007/s00521-022-08066-8
35. Ramos Rojas, J.A., Beth Kery, M., Rosenthal, S., Dey, A.: Sampling techniques to improve big data exploration. In: 2017 IEEE 7th Symposium on Large Data Analysis and Visualization (LDAV), pp. 26–35 (2017). https://doi.org/10.1109/LDAV.2017.8231848

36. Santos, A., Bessa, A., Chirigati, F., Musco, C., Freire, J.: Correlation sketches for approximate join-correlation queries. In: Proceedings of the 2021 International Conference on Management of Data, pp. 1531–1544. SIGMOD/PODS '21, ACM (2021). https://doi.org/10.1145/3448016.3458456

37. Wang, F., Sun, K., Chen, M., Pujara, J., Szekely, P.: Retrieving complex tables with multi-granular graph representation learning. In: Proceedings of the 44th International ACM SIGIR Conference on Research and Development in Information Retrieval, pp. 1472–1482. SIGIR '21, ACM (2021). https://doi.org/10.1145/3404835.3462909

38. Wang, Q., Castro Fernandez, R.: Solo: data discovery using natural language questions via a self-supervised approach. Proc. ACM Manag. Data 1(4) (2023). https://doi.org/10.1145/3626756

39. Xu, P., Ji, X., Li, M., Lu, W.: Virtual sample generation in machine learning assisted materials design and discovery. J. Mater. Inf. 3(3) (2023). https://doi.org/10.20517/jmi.2023.18

40. Zhang, S., Balog, K.: Ad hoc table retrieval using semantic similarity. In: Proceedings of the 2018 World Wide Web Conference on World Wide Web - WWW '18, pp. 1553–1562. WWW '18, ACM Press (2018). https://doi.org/10.1145/3178876.3186067

Architecture of Multi-agent System for Automatic Code Template Maintenance

Elena Akik$^{(\boxtimes)}$ ⓘ, Marko Vještica ⓘ, Vladimir Dimitrieski ⓘ, Slavica Kordić ⓘ, and Sonja Ristić ⓘ

Faculty of Technical Sciences, University of Novi Sad, Trg Dositeja Obradovića 6, 21000 Novi Sad, Serbia
{elena,marko.vjestica,dimitrieski,slavica,sdristic}@uns.ac.rs

Abstract. In modern software engineering, template-based code generation, and in particular Large Language Model (LLM) driven code synthesis, have been identified as key enablers of developer productivity in environments in which complex, predefined components are subject to continuous adaptation to evolving requirements. However, the maintenance of code generator templates is frequently performed as a manual, error prone task when dependencies on external specifications or components are present, especially when they change often. To address this gap, a Multi-Agent System (MAS) is proposed, in which LLMs are relied upon for the interpretation of template code, and Model-Driven Software Development (MDSD) principles are applied for the structured representation of templates. In this system, specialized agents are tasked with detecting and interpreting changes in external specifications, performing targeted updates to template code, and validating consistency against a canonical model. Communication among agents, and the storage of semantic embeddings of template fragments and domain models, are managed through a vector database, thereby enabling retrieval of prior template versions and reasoning over them. The proposed solution is to be integrated into an existing MDSD solution for uniform access to vector databases as proof of concept.

Keywords: Multi-Agent System · Large Language Model · Model-Driven Software Development · Code Generator · Vector Database

1 Introduction

In recent decades, escalating system complexity and rapid iteration cycles have raised software development costs, particularly in programming and debugging, thereby increasing interest in advanced engineering paradigms and real-time code generation [1]. Consequently, research efforts have been directed toward automatic code generation pursued to streamline development by reducing manual coding and enhancing reliability. Generating code from high-level specifications has thus become a central objective.

Two approaches to automatic code generation are distinguished [2]. In the first, search-based techniques are used to explore a program space defined by Domain-Specific

P. K. Chrysanthis et al. (Eds.): ADBIS 2025, CCIS 2676, pp. 296–310, 2026.
https://doi.org/10.1007/978-3-032-05727-3_26

Languages (DSLs) and target programs are produced via code generators [3]. However, scalability is constrained as DSL is expanded, and dependencies on evolving external specifications and solutions are placed upon generator templates. In the second, Deep Learning (DL) models are used to synthesize code from natural-language descriptions or incomplete code fragments, with supervisory mechanisms introduced to improve output fidelity, yet practical challenges remain unresolved [4].

Advances in automated generation arise from DL models, such as Large Language Models (LLMs), capable of synthesizing various software artifacts, like source code and documentation [5]. However, structural inconsistencies and syntactic variability hinder reliable integration and maintenance [6]. These challenges limit a development lifecycle consistency, and their resolution is not yet achieved through LLM-based approaches alone [2]. To address these limitations, recent research explored integrating LLMs within Multi-Agent System (MAS) frameworks, where autonomous agents are assigned to generation, validation, and refinement tasks, thereby improving coherence, scalability, and adaptability across complex software engineering workflows [7, 8].

Building upon software artifact management, our previous work identified challenges in syntactic heterogeneity and lack of standardization among vendor-specific vector databases [9]. Therefore, a Model-Driven Software Development (MDSD) solution was proposed, featuring a DSL named vecDSL [10], which transforms platform-independent models into code for the selected Vector Database Management System (VDBMS). Due to dynamic and evolving nature of query language syntaxes and documentation, frequent manual updates to code generators became a burdensome and time-consuming task. Consequently, LLMs and MASs are explored to automate and enhance code-generator template maintenance for greater autonomy and adaptability.

In this paper, we propose the Multi-Agent System for Automatic Code Template Maintenance (MAS4ACTM) architecture, to enable adaptive maintenance of diverse software artifacts in response to changes in external specifications and documentation. MAS4ACTM aims to reduce manual overhead and implementation errors arising from maintaining compatibility across heterogeneous reference materials. The system's objectives include: (i) minimizing engineering workload; (ii) enhancing adaptability to documentation changes; and (iii) improving maintainability across varied deployment contexts. As proof of concept, MAS4ACTM is to be integrated within the existing MDSD solution for uniform access to vector databases. Through this approach, a foundation is established for replacing reactive and error-prone maintenance processes with proactive, intelligent, and context-aware adaptation mechanisms.

Following Introduction, this paper proceeds as follows. In Sect. 2, the evolution of automatic code generation and relevant studies are reviewed. The MAS4ACTM architecture is described in Sect. 3, while proof of concept is presented in Sect. 4. MAS4ACTM's advantages, limitations, ethical considerations, and integration requirements are outlined in Sect. 5. Conclusions and future work are given in Sect. 6.

2 Background and Related Work

With the development of Artificial Intelligence (AI) and Machine Learning (ML), data-driven methods were adopted for code generation, by using DL to predict code tokens from extensive datasets. In recent years, LLMs have been introduced, expanding automated code generation through the translation of natural-language input into code [11]. As software systems grew in complexity and documentation evolved, documentation-driven development was adopted to align generated code with changing requirements.

In parallel, MASs emerged as computational frameworks of multiple autonomous agents deployed in a shared environment and suitable for perceptual reasoning [12]. In such systems, task decomposition is achieved through the assignment of subtasks to agents, context discovery is automated, and iterative refinement and self-healing mechanisms are enabled [13]. Such advances set the stage for applying LLMs to update code generator templates, especially in rapidly changing environments. This combination of perception provided by LLMs and orchestration provided by MASs was positioned as central to the realization of robust and adaptive code generation workflows [14].

In this section, recent studies utilizing LLMs and MAS for code generation are reviewed, primarily in Sect. 2.1. A brief outline of the integration of MAS and MDSD is provided in Sect. 2.2, accompanied by a summary of described studies.

2.1 Application of LLMs and MAS in Code Generation

In this section, an overview of prior work on LLM-based MAS and code-synthesis methodologies is provided, with emphasis on agent capabilities, collaboration patterns, and automation frameworks. It was found that potential efficiency was demonstrated by LLMs in various contexts when applied to specific development tasks [5].

In the related study [15], an integration of LLMs into automation frameworks within agile development pipelines is examined. Approaches such as Chat2Code utilize conversational interfaces to guide model specification and code generation, enabling nontechnical users to interact with code synthesis systems. This integration enhances system usability and creates a feedback loop aligned with evolving requirements.

With respect to context-aware collaboration, AgentFlow [8] establishes an MAS in which multiple LLM-based agents collaborate through dynamic role switching, leveraging structured workflows and a perception loop to ensure task coherence and creativity. Empirical evaluations on HumanEval [16] and MMLU [17] benchmarks demonstrated that this MAS consistently outperformed single-agent baselines in terms of accuracy, coherence, and creativity of generated code and text responses.

Within iterative debugging workflows, the Synthesize, Execute, Instruct, Debug, and Repair (SEIDR) framework [18] is constituted as an MAS. Dedicated LLM agents generate, execute and test initial code drafts, and produce debugging instructions to guide subsequent code repairs. It was empirically shown that SEIDR outperformed both basic LLM code generation and traditional genetic programming approaches.

To support continuous maintenance and iterative code updates, a framework for development pipelines was proposed [19], utilizing LLMs for ongoing code generation and refinement. Automated processes were established to monitor and update codebases, reducing manual intervention and ensuring that software artifacts remain current over extended periods.

Finally, LLMs were employed to generate initial high-level code generation plans, which were then refined via fine-tuning and validated through automated processes in continuous integration pipelines [20]. This approach was designed to ensure that the code generation process remained consistent and adaptable to dynamic environments.

Based on these studies, efficiency gains and improved error handling have been demonstrated through LLM-based synthesis from natural language fragments. Iterative refinement has been enhanced by conversational interfaces and prompt chaining. Varied communication paradigms, multi-agent coordination, memory mechanisms, and retrieval-augmented generation have been employed in LLM-based MAS architectures to support code understanding, generation, and debugging [12].

However, several notable considerations have been identified across these approaches. It has been observed that hallucination risks and inconsistency in agent outputs can undermine overall system reliability, and the computational overhead associated with multi-agent orchestration has been shown to impede scalability in real-world deployments. Furthermore, insufficient mechanisms for traceability and error provenance have limited the ability to audit and refine agent interactions post hoc.

2.2   Enhancing MDSD with LLMs and MAS

Code generation has been automated by integrating model transformation techniques and template-based approaches in MDSD. Approaches such as Models2Code [21] have transformed models into code artifacts via plugins and automated workflows, reducing manual coding efforts and promoting consistency and maintainability. Continuous integration and automated validation frameworks have been devised to ensure that models remain synchronized with rapidly evolving system requirements [22]. The potential for engineering cost reduction and reliability improvement has thus been highlighted.

The integration of LLMs with MDSD frameworks has been explored to enable adaptive code generation. Model transformation techniques have been combined with generative abilities of LLMs to autonomously update code templates in response to evolving design specifications and Application Programming Interface (API) landscapes [23]. Moreover, research on modeling and model transformation as a service provided an agile approach to integrating LLMs into MDSD pipelines [24].

Early fault detection capabilities during the model-to-code transformation process in MDSD environments were demonstrated by MASDebugFW [25]. Design flaws were thereby identified and resolved before propagation to generated implementations, and post-deployment error correction costs were substantially reduced.

Incorporating world-model-driven learning paradigms from Multi-Agent Reinforcement Learning (MARL) into MDSD frameworks offers a promising approach [7]. By applying transition-informed representations to model transformation processes, systems can adapt from previous transformation patterns and improve code generation quality over time. This approach is effective when system requirements frequently evolve, enabling resilient software architectures with minimal manual intervention.

It has been demonstrated by reviewed studies that LLMs and MAS have been incorporated into code synthesis to automate maintenance driven by design artifacts. Consequently, MDSD-based systems are being transformed into AI-supported systems, manual coding is reduced, and consistency is enhanced, while continuous integration and automated validation ensure alignment of evolving requirements and implementations. However, prior MDSD solutions have been predicated on statically defined templates without automated synchronization with evolving specifications, causing template drift and necessitating manual interventions, thus impeding scalability and consistency across software lifecycle.

To address the identified limitations in existing LLM-MAS integrations, MAS4ACTM is proposed as a unified solution for MDSD environments. In this architecture, agent roles are formally specified and coordinated via a lightweight protocol, interaction histories are persistently logged for traceability, and a validation component is interposed to verify generated templates against external specifications before deployment.

3   Architecture of MAS4ACTM

The proposed MAS4ACTM architecture, as presented in Fig. 1, is designed to ensure that the software and its documentation are automatically synchronized. The main purpose of the proposed system is to automatically maintain code templates, based on the identified documentation changes. The MAS4ACTM is comprised of dedicated agents which are organized into four main modules. Each module is assigned a specific role in documentation-driven maintenance.

First, **Documentation Change Detection Module** is tasked with the ingestion and normalization of heterogeneous source formats, the segmentation of content into semantically meaningful units, and the flagging of substantive edits while minor revisions are filtered out. Next, in **Software Artifact Generation Module**, the changes detected are translated into concrete code modifications, ranging from simple example patches to complete stub and test scaffolding. Then, **Code Validation Module** ensures that all generated artifacts are subjected to automated checks (including syntax, functionality, security, and performance) and to human review, with any failures being routed back for automated rework. Finally, **System Improvement Module** ensures that continuous refinement of MAS4ACTM is driven by linking code updates to their documentation origins, with usage metrics and developer feedback being analyzed to detect recurring patterns. Model and parameter updates are being applied so that accuracy and efficiency are incrementally enhanced over time. When **Task Orchestration Agent** is considered, the orchestration of the system's operations is regarded as its primary duty, and no specific module is designated as its owner.

Each agent is maintained with its own private state while access is granted to shared information. Autonomous operations are carried out, with decisions being made via internal logic, communication being conducted with peers to achieve goals, and behavioral adaptations being made based on system conditions and interactions. Agents may be LLM-based, algorithm-based, or human-in-the-loop, with each type being tailored to leverage specific reasoning capabilities and expertise.

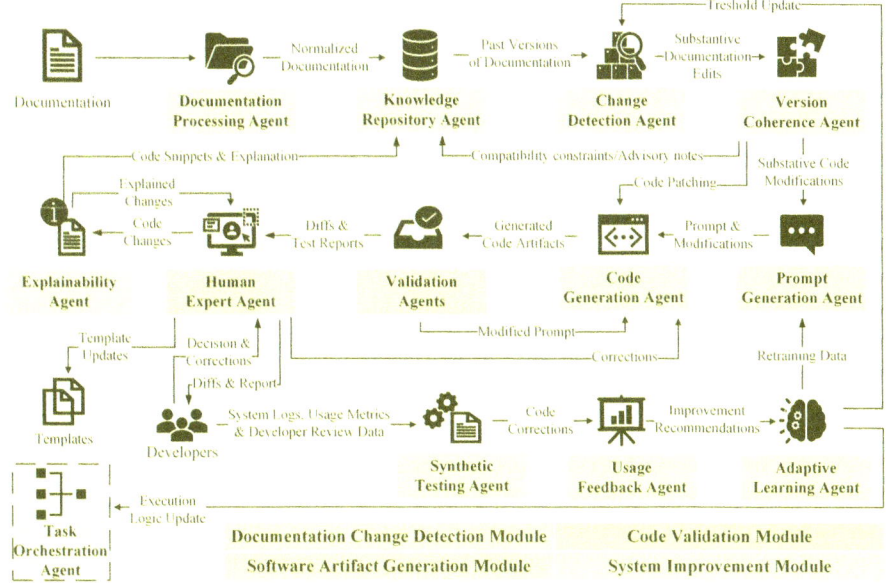

Fig. 1. The MAS4ACTM architecture

In the rest of this section, each module is outlined in a more detailed manner, followed by a discussion on the potential and limitations of the proposed MAS4ACTM.

3.1 Documentation Change Detection Module

Raw documentation referring to the definition of software components that are generated by code generators, in the form of web pages or documents of various types, are processed and structured, and meaningful changes are identified by **Documentation Change Detection Module**.

Raw documents are structured by **Documentation Processing Agent**, an algorithmic system incorporating Natural Language Processing (NLP) components. Document sections are extracted and annotated with labels such as "constraint", "blog post", "image", or "hands-on example", which are employed in lieu of generic tags to enable downstream mapping to code templates, such as stub generation, input validation logic, and exception handling routines, with greater precision. An evolving knowledge graph of document versions, labeled sections, and inter-section dependencies is maintained by **Knowledge Repository Agent**, which serves to detect and defer incompatible revisions of code templates. The latest content extracted from the documents is compared to past versions by **Change Detection Agent**, a hybrid LLM–algorithm system, so that substantive edits, such as added functions or revised parameters, are flagged while minor rewordings are ignored. Potential effects of individual edits on other software artifacts are analyzed by **Version Coherence Agent**, which employs rule-based algorithms with ML, and compatibility constraints or advisory notes are inserted into the knowledge repository graph accordingly.

Documentation Processing Agent and **Change Detection Agent** are executed in parallel on heterogeneous document types (e.g., API reference pages and design diagrams) so that detection is accelerated without requiring full sequential normalization. By processing and comparing API reference pages upon extraction and independently transcribing and analyzing design diagrams, full normalization is avoided, and throughput is increased.

3.2 Software Artifact Generation Module

Detected documentation changes are transformed into code modifications by **Software Artifact Generation Module**. First, an optimal execution plan is devised by **Task Orchestration Agent** in which decision-tree logic is employed to determine the sequencing and activation of agents within the module. Execution paths are selected based on defined criteria, including the type and scope of the detected change, the required agent capabilities, and historical performance metrics associated with similar update patterns. When only template comments or example snippets are identified for patching, **Code Generation Agent** (a specialized LLM) is invoked in a "light mode", during which examples are updated without full templates being regenerated. For substantive modifications, **Prompt Generation Agent** (a fine-tuned LLM) is sequenced to build structured prompts from the identified document changes. Prompts are constructed to include: (i) the original template fragment or code stub affected; (ii) a concise summary of the detected documentation changes; (iii) contextual metadata, such as target programming language, framework version, and coding standards; and (iv) exemplar input–output pairs when applicable. The prompts generated are then sent to **Code Generation Agent**, producing both stub implementations and accompanying unit-test scaffolds, with the agent's responsibilities extended to include test generation.

Finally, these outputs are routed to the following **Code Validation Module**. In complex, multi-part updates, such as simultaneous schema and endpoint alterations, parallel processing paths are launched by **Task Orchestration Agent**, and overlapping edits are reconciled internally before the combined patch set is forwarded for validation.

3.3 Code Validation Module

Generated code artifacts are evaluated across multiple dimensions by **Code Validation Module**. They are subjected to quality checks through two parallel streams, rule-based and model-driven analyses.

The code analyses are performed by **Validation Agents** (rule-based algorithms with specialized models) to verify the syntax, enforce coding standards, and execute integration tests. Whenever a check fails, the corresponding patch with error metadata is returned to the **Task Orchestration Agent** for automated rework, whereupon the modified generation prompt is resent to **Code Generation Agent** to produce an updated implementation. The initially received or reworked code is examined for known vulnerabilities within **Validation Agents** by security-scanning components, and quick benchmarks are conducted on critical paths by performance-profiling components to detect regressions. Once all automated checks are successfully passed, code diffs and test reports are presented to developers in a unified interface by **Human Expert Agent** (human-in-the-loop interface). At this stage, potential updates to code generation templates are identified. Human intervention is introduced at this stage of the workflow to enable manual oversight prior to deployment. Developers, acting through the **Human Expert Agent** interface, are prompted to inspect validation outputs, verify correctness in ambiguous cases, and assess whether template updates may be required. Based on this assessment, they may approve the proposed changes, annotate the artifacts with context-specific insights, or request further modifications.

All approvals or requested corrections are recorded for subsequent system refinement. If corrections are requested by developers, those modifications are recorded and incorporated as feedback into **System Improvement Module** for subsequent model refinement before being resubmitted to **Code Generation Agent**. Any manually applied changes are captured and used to guide future automated rework.

3.4 System Improvement Module

To ensure ongoing MAS4ACTM refinement, **System Improvement Module** is continually informed by its own activity. Clear, human-readable justifications that trace each code change back to its source in the documentation are produced by **Explainability Agent** (a fine-tuned LLM) and are dispatched to **Human Expert Agent**, regardless of whether the proposed change is accepted or rejected, to guarantee transparency at every stage. These explanations are supplementary to those provided during initial review, and links, document snippets, and the generated explanatory notes are all stored in **Knowledge Repository Agent**'s evolving knowledge-graph database for traceability and future analysis. For example, an update to database schema documentation, such as "changing the description column data type from VARCHAR(255) to TEXT to support longer descriptions as specified in schema guide v2.1", is recorded alongside the relevant snippet and is reviewed by a human through **Human Expert Agent**.

Unlike **Validation Module**, which operates earlier in the pipeline to verify correctness and structural conformity of generated templates, robustness is evaluated by **Synthetic Testing Agent**, through algorithm-based simulation. The product resulting from the combined outputs of all previous agents is subjected to testing by this agent. Simulated adversarial conditions and edge cases are introduced so that weaknesses, such as silent failures in schema propagation or misinterpretation of ambiguous documentation, are exposed. For instance, conflicting version updates or malformed parameter definitions are injected so that the system's reliability in handling unexpected scenarios can be assessed. **Synthetic Testing Agent** focuses specifically on failure modes under controlled stress conditions that may not be captured during routine validation.

Continuous adaptation of system behavior based on real-world usage is ensured by **Usage Feedback Agent** (a telemetry and analytics system). System logs, usage metrics, and developer review data, including approvals, rejections, and manual code corrections, are collected and analyzed by this agent. Emerging trends are identified; for instance, prompt templates that frequently result in off-by-one errors, or test name generators that yield unhelpful outputs. Performance bottlenecks and error patterns are detected through these analyses. From these insights, actionable recommendations, such as prompt-parameter adjustments, validation-threshold tuning, and knowledge-graph refinements, are generated by **Usage Feedback Agent** and forwarded directly to **Adaptive Learning Agent** for implementation to support ongoing system optimization and enhance user satisfaction.

System parameters are refined through continuous updates applied by **Adaptive Learning Agent** (an ML model incorporating reinforcement learning techniques). In practice, this entails retraining or fine-tuning components: (i) **Prompt Generation Agent**'s LLM is retrained on the latest code-review feedback; (ii) **Change Detection Agent**'s threshold for flagging edits is recalibrated; and (iii) decision-logic rules in **Task Orchestration Agent** are adjusted to favor parallel execution where speed improvements have been shown in past runs. Updates are applied in scheduled batches (e.g., nightly retraining on feedback collected that day) or incrementally (with a rule adjusted as soon as a new pattern is detected), so that system behavior has evolved fluidly in response to real-world use. Long-term stability is supported, and improvements in accuracy, efficiency, and alignment with developer expectations are ensured by this agent through traditional ML and transformer-based adaptation.

The proposed MAS4ACTM is to be integrated into the MDSD solution for vector database access, and use-case details are provided in the following section.

4   Integration of MAS4ACTM into MDSD Solution for Vector Database Access

We observed that MAS4ACTM could be effectively integrated into our MDSD solution for uniform access to vector databases in order to enhance its adaptability and reduce documentation-related overhead. In our previous research [9, 26], the MDSD solution has been developed to address challenges related to vector database access. Vector databases are typically accessed through proprietary APIs and query languages that vary significantly between vendors, which results in tight coupling, interoperability challenges, and transition difficulties that negatively impact usability. To overcome these issues, a DSL named vecDSL [10] has been implemented as the central component of the MDSD solution, as well as code generators through which DSL models should be transformed into executable scripts for various databases and programming languages. Through this solution, vector database management is simplified, and interactions are streamlined, which leads to enhanced end-user efficiency.

However, the challenge of frequent manual template modifications has been noted, since the vector database field remains nascent and both VDBMSs and their access languages evolve rapidly. In that manner, as VDBMSs version updates often entail syntax changes, recurrent adjustments to code generator templates are required.

Vector databases were selected as the initial test environment for MAS4ACTM because they are characterized by structural complexity, frequent documentation updates, and significant vendor-specific variability. These characteristics reflect the key challenges targeted by the proposed system. Due to their evolving nature and lack of standardization, vector databases provide a high-variance context in which the effectiveness of autonomous maintenance, multi-agent coordination, and documentation-aware code generation can be systematically evaluated.

MAS4ACTM is proposed to be integrated with the existing MDSD solution, as presented in Fig. 2. The MDSD solution is comprised of the vecDSL tool and code generators. VecDSL statements are inputted by users to the code editor, where they are compiled, validated and transformed into models. These models are fed into specialized generators, by which executable scripts for vector databases are produced. MAS4ACTM monitors VDBMSs documentation and code generators, automatically detecting updates through event-driven notifications (e.g. GitHub/GitLab webhooks) and scheduled API polling (e.g. Confluence). Detected documentation changes are analyzed, template modifications are generated, validated, and then integrated into the code generators. Communications with vecDSL developers are facilitated when template updates are necessitated. Critical changes are reviewed by developers who serve as human supervisors within this multi-agent workflow, by whom expert guidance is provided when human judgment is required. The established MDSD pipeline, consisting of models, transformation rules, and code generation templates, is preserved while intelligent maintenance automation is included.

Several strategic advantages are to be realized through this integration. It is intended to minimize manual documentation efforts for developers and to enhance user experience by aiming to prevent code failures that can arise from VDBMS version changes. Knowledge pertaining to vector database APIs is preserved within the Knowledge Repository Agent's semantic network, through which the risk of loss of information during developer turnover is mitigated. The onboarding process for developers of the new MDSD solution is accelerated through contextual reasoning capabilities that are provided by specialized agents. Error prevention is facilitated proactively through the simulation of potential changes, although human oversight is recognized as necessary for specialized domains, and cross-vendor standardization is enabled, which creates opportunities for more flexible technology selection strategies. Through these mechanisms, the maintenance process shall be transformed into an operation that is intelligent, scalable, and efficient, through which the long-term sustainability of the MDSD solution is enhanced.

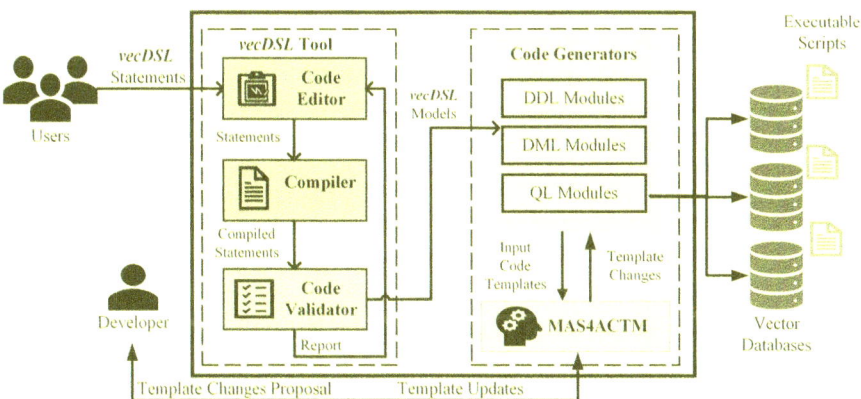

Fig. 2. The architecture of MDSD solution integrated with MAS4ACTM

5 Discussion

In this section, the MAS4ACTM possibilities and advantages are discussed, as well as potential limitations, ethical considerations, and integration requirements.

End-to-end automation of the documentation-to-code maintenance lifecycle is provided by MAS4ACTM, differing from conventional solutions. While code is generated from prompts by tools like GitHub Copilot [27], synchronization between evolving documentation and dependent artifacts is addressed by this system, with a persistent understanding of the documentation landscape being maintained. Value is delivered by MAS4ACTM where documentation serves as a true source, such as in an API-first development approach where implementation is preceded by documentation. Manual translation of documentation changes may be error-prone and resource-intensive, making benefits possible through this automated approach.

Modularity and adaptability are key advantages for MAS4ACTM. Unlike monolithic systems functioning as black boxes, transparent reasoning is enabled by this architecture, targeted improvements are possible, and incremental adoption is facilitated. Understanding documentation patterns and code modifications is continuously improved by the system through developer feedback. System evolution is shaped by developer expertise through feedback that informs learning processes, and system accuracy is enhanced over time by developer input.

Potential limitations may exist despite comprehensive design, particularly with specialized or visual-heavy documentation, ambiguous specifications, and rapidly evolving requirements. Nevertheless, human oversight in specialized domains is acknowledged as complementing automated processes. Substantial computational resources are required for implementation, particularly Graphics Processing Unit (GPU) acceleration for semantic processing of documentation. Standardized APIs for documentation ingestion and repository access are needed for integration with development environments.

The integration of MAS4ACTM raises several ethical considerations around the evolving role of developers and the integrity of the development process. Automating routine tasks can liberate developers to focus on higher-value work, but over-reliance on generated code risks deskilling and eroding deep understanding of how documentation maps to implementation. Moreover, biases present in training data or LLM outputs can be inadvertently propagated into production code, creating fairness, security, or reliability issues. To mitigate these risks, developers must retain responsibility for validating and auditing all generated artifacts, exercising human judgment in complex change scenarios, and guiding model retraining with clear feedback. In this way, the system enhances rather than replaces human expertise, preserving transparency, accountability, and ongoing knowledge transfer.

An extension of the autonomous software maintenance landscape is achieved by bridging documentation analysis and code generation. A documentation-driven, end-to-end maintenance approach is represented by the MAS4ACTM architecture, in which maintenance is framed as a multi-agent collaborative problem rather than a sequential process. The proposed hybrid multi-agent architecture aims to improve upon traditional sequential approaches by: (i) adapting processing paths based on the characteristics of

detected changes; (ii) enabling parallel processing; (iii) facilitating efficient information sharing; (iv) integrating human expertise at strategic points; and (v) incorporating
continuous feedback for ongoing refinement.

Furthermore, an evaluation framework is proposed to facilitate assessment of agent
knowledge alignment within MAS4ACTM by applying four symbolic comparison metrics [28], because it is enabling objective measurement of update quality and reliability.
Consistency is assessed by measuring overlap in agents' proposed code changes to
confirm uniform application of documentation edits. Consensus on complex scenarios
is evaluated by checking agreement on critical edge cases to ensure robust handling of
ambiguous specifications. Containment of detailed patches within broader update scopes
is verified to guarantee alignment with maintenance policies. Completeness is confirmed
by ensuring that every flagged code element is included in the final update plan. By applying these metrics to MAS4ACTM, a clearer picture of the system's capabilities should
be obtained.

6 Conclusion

In this paper, MAS4ACTM for addressing documentation-dependent software maintenance challenges was proposed. Through this system, the substantial maintenance
overhead that is associated with continually evolving documentation sources could be
addressed, as components that rely on external specifications could be automatically
updated when changes are detected.

A sophisticated integration has been constructed encompassing LLM-based components for semantic understanding and code generation, algorithm-based agents for
change detection and verification, and human-in-the-loop mechanisms for final approval
and edge-case resolution. By combining NLP capabilities of LLMs with traditional computational methods and human expertise, a robust maintenance automation system is
established. Projected benefits are manifested as: (i) enhanced developer productivity
through reduced manual maintenance; (ii) improved code reliability via synchronization
of updates with documentation changes; and (iii) preservation of institutional knowledge
within the system's semantic repository.

As the architecture presented in this paper and the first agents implementation
demonstrated promising results for agent-based automatic maintenance of code generators supporting uniform vector database access, a complete MAS4ACTM prototype
will be developed and evaluated for such a use case. Future work will be devoted to
the integration of the proposed system into existing development pipelines and to the
evaluation of its performance across diverse platforms and documentation change scenarios. Both conventional performance metrics and symbolic knowledge comparison
techniques will be applied. By so doing, a holistic assessment of the system's capacity
to meet access requirements and to function effectively beyond vector database use cases
will be provided.

Future evaluation of generated code templates should combine established benchmarks like HumanEval [16] and MMLU [17] alongside knowledge-alignment metrics
to assess consensus, containment, and completeness of agent-driven updates. Accuracy
and API efficiency will be measured in parallel to evaluate system behavior across

documentation-change scenarios. This combined approach will reveal correlations between agent alignment and overall maintenance performance.

While MAS4ACTM is built on domain-agnostic design principles, its transferability to other documentation-driven software development contexts remains to be empirically evaluated. To facilitate this, the system's effectiveness and potential optimization opportunities should be systematically validated using consistent evaluation protocols that incorporate both performance metrics and knowledge-alignment measures, thereby laying the foundation for deployment across diverse development domains.

Acknowledgments. This research has been supported by the Ministry of Science, Technological Development and Innovation (Contract No. 451–03-137/2025–03/200156) and the Faculty of Technical Sciences, University of Novi Sad through project "Scientific and Artistic Research Work of Researchers in Teaching and Associate Positions at the Faculty of Technical Sciences, University of Novi Sad 2025" (No. 01–50/295).

Disclosure of Interests. The authors have no competing interests to declare that are relevant to the content of this article.

References

1. Antero, U., Blanco, F., Oñativia, J., Sallé, D., Sierra, B.: Harnessing the power of large language models for automated code generation and verification. Robotics **13**, 137 (2024). https://doi.org/10.3390/robotics13090137
2. Vaithilingam, P., Zhang, T., Glassman, E.L.: Expectation vs. experience: evaluating the usability of code generation tools powered by large language models. In: Proceedings of the CHI Conference on Human Factors in Computing Systems Extended Abstracts, pp. 1–7. ACM, USA (2022). https://doi.org/10.1145/3491101.3519665
3. Zhang, T., Lowmanstone, L., Wang, X., Glassman, E.L.: Interactive program synthesis by augmented examples. In: Proceedings of the 33rd Annual ACM Symposium on User Interface Software and Technology, pp. 627–648. ACM, Virtual Event USA (2020). https://doi.org/10.1145/3379337.3415900
4. Sun, Z., Zhu, Q., Xiong, Y., Sun, Y., Mou, L., Zhang, L.: TreeGen: a tree-based transformer architecture for code generation. In: Proceedings of the AAAI Conference on Artificial Intelligence, vol. 34, pp. 8984–8991 (2020). https://doi.org/10.1609/aaai.v34i05.6430
5. Liu, Z., Tang, Y., Luo, X., Zhou, Y., Zhang, L.F.: No need to lift a finger anymore? assessing the quality of code generation by ChatGPT. IEEE Trans. Softw. Eng. **50**, 1548–1584 (2024). https://doi.org/10.1109/TSE.2024.3392499
6. Jin, H., Huang, L., Cai, H., Yan, J., Li, B., Chen, H.: From LLMs to LLM-based agents for software engineering: a survey of current, challenges and future (2025). https://doi.org/10.48550/arXiv.2408.02479
7. Feng, M., Yang, Y., Zhou, W., Li, H.: TIMAR: transition-informed representation for sample-efficient multi-agent reinforcement learning. Neural Netw. **184**, 107081 (2025). https://doi.org/10.1016/j.neunet.2024.107081
8. Nettem, G., Disha, M.J.A., Prasad, S., Natarajan, S.: AgentFlow: a context aware multi-agent framework for dynamic agent collaboration: In: Proceedings of the 17th International Conference on Agents and Artificial Intelligence, pp. 687–693. SCITEPRESS - Science and Technology Publications, Porto, Portugal (2025). https://doi.org/10.5220/0013375700003890

9. Akik, E., Vještica, M., Dimitrieski, V., Kordić, S., Ristić, S.: Towards a model-driven approach to enable uniform access to vector databases. In: Tekli, J., et al. (eds.) New Trends in Database and Information Systems, pp. 225–237. Springer Nature Switzerland, Cham (2024). https://doi.org/10.1007/978-3-031-70421-5_19

10. Akik, E., Vještica, M., Dimitrieski, V., Čeliković, M., Kordić, S., Ristić, S.: Interacting with vector databases by means of domain-specific language. Open Comput. Sci. (2025). accepted for publication

11. Ren, X., Ye, X., Zhao, D., Xing, Z., Yang, X.: From misuse to mastery: enhancing code generation with knowledge-driven AI chaining. In: Proceedings of the 38th IEEE/ACM International Conference on Automated Software Engineering (ASE), pp. 976–987. IEEE, Luxembourg, Luxembourg (2023). https://doi.org/10.1109/ASE56229.2023.00143

12. Li, X., Wang, S., Zeng, S., Wu, Y., Yang, Y.: A survey on LLM-based multi-agent systems: workflow, infrastructure, and challenges. Vicinagearth. 1, 9 (2024). https://doi.org/10.1007/s44336-024-00009-2

13. Hu, J., Bhowmick, P., Jang, I., Arvin, F., Lanzon, A.: A decentralized cluster formation containment framework for multirobot systems. IEEE Trans. Robot. 37, 1936–1955 (2021). https://doi.org/10.1109/tro.2021.3071615

14. Bassamzadeh, N., Methani, C.: Plan with Code: Comparing approaches for robust NL to DSL generation. (2024). https://doi.org/10.48550/ARXIV.2408.08335

15. Qasse, I., Mishra, S., Jónsson, B.Þ., Khomh, F., Hamdaqa, M.: Chat2Code: a chatbot for model specification and code generation, the case of smart contracts. In: Proceedings of the IEEE International Conference on Software Services Engineering (SSE), pp. 50–60. IEEE, Chicago, IL, USA (2023). https://doi.org/10.1109/SSE60056.2023.00018

16. Chen, M., et al.: Evaluating Large Language Models Trained on Code (2021). https://doi.org/10.48550/arXiv.2107.03374

17. Hendrycks, D., et al.: Measuring Massive Multitask Language Understanding (2021). https://doi.org/10.48550/arXiv.2009.03300

18. Grishina, A., Liventsev, V., Härmä, A., Moonen, L.: Fully autonomous programming using iterative multi-agent debugging with large language models. ACM Trans. Evol. Learn. Optim. 5, 1–37 (2025). https://doi.org/10.1145/3719351

19. Adnan, M., Xu, Z., Kuhn, C.C.N.: Large Language Model Guided Self-Debugging Code Generation (2025). https://doi.org/10.48550/arXiv.2502.02928

20. Yan, K., Guo, H., Shi, X., Xu, J., Gu, Y., Li, Z.: CodeIF: Benchmarking the Instruction-Following Capabilities of Large Language Models for Code Generation (2025). https://doi.org/10.48550/arXiv.2502.19166

21. Paniagua, C., Caso, F.L.: Models2Code: autonomous model-based generation to expedite the engineering process. Syst. Eng. 28, 224–237 (2025). https://doi.org/10.1002/sys.21789

22. Saini, R., Mussbacher, G., Guo, J.L.C., Kienzle, J.: DoMoBOT: an AI-empowered bot for automated and interactive domain modelling. In: Proceedings of the ACM/IEEE International Conference on Model Driven Engineering Languages and Systems Companion (MODELS-C), pp. 595–599. IEEE, Japan (2021). https://doi.org/10.1109/MODELS-C53483.2021.00090

23. Buchmann, T., Bank, M., Westfechtel, B.: BXtendDSL: a layered framework for bidirectional model transformations combining a declarative and an imperative language. J. Syst. Softw. 189, 111288 (2022). https://doi.org/10.1016/j.jss.2022.111288

24. Vahdati, A., Ramsin, R.: Modeling and model transformation as a service: towards an agile approach to model-driven development. In: Przybyłek, A., Jarzębowicz, A., Luković, I., Ng, Y.Y. (eds.) Lean and Agile Software Development, pp. 116–135. Springer International Publishing, Cham (2022). https://doi.org/10.1007/978-3-030-94238-0_7

25. Tezel, B.T., Kardas, G.: Debugging in the domain-specific modeling languages for multi-agent systems. J. Comput. Lang. 83, 101325 (2025). https://doi.org/10.1016/j.cola.2025.101325

26. Akik, E., Vještica, M., Dimitrieski, V., Čeliković, M., Ristić, S.: Prototype of domain-specific language for uniform access to vector databases. In: Proceedings of the IEEE 17th International Scientific Conference on Informatics (Informatics), pp. 11–16. IEEE, Poprad, Slovakia (2024). https://doi.org/10.1109/Informatics62280.2024.10900871
27. GitHub Copilot. https://github.com/features/copilot. Accessed 22 Mar 2025
28. Sabbatini, F., Sirocchi, C., Calegari, R.: Symbolic knowledge comparison: metrics and methodologies for multi-agent systems. In: Proceedings of the 25th Workshop "From Objects to Agents.", pp. 221–235, Italy (2024)

Sequence Management in the Relational Database System Oracle – Case Study

Michal Kvet[✉]

Faculty of Management Science and Informatics, University of Žilina, Univerzitná 8215/1, 010 26 Žilina, Slovak Republic
Michal.Kvet@fri.uniza.sk

Abstract. Sequence data management is a fundamental aspect of database design and operation, particularly in environments requiring the generation of unique identifiers and orderly numerical progression. Oracle Database provides a robust implementation of sequences—schema-level objects designed to generate unique numeric values autonomously. This paper examines the architecture and operational semantics of Oracle sequences, reflecting their roles in ensuring data consistency, integrity, and performance in multi-user, transactional systems. It focuses on various implementation details and impact on the performance and applicability, delimited by the context switches, obtaining and setting values. Practical deployment scenarios are analyzed, as well. Additionally, the study addresses issues related to concurrency, sequence value gaps, and the implications of caching on system throughput and reliability. Through theoretical discussion and applied examples, this work highlights best practices and optimization strategies for sequence data management, positioning it as a critical component in the design of scalable and efficient Oracle-based information systems.

Keywords: sequences · data optimization · Oracle · relational database · methodology

1   Introduction

Relational databases have fundamentally database structures and organization of the data in terms of storage and query management in the modern computing systems. They were introduced by Edgar F. Codd in 1970. The relational model provided a theoretical framework for organizing data into tables interconnected by relationships by establishing a foundation for data management that has remained relevant for over five decades and is still widespread [5, 8, 12]. The advent of the relational database management system (RDBMS) marked a significant departure from earlier, more rigid hierarchical and network models, offering a flexible and scalable way to manage structured data. Central to the relational paradigm is the use of Structured Query Language (SQL) for data manipulation, as well as principles of data normalization, ensuring data integrity, minimizing redundancy and transaction management [14, 17, 18].

© The Author(s), under exclusive license to Springer Nature Switzerland AG 2026
P. K. Chrysanthis et al. (Eds.): ADBIS 2025, CCIS 2676, pp. 311–324, 2026.
https://doi.org/10.1007/978-3-032-05727-3_27

Over the years, relational databases have evolved to meet the growing demands of both transactional and analytical processing in increasingly complex computing environments. Based on the study, the annual growth in the amount of data has exceeded 11% [10]. However, it is not only about the amount itself, but also about the complexity. It goes with the requirement for temporal data processing by handling individual versions of the objects by storing states of the objects bordered by the time frame, commonly expressing the validity [9, 13]. RDBMSs have witnessed continuous enhancements in terms of performance, scalability, and functionality [11, 14, 17]. This evolution is largely driven by technological advances, such as the advent of distributed systems, the need for greater transaction throughput, and the desire for seamless integration with emerging technologies like cloud computing [1, 2, 6], big data [3, 16, 19], and machine learning [15].

As the landscape of data management has become more diverse, relational databases have also faced competition from NoSQL databases, which are designed for unstructured data and huge horizontal scalability [17, 18]. Despite this, relational databases remain integral to enterprise data architectures due to their robustness, relational algebra principles and well-defined standards, supervised by the transactions. Current cloud-native relational databases even more specifically demonstrate the ongoing adaptability of the relational model.

When dealing with the complex data set and operational data in the environment of massive parallelism and huge performance requirements, it is necessary to optimize the data management and internal processes to serve the data in a proper way, reliable format, but mostly on time.

In relational databases, sequences form a significant concept for generating unique, automatically incrementing numbers, often used for creating primary keys. These sequences ensure data integrity, efficiency, and scalability in several ways to adopt the following requirements [11]:

- *Data identity* – limiting manual management of the primary key values.
- *Data integrity* – prevents accidental overwriting of the primary key values.
- *Performance* – related to the bulk operations.
- *Consistency* – ensured by the automatic value assignment.
- *Scalability* – sequences can continue to generate unique values without the need for redesign or additional logic.
- *Parallelism and concurrency* – managed by the database engine ensuring proper management in concurrent transaction environment.
- *Flexibility* – they are not strictly related to the numerical representation, but can be adopted for textual or date values, as well.

This paper deals with the sequences by proposing a case study forming the methodology, which deals with the creation of the performance effective database approach. It focuses on the value assignment, identity columns, approaches for getting (currval vs. returning) and setting values (direct vs. trigger). Besides, the style for referencing Select statements is evaluated. By analyzing the core principles and paradigms of the relational databases, this study highlights the performance and relevance of the data processing related to the sequential value assigned to the primary keys in a distributed environment.

The evaluation study points to the environmental data [23] regarding the field of aviation [4, 7, 20–22].

The proposed paper is organized as follows: Sect. 2 deals with the sequences and associated identity columns. Section 3 discusses approaches for assignment and getting value and focuses on the value setting using direct approach and trigger definition. The performance evaluation study is present in Sect. 4, followed by the discussion of the results. The novelty of the paper is covered by considering performance of the new features introduced in Oracle Database 23ai. While new technologies and methodologies continually emerge, it remains essential to prioritize performance, robustness, and usability within the context of specific application scenarios.

2 Sequences and Identity Columns

A sequence is a database object that generates a sequence of unique values, based on the set parameters. As stated, they are often used for generating and assigning primary key values. The processing is session specific and offers two functions – *currval* for getting currently assigned and processed value and *nextval* applying the set increment value (at least one *nextval* function call must be present to enable *currval* function for the sequence) [11]. The syntax of the sequence is defined by the following code block. Only the name of the sequence is mandatory, all rest clauses have default values. The meaning of the clauses and the default values are shown in Table 1

```
CREATE SEQUENCE sequence_name
  [ START WITH start_value ]
  [ INCREMENT BY increment_value ]
   [ MINVALUE min_value ]
   [ MAXVALUE max_value ]
     [ CYCLE | NOCYCLE ]
     [ CACHE cache size | NOCACHE ]
```

Table 1. Sequence clause definition

Clause	Meaning	Default value
Sequence_name	The name of the sequence	
Start with	The first generated value	1
Increment by	The number by which the sequence value will be incremented each time	1
Minvalue	Left limit (border) for the sequence	1
Maxvalue	Right limit (border) for the sequence	10^{29}
Cyclicality	Ability to restart the sequence if the border value is reached	nocycle
Caching	Storing a specified number of pre-fetched values in the instance memory to improve performance	20

RDBMS Oracle, which will be also used for the performance evaluation study, offers the ability to define identity columns, which handles values automatically, similarly to other RDBMS, like MySQL or SQL Server. The values in an identity column are automatically assigned by Oracle when new rows are inserted, eliminating the need for manual insertion of values into that column. So, in principle, trigger definition necessity is limited.

An identity column automatically generates a unique value for each new row inserted into a table. This value can be defined with a starting point and an incremental value, just like a sequence. The available clauses are the same as sequences (stated in Table 1). The definition of the identity column was introduced in Oracle 12c by offering two core principles [5]:

- *Generated always* – the value for the attribute cannot be specified (if defined, exception is raised), it is always set automatically.
- *Generated by default* – the value is assigned automatically unless a value is explicitly provided during the Insert operation.

The identity column is associated with the attribute. The syntax is shown in the following snippets:

```
column_name datatype GENERATED ALWAYS AS IDENTITY
column_name datatype GENERATED BY DEFAULT AS IDENTITY
```

Advantages of using identity columns include simplified table definition (sequences do not need to be explicitly created, nor handled by the triggers), values are automatically assigned. Since the management is analogous to the sequences, consistency and concurrency are always ensured.

Limitations of the identity column include the impossibility of changing parameters. It would be necessary to recreate a table or alter the column as a workaround.

The parameters of the sequences can be obtained from the data dictionary by considering the following column elements – *sequence_name, min_value, max_value, increment_by, cycle_flag, order_flag, cache_size, last_number, scale_flag, extend_flag, sharded_flag, session_flag*, and *keep_value* obtained from the {user | all | dba}_sequences data dictionary view. However, what about the identity columns? There is no specific data dictionary for handling those parameters, right? By looking to the aforementioned data dictionary, it is evident, that identity columns create internal sequences that can be directly referenced, like „*ISEQ$$_73321*" for the sequence name.

3   Techniques for Assignment, Getting and Setting Values

The assignment of the value to the primary key column can be done directly by the user embedding sequence call in the statement or automatically by firing a *row-level trigger*. The direct assignment looks like the following. The limitation is reliability, since the call is directly up to the user and not specifically protected by the database environment:

```
INSERT INTO table_name(ID, …)
  VALUES(sequence_name.NEXTVAL, ...);
```

By using a trigger, assignments are always precisely used, irrespective of the user definition or selection by changing the *NEW* record, which is them used for the Insert statement execution. Therefore, the timing of the trigger is *BEFORE*:

```
CREATE OR REPLACE TRIGGER trigger_name
 BEFORE INSERT ON table_name
  FOR EACH ROW
  IS
BEGIN
  SELECT sequence_name.NEXTVAL into :new.ID FROM DUAL;
END;
/
```

The performance evaluation study emphasizes the impact on the performance by firing a trigger. However, by looking at the trigger body, Select statement is present, which is SQL environment related, isn´t it? Does it require context switch between SQL and PL/SQL environment? How does it impact the processing time demands? So, another way is to use direct PL/SQL assignment. The performance differences are evaluated in Sect. 5:

```
SELECT sequence_name.NEXTVAL
  FROM user_one_row_table_name;

SELECT sequence_name.NEXTVAL
  FROM (select 1
          FROM user_one_row_table_name
           FETCH FIRST 1 row only);
```

Besides, when dealing with the SQL, there are many ways, how to implement *Currval* function call embedded by the Select statement. Namely, the function can be called by the statement associated with the *DUAL* table. In Oracle database, *DUAL* table represents a special one-row, one-column table owned by the *SYS* user. It is primarily used for selecting pseudo rows in scenarios where a physical table is not required. This table serves as a convenient mechanism for executing queries that involve functions, constants, or expressions, particularly when the retrieval of data from actual user tables is unnecessary:

```
SELECT sequence_name.NEXTVAL
   FROM user_one_row_table_name
   FETCH FIRST 1 row only;
    ==> ORA-02287: "sequence number not allowed here"
```

In Oracle Database 23ai, released on May 2, 2024, a new concept has been introduced, allowing us to omit *FROM* clause of the Select statement, if *DUAL* table is referenced. It is based on the AI-powered SQL tools, which involve automatic recognition of expressions and system queries, so no explicit table reference is required:

```
SELECT sequence_name.NEXTVAL
   FROM (select 1
           FROM data_dictionary_table_name
            FETCH FIRST 1 row only);
```

In addition, the next value obtained from the sequence can also be provided from user-defined tables. Two approaches are analyzed – user defined 1-row table and general table by limiting the result set to provide just one value:

```
DECLARE
 val integer;
BEGIN
  val:=sequence_name.NEXTVAL;
    INSERT INTO table_namePK VALUES(val, …);
    INSERT INTO table_nameFK VALUES(val, …);
END;
/
```

Please note that the inner statement is mandatory, if placed in one query, exception would be raised:

```
DECLARE
 val integer;
BEGIN
    INSERT INTO table_namePK
      VALUES(sequence_name.NEXTVAL, …)
        RETURNING id INTO val;
    INSERT INTO table_nameFK VALUES(val, …);
END;
/
```

The last evaluated solution related to the query processing is based on referencing a data dictionary instead of user defined query. Since generally it takes multiple rows, *FETCH FIRST* clause will be used to reduce the result set to just a single row:

```
BEGIN
    INSERT INTO table_namePK
      VALUES(sequence_name.NEXTVAL, …)
    INSERT INTO table_nameFK
      VALUES(sequence_name.CURRVAL, …);
END;
/
```

Many times, it is necessary to store the assigned value from the sequence to multiple destinations, typically represented by the referential integrity, by which the core object is then referenced in a child record. To do that, assigned value from the sequence must be somehow preserved for the consecutive processing. There are also multiple ways to do that. The performance of individual solutions is then evaluated. The following strategies are considered:

Strategy 1 – prefetching value into local variable
The value from the sequence is stored in a local variable and then referencing multiple times. This approach requires additional memory storage for the variable:

```
DECLARE
 val integer;
BEGIN
  val:=sequence_name.NEXTVAL;
    INSERT INTO table_namePK VALUES(val, …);
    INSERT INTO table_nameFK VALUES(val, …);
END;
/
```

Strategy 2 – using returning clause of the Insert statement
The second option is to postpone the assignment of the variable just after the first Insert operation by invoking RETURNING clause storing the result value into an local variable:

```
DECLARE
 val integer;
BEGIN
   INSERT INTO table_namePK
     VALUES(sequence_name.NEXTVAL, …)
       RETURNING id INTO val;
   INSERT INTO table_nameFK VALUES(val, …);
END;
/
```

Strategy 3 – Using CURRVAL Function Call
Both above solutions require a variable to be defined, filled either before the Insert operation related to the primary key definition or obtained as a result of the operation. One way or another, local variable needs to be defined. It is simply impossible to shift the *RETURNING* clause output directly as a parameter of another operation. Furthermore, there could be multiple child records, so multiple returning clauses would be necessary to be defined. Strategy 3 is based on another approach by listing the value from the sequence definition directly. Namely, *CURRVAL* function gets the current value of the session and is session specific, so the reliability and consistency are ensured:

```
BEGIN
   INSERT INTO table_namePK
     VALUES(sequence_name.NEXTVAL, …)
   INSERT INTO table_nameFK
     VALUES(sequence_name.CURRVAL, …);
END;
/
```

4   Performance Evaluation Study

The performance evaluation study was conducted in a controlled testing environment using a server configured with the Oracle Database system. The server specifications are as follows:

- **Processor**: AMD Ryzen 5 PRO 5650U with Radeon Graphics, 2.30 GHz

- **Memory**: 64 GB DDR4 (2 × 32 GB), 3200 MHz, CL20
- **Storage**: 2 TB NVMe SSD, with read/write speeds of up to 3500 MB/s
- **Operating System**: Windows 11 Pro, version 24H2
- **Database System**: Oracle Database 23ai Free, Release 23.0.0.0.0 – Production Version 23.4.0.24.05

For the computational study, a real-world dataset from the aviation domain was used, comprising three distinct components:

- *planned routes* for the aircraft operations,
- *real routes* of the planes by reflecting positional data of other objects, planes, as well as statuses of the flight regions,

flight monitoring by assigning airplanes to the Flight Information Regions (*FIRs*) delimited by the entry and exit time. Figure 1 shows an example of the input data related to the FIR assignment. It consists of the identifier (*ECTRL_ID*), FIR assignment reference (*AUA_ID*), temporal borders (*ENTRY_TIME* and *EXIT_TIME*). The data are sequentially stored based on the *SEQUENCE_NUMBER* attribute.

```
"ECTRL ID","Sequence Number","AUA ID","Entry Time","Exit Time"
"186858226","1","EGGXOCA","01-06-2015 04:55:00","01-06-2015 05:57:51"
"186858226","2","EISNCTA","01-06-2015 05:57:51","01-06-2015 06:28:00"
"186858226","3","EGTTCTA","01-06-2015 06:28:00","01-06-2015 07:00:44"
"186858226","4","EGTTTCTA","01-06-2015 07:00:44","01-06-2015 07:11:45"
"186858226","5","EGTTICTA","01-06-2015 07:11:45","01-06-2015 07:15:55"
"186858227","1","EEGXOCA","01-06-2015 04:08:00","01-06-2015 05:01:00"
```

Fig. 1. Input data source

A Flight Information Region (*FIR*) is a designated area of airspace where a specific country's air traffic control (*ATC*) authority is responsible for providing Flight Information Services (*FIS*) and alerting services. Defined by the International Civil Aviation Organization (*ICAO*), FIRs encompass the entire globe, including both land and oceanic regions. The primary functions of FIRs include:

- ensuring the safety and efficiency of air traffic by delivering critical information and coordinating emergency responses,
- managing and controlling airspace within their boundaries,
- supporting route planning and optimization to reduce environmental impact and operational consequences.

The borders of the FIR are temporally oriented and can evolve. The complete dataset comprised 1,000 flights. On average, each flight was represented by approximately 1,000 rows for both the planned and actual routes. Four experiments were done in this study.

Experiment 1 – SQL vs. PL/SQL

The first part of the evaluation study points to the context switches and impact of the SQL call inside PL/SQL environment. It takes one million calls. Tab. 2 shows the results. The processing time difference is 10.82% in favor of the first solution, in which the entire implementation is done in the PL/SQL environment. The value is set by a direct assignment command (Table 3).

Table 2. Processing time results – SQL vs. PL/SQL call

PL/SQL assignment	SQL assignment
x: = seq.nextval;	**select seq.nextval into x from dual;**
00:00:22.891	00:00:25.667

Table 3. Processing time results – Value assignment

Strategy		
1	2	3
Local variable	**Returning**	**Currval**
00:01:18.895	00:00:55.120	00:00:55.913

Experiment 2 – Referencing Assigned Value Multiple Times

The second evaluation study considers three strategies for the value assignments. The first strategy takes the assignment before the first statement processing, while the second strategy uses *RETURNING* clause of the Insert statement. The last third strategy is based on calling *CURRVAL* for the later sequence reference, without the necessity to declare and manage local variable. For the evaluation study, the obtained sequence number is twice referenced, for the master table (primary key) and child record as foreign key (Table 3).

The rapid performance improvement is obtained by postponing the variable management to the second operation by reaching 30.31% improvement for strategy 2 and 29.12% for strategy 3. The difference between strategies 2 and 3 is negligible, below the level of recognizability and statistical error (below 1.5%).

Graphical representation of the results is depicted in Fig. 2.

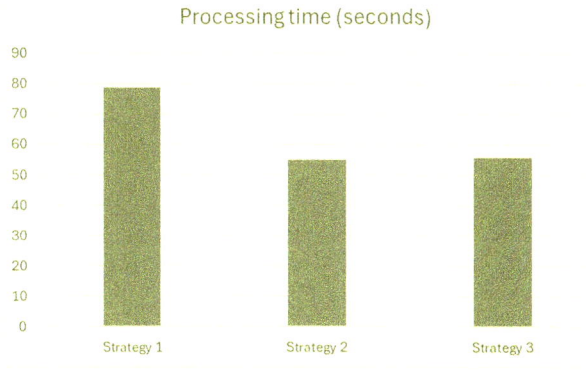

Fig. 2. Experiment 2 results

Experiment 3 – Impact of the Select Statement Definition and Table Reference

This evaluation experiment provides the highest number of result range. It emphasizes Select statement by considering various data source references. Namely, to get the sequence function call result, one row should be provided as a reference. To consider that, five solutions were used. In solution 1, *DUAL* table is explicitly referenced, while solution 2 deals with the Select statement with no *FROM* clause, requesting to "add" it automatically by the database system. Solution 3 references data dictionary view reduced to one row using *FETCH FIRST* clause. The used system table was *user_tables*, just tiny differences are obtained if another system table is referenced. Solutions 4 and 5 use a user defined table. Solution 4 uses FETCH FIRST clause, so there is no limit for the cardinality of the table (naturally, it must consist of at least one row to produce a non-empty result set. Solution 5 is based on one row user-defined table. The results are shown in Table 4, enhanced by the chart in Fig. 3.

Table 4. Processing time results – Various table references

	1	**DUAL**	00:00:25.872
SOLUTION	2	**No FROM clause**	00:00:28.385
	3	**Data dictionary view, fetch first**	00:4:24.289
	4	**User defined, fetch first**	00:00:46.242
	5	**One row user table**	00:00:51.167

Table 4 shows the results. The best solution was obtained by the original solution referencing *DUAL* table. If *FROM* clause is omitted (available in Oracle 23ai), additional demands can be identified. Precisely, additional processing time costs are 2.513 s, which reflects 8.85%. Regarding the user defined table, surprisingly, the better solution is provided by the table with unlimited number of rows, compared to just one row table. The difference is denoted by 9.63%. It is caused by the pre-fetching and indexing availability. The clearly worst solution is related to system tables (solution 3). It is evident that these tables are not suitable ways for referencing standard conventional functions, nor sequences. They are specifically defined for a different purpose and their structure optimization is limited by that. Moreover, from an infrastructure point of view they are typically stored in a different table space, optionally enhanced by the different table blocks. Compared to the best solution (solution 1), data dictionary view reference (solution 3) requires more than 8.44 times for the processing time. Currently introduced no *FROM* clause solution requires only 11.62%. Results in the form of chart are in Fig. 4.

Experiment 4 – Sequence Number Assignment

In Sect. 3, two approaches for the sequence number assignment were discussed. The first approach is based on a direct value assignment embedded to the Insert statement, which is, however, error-prone, as the user is not forced the next value of the sequence and in general, any value can be used as a reference. The second solution invokes a

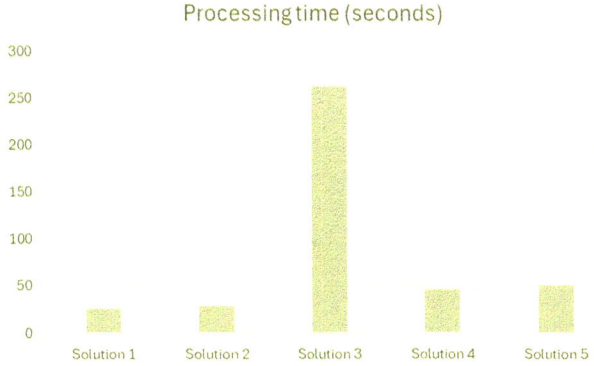

Processing time (seconds)

Fig. 3. Experiment 3 results

trigger, which assigns the primary key value automatically inside the body of the row-level trigger. Its limitation is the PL/SQL environment, which requires context switch for each processed row. Table 5 shows the results for one million row manipulation.

Table 5. Processing time results – Various table references

Direct assignment	Trigger
00:00:34.821	00:01:28.351

Reflecting the results, significant processing time changes and differences can be identified. Namely, In the case of using triggers, the processing costs have increased significantly, up to 2.54 times the original processing time. However, this is a "price" for the consistency and sequential processing of values without gaps. Values are assigned automatically by the database system. This increase itself is caused by two factors - a change between the SQL and PL/SQL language environment (context switch), but mainly by separate record processing – row by row. A trigger is called for each row.

The results in form of chart are in Fig. 4.

5 Discussion

Relational databases are still hugely used, mainly due to the unambiguous structure, transaction support, and relational algebra processing individual records and operations. It maintains efficiency even if the data number is significantly rising. Currently, the main part of the relational databases focuses on the transformation between conventional data into the temporal paradigm. In the sensorical database environment, the stream is even more significant. This paper deals with the sequences allowing to manage individual versions by assigning object identifiers from the sequence. It describes the syntactical part, available functions and existing principles. The main contribution of the paper is the methodology based on the evaluation study. The available options are described

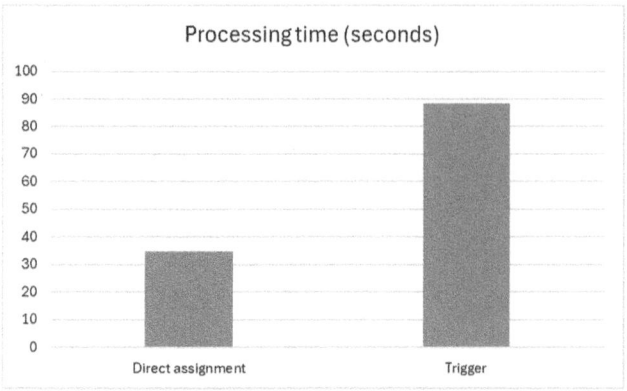

Fig. 4. Experiment 4 results

in the third section, followed by the evaluation study present in Sect. 4. Namely, this paper focuses on four parts depicted by the experiments. The first experiment deals with the SQL call inside the PL/SQL, which is expressed by the context switch. It takes approximately an additional 10%. The second experiment focuses on the assignment of the sequence value multiple times forming the object references. There are two suitable ways, by using *RETURNING* clause or by calling *CURRVAL* function of the sequence, which is session specific. Explicit management using local variable does not provide sufficient power and takes an additional 43%.

When dealing with the Select statement providing sequences, the most efficient way is to use existing conventional *DUAL* table reference. Introduced solution allowing to omit *FROM* clause introduced in Oracle 23ai brings user experience and simplifies coding, but only from the developer point of view. Reflecting the performance, the solution introduced in Oracle 23ai requires an additional 9.71%. All other solutions, like referencing user defined tables or data dictionary views are significantly worse. Practically, referencing data dictionary can extend the processing time demands almost up to 9 times.

Finally, the last experiment 4 reflected the processing time demands by using a trigger, which assigns that. Although the costs are risen, the final solution is protected by the consistency.

In the context of Oracle Database, *SEQUENCE* objects are commonly used to generate unique numeric values, often serving as surrogate keys in relational schemas. However, several threats to validity can compromise their intended use. One major concern is the presence of *gaps in sequence values*, which may arise due to transaction rollbacks, system crashes, or the use of caching mechanisms. While sequences are designed for uniqueness, they do not guarantee continuity, which can be problematic in applications requiring strictly sequential numbering. Additionally, *concurrency issues* may occur when multiple sessions access a sequence simultaneously, leading to non-deterministic increments. *Caching behavior*, while improving performance, introduces risks of lost values in the event of system failure. Further threats include *misconfiguration*—such as inappropriate increment values or maximum limits—and *security risks* stemming from unauthorized access or manipulation of sequence objects. Finally, incorrect assumptions in application logic, such as relying on sequential or timestamp-like ordering of sequence values, can lead to semantic mismatches and flawed data processing. These validity threats must be carefully considered when designing systems that depend on Oracle SEQUENCE objects to ensure correctness, robustness, and alignment with business requirements.

6 Conclusions

This paper examines the Oracle database system, recognized as one of the most complex and feature-rich relational database management systems (RDBMS). Oracle provides an extensive set of functionalities, supporting a wide range of development and deployment scenarios. Core features include the use of sequences, auto-increment columns, and PL/SQL-based mechanisms for automated value generation.

Although certain implementation strategies in Oracle are system-specific and optimized at the database engine level, many concepts are transferable to other relational databases, such as MySQL, PostgreSQL, and Microsoft SQL Server. Most of these systems support sequence definition and procedural language extensions, although the degree of integration and performance characteristics may vary—particularly with regard to context switching between SQL and procedural execution layers.

Auto-increment functionality is now a standard feature across most RDBMS platforms. Notably, Oracle was among the last major systems to introduce native support for auto-increment columns. However, this functionality in Oracle is still fundamentally implemented through sequences and triggers. As such, its primary advantage lies in reducing development complexity rather than offering significant performance enhancements.

Our future research will focus on using the sequences in temporal versioned system related to the consistency of the time frames. It aims to assign sequentially order to the object state versions directly in the relation of the temporal frames. So, the order of states will directly correspond to the order in the timeline.

Acknowledgment. This paper was also supported by the **VEGA 1/0192/24** project - *Developing and applying advanced techniques for efficient processing of large-scale data in the intelligent transport systems environment.*

References

1. Abhinivesh, A., Mahajan, N.: The Cloud DBA-Oracle. Apress (2017)
2. Anders, L.: Cloud Computing Basics. Apress (2021)
3. Bhat, A.Z., Ahmed, I.: Big data for institutional planning, decision support and academic excellence. In: 2016 3rd MEC International Conference on Big Data and Smart City (ICBDSC), pp. 1–5. Muscat, Oman (2016). https://doi.org/10.1109/ICBDSC.2016.7460353
4. Cheng, Y., Jiao, Y., Wei, W., Wu, Z.: Research on construction method of knowledge graph in the civil aviation security field. In: 2019 IEEE 1st International Conference on Civil Aviation Safety and Information Technology (ICCASIT), pp. 556–559. Kunming, China, (2019). https://doi.org/10.1109/ICCASIT48058.2019.8973190
5. Hansen, K.B.: Practical Oracle SQL, Mastering the Full Power of Oracle Database. Apress (2020)
6. Jakóbczyk, M.: Practical Oracle Cloud Infrastructure: Infrastructure as a Service, Autonomous Database, Managed Kubernetes, and Serverless. Apress (2020)

7. Jiao, Y., Han, J., Xu, B., Xiao, M., Shen, B., Sun, H.: Research on domain entity extraction in civil aviation safety. In: 2021 IEEE 3rd International Conference on Civil Aviation Safety and Information Technology (ICCASIT), pp. 384–388. Changsha, China (2021). https://doi.org/10.1109/ICCASIT53235.2021.9633439

8. Kuhn, D., Kyte, T.: Expert Oracle Database Architecture: Techniques and Solutions for High Performance and Productivity. Apress (2021)

9. Kvet, M.: Developing Robust Date and Time Oriented Applications in Oracle Cloud: A comprehensive guide to efficient Date and time management in Oracle Cloud. Packt Publishing (2023). ISBN: 978–1804611869

10. Kvet, M., Baggia, A., Borkovcová, M., Urem, F.: Environmental data analysis. EDIS UNIZA (2024)

11. Malcher, M., Kuhn, D.: Pro Oracle Database 23ai Administration: Manage and Safeguard Your Organization's Data. Apress (2024)

12. Mukherjee, N., Kulkarni, K., Jin, H., Kamp, J., Lahiri, T.: How does oracle database in-memory scale out?. In: 2015 10th International Joint Conference on Software Technologies (ICSOFT), pp. 1–6. Colmar, France (2015)

13. Mukherjee, N., et al.: Fault-tolerant real-time analytics with distributed oracle database in-memory. In: 2016 IEEE 32nd International Conference on Data Engineering (ICDE), pp. 1298–1309. Helsinki, Finland (2016). https://doi.org/10.1109/ICDE.2016.7498333

14. Nuijten, A., Barel, P.: Modern Oracle Database Programming: Level Up Your Skill Set to Oracle's Latest and Most Powerful Features in SQL, PL/SQL, and JSON. Apress (2023)

15. Pastierik, I.: Deploying oracle machine learning AutoML models for oracle APEX analytics. In: 2024 IEEE 17th International Scientific Conference on Informatics (Informatics), pp. 499–506. Poprad, Slovakia (2024). https://doi.org/10.1109/Informatics62280.2024.10900807

16. Patel, J.: An effective and scalable data modeling for enterprise big data platform. In: 2019 IEEE International Conference on Big Data (Big Data), pp. 2691–2697. Los Angeles, CA, USA (2019). https://doi.org/10.1109/BigData47090.2019.9005614

17. Pendse, S., et al.: Oracle database in-memory on active data guard: real-time analytics on a standby database. In: 2020 IEEE 36th International Conference on Data Engineering (ICDE), pp. 1570–1578. Dallas, TX, USA (2020). https://doi.org/10.1109/ICDE48307.2020.00139

18. Rosenzweig, B., Rakhimov, E.: Oracle PL/SQL by Example (The Oracle Press Database and Data Science). Oracle Press (2023)

19. Shi, Z., Jiang, H., Ding, X.: Research on the impact of big data analysis and integration capability on enterprise innovation performance—the intermediary effect of supply chain collaborative innovation. In: 2023 IEEE 3rd International Conference on Information Technology, Big Data and Artificial Intelligence (ICIBA), pp. 138–142. Chongqing, China (2023). https://doi.org/10.1109/ICIBA56860.2023.10165596

20. Wang, C., Sun, H., Chen, Q.: Analysis of China civil aviation turbulence index eddy dissipation rate. In: 2019 IEEE 1st International Conference on Civil Aviation Safety and Information Technology (ICCASIT), pp. 607–610. Kunming, China (2019). https://doi.org/10.1109/ICCASIT48058.2019.8972995

21. Zhou, M., Liu, M., Zhang, R., Wang, X.: Development status and suggestions for sustainable aviation biofuel. In: 2022 International Conference on Environmental Science and Green Energy (ICESGE), pp. 187–191. Shenyang, China (2022). https://doi.org/10.1109/ICESGE56040.2022.10180375

22. Zhuang, Q., Zhou, T.: International development of china civil aviation aircraft tracking and monitoring system. In: 2020 IEEE 2nd International Conference on Civil Aviation Safety and Information Technology (ICCASIT, pp. 1099–1103. Weihai, China (2020). https://doi.org/10.1109/ICCASIT50869.2020.9368807

23. Erasmus+ project EverGreen dealing with the complex data analytics: https://evergreen.uniza.sk/

DOING 2025: 6th Workshop on Intelligent Data - From Data to Knowledge

From Flows to Graphs: Data-Driven Insights on Latent Overtourism with Frequent Pattern Mining

Hugo Alatrista-Salas[1], Gaël Chareyron[1,2], Sonia Djebali[1],
Imen Ouled-Dlala[1], and Nicolas Travers[1(✉)]

[1] De Vinci Higher Education, De Vinci Research Center, Paris, France
{hugo.alatrista_salas,gael.chareyron,sonia.djebali,
imen.ouled_dlala,nicolas.travers}@devinci.fr
[2] EA EIREST, Université Paris 1 Panthéon Sorbonne, Paris, France

Abstract. Overtourism presents complex and often hidden challenges for urban environments, impacting residents, infrastructure, and visitor satisfaction. This study proposes a novel, data-driven methodology to detect and analyze latent overtourism— the early, subtle warning signs of excessive tourism—before visible breakdowns occur. By leveraging user-generated content from *Tripadvisor*, a temporal circulation multidigraph is modeled to capture tourist mobility. Using frequent subgraph mining algorithms, the approach identifies recurring tourist movement patterns across different urban scales. These patterns are then analyzed in both spatial and temporal dimensions to detect hotspots and evaluate dynamic attractiveness through a Huff-based probabilistic model. The approach is applied to three cities of varying sizes revealing consistent tourist flows and areas under increasing pressure, suggesting early overtourism.

Keywords: Overtourism · Social Media · Graph mining · Frequent Pattern Mining

1 Introduction

Tourism is not only a powerful economic sector but also a strategic tool for advancing the Sustainable Development Goals (SDGs), particularly those related to decent work, gender equality, and sustainable communities. Artificial intelligence presents new opportunities to support the Sustainable Development Goals in tourism, yet its implementation involves complex and multidimensional challenges that require informed governance, especially to detect hidden patterns [1].

However, the unchecked growth of tourism can lead to significant socio-environmental challenges that extend beyond immediate economic gains. As defined by Phil Butler [2], Overtourism has adverse effects on three key stakeholder groups. Local residents face increased housing and food prices, community disruption, and growing anti-tourism sentiment. At the same time, local

---

H. Alatrista-Salas, G. Chareyron, S. Djebali, I. Ouled-Dlala—Contributed equally.

P. K. Chrysanthis et al. (Eds.): ADBIS 2025, CCIS 2676, pp. 327–342, 2026.
https://doi.org/10.1007/978-3-032-05727-3_28

infrastructure, natural ecosystems, and cultural heritage sites are suffering from overuse and degradation. Tourists experience reduced satisfaction due to over-crowding, which limits photo opportunities and meaningful engagement, thereby reducing the overall authenticity of their visit.

Recognizing overtourism is inherently difficult, as it is context-dependent with no universal thresholds. Its impacts—sociocultural, economic, and environmental—vary by destination; for example, a visitor-resident ratio unsustainable in Venice may be manageable in Tokyo. Data gaps, like unrecorded rentals or unofficial tourism distort assessments, and resident discontent often emerges too late. Given this complexity, some authors explore early indicators of overtourism [3]. *Latent overtourism* refers to destinations showing early warning signs before tipping points. It differs from overtourism by relying on predictive indicators rather than visible disruption.

Our approach aims to detect latent overtourism using data-driven techniques. The key idea is to analyze tourist movements over the city and their negative impact on areas denoting a premise to overtourism. Our methodology relies on the modelization of a circulation multidigraphs [4] to represent tourist movement, where nodes represent locations and directed edges denote trips between them. The multidigraph integrates the evolution of relationships to enable longitudinal analysis. Then, to focus on relevant nodes, frequent subgraph mining is applied on this circulation multidigraph, producing a ranking of the most frequent places and trips between them. We can finally exploit these frequent patterns which correspond to recurrent patterns of tourist visits and travel routes by combining metrics such as attractivity, ratings, and number of visitors. This step helps to target subgraphs analysis to characterize overtourism dynamics. To validate its scalability and versatility, we tested the approach on datasets from three differently sized cities.

This paper is structured as follows: Sect. 2 reviews relevant state-of-the-art contributions. Section 3 outlines the proposed methodology. Section 4 presents the dataset, the experimental results, and a discussion of the key findings. Finally, Sect. 5 concludes and suggests potential directions for future research.

2   Related Works

The tourism sector is constantly evolving, as well as sophisticated tools to understand traveler behavior and manage destinations effectively like overtourism. This overview explores the latest advancements in the tourism analysis, focusing on three key areas: Graph-based techniques for modeling relationships and movements; Trajectory mining for extracting patterns from sequential location data and diverse approaches to identify, measure, and mitigate the negative impacts of overtourism.

2.1   Graph-Based Techniques for Analyzing Tourism

Graph-based methods provide a robust framework for understanding complex tourist behavior. Geo-tweets can be transformed into tourism graphs to pinpoint popular attractions and routes for tourism management [5]. Graph-data and provenance can be used to recommend contextual tourist management [6]. Geo-tagged photos can also be used to analyze travel habits, modeling daily itineraries, identifying regions of attraction and examining movement using Markov chains and sequence clustering [7]. Mobile positioning datasets from cell tower connections analyze international traveler mobility patterns and identify travel motives [8]. By integrating TELCO data with *Tripadvisor* reviews and applying Social Network Analysis, we can better understand tourist behavior, helping to promote sustainable tourism and reduce overcrowding [9].

Finally, license plate recognition data distinguishes tourists from residents and uncovers frequent travel patterns using `gSpan` frequent subgraph mining [10].

2.2   Trajectory Mining

Trajectory mining focuses on extracting sequential patterns and insights from various forms of mobility data, offering a detailed understanding of tourist movements over time. A four-step methodology uses mobile roaming datasets to extract location points, apply density-based clustering to merge traces into sequences, which are then analyzed using the SPADE algorithm to extract sequential patterns [11]. Open data from GPS trajectories and geo-tagged photos are analyzed using spatial gridding, Markov chain models to identify movement patterns, and K-means clustering to uncover spatiotemporal behavioral patterns [12].

Another approach focuses on extracting disjoint sequential patterns from trajectory data, where routes begin and end at the same locations but differ in the paths taken between them [13]. While graph-based techniques and trajectory mining methods provide valuable insights into tourism trends, they often face limitations in data consistency, scalability for dynamic analysis, and the ability to infer causality–challenges that become especially significant when tackling complex problems like overtourism.

2.3   Analyzing Overtourism

Addressing the impacts of mass tourism has become critical, and this requires a mix of qualitative, textual and sophisticated, data-driven techniques to study overtourism.

Qualitative and Survey-Based Approaches. Qualitative studies synthesize existing literature to understand the impact of overtourism on society, tourists. Tourist surveys are essential for recognizing the effects of overtourism and for providing information tailored to tourist attitudes, preferences, and behaviors. They help us understand overuse, assess the quality of the tourist experience, and evaluate the impact of tourism on the community. These surveys often explore social carrying capacity and apply social exchange theory [14].

Textual Data Analysis. Analyzing textual data, such as online reviews and comments, is a popular way to understand how the public perceives overtourism.

Researchers use text analysis techniques to extract insights from social media sources and assess the opinions of tourists and local residents. For instance, the authors apply sentiment analysis and topic modeling to *Tripadvisor* reviews of popular attractions to identify key discussion topics and associated sentiments [15].

Quantitative and Machine Learning Approaches. Researchers often combine extensive literature reviews, statistical analyses, and numerous case studies with diverse data sources to define overtourism, its causes, and its impacts. However, they also note that a lack of reliable, detailed data often hinders the effective determination of overtourism status [16]. Machine Learning provides powerful tools for analyzing large datasets, predicting overcrowding, and optimizing resource allocation. In this context, Perles-Ribes et al. [17] use binary classification methods, to predict overtourism based on variables such as competitiveness, hotel/*Airbnb* supply ratios, and dependency. Statistical methods are employed to examine overtourism, using multi-criteria decision-making methods to create a dynamic ranking model based on key indicators like paid overnight stays per capita, trips per capita, and air passenger flows. Techniques such as principal component analysis, cluster analysis are employed to analyze trends and classify cities by their levels of overtourism [18].

Spatiotemporal and Sensor-Based Analysis. Analyzing spatiotemporal data is crucial for understanding tourists flow evolution and across different locations and so managing overtourism. The Grid-Level Overtourism Model quantifies overtourism at a granular level using taxi drop-off and traffic congestion data. This model uses Mahalanobis distance to detect overtourism and employs a Temporal Convolutional Network to predict trends by learning spatial and temporal dependencies [19]. Other studies differentiate overtourism from overcrowding by emphasizing its impact on ecological and social capacities. They use key indicators like tourism density and tourism intensity [20]. The integration of specialized technologies, including IoT and Geographic Information Systems (GIS), enhances the efficiency of tourist flow management in historic city centers.

These technologies use sensor networks to monitor pedestrian movements, calculate KPIs, and discover patterns in visitor behavior using advanced data analysis methods. The goal is to optimize management and mitigate overtourism [21].

Though graph-based techniques and trajectory mining offer valuable insights into tourism in a global vision of the city and POIs, they are limited by inconsistent data, dynamic scalability challenges, border effects on POIs' area (*i.e.* inferring causality), especially regarding complex issues like overtourism.

3   Methodology

Our approach aims at identifying latent overtourism spatially and temporally using circulation multidigraphs. It is based on two main components: (1) identifying the most frequently visited tourist destinations using subgraph pattern

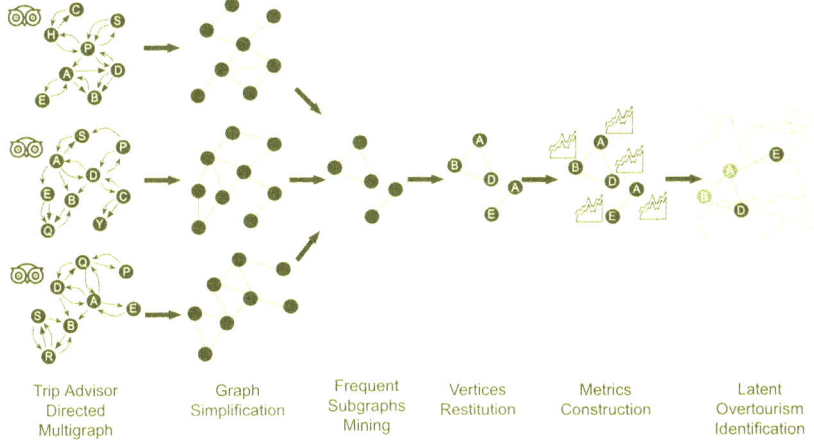

Fig. 1. Methodology for identifying the latent overtourism.

mining algorithms and (2) analyzing the temporal behavior of locations susceptible to overtourism.

Figure 1 illustrates the overall methodological framework proposed in this
study. i) It begins by constructing directed graphs where nodes represent visited locations and directed edges reflect sequential visits by individual users,
resulting in multidigraphs potentially containing multiple edges and self-loops.
ii) To manage complexity while maintaining core topological properties, duplicate edges and self-loops are removed in a graph simplification step. iii) Then,
the gSpan algorithm [22], a depth-first frequent subgraph mining technique, is
employed to detect commonly visited location sets, interpreted as premises of
overtourism hotspots, informed by prior research. iv) Following this, original
node attributes are reintegrated into these subgraphs for temporal analysis. v)
Finally, to identify areas vulnerable to overtourism and vi) visualize temporal
trends, the *Huff model* is implemented to calculate location attractiveness scores
based on visit frequency and geodesic distance.

3.1   Circulation Multidigraph

Definition 1 Circulation Multidigraph. A directed multigraph [23] with
associated data for tourist mobility analysis is a structure $\mathcal{G} = (\mathcal{V}, \mathcal{E}, \mu, \mathcal{L}_\mathcal{V}, \mathcal{L}_\mathcal{E})$
where:

- $\mathcal{V}$ is a finite *set of vertices* representing tourist location.
- $\mathcal{E}$ is a finite *set of edges* representing sequential visits between locations.
- $\mu : \mathcal{E} \to \mathcal{V} \times \mathcal{V}$, is a function that maps each edge $e \in \mathcal{E}$ to an ordered pair
  of vertices (u, v); representing a direct trip from u to v, allowing for multiple
  such trips.

- $\mathcal{L}_\mathcal{V}$ is a mapping that assigns a set of discrete labels (continuous data) to each vertex in $\mathcal{V}$. These labels can represent initial properties of the locations and are enriched after the subgraph mining process.
- $\mathcal{L}_\mathcal{E}$ is a mapping that assigns a set of discrete labels (continuous data) to each edge in $\mathcal{E}$. The primary label denotes a sequential visit, with its frequency (edge count) between locations analyzed in frequent subgraph mining.

As shown in the first step of Fig. 1, multiple edges may exist between a given pair of vertices u and v, resulting in a circulation multidigraph structure (see Definition 1). Furthermore, the graph is labeled: the sets $\mathcal{L}_E$ and $\mathcal{L}_V$ store metadata associated with edges and vertices, respectively. Edge attributes may include variables such as the elapsed time between reviews (measured in days) or the nationality of the tourist. Vertex attributes may comprise the location's name, its geographic coordinates (*e.g.* Lambert coordinates), or user-assigned ratings.

3.2 Subgraph Pattern Mining

Extracting frequent subgraphs (see Definition 2) from a set of graphs can reveal recurring structural configurations that may represent underlying spatial, behavioral, or relational patterns. The gSpan algorithm is one of the most effective approaches to identify all subgraphs that frequently appear across a given set of graphs, based on a user-defined support threshold. The gSpan algorithm's methodology is grounded in a rigorous formalization of graph structures and their relationships.

Definition 2. Subgraph Isomorphism. Give a graph database $\mathcal{D} = \{G_1, G_2, \ldots, G_n\}$ where $\mathcal{G}_i = (\mathcal{V}_i, \mathcal{E}_i, \mu_i, \mathcal{L}_{\mathcal{V}i}, \mathcal{L}_{\mathcal{E}i})$, a graph $g = (\mathcal{V}_g, \mathcal{E}_g, \mu_g, \mathcal{L}_{\mathcal{V}g}, \mathcal{L}_{\mathcal{E}g})$ is subgraph-isomorphic to a graph $\mathcal{G}_i \in \mathcal{D}$ if there exists an bijective function $f : \mathcal{V}_g \to \mathcal{V}_i$ that preserves the edge structure and the vertex and edge labels. Specifically, for ever:

- Edge $(u, v) \in \mathcal{E}_g$, the corresponding edge $(f(u), f(v))$ exists in $\mathcal{E}_i$,
- Vertex $v \in \mathcal{V}_g$ its label $\mathcal{L}_{\mathcal{V}g}(u)$ is equal to the label of its mapped vertex $\mathcal{L}_{\mathcal{V}i}(f(u))$,
- Edge $(u, v) \in \mathcal{E}_g$ its label $\mathcal{L}_{\mathcal{V}g}(u, v)$ is equal to the label of the corresponding edge $\mathcal{L}_{\mathcal{E}i}(f(u), f(v))$.

Given a minimum support threshold $\sigma \in \mathbb{N}$, a subgraph g is said to be *frequent* if $support(g) \geq \sigma$. gSpan's goal is to efficiently discover all frequent subgraphs in a database $\mathcal{D}$, without redundancy, which it achieves using a depth-first strategy that explores the search space in a structured and constrained manner. The key idea is to map each graph to a canonical string representation called a *DFS code* derived from a depth-first traversal of the graph. This encoding captures structural and label information, providing a consistent way to represent and compare graphs. In addition, the rightmost path extension strategy further constrains pattern growth, ensuring deterministic and compact exploration. Together, these mechanisms guarantee completeness and non-redundancy. Early pruning based on minimum support thresholds also discards infrequent subgraphs before further exploration.

It is worth nothing that the structure of a frequent subgraph is formally equivalent to the graphs contained in the dataset $\mathcal{D}$. Specifically, frequent subgraphs are graph-theoretic patterns that have the same structural properties as individual graphs in the dataset $\mathcal{D}$, but they are distinguished by their frequent appearance in multiple graphs. This characteristic helps us identify frequently visited places while taking into account the topology of tourist visits.

3.3   Latent Overtourism Identification

Once frequent patterns are extracted from the circulation digraph, we focus the analysis of involved nodes in those patterns. For this, we rely on an attractiveness model which serves as a witness of overtourism on involved frequent patterns. This measure must be confronted to activity data on those nodes to highlight the impact of attractiveness.

3.3.1   Huff Model

The *Huff* model [24], a gravity-based spatial interaction model, analyzes shopping center trade area, *i.e.* the regions from which a store attracts customers. The model considers two main factors: store attractiveness (product variety) and travel cost (distance or time). As travel costs increase, the likelihood of a visit decreases. Huff probability that a visitor at location i selects destination j is defined as:

$$p_{i,j} = \frac{A_j^{\alpha}/D_{ij}^{\beta}}{\sum\limits_{j \neq i} \frac{A_i^{\alpha}}{D_{ij}^{\beta}}} \qquad (1)$$

where $i, j \in \mathcal{V}_g$: vertices are among frequent patterns, $p_{i,j}$: probability that individual i visit site j. A_j: attractiveness of site j (represented by a node's feature: $A_j \in \mathcal{L}_{\mathcal{V}_j}$). D_{ij}: distance between site i and site j. α, β: sensitivity parameters to A_j and D_{ij}.

For identifying latent overtourism, the Huff probability can serve as a proxy for measuring the attractiveness of a destination compared to its competitors, taking into account both the Haversine distance and the number of visitors (node's in-degree) as the attractivity.

Huff attractiveness scores, derived from visit frequency and spatial proximity, can vary widely across locations and over time. Without normalization, these scores reflect scale disparities or data sparsity rather than genuine attractiveness trends, thereby distorting the interpretation of temporal dynamics. To mitigate this issue, the *rolling average* normalization method was applied to monthly Huff attractiveness scores.

3.3.2   Normalize Reviews and Rating

To compare the attractiveness with tourist activity, a normalization procedure is applied to reviews data. This step enables accurate and interpretable comparisons of user ratings across locations and over time. In fact, due to volume fluctuation of reviews over time, direct visualization can be misleading.

This cumulative normalization corrects for fluctuations in review volume and density.

- **Cumulative Review Count**: for each location i at time τ the cumulative number of reviews up to time is calculated as: $NB_{i,\tau}^{\mathrm{acc}} = \sum_{\tau=1}^{t} NB_{i,\tau}$
- **Cumulative Weighted Rating**: aggregate influence of each review weighted by its corresponding rating : $R_{i,t}^{\mathrm{add}} = \sum_{\tau=1}^{t} NB_{i,\tau} \cdot \mathrm{rating}_{i,\tau}$
- **Normalized Rating Score**: these two cumulative measures are used to define the normalized rating score $R_{i,t}^{\mathrm{norm}} = R_{i,t}^{\mathrm{add}}/NB_{i,t}^{\mathrm{acc}}$.
- **Reference Review Volume (per month/year)**: total reviews in a given month and year: $R_{y,m}^{ref} = \sum_{i} NB_{i,(y,m)}$
- **Normalized Review Count**: normalize each observation's review count: $NB_{i,t}^{\mathrm{norm}} = NB_{i,t}/R_{y_t,m_t}^{ref}$

The final output consists of the two normalized variables $R_{i,t}^{\mathrm{norm}}$ and $NB_{i,t}^{\mathrm{norm}}$, that are visualized over time to compare trends across locations. This approach smooths out short-term fluctuations and allows for fair temporal and spatial comparisons, even when review data is sparse or unevenly distributed.

4   Results and Discussion

This study pursues two primary objectives: (1) to identify urban areas that may be exhibiting early indicators of latent overtourism, and (2) to examine how the defining characteristics of these areas change over time. To address these goals, this section first provides a comprehensive overview of the dataset employed, followed by a presentation and interpretation of the analytical results for the selected cities.

4.1   Dataset Description

The methodological pipeline begins by collecting 96 months of *Tripadvisor* review data from January 2013 to December 2020. Tourist mobility is modeled as monthly circulation multidigraphs, where the nodes represent reviewed locations and the directed edges represent sequential visits by users within the same month. These time series graphs capture actual behavior and offer a basis for identifying latent overtourism.

To validate the method, we analyzed three cities with varying levels of tourist activity: *Tampere* (688 nodes), *Lille* (2,531 nodes), and *Barcelona* (18,514 nodes). The graphs were simplified by merging duplicate edges and removing self-loops. *Tampere* had small graphs (2128 nodes, averaging 55 nodes and 51 edges), *Lille* showed steady activity (38646 nodes, averaging 419 nodes and 514 edges), and *Barcelona* exhibited high complexity (4674,672 nodes and up to 11,767 edges), reflecting intense tourism.

4.2 Latent Overtourism Detection

The temporal graphs representing 96 monthly instances per city are used as input to the gSpan algorithm to extract frequent subgraphs. To account for differences in city size (node counts ranging from 688 to 18,514), a relative minimum support threshold of 30% is applied to enable pattern discovery in small and medium-sized graphs and 90% for large graphs. To ensure computational efficiency while preserving pattern relevance, the algorithm is also limited to subgraphs with at most five edges.

Using gSpan with the same parameters (minimum support count equal to 0.95 and a maximal number of edges equal to 5) we identified frequent subgraphs with 4 in *Tampere*, 74 in *Lille*, and 1,010 in *Barcelona*. It is important to note that the number of frequent subgraphs extracted for each city varies significantly due to the density of the graphs. The number of nodes in the graph representing *Barcelona* is 172 times the number of nodes in *Tampere*.

Our analysis of the *Tampere* Finland dataset revealed three distinct two-length frequent subgraphs. It represents significant tourist movement patterns between key attractions in the city. These patterns involve visits between four major landmarks: *The Pyynikki Observation Tower* (#661108), which offers panoramic views and is near culinary destinations like *Pizzeria Napoli* (#777510) and the *Pyynikin Nakotornin Kahvila* restaurant (#941703). Meanwhile, the *Tampere Cathedral* (#604970), as the city's spiritual and architectural centerpiece, is another frequently visited place.

In the same spirit, Table 1 highlights the five most prevalent patterns found in *Lille*, France. The second pattern (#1) shows tourists traveling between two cultural landmarks: the *Grande Place* (#269814) and the *Palais des Beaux-Arts* (#269820). This sequence occurred in 75% of the months (72/96). A more complex four-node pattern is needed to illustrate behavioural trends. Tourists frequently visited the *Grande Place* (#269814), then the *Vieille Bourse* (#269817), followed by the *Palais des Beaux-Arts* (#269820). This pattern appeared in 66 out of 96 months of months.

Table 1. Lille's Most Frequent Subgraphs

Patt_ID	Vertices	Supp
0	[269814, 269817]	85
1	[269817, 269820]	72
2	[269814, 269817, 269820]	71
3	[269814, 269820]	70
4	[269814, 269820, 269817]	66

Table 2. Barcelona's Most Frequent $\mathcal{V}_g$

Patt_ID	Vertices	Supp
0	[190166, 190624]	95
1	[190162, 190166]	92
2	[190624, 190629]	91
3	[190166, 190624, 190629]	91
4	[190166, 1059712]	91

Similarly, Table 2 presents the five most frequent subgraphs identified in *Barcelona*'s dataset. The first pattern (Pattern #0) shows tourists traveling between the *Sagrada Família*, represented by #190166, and the *Güell Park* (#190624), both of which are among the most famous places in *Barcelona*.

Using gSpan, we uncover hidden movement patterns that concentrate visitor flows—potential early indicators of overtourism by highlighting local impacts. Certain location sequences emerge regularly across multiple months and years, highlighting persistent pressure points where local infrastructure and visitor experiences are most susceptible to strain. For example, the pattern #4 in Table 1 comprises three locations: *Grande Place* (#269814), *Vieille Bourse* (#269817), and *Palais des Beaux-Arts* (#269820). This pattern manifests in 68.7% of monthly graphs. This phenomenon is not merely a conventional tourist route; rather, it is a mathematical signature of systemic overload. Tourist attractions are packed with visitors, and queues at eateries stretch out onto the pavements. Such persistent patterns are typically referred to as carrying capacity breaches, particularly when they pertain to bottleneck locations and demonstrate increasing edges' density over time.

Furthermore, the presence of frequent subgraphs can facilitate the identification of locations that are particularly vulnerable to overtourism. To accomplish this objective, we compute the frequency of appearance of all nodes that are part of each frequent subgraph. Tables 3, 4 and 5 illustrate the most popular locations in the three cites

4.3   Temporal Analysis of Places Subject to Latent Overtourism

User-generated data from platforms like *Tripadvisor* provide a valuable lens for exploring latent overtourism, early tourism pressure that is not yet evident

Table 3. Top 5 frequent places in *Lille*

ID	Name	Freq.
269820	Palais des Beaux-Arts de *Lille*	45
269814	Grande Place	44
269817	Vieille Bourse	40
793722	La Chicoree	25
247498	LaM *Lille* Métropole Musée d'Art Moderne	13

Table 4. Top 5 frequent places in *Barcelona*

ID	Name	Freq.
190166	Sagrada Família	926
190624	Güell Park	866
246171	église Sainte-Marie-de-la-Mer	483
190629	Casa Milà	431
271009	Camp Nou	397

Table 5. Frequent places found by gSpan in *Tampere*

ID	Name	Frequency
661108	Park and Pyynikki observation tower	2
777510	Pizzeria Napoli	2
604970	*Tampere* Cathedral	1
941703	Pyynikin Nakotornin Kahvila	1

through traditional signs, such as overcrowding or infrastructure saturation. A detailed examination of the normalized review counts $NB_{i,t}^{norm}$ and average ratings over time $R_{i,t}^{norm}$ for selected locations, as outlined in Sect. 4.2, facilitates the identification of subtle yet significant shifts in visitor dynamics. The normalized review count captures shifts in user activity relative to seasonal and temporal changes. It helps detect sustained or rising interest in a site. The normalized rating reflects user satisfaction; declines may indicate discomfort due to overcrowding or unmet expectations, even with steady or growing visit numbers. Together, these indicators offer a comprehensive view of evolving perceptions and pressures on a location, serving as early indicators of potential overtourism. Figures 2, 3 and 4 depict the normalized number of comments and the normalized rating for *Tampere*, *Lille*, and *Barcelona*, respectively. We observe the evolution of monthly reviews (red lines) and cumulative ratings (dark lines) over time. The data comes from TripAdvisor, a platform that has declined in recent years due to changing user behaviour and rising competition. The 2020 pandemic also disrupted tourism, reducing reviews and travel. For each city studied, two locations from $\mathcal{V}_g$ were selected. Overall, review numbers are decreasing as tourists increasingly rely on existing reviews. The cumulative ratings are converging, however some variations of tendency can be seen in Figs. 2.B, 3.B and 4.B with a decrease of the rating which can be a witness of consequences of overtourism.

Further, applying the Huff model over time provides a dynamic measure of site attractiveness. This measure can be analyzed to identify persistent increases or decreases in tourist attention, which is a key indicator of latent overtourism. To highlight these trends, a logarithmic regression was applied to the smoothed attractiveness curve. Sustained growth or decrease of probability $p_{i,j}$ or attractiveness A_j over time may reveal hidden dynamics that indicate emergent overtourism. When employed in conjunction with rolling average smoothing techniques, the Huff model offers a robust and interpretable approach to monitoring spatial pressure dynamics.

Figures 5, 6, and 7, show the normalized Huff attractiveness undergoes a temporal evolution for the four most visited locations in *Tampere*, *Lille*, and *Barcelona*, respectively. The monthly values (red dotted lines), based on node in-degree as a proxy for appeal – tourists moving to this specific place – reveal fluctuations over time. A logistic regression curve (solid black line) smooths out short-term variations to highlight long-term trends.

Figure 5.A shows a gradual decline in the Huff-based attractiveness score of the Pyynikki Observation Tower over time. In contrast, Fig. 5.B shows that the attractiveness of the Pyynikin Näkötornin Kahvila restaurant has steadily increased over time. This growth is likely due to rising tourist interest, positive word-of-mouth, and greater visibility on platforms like *Tripadvisor*. This systematic augmentation in its perceived appeal potentially reflects the onset of latent overtourism.

Interestingly, we studied the evolution of reviews overtime for those specific locations. The first tourists identified the *Pyynikin Nakotornin Kahvila* restaurant (B) as "*A must-visit for tourists* (2013)" after visiting the observation tower

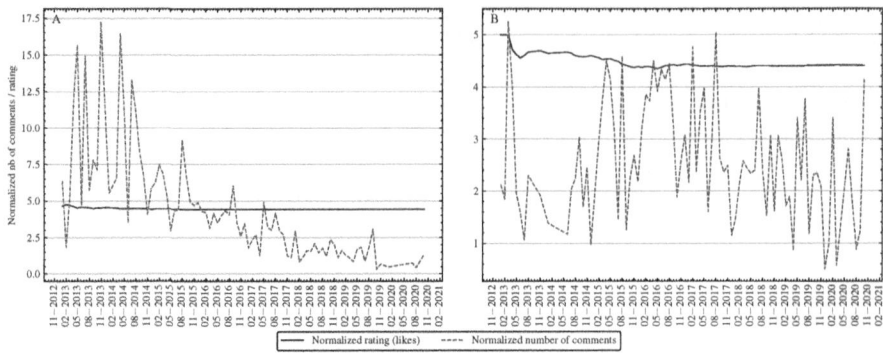

Fig. 2. Normalized visitors Nb vs. normalized rating of 2 frequently visited places in *Tampere*: A) Park and Pyynikki observation Tower, B) Pyynikin Nakotornin Kahvila. (Color figure online)

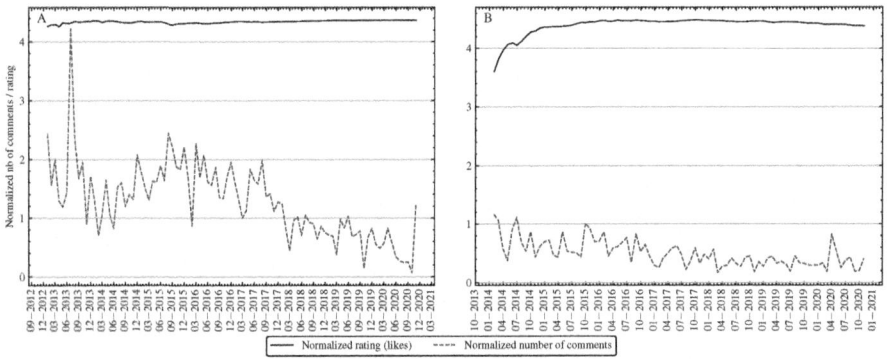

Fig. 3. Normalized visitors Nb vs. normalized rating of 2 frequently visited places in *Lille*: A) Palais des Beaux-Arts de *Lille*, B) Bloempot restaurant. (Color figure online)

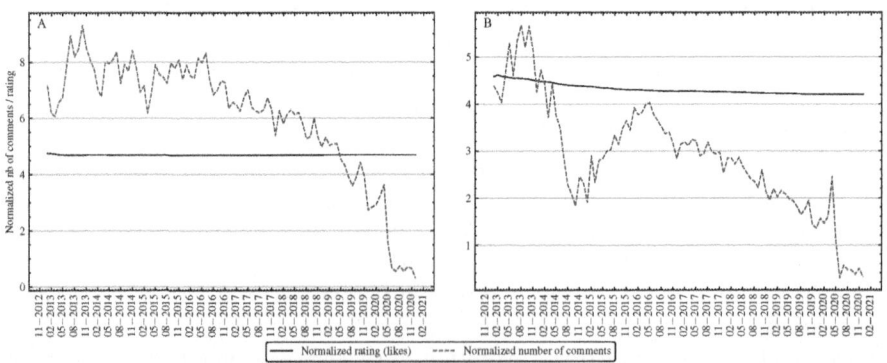

Fig. 4. Normalized visitors Nb vs. normalized rating of 2 frequently visited places in *Barcelona*: A) Sagrada Família, B) Güell Park. (Color figure online)

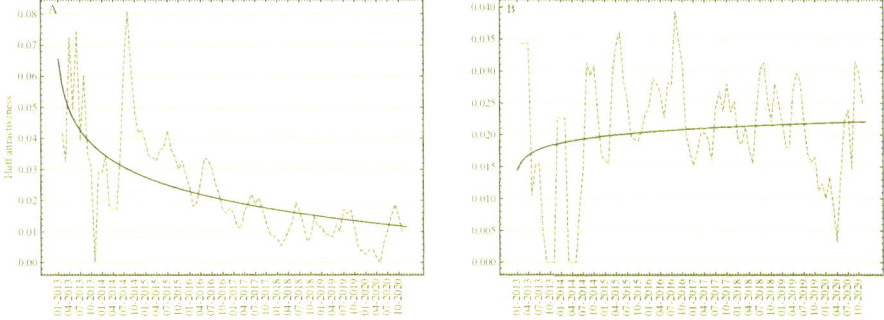

Fig. 5. Normalized Huff Attractiveness of two of the most frequently visited places in *Tampere*, Finland: A) Park and Pyynikki Observation Tower, B) Pyynikin Nakotornin Kahvila. (Color figure online)

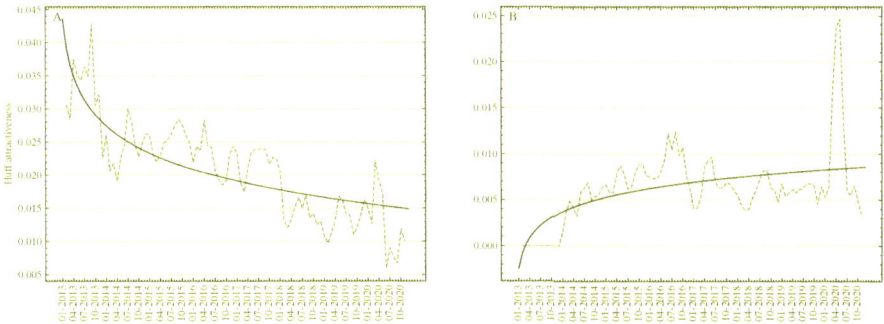

Fig. 6. Normalized Huff Attractiveness of two of the most frequently visited places in *Lille*, France: A) Palais des Beaux-Arts de *Lille*, B) Bloempot restaurant. (Color figure online)

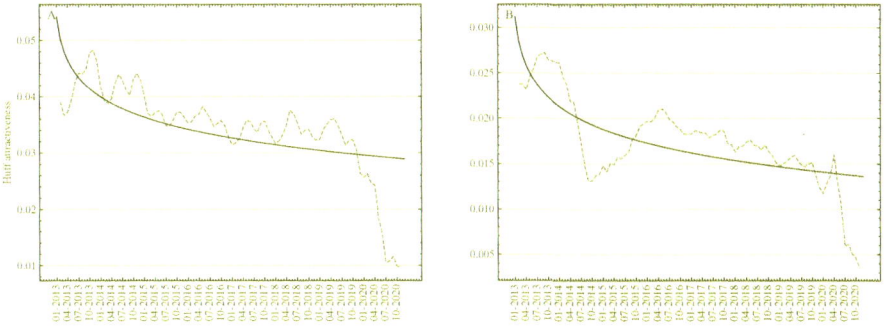

Fig. 7. Normalized Huff Attractiveness of two of the most frequently visited places in *Barcelona*, Spain: A) Sagrada Família, B) Güell Park. (Color figure online)

(A). This trend has implied more visits as visible in Fig. 5.B, however the quality of reviews and visits decreased over time (as seen in Fig. 2.B) with comments witnessing overtourism effect (c1-2015, c2-2016, c3-2017, and especially c4-2017). A similar behavior is witnessed for Bloempot restaurant (Fig. 6): same positive Huff attractiveness regression value and average rating decrease (conversely with the *Palais des Beaux-Arts*). This effect was not present in Barcelona (Fig. 7) where all Huff regressions were negative and not witnessing reviews on overtourism.

As previously discussed, the augmentation in attractiveness that occurs before overt manifestations of overcrowding or degradation provide empirical support for the concept of latent overtourism. In other words, the Huff attractiveness and rating evolution on frequent patterns in the circulation digraph can be considered a subtle yet significant early warning signal of overtourism. These findings suggest that digital traces from travel platforms, when integrated with spatial interaction models such as Huff, can be effectively leveraged to anticipate shifts in the desirability of tourist sites.

5   Conclusion and Future Directions

This study presents a data-driven approach to the early detection of latent overtourism. This approach combines temporal graph modeling, frequent subgraph mining, and spatial attractiveness metrics. Monthly circulation multidigraphs were constructed using user-generated content from *Tripadvisor*, and the gSpan algorithm identified frequently visited locations over time. The algorithm assessed the locations' susceptibility to latent overtourism based on visitor numbers, user rankings, and attractiveness. Applying this approach to three cities of varying sizes demonstrates its effectiveness in uncovering recurrent visitation patterns and detecting emerging hotspots experiencing growing tourism pressure.

Future work may improve the model by incorporating contextual variables, such as local events, weather, and seasonality, to distinguish between short-term fluctuations and long-term trends more effectively. Using a hierarchical spatial indexing system, such as the H3 grid, could improve spatial resolution and integrate factors such as accommodation density, transit access, and cultural asset distribution to detect hotspots more precisely.

References

1. Gössling, S., Mei, X.Y.: AI and sustainable tourism: an assessment of risks and opportunities for the SDGs. Curr. Issues Tourism, 1–14 (2025). https://doi.org/10.1080/13683500.2025.2477142
2. Dodds, R., Butler, R. (eds.): Overtourism: Issues, Realities and Solutions. Walter De Gruyter Oldenbourg, Berlin, Boston (2019). https://doi.org/10.1515/9783110607369
3. Adie, B.A., Falk, M., Savioli, M.: Overtourism as a perceived threat to cultural heritage in Europe. Curr. Issues Tourism **23**(14), 1737–1741 (2020). https://doi.org/10.1080/13683500.2019.1687661

4. Chareyron, G., Quelhas, U., Travers, N.: Tourism analysis on graphs with Neo4Tourism. In: U, L.H., Yang, J., Cai, Y., Karlapalem, K., Liu, A., Huang, X. (eds.) WISE 2020. CCIS, vol. 1155, pp. 37–44. Springer, Singapore (2020). https://doi.org/10.1007/978-981-15-3281-8_4

5. Hu, F., Li, Z., Yang, C., Jiang, Y.: A graph-based approach to detecting tourist movement patterns using social media data. Cartography GIS **46**(4), 368–382 (2019). https://doi.org/10.1080/15230406.2018.1496036

6. Costa, U.S., Espinosa-Oviedo, J., Musicante, M.A., Vargas-Solar, G., Zechinelli-Martini, J.: Using provenance in data analytics for seismology: challenges and directions. In: ADBIS'22 Short Papers, vol. 1652, pp. 311–322. Springer, Italy (2022). https://doi.org/10.1007/978-3-031-15743-1_29

7. Zheng, Y.-T., Zha, Z.-J., Chua, T.-S.: Mining travel patterns from geotagged photos. ACM TIST **3**(3), 1–18 (2012). https://doi.org/10.1145/2168752.2168770

8. Park, S., Zhong, R.R.: Pattern recognition of travel mobility in a city destination: application of network motif analytics. J. Travel Res. **61**(5), 1201–1216 (2022). https://doi.org/10.1177/00472875211024739

9. Confente, I., Mazzoli, V., Camatti, N., Bertocchi, D.: Integrating tourists' walk and talk: a methodological approach for tracking and analysing tourists' real behaviours for more sustainable destinations. J. Sustain. Tour. **32**(11), 2323–2343 (2024). https://doi.org/10.1080/09669582.2024.2322120

10. Bing, H., et al.: Discovering the graph-based flow patterns of car tourists using license plate data: a case study in Shenzhen. Chin. J. Adv. Transp. **2020**(1), 4795830 (2020). https://doi.org/10.1155/2020/4795830

11. Park, S., Xu, Y., Jiang, L., Chen, Z., Huang, S.: Spatial structures of tourism destinations: a trajectory data mining approach leveraging mobile big data. Ann. Tourism Res. **84**, 102973 (2020). https://doi.org/10.1016/j.annals.2020.102973

12. Liu, W., et al.: Cluster analysis of microscopic spatio-temporal patterns of tourists' movement behaviors in mountainous scenic areas using open GPS-trajectory data. Tour. Manage. **93**, 104614 (2022). https://doi.org/10.1016/j.tourman.2022.104614

13. Peng, S., Yamamoto, A.: Mining disjoint sequential pattern pairs from tourist trajectory data. In: Appice, A., Tsoumakas, G., Manolopoulos, Y., Matwin, S. (eds.) Discovery Science, pp. 645–658. Springer, Cham (2020)

14. Sansone, M., Colamatteo, A., Pagnanelli, M.A., D'Agostini, M., et al.: How can artificial intelligence tools help mitigate the phenomenon of overtourism? In: Brands and Purpose in a Changing Era, Atti della XXI SIM Conference (2024)

15. Singgalen, Y.A.: Sentiment classification of over-tourism issues in responsible tourism content using naïve bayes classifier. JoSYC **5**(2), 275–285 (2024)

16. Peeters, P., et al.: Research for TRAN committee-overtourism: impact and possible policy responses. University of Brighton (2021)

17. Perles-Ribes, J., Ramón-Rodríguez, A., Moreno-Izquierdo, L., Such-Devesa, M.: Competitiveness and overtourism: a proposal for an early warning system in Spanish urban destinations. Euro. J. Tourism Res. **27**, 2707 (2021). https://doi.org/10.54055/ejtr.v27i.2137

18. Nádasi, L., Kovács, S., Szöllős-Tóth, A.: The extent of overtourism in some European locations using multi-criteria decision-making methods between 2014 and 2023. Int. J. Tourism Cities (2024). https://doi.org/10.1108/IJTC-05-2024-0103

19. Kong, X., Huang, Z., Shen, G., Lin, H., Lv, M.: Urban overtourism detection based on graph temporal convolutional networks. IEEE TOCSS **11**(1), 442–454 (2024). https://doi.org/10.1109/TCSS.2022.3226177

20. Vourdoubas, J.: Evaluation of overtourism in the island of Crete. Greece. EJASET **2**(6), 21–32 (2024)

21. Zubiaga, M., Izkara, J.L., Gandini, A., Alonso, I., Saralegui, U.: Towards smarter management of overtourism in historic centres through visitor-flow monitoring. Sustainability **11**(24), 7254 (2019). https://doi.org/10.3390/su11247254
22. Yan, X., Han, J.: gSpan: graph-based substructure pattern mining. In: ICDM'22, pp. 721–724 (2002). https://doi.org/10.1109/ICDM.2002.1184038
23. Abello, J., Korn, J.: Visualizing massive multi-digraphs. In: IEEE Symposium on Information Visualization 2000. INFOVIS 2000. Proceedings, pp. 39–47 (2000). https://doi.org/10.1109/INFVIS.2000.885089
24. Huff, D.L.: Defining and estimating a trading area. J. Mark. **28**(3), 34–38 (1964). https://doi.org/10.1177/002224296402800307

Using a Model-Agnostic Meta-model as a Flexible Database Migration Tool

Baptiste Haudebourg and Nicolas Hiot[✉] [iD]

Université d'Orléans, INSA CVL, LIFO, UR 4022, Orléans, France
baptiste.haudebourg@etu.univ-orleans.fr, nicolas.hiot@univ-orleans.fr

Abstract. This work is an extension of the ArchiTXT structuration algorithm. Building on its model-agnostic data representation, we propose a translation algorithm that can convert any instance in this meta-model into either a relational or a property graph database. Furthermore, we demonstrate the model's versatility by providing translations from documents, relational databases, and graph databases towards the meta-model. This provides a fully bidirectional translation algorithm that allows databases to be converted between other models. This also paves the way for future work involving the application of the ArchiTXT structuration algorithm to database instances for automatic data integration.

Keywords: database translation · data integration · interoperability

1 Introduction

With the continuous growth of data production, both organisations and researchers face the challenge of integrating and analysing large volumes of heterogeneous information. Data is distributed across a wide range of models, including documents, relational databases, and graph structures, each optimised for particular storage, retrieval, or processing tasks. Nevertheless, the extraction of meaningful insights frequently requires exploring multiple databases simultaneously. Despite recent advances in multi-paradigm database systems and ongoing research on data lakes, data integration remains a complex and resource-intensive task. It often requires substantial manual intervention, including schema alignment and the design of supervised transformation pipelines, which significantly constrain both scalability and adaptability. In response to these limitations, we propose an unsupervised, lightweight, flexible, and format-agnostic approach that enables seamless exploration of heterogeneous data sources, reducing the cost and complexity of integration.

This work is a direct extension of ArchiTXT [6], a tool designed for easy exploration of textual data by automatically structuring it without the need for supervision. At its core, it uses a simple tree-based meta-model that provides a uniform representation of the data. Its structuration algorithm relies on tree

P. K. Chrysanthis et al. (Eds.): ADBIS 2025, CCIS 2676, pp. 343–353, 2026.
https://doi.org/10.1007/978-3-032-05727-3_29

rewriting, enabling it to operate on any dataset while remaining transparent and auditable. Unfortunately, ArchiTXT misses the final step of the process and does not convert the intermediate representation into a concrete database model. This paper aims to address that gap by proposing a method to transform any instance of the intermediate model into relational and graph database formats. Furthermore, we introduce a set of simple parsing algorithms that enable the conversion of documents, relational databases, and graph databases into this intermediate model, facilitating translation between different modelling paradigms. This work is part of an ongoing effort, and while the proposed methods are implemented and functional, further evaluation and refinement remains to be conducted.

Paper Organisation. The next section provides a review of related work on data translation and existing meta-models. Section 3 introduces the ArchiTXT meta-model and key definitions. Section 4 explains how to export a meta-model instance to relational or property-graph databases, while Sect. 5 covers conversion from databases to the meta-model. Finally, Sect. 6 concludes the paper and outlines future work.

2   Related Works

Since the introduction of the relational model, alternative database models have emerged, each designed to address limitations of its predecessors or to optimise for particular use cases such as enhanced flexibility, efficient recursive queries, or large-scale distributed storage. Over time, organisations have collected vast silos of information in a variety of formats, chosen for their alignment with initial requirements, for developer familiarity, or simply because a given technology was in trend. Today, getting valuable insights often requires aggregating and analysing multiple data sources at once. Although schema and data migration across relational, object-oriented, document, graph, and other database paradigms has been studied for some time [3,4,7], existing approaches typically address only one-to-one conversions. We still lack a general framework for integrating multiple heterogeneous databases together automatically. This gap is particularly important in modern data lake architectures and ETL pipelines, where heterogeneous data must be ingested at scale. Furthermore, querying across diverse data formats in a data lake, such as combining JSON, CSV, and relational tables in a single SQL-like query, remains an open challenge [8,9].

In the literature, we saw a focus on relational-to-graph conversion, especially since the rise of Linked Open Data [1,4,12] and property-graph (PG) systems [7,13]. In these approaches, each relation tuple is often promoted to a graph node whose columns become node properties, while foreign keys and join tables are represented as edges, with any associated attributes attached as edge properties.

[3] shows an overview of query processing in multi-store, which mostly relies on the mediator-wrapper pattern. In this architecture, the mediator acts as the query planner, similar to a distributed database query processor, and decomposes

a query into sub-queries. The wrappers are bridges that translate the sub-query into the designated database query language. Despite advances in optimisation and heterogeneous query planning, they do not fully cover data integration and often rely on specialised SQL extensions for querying.

ArchiTXT [6] is a tool built on an intermediate, model-agnostic data representation originally developed for structuring textual data. It relies on a similarity metric to identify structurally equivalent parts and tries to minimise the schema size while preserving the data. The objective was to facilitate the exploration and analysis of textual data by persisting it in a structured database, thereby leveraging the optimisation capabilities of modern DBMS. It relies on a model-agnostic meta-model as its internal representation, which closely follows the Entity-Relationship (E/R) model using a tree-like structure.

On one hand, a similar approach is described in [11], though with a notable difference in how relationships are represented. While the ArchiTXT meta-model adheres to the E/R paradigm by explicitly modelling relationships between individual entities, the canonical model in [11] is more aligned with the relational model, where data is represented as tuples grouped into tables, without explicitly capturing the semantic relationships implied by foreign keys.

On the other hand, Abstra [2] is a tool designed to support exploration of structured and semi-structured data, using concepts of the E/R model as employed by ArchiTXT to achieve a uniform representation of data. It is particularly useful for navigating unfamiliar datasets and identifying those of potential value. While the underlying motivation aligns with ours, their approach does not construct a unified data representation; instead, it remains focused solely on data exploration rather than integration or transformation.

This paper proposes extending ArchiTXT into a data migration tool capable of ingesting diverse semi-structured and textual data to produce valid database instances so they can be queried using standard languages. By routing all conversions through a unified intermediate abstraction, we avoid brittle one-to-one mappings, simplify support for new formats, and enable seamless merging of heterogeneous databases. This shared layer also allows universal transformations, like normalization, to be implemented once across formats. While beyond the paper's scope, ArchiTXT's automatic structuration algorithm could be adapted to allow fully automated end-to-end integration.

3   Intermediate Meta-model

The model proposed by ArchiTXT is built upon three fundamental components. Property (Prop), is the most basic element, representing a named data value that captures a specific characteristic of an object (e.g., the name of a person). Group (Grp) represents a named collection of properties that together represent an instance of an object. Finally, we

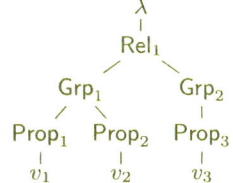

Fig. 1. Example of ArchiTXT's meta model.

have relation (Rel) between groups, capturing higher-level connections between object instances. An example instance of this meta-model is shown in Fig. 1.

Collections (Coll) are also represented in the model, but their presence is mostly implicit. Groups or relations that share the same name are considered part of the same collection. This aligns with the property graph paradigm, where collections are implicitly defined through node labels. While Coll elements are generally omitted in the final representation, they can be useful during intermediate steps to explicitly mark nested collections.

Formally, we can define an instance and its schema as follows:

Definition 1 (Instance). *An instance I is represented as a collection of ordered tree $T = (D, l, i)$, where $D \subseteq \mathbb{N}^*$ is a prefix-closed domain of positions. $l : D \to \mathbb{T} \times (\Sigma \cup \{\lambda\})$ is a labelling function that assigns to each position a pair consisting of a node type from $\mathbb{T} = \{Rel, Grp, Prop, \epsilon\}$, and a label from the set Σ, or the special root symbol λ. i is a function that maps a position to an OID where $i(u)$ is the unique OID assigned to the node at the position u. A node at the position $u.k \in D$ is called a child of u and u is the parent of $u.k$.*

Definition 2 (Schema). *A schema $S = (N, T, P, \lambda)$ is a condensed context-free grammar (CFG) [6], where N is the set of non-terminal symbols, T the set of terminal symbols, P the set of production rules (which may include rules of the form $A \to B^*$ for repetitions), and λ the start symbol. A schema conforms to a meta-grammar $\mathcal{G}$ [6], which defines its semantics. Following $\mathcal{G}$, the rules in P has one of the following shapes:*

$$Prop \to \langle data \rangle \quad \text{(properties)} \qquad Rel \to Grp_i \; Grp_j \qquad \text{(relations)}$$
$$Grp \to Prop_0 \ldots Prop_i \quad \text{(groups)} \qquad Coll \to Rel_i^+ \mid Grp_i^+ \quad \text{(collections)}$$

For each type $t \in \mathbb{T}$, we note $N_t \subseteq N$ the set of non-terminal of the type t and $P_t = \{A \to \alpha \in P \mid A \in N_t\}$ the set of production rules. We note $leftR \subseteq P_{Rel}$ the one-to-many relations and $rightR \subseteq P_{Rel}$ with the many-to-one relations, such that, $leftR \cup rightR = P_{Rel}$ and $leftR \cap rightR = \emptyset$. $leftR$ and $rightR$ can be determined from an instance I of S by counting examples of each relation.

Identity and De-duplication. The meta-model used by ArchiTXT is structured as a tree rather than a directed graph to avoid cycles. This means that if an object A is related to both object B and object C, we need to represent it as two separate relations: one between A and B, and another between A and C. As a result, A must be duplicated in the structure. While this duplication is not an issue within ArchiTXT itself, it becomes problematic when translating the instance into a concrete database model. To avoid redundancy in the target database, we need a way to identify equivalent group instances. However, the original ArchiTXT model does not include a concept of keys or unique identifiers for groups, making it difficult to recognise that the duplicated A instances refer to the same object. Ideally, A should be represented only once in the resulting database.

In this work, we address this issue by assigning a unique object identifier (OID) to each Grp node in the trees. These OIDs serve as keys to track object identity across duplicated branches. When two nodes are considered semantically equivalent, they are assigned the same OID. This provides a consistent mechanism for identity resolution across representations, bridging the gap between different paradigms: relational (primary keys), property graphs (node IDs), and documents (hierarchical paths). In practice, OIDs are either derived directly from existing identifiers in the source database when available, or else generated uniquely per group instance. The OIDs may also be determined using techniques such as clustering or similarity scoring to identify equivalent Grp instances. This paper does not explore such techniques in detail, as they depend heavily on domain-specific constraints and data availability. Instead, we treat OIDs as given, focusing on how they enable consistent object tracking throughout the translation pipeline.

4   Export to Database

The first contribution of this paper is enhancing ArchiTXT to export its intermediate data representation directly into concrete databases. As the meta-model is designed to accommodate the limitations of diverse source formats, transforming data into concrete database models becomes relatively straightforward. In particular, we target relational databases and property-graph databases, as these are the two most widely adopted paradigms in industry. While for RDF a similar process to property-graph can be used, it would require mapping extracted facts to established ontologies.

Relational Database. To translate the meta-model to a relational database, Grp trees are promoted as relation tuples, with Prop as attributes. The leaf of the property always defines the value of the attribute. A relational database $\mathcal{R} = (\mathcal{S}, \mathcal{D})$ is defined as a pair where $\mathcal{S}$ is the schema or set of relations $R = (name, attrs, PK)$ with $name$ the name of the relation, $attrs$ a finite set of attributes, and $PK \subseteq attrs$ the set of primary key attributes. $\mathcal{D}$ is a mapping between each $R \in \mathcal{S}$ and a finite set of tuples over $attrs(R)$. We denote R_x as the relation with the name x and $\forall y \in attrs(R_x)$ $R_x.y$ the attribute y of the relation x. As defined in Sect. 3, each group is given a unique object identifier; we denote its attribute by oid.

In a relational schema, any one-to-many relation $\mathsf{Grp_x} \xrightarrow{\mathsf{Rel}} \mathsf{Grp_y}$ is represented by adding a foreign-key column $R_x.oid$ to R_y. If Rel is many-to-many, we create a join table instead. For each $\mathsf{Grp_{name}}$, we define its foreign-key attribute set as:

$$FK_{name} = \{R_x.oid \mid \forall(\mathsf{Rel} \to \mathsf{Grp_x}\ \mathsf{Grp_y}) \in leftR \quad \text{such that } y = name\}$$
$$\cup \{R_y.oid \mid \forall(\mathsf{Rel} \to \mathsf{Grp_x}\ \mathsf{Grp_y}) \in rightR \quad \text{such that } x = name\}$$

Considering a schema S, the mapping from $S \mapsto \mathcal{S}$ is given by:

$$\mathcal{S} = \{(name, \alpha \cup \{oid\} \cup FK_{name}, \{oid\}) \mid \forall (\mathsf{Grp}_{name} \to \alpha) \in P_{\mathsf{Grp}}\}$$
$$\cup \{(name, \{R_{g1}.oid, R_{g2}.oid\}, \{R_{g1}.oid, R_{g2}.oid\}) \mid$$
$$\forall (\mathsf{Rel}_{name} \to \mathsf{Grp}_{g1}\ \mathsf{Grp}_{g2}) \in P_{\mathsf{Rel}} \setminus (leftR \cup rightR)\}$$

Given $\mathcal{S}$, the final step is to insert an instance I that corresponds to $\mathcal{S}$ into the database. The process is similar to the schema construction, but instead of using the meta-model schema, it operates on each tree $T \in I$ that contains the data.

Property Graph. Converting the metamodel to a property graph is analogous to the relational database approach; each Grp becomes a node with Prop children turned into properties. Rel, on the other hand, is almost always represented as an edge without the need to introduce an extra node. In this direction, we identify and collapse any Grp that serves solely as a connection between two others into an attributed edge.

Definition 3 (Collapsible relations). *A Grp M is collapsible if and only if it occurs exactly twice as the "one" side of two distinct one-to-many relations. In other terms, there exist relations $A \xrightarrow[n-1]{R_1} M$ and $M \xleftarrow[1-n]{R_2} B$. In that case, M can be removed and replaced by a direct many-to-many edge $A \xleftrightarrow{M} B$, whose attributes are those originally carried by M. If M appears in a third relation or in the "many" side of a relation, it will be preserved. Given a tree $T = (D, l, i)$, "collapsible" is the set of all tree positions that can be collapsed.*

$$oidrel = \{\!\!\{i(u.0) \mid \forall u \in D, l(u) \in leftR\}\!\!\} \cup \{\!\!\{i(u.1) \mid \forall u \in D, l(u) \in rightR\}\!\!\}$$
$$collapsible = \{u \in D \mid i(u) \in oidrel \wedge m(i(u)) = 2\}$$

To convert an instance I, the procedure toPG is applied for each tree $T \in I$. It depends on two Cypher-based routines: **createNode**, which uses MERGE to ensure each node is unique by its OID, and **createEdge**, which similarly uses MERGE to create a unique edge for the pair of nodes and a given set of properties.

```
MERGE (n:'$label' { oid: $oid })
ON CREATE SET n += $props

      createNode(oid, label, props)
```

```
MATCH (src  { oid: $src$ })
MATCH (dest { oid: $dst$ })
MERGE (src)-[r:'$label' $props]->($dst$)

      createEdge(src, dst, label, props)
```

5   Translate a Database Instance

As shown previously, we can translate any meta-model instance into relational or graph databases. We now look at the reverse process: mapping various concrete database models into this intermediate representation. By first converting a

---

Procedure $\text{toPG}(T = (D, l, i))$

1 **foreach** $u \in D$ *where* $l(u) \in N_{\text{Grp}}$ *and* $u \notin collapsible$ **do**
2 | $\texttt{createNode}(i(u), l(u), \{l(u.i.0) \mid \forall u.i \text{ such that } l(u.i) \in N_{\text{Prop}}\})$;
3 **end**
4 **foreach** $u \in D$ *where* $l(u) \in N_{\text{Rel}}$ *and* $\{u.0, u.1\} \cap collapsible = \emptyset$ **do**
5 | $\texttt{createEdge}(i(u.0), i(u.1), l(u), \{\})$;
6 **end**
7 **foreach** $u.i \in collapsible$ **do**
8 | find $r \in D$ such that $l(r) \in N_{\text{Prop}}$ and $i(u.i) \in \{i(r.0), i(r.1)\}$;
9 | $v \leftarrow u.(1 - i)$;
10 | $w \leftarrow r.0$ if $i(r.1) = i(u.i)$ else $r.1$;
11 | $props \leftarrow \{l(u.i.j.0) \mid \forall u.i.j \text{ such that } l(u.i.j) \in N_{\text{Prop}}\}$;
12 | $\texttt{createEdge}(i(v), i(w), l(u), props)$;
13 **end**

---

source database into the meta-model, and then translating from the meta-model into a target database format, we effectively decouple source and target schema. Using this two-step approach, adding support to a new database model only requires two translators (into and out of the meta-model) rather than building direct converters for every pair of database paradigms. As a result, our framework remains model-agnostic, and fully extensible.

Document Database. In this work, we primarily focus on XML, JSON, and YAML documents, although the method applies to any nested data structure. The document schema is always inferred from the data to accommodate any document, but it can, instead, take into account a JSON schema or an XSD description. The transformation occurs in two phases:

First, the documents are standardised into a uniform nested structure composed of mappings and homogeneous collections. JSON and YAML are semantically similar, both being composed of mappings (key-value pairs) and collections (lists or arrays). In contrast, XML lacks explicit markers to distinguish collections. To infer collections in XML, we apply the following heuristic: if an element contains multiple child elements with the same tag name, we treat the element as a collection. When a collection contains heterogeneous children (e.g., different tag names), we can rewrite the document tree to split the collection, grouping homogeneous elements together. Additionally, for any collection whose items are literals, we introduce an intermediate node carrying the parent's label so that collections hold only non-leaf nodes. For instance, the tree $(A\ 1\ 2)$ becomes $(\text{Coll}_A\ (A\ 1)\ (A\ 2))$.

In a second phase, the nested trees are flattened to match the ArchiTXT meta-model. Any node whose children are only literal leaves becomes a Prop, and all other nodes become Grp elements. A Grp is added for Prop that appear in a collection, creating one-property groups. For every parent-child pair of Grps, we create a Rel linking the parent to each child. To accommodate collections, parent-child relations are distributed over all elements of the collection.

For example, the JSON document in Fig. 2a will be converted to the meta-model as shown in Fig. 2b and can be exported to get a PG (Fig. 2c). We name the document root R in this example.

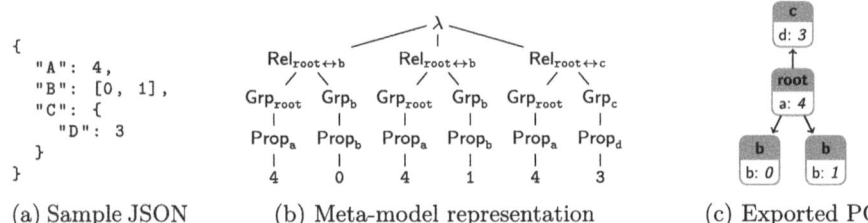

(a) Sample JSON   (b) Meta-model representation   (c) Exported PG

Fig. 2. JSON to PG conversion example.

Tabular data, such as CSV files, can be naturally represented as a set of Grp elements, where each instance (or sub-tree) corresponds to a row, all sharing a common group name.

Relational and Graph Databases. A relational database can be modelled as a directed graph, with each tuple represented as a node and each foreign-key relationship as a directed edge, which makes the computation of relational and graph databases pretty similar. For relational databases, join tables that do not carry extra attributes other than the foreign-keys are condensed into only one edge instead of using an intermediate node. For PGs, attributed edges can be replaced by an intermediate node containing the attributes to make a graph without attributed edges.

As the meta-model is a tree-based representation, converting this graph into our meta-model is analogous to the graph arboricity problem: we need to select a set of root nodes from which to launch breadth-first searches (BFS) to cover every edge and arc of the graph. To ensure scalability and parallelism, ArchiTXT represents a database instance as a forest of moderate-sized trees. During conversion, we choose to balance two objectives: limiting tree size for efficient processing and maximising schema coverage so that each tree covers as much as possible of the database schema to reflects the original context.

We address root node selection by leveraging the natural directionality of the edges. Node labels that do not have entering edges, such as the join tables, naturally become the root of the BFSs. For each instance of these root nodes, we perform a BFS that follows outgoing edges, generating one tree per traversal. Each edge is traversed at most once, mitigating tree growth and preventing infinite loops. Each visited edge yields a Rel at the root of the tree between the two node that are represented as Grp in the tree. Entities not reached by the initial BFS runs are handled in a cleanup pass: after all root-based traversals finish, we revisit each table in BFS order from the previous roots and identify any unprocessed rows. Each of these rows then becomes the root of a new BFS, creating its own sub-tree. We repeat this scan-and-build cycle until every tuple in the database has been converted.

In the presence of cycles in the database schema, and if no obvious root table can be selected, we break the cycle by selecting as the root the table in that cycle with the most foreign-keys and the most rows, making the assumption that increasing the number of initial trees will lead to more reasonably sized trees.

Large Database Handling. In this work, we adopt a breadth-first search (BFS) strategy to traverse source databases and convert them into the meta-model tree representations. However, this naive BFS can produce extremely wide trees that exceed available memory and reduce the effectiveness of parallel processing. While some strategies have been proposed to avoid infinite cycles, we plan to explore additional strategies in an extensive study:

First, instead of traversing all outgoing edges of the same label concurrently (which expands tree width), we can traverse only one per BFS pass. For example, if an Order node links to multiple Products, we can partition traversal into separate trees, each containing exactly one OrderProduct pair. This reduces tree width while preserving a context sufficient for structural equivalence computation (however, we will lose the proximity of products).

A second approach is to limit the BFS depth, effectively segmenting the tree. This is motivated by the fact that elements that are far apart are less related to each other and can be separated without too much context loss. Alternatively, random-walk sampling can be explored to extract small examples from the graph. This approach is in a way similar to ArchiTXT's original handling of textual input, where sentences were interpreted as partial descriptions of the underlying data.

6   Conclusion

In this paper, we have introduced a comprehensive methodology for translating concrete database instances such as relational, document-based (XML, JSON, YAML), and graph into and out of the ArchiTXT meta-model. This work fills the gap in integrating textual data into a unified database representation as originally presented in [6]. Our investigation is preliminary, and future work must evaluate the completeness and cohesion of each translation path. Nevertheless, it demonstrates the meta-model's versatility in accommodating diverse data structures with minimal ad hoc rules. However, we saw that the strictness of the binary, non-attributed relation model does limit direct representation of certain n-ary or richly attributed relationships permitted by other data models, necessitating additional steps to preserve the semantic nuance.

We have started to experiment with the application of the simplification algorithm from [6] directly to multiple database instances. Initial tests indicate that it successfully merges, simplifies, and integrates heterogeneous databases automatically, indicating that the algorithm is truly data-agnostic rather than limited to textual inputs. An early exploration in [10] showed that augmenting textual datasets with synthetic instances further enhances simplification by providing structuration hints. With this work, it is possible to integrate textual data with an existing database instead of relying on synthetic data.

While this work can be an important step toward automatic data integration, measuring the quality of the simplification remains a challenge. Moreover, our current tree extraction strategy does not scale well to very large databases. In particular, as described in Sect. 5, additional strategies remain to be explored to mitigate tree size and ensure they would fit into memory without too much loss of contextual information. Despite these limitations, our methodology serves both as a complementary step for ArchiTXT and as a general-purpose tool for converting data models.

Implementation. All the translations described in this paper have been implemented and are publicly available in the ArchiTXT GitHub repository [5].

Acknowledgment. This work was partially supported by the *Ambition Recherche Développement Centre-Val de Loire* (ARD) JUNON-DATA project.

References

1. A Direct Mapping of Relational Data to RDF (2012). https://www.w3.org/TR/rdb-direct-mapping/
2. Barret, N., Manolescu, I., Upadhyay, P.: Abstra: toward generic abstractions for data of any model. In: Proceedings of the 31st ACM International Conference on Information & Knowledge Management, CIKM '22, pp. 4803–4807. Association for Computing Machinery, Atlanta (2022). https://doi.org/10.1145/3511808.3557179
3. Bondiombouy, C., Valduriez, P.: Query Processing in Multistore Systems: An overview. Research Report RR-8890, INRIA Sophia Antipolis - Méditerranée (2016)
4. Cerbah, F.: Learning highly structured semantic repositories from relational databases. In: Bechhofer, S., Hauswirth, M., Hoffmann, J., Koubarakis, M. (eds.) ESWC 2008. LNCS, vol. 5021, pp. 777–781. Springer, Heidelberg (2008). https://doi.org/10.1007/978-3-540-68234-9_57
5. Chabin, J., Halfeld-Ferrari, M., Hiot, N.: ArchiTXT (2024). https://github.com/Neplex/ArchiTXT
6. Chabin, J., Halfeld-Ferrari, M., Hiot, N.: From Text to Databases: Attribute grammar as database meta-model (2024). https://doi.org/10.48550/arXiv.2410.09441
7. De Virgilio, R., Maccioni, A., Torlone, R.: Converting relational to graph databases. In: First International Workshop on Graph Data Management Experiences and Systems, pp. 1–6 (2013)
8. Gu, Z., et al.: A systematic overview of data federation systems. Semant. Web **15**(1), 107–165 (2024)
9. Hai, R., Koutras, C., Quix, C., Jarke, M.: Data lakes: a survey of functions and systems. IEEE Trans. Knowl. Data Eng. **35**(12), 12571–12590 (2023)

10. Hiot, N.: Construction automatique de bases de données pour le domaine médical : Intégration de texte et maintien de la cohérence. Ph.D. thesis, Orléans (2024). https://theses.fr/s265149
11. Schreiner, G.A., Duarte, D., Santos Mello, R.: Bringing SQL databases to key-based NoSQL databases: a canonical approach. Computing **102**(1), 221–246 (2020). https://doi.org/10.1007/s00607-019-00736-1
12. Sequeda, J.F., Arenas, M., Miranker, D.P.: On directly mapping relational databases to rdf and owl. In: Proceedings of the 21st International Conference on World Wide Web, pp. 649–658 (2012)
13. Unal, Y., Oguztuzun, H.: Migration of data from relational database to graph database. In: Proceedings of the 8th International Conference on Information Systems and Technologies, pp. 1–5 (2018)

DGP: Towards A Distributed Graph Programming

Alpha Mouhamadou Diop⬛ and Cheikh Ba$^{(\boxtimes)}$⬛

LANI - Université Gaston Berger, Saint-Louis, Senegal
{diop.alpha-mouhamadou,cheikh2.ba}@ugb.edu.sn

Abstract. Nowadays, Graph data are very important and omnipresent. Due to the fast growth of Internet, very large data are being more and more produced. In such a case, a cluster of machines is commonly used to store data in a distributed manner, and to process them in parallel. Many platforms and solutions exist that can process these data transparently. However, the corresponding paradigms suffer from their low-level. The purpose of this work-in-progress is to start from a (non-distributed) graph programming language, and translate its buildings blocks/constructs into lower level programming platforms. The graph programming language depends on graph rewriting, and its constructs are proved to cover any graph computable program.

Keywords: Big Data · Graph programming · Graph transformation · Pregel

1 Introduction

Due to the development of data processing technologies and social networks, graph data become more and more omnipresent. In fact, objects of interest form graphs since they are relationally linked to each other. For instance, we can cite well known knowledge graphs that are often represented by using RDF, and many other graph related domains such as machine learning, knowledge bases, social networks, biological networks, graph databases, etc.

Nowadays, in our digital world, data are more and more digitized, and generated much more than ever. This exponential growth is due to the ubiquitousness of Internet and the fact that humans are data generators.

In cases where the data are so large that conventional solutions cannot process them in a reasonable time, the best solution is the use of a cluster of machines in order to distribute the storage and to process in parallel.

Several platforms and solutions, developed from the ground up, have existed [36,41], but they required knowledge in distributed systems, as well as in low-level tasks such as computer communication and network configuration.

Fortunately, higher level solutions for a transparent and distributed data processing have been proposed. One of the most known is Hadoop [2], inspired

P. K. Chrysanthis et al. (Eds.): ADBIS 2025, CCIS 2676, pp. 354–363, 2026.
https://doi.org/10.1007/978-3-032-05727-3_30

by Google's paper [10] on a programming model (called MapReduce) and a File System (called Google File System).

Since MapReduce is not so suitable for graph algorithms, known to be often iterative, several in-memory solutions have been introduced in order to accelerate the processing. We can cite Google's Pregel [29], GraphLab [27], Power-Graph [18] and Spark/GraphX [19]. Consequently, some traditional algorithms on graphs such as connected components, PageRank, shortest path, and so on, have been implemented [4,9,14,20,21,26,37,38], and sometimes accompanied by comparative studies and experiments [3,23,24,45,46]. Furthermore, to ensure completeness, novel approaches have been proposed [5,6,11] and tested [12] for automata-based abstract machines.

However, when we consider programming models, the aforementioned platforms are still low-level since they lack intuitiveness. For instance, the programming model of Pregel [29] is called "Think Like a Vertex" and determines a graph processing by what each graph node or vertex will have to process. Programmers have to handle low-level aspects such as when a node sends a message, to whom, what is the present superstep and when the whole processing finishes. Graph algorithms may therefore be hard to implement and to maintain.

With this in mind, we present our proposed approach. Our long-term goal is to develop a high-level language dedicated to distributed graph programming that conceals the graph's distributed nature, enabling programs to operate as if the graph were not distributed. Programs with this language will be automatically or semi-automatically translated into a low-level programming model, such as the "Think Like a Vertex" framework. To achieve this, we leverage a language based on graph transformation, designed for non-distributed graphs [8,31–35]. Subsequently, its constructs will be translated into the Pregel model.

The document is structured as follows. Section 2 gives some definitions and platforms for distributed graph processing. Section 3 recalls graph rewriting and a related graph programming language. Section 4 presents our proposal. Section 5 gives some related works while Sect. 6 concludes the document with some remarks and perspectives.

2 Preliminaries

This section introduces key concepts related to graphs and distributed graph processing, focusing on relevant frameworks and techniques.

2.1 Graph Morphism

A graph G generally consists of a set V of vertices and a set E of edges connecting vertices $\{(v_1, v_2) \mid v_1, v_2 \in V\}$. It may be useful, for some reason, to label edges and vertices in order to represent weight, symbol, distance, and so on, depending on field of interest. Furthermore, considering characteristics of edges, several topologies can be considered like directed graphs, undirected graphs, multigraphs or hypergraphs. For now, we only consider labeled and directed graphs (Definition 1).

Definition 1. *Let $\mathcal{L}$ be an alphabet of labels. A graph G is defined as (V, E, s, t, l, m), where V is the set of vertices, E the set of edges, functions $s, t : E \to V$ are source and target functions for edges, function $l : V \to \mathcal{L}$ labels vertices and function $m : E \to \mathcal{L}$ labels edges.* □

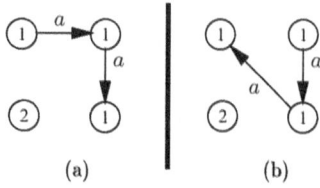

(a)   (b)

Fig. 1. (a) A directed and labeled graph G. (b) A graph H isomorphic to graph G.

Figure 1-(a) shows graph G, over the alphabet $\mathcal{L} = \{a, b, 1, 2, 3\}$, and defined by $V = \{v_1, \cdots, v_4\}$; $l : l(v_1) = l(v_2) = l(v_3) = 1, l(v_4) = 2$; $E = \{e_1, e_2\}$; $m : m(e_1) = m(e_2) = a$; $s : s(e_1) = v_1, s(e_2) = v_2$; $t : t(e_1) = v_2, t(e_2) = v_3$.

Definition 2. *Let $G = (V_G, E_G, s_G, t_G, l_G, m_G)$ and $H = (V_H, E_H, s_H, t_H, l_H, m_H)$ be directed graphs with vertex sets V_G, V_H, edge sets E_G, E_H, source and target functions s_G, t_G, s_H, t_H, and vertex and edge labeling functions l_G, m_G, l_H, m_H. A graph morphism $g : G \to H$ is a pair of mapping functions $g_V : V_G \to V_H$ and $g_E : E_G \to E_H$ such that the following conditions hold:*

$s_H \circ g_E = g_V \circ s_G$, $t_H \circ g_E = g_V \circ t_G$, $m_H \circ g_E = m_G$, $l_H \circ g_V = l_G$ for all vertices in the domain of l_G. Here, $\circ$ denotes function composition. These conditions ensure that g preserves the graph's structure and labeling. □

2.2 Distributed Graph Processing

A graph is distributed if its edges and vertices are stored across a cluster of computers. Below, we briefly present some high-level platforms for distributed data, noting that solutions developed from the ground up were discussed in Sect. 1.

MapReduce When it comes to processing big data, Google's MapReduce framework [10], with its two functions *map* and *reduce*, is actually a well-known paradigm. Therefore we assume familiarity and, as a reminder, we just recall that MapReduce is not suitable for iterative programs due to excessive *shuffle & sort* phases and disk-based input/output operations.

Memory-Based Solutions As we recall above, MapReduce is unsuitable for programs that are iterative. Therefore, in order to enhance efficiency of data processing, several in-memory solutions have been proposed. One of the well-known solutions is Spark [47], which can be considered as an "in-memory MapReduce".

Sadly, neither Spark's nor MapReduce's paradigm is suitable to graph computation. We have theoretically [5,6,11] and experimentally [12] proven that the use of an in-memory *Think Like a Vertex* paradigm has at least two pros: (1) it is easier to write graph algorithms, and it prevents from using counterintuitive structure and generating intermediary and useless data; (2) naturally, execution is faster thanks to the in-memory aspect.

In this manner, in order to take into account the intrinsic nature of graph, some in-memory solutions have been proposed, among which Spark-GraphX [19] and Giraph(Pregel) [29]. We presently use he latter in this work and it is based on BSP. BSP (Bulk Synchronous Parallel [44]) is a model for parallel programming with a *Message Passing Interface*. The computation is a sequence of supersteps. Pregel [29] is a BSP implementation, and hides details of communications between workers. The computing model is *"Think Like a Vertex"* for which what each graph vertex has to execute determines the whole graph computation. In a superstep, active vertices run `compute()`, a user-defined function, may send and/or receive messages to/from some other vertices, and become inactive. Changes on edges and vertexes will be available in the following superstep. Supersteps are delimited by synchronization barriers. These barriers ensures that a message sent in one superstep will be delivered to the target vertices during the following superstep. In a superstep, an active vertex can call `voteToHalt()` function in order to become inactive in the next superstep, unless it receives a new message. The whole computation will end when all vertices are inactive, and no message is to be sent. Our implementation is based on an open-source implementation of Pregel, named Giraph [1].

Even though this programming model can express lots of graph algorithms, it is not intuitive enough and easy to use in many cases since programmers still have to consider low-levels aspects and have to find one single `compute()` (vertex) function to process the whole graph. In this way, graph algorithms are often hard to implement and to maintain. We therefore aim to raise the programming level by using a language that, as far as possible, hides the distributed aspect.

3   High-Level Graph Programming

Considering that our proposal is based on graph transformation, we briefly introduce this notion in this section. The focus is on algebraic approach, precisely the *double-pushout* approach [15].

3.1   Graph Rewriting

Graph transformation or rewriting consists in the obtaining of a new graph from another graph by the use of transformation rules of the form $r = (L \leftarrow I \rightarrow R)$.

Part L is the *left-hand side* of r, part R is the *right-hand side* and part I is the *interface*. The rule r is composed of two graph morphisms $I \to L$ and $I \to R$. Informally, the application of the rule to an input graph G consist in finding a matching of L, then deleting from that matching elements in $L \setminus I$, and adding elements in $R \setminus I$ to the result.

In Fig. 2, graph morphism are represented by numbers decorating nodes.

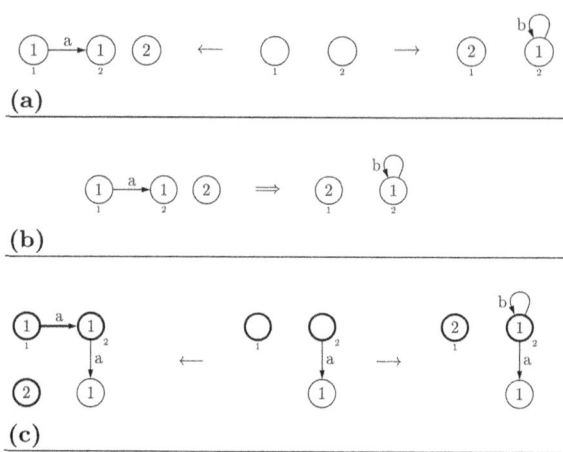

(a)

(b)

(c)

Fig. 2. (a) A rule. (b) An equivalent notation. (c) An example of application.

3.2   Distributed Graph Programming

We recall that our ultimate goal is to have high-level language for distributed graph programming. This language should conceal the graph's distributed nature, that is, it would act as if the graph was not distributed. A program in this language would be automatically or semi-automatically translated into a low-level programming model such as "Think Like a Vertex". For this purpose, our idea is to build upon a language (GP/GP2) based on graph transformation [8, 31–35] and designed for graphs that are non-distributed. Then we translate its constructs into the Pregel model. The set of constructs in GP/GP2 has been proven to be not only minimal, but is also able to process any program on graphs [22].

Let $\mathcal{L}$ be a alphabet of labels. A graph ***program*** over $\mathcal{L}$ is based on graph transformation rules, and is defined by three specifications: (i) a set $\mathcal{R}$ of rules over $\mathcal{L}$ is a program. (ii) *sequential composition*: if P_1 and P_2 are programs, the composition $\langle P_1; P_2 \rangle$ is also a program. (iii) *iteration*: if P is a program, so $P{\downarrow}$ is also a program.

Figure 3 shows a Dijkstra's Shortest Path Algorithm in GP2, a revised version of the GP. The algorithm starts from a graph with positive numbers as edge

Simple_Dijkstra = S_Prepare ↓; S_Start: S_Reduce ↓

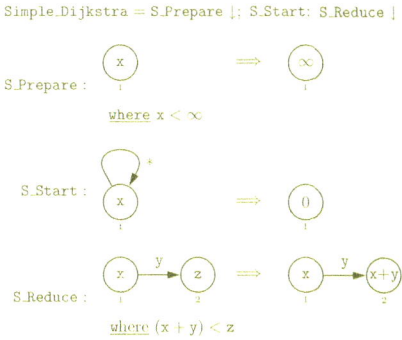

S_Prepare :

where x < ∞

S_Start :

S_Reduce :

where (x + y) < z

Fig. 3. Simple program for Dijkstra's SPA

labels. A loop with * labels the starting node. The rule S_Prepare relabels all nodes with ∞. The rule S_Start removes the loop of the starting node and relabels it with 0, and the rule S_Reduce will change a current distance when a shorter path is detected.

4 From High-Level to Low-Level

4.1 Soundness of the Approach

One of the advantages of GP/GP2 is its formal semantics for reasoning about graph transformations. Nevertheless, our most important concern is the translation of GP programs into a lower programming paradigm like Pregel. In fact, we target the possibility to write programs for distributed graphs in a transparent manner, that is, as if the graphs were not distributed. This approach is sound [1,13,43]. In fact, the idea of graph rewriting is a graph modification based on rules. Each rule application is a step of the graph transformation. When we consider the definition in Sect. 3.1, we can note that a rule application - mostly subgraphs adding or deleting - is about graph atomic modifications such as nodes and edges adding, relabeling and removing. These operations are natively present in platforms like Spark/GraphX, Giraph and other similar solutions.

4.2 Discussions

The content of the previous subsection is enough to prove the practicability of our proposal. Nevertheless, instead of atomic operations on graphs, it may be more convenient and interesting to have high-level design patterns or reusable building blocks. Naturally, this would prevent from having one superstep for each elementary operation. Otherwise, the execution would be computationally expensive.

However, this task is challenging. Besides the NP-completeness of subgraph isomorphism [17], its strict classical constraints make it unwieldy for graph pattern matching. Consequently, several relaxed models have appeared as they produce good results, in terms of execution time, with "Think Like a Vertex" model.

Unfortunately, our work is currently based only on the Ullmann's algorithm (subgraph isomorphism) proposed in [43], whose approach is recursive backtracking. The implementation leads to many messages and supersteps. As a result, we are presently investigating surveys on distributed graph pattern matching in big and distributed graphs [7], in order to efficiently implement GP constructs.

5   Related Work

So far, to the best of our knowledge, we found no work exactly aligning with our purpose. Our main goal is to build upon a graph language designed for non-distributed graphs, and to use it for distributed graphs. The main motivation is that distributed data platforms are still low-level. However, for instance, works have been proposed to implement graph transformation on distributed and large-scale graphs. Thus, the next steps of our work will take into account some of these proposals.

In [39], authors present a sub-graph matching on big graphs stored on a distributed memory. Their approach is limited on vertex-labeled but. In addition, they don't support conditions such as negative application, as well as transformations. The remark is similar for work on distributed graph pattern matching in [16,28]. Taentzer [40] proposed an approach for distributed graph rewriting, but it only concerns distributed graph and rewriting modeling, but not about physical distribution over a cluster. A refinement of CoqTL specification is presented in [30], which is intended to optimize a Spark-based execution of model transformations. In [42], authors propose an evaluation, unfortunately, only of structural recursion on graphs. Based on the use of a DSL (domain-specific language), they design a high-level framework for parallel programming, also translated into Pregel. In [25], authors introduce a distributed and parallel graph transformation by using the BSP model. Their modeling relies on typed directed graphs. Transformations are defined by considering procedural rewriting units and declarative rewriting rules, and their graphs are extended with stereotypes for edges and vertices.

6   Conclusion and Perspectives

In the present work-in-progress, we base ourselves on a graph-transformation-based language, named GP, and that was designed for graphs that are not distributed, to show its translation into a low-level programming model. The set of constructs of GP is minimal and can compute any program on graphs. In this way, it is sufficient to assure the soundness of the proposal. In fact, a primary benefit is the ability to transparently process distributed graphs as if they were stored in one single repository.

Nevertheless, this translation is challenging regarding distributed graph pattern matching, with "Think Like a Vertex" model. Consequently, our next work will focus on the efficiency of translating from graph transformation rules to Pregel programs, with a minimum number of supersteps and communications between vertices.

References

1. The Apache Software Foundation: Apache giraph. https://giraph.apache.org/
2. The Apache Software Foundation: Apache hadoop. https://hadoop.apache.org/
3. Ammar, K., Özsu, M.T.: Experimental analysis of distributed graph systems. PVLDB **11**(10), 1151–1164 (2018)
4. Aridhi, S., Lacomme, P., Ren, L., Vincent, B.: A MapReduce-based approach for shortest path problem in large-scale networks. Eng. Appl. Artif. Intell. **41** (2015)
5. BA, C., GUEYE, A.: On the distributed determinization of large NFAS. In: 2020 IEEE 14th International Conference on Application of Information and Communication Technologies (AICT), pp. 1–6 (2020)
6. Ba, C., Gueye, A.: A BSP based approach for NFAs intersection. In: Qiu, M. (ed.) Algorithms and Architectures for Parallel Processing-20th International Conference, ICA3PP 2020, New York City, NY, USA, October 2–4, 2020, Proceedings, Part I. LNCS, vol. 12452, pp. 344–354. Springer (2020)
7. Bouhenni, S., Yahiaoui, S., Nouali-Taboudjemat, N., Kheddouci, H.: A survey on distributed graph pattern matching in massive graphs. ACM Comput. Surv. **54**(2) (2021). https://doi.org/10.1145/3439724
8. Campbell, G., Courtehoute, B., Plump, D.: Fast rule-based graph programs. Sci. Comput. Program. **214**, 102727 (2022)
9. Cohen, J.: Graph twiddling in a MapReduce world. Comput. Sci. Eng. **11**(4), 29–41 (2009)
10. Dean, J., Ghemawat, S.: MapReduce: simplified data processing on large clusters. Commun. ACM **51**(1), 107–113 (2008)
11. Diop, A.M., Ba, C.: A distributed memory-based minimization of large-scale automata. In: Faye, Y., Gueye, A., Gueye, B., Diongue, D., Nguer, E.H.M., Ba, M. (eds.) Research in Computer Science and Its Applications, pp. 3–14. Springer, Cham (2021)
12. Diop, A.M., Ba, C.: On large automata processing: towards a high level distributed graph language. Int. J. Big Data Intell. **8**(2), 100–109 (2024)
13. Diop, A.M., BA, C.: On the soundness of a language for large and distributed graph processing. In: 2024 IEEE 18th International Conference on Application of Information and Communication Technologies (AICT), pp. 1–6 (2024). https://doi.org/10.1109/AICT61888.2024.10740415
14. Diop, L., Ba, C.: Parallelization of sequential pattern sampling. In: 2021 IEEE International Conference on Big Data (Big Data), Orlando, FL, USA, December 15–18 (2021)
15. Ehrig, H., Pfender, M., Schneider, H.J.: Graph-grammars: an algebraic approach. In: 14th Annual Symposium on Switching and Automata Theory (SWAT 1973), pp. 167–180 (1973)
16. Fard, A., Abdolrashidi, A., Ramaswamy, L., Miller, J.A.: Towards efficient query processing on massive time-evolving graphs. In: 8th International Conference on Collaborative Computing: Networking, Applications and Worksharing (CollaborateCom), pp. 567–574 (2012)
17. Garey, M.R., Johnson, D.S.: Computers and Intractability: A Guide to the Theory of NP-Completeness. W.H. Freeman, USA (1979)
18. Gonzalez, J.E., Low, Y., Gu, H., Bickson, D., Guestrin, C.: Powergraph: distributed graph-parallel computation on natural graphs. In: Thekkath, C., Vahdat, A. (eds.) 10th USENIX Symposium on Operating Systems Design and Implementation, OSDI 2012, Hollywood, CA, USA, October 8–10, pp. 17–30 (2012)

19. Gonzalez, J.E., Xin, R.S., Dave, A., Crankshaw, D., Franklin, M.J., Stoica, I.: GraphX: graph processing in a distributed dataflow framework. In: Flinn, J., Levy, H. (eds.) 11th USENIX Symposium on Operating Systems Design and Implementation, OSDI '14, Broomfield, CO, USA, October 6–8, pp. 599–613 (2014)
20. Grahne, G., Harrafi, S., Hedayati, I., Moallemi, A.: DFA minimization in mapreduce. In: Afrati, F.N., Sroka, J., Hidders, J. (eds.) Proceedings of the 3rd ACM SIGMOD Workshop on Algorithms and Systems for MapReduce and Beyond, BeyondMR@SIGMOD 2016, San Francisco, CA, USA, July 1, 2016, p. 4. ACM (2016)
21. Grahne, G., Harrafi, S., Moallemi, A., Onet, A.: Computing NFA intersections in map-reduce. In: Fischer, P.M., Alonso, G., Arenas, M., Geerts, F. (eds.) Proceedings of the Workshops of the EDBT/ICDT 2015 Joint Conference (EDBT/ICDT), Brussels, Belgium, March 27th, 2015. CEUR Workshop Proceedings, vol. 1330, pp. 42–45 (2015)
22. Habel, A., Plump, D.: Computational completeness of programming languages based on graph transformation. In: Honsell, F., Miculan, M. (eds.) Foundations of Software Science and Computation Structures, 4th International Conference, FOSSACS 2001 Held as Part of the Joint European Conferences on Theory and Practice of Software, ETAPS 2001 Genova, Italy, April 2–6, Proceedings (2001)
23. Han, M., Daudjee, K., Ammar, K., Özsu, M.T., Wang, X., Jin, T.: An experimental comparison of Pregel-like graph processing systems. PVLDB 7(12), 1047–1058 (2014)
24. Koch, J., Staudt, C.L., Vogel, M., Meyerhenke, H.: An empirical comparison of big graph frameworks in the context of network analysis. Soc. Netw. Anal. Mining 6(1), 84:1-84:20 (2016)
25. Krause, C., Tichy, M., Giese, H.: Implementing graph transformations in the bulk synchronous parallel model. In: Gnesi, S., Rensink, A. (eds.) Fundamental Approaches to Software Engineering, pp. 325–339. Springer, Berlin, Heidelberg (2014)
26. Lattanzi, S., Mirrokni, V.S.: Distributed graph algorithmics: theory and practice. In: WSDM, pp. 419–420 (2015). http://dl.acm.org/citation.cfm?id=2697043
27. Low, Y., Gonzalez, J., Kyrola, A., Bickson, D., Guestrin, C., Hellerstein, J.M.: Distributed GraphLab: a framework for machine learning in the cloud. PVLDB 5(8), 716–727 (2012)
28. Ma, S., Cao, Y., Huai, J., Wo, T.: Distributed graph pattern matching. In: Proceedings of the 21st International Conference on World Wide Web, pp. 949–958. WWW '12, Association for Computing Machinery, New York, NY, USA (2012). https://doi.org/10.1145/2187836.2187963
29. Malewicz, G., et al.: Pregel: a system for large-scale graph processing. In: Proceedings of the ACM SIGMOD International Conference on Management of Data, SIGMOD 2010, Indianapolis, Indiana, USA, June 6–10, pp. 135–146. ACM (2010)
30. Philippe, J., Tisi, M., Coullon, H., Sunyé, G.: Executing certified model transformations on apache spark. In: Proceedings of the 14th ACM SIGPLAN International Conference on Software Language Engineering, pp. 36–48. SLE 2021, Association for Computing Machinery, New York, NY, USA (2021). https://doi.org/10.1145/3486608.3486901
31. Plump, D.: The graph programming language GP. In: Bozapalidis, S., Rahonis, G. (eds.) CAI 2009. LNCS, vol. 5725, pp. 99–122. Springer, Heidelberg (2009). https://doi.org/10.1007/978-3-642-03564-7_6

32. Plump, D.: The design of GP 2. In: Escobar, S. (ed.) Proceedings 10th International Workshop on Reduction Strategies in Rewriting and Programming, WRS 2011, Novi Sad, Serbia, 29 May 2011. EPTCS, vol. 82, pp. 1–16 (2011). https://doi.org/10.4204/EPTCS.82.1

33. Plump, D., Steinert, S.: Towards graph programs for graph algorithms. In: Ehrig, H., Engels, G., Parisi-Presicce, F., Rozenberg, G. (eds.) ICGT 2004. LNCS, vol. 3256, pp. 128–143. Springer, Heidelberg (2004). https://doi.org/10.1007/978-3-540-30203-2_11

34. Plump, D., Steinert, S.: The semantics of graph programs. In: Mackie, I., Moreira, A.M. (eds.) Proceedings Tenth International Workshop on Rule-Based Programming, RULE 2009, Brasília, Brazil, 28th June 2009. EPTCS, vol. 21, pp. 27–38 (2009)

35. Poskitt, C.M., Plump, D.: Monadic second-order incorrectness logic for GP 2. J. Log. Algebraic Methods Program. **130**, 100825 (2023). https://doi.org/10.1016/J.JLAMP.2022.100825

36. Ravikumar, B., Xiong, X.: A parallel algorithm for minimization of finite automata. In: Proceedings of IPPS '96, The 10th International Parallel Processing Symposium, April 15-19, Honolulu, USA, pp. 187–191. IEEE Computer Society (1996)

37. Slavici, V.: Scaling up scientific computations by using map-reduce-like control flow on Numa architectures, Ph.D. thesis, USA (2013)

38. Srirama, S.N., Jakovits, P., Vainikko, E.: Adapting scientific computing problems to clouds using MapReduce. Future Gener. Comput. Syst. **28**(1), 184–192 (2012)

39. Sun, Z., Wang, H., Wang, H., Shao, B., Li, J.: Efficient subgraph matching on billion node graphs **5**(9), 788–799 (2012)

40. Taentzer, G.: Distributed graphs and graph transformation. Appl. Categorical Struct. **7**(4), 431–462 (1999). https://doi.org/10.1023/A:1008683005045

41. Tewari, A., Srivastava, U., Gupta, P.: A parallel DFA minimization algorithm. In: Sahni, S., Prasanna, V.K., Shukla, U. (eds.) High Performance Computing - HiPC 2002, 9th International Conference, Bangalore, India, December 18-21. LNCS, vol. 2552, pp. 34–40. Springer (2002)

42. Tung, L., Hu, Z.: Towards systematic parallelization of graph transformations over Pregel. Int. J. Parallel Prog. **45**(2), 320–339 (2017)

43. Ullmann, J.R.: An algorithm for subgraph isomorphism. J. ACM **23**(1), 31–42 (1976). https://doi.org/10.1145/321921.321925

44. Valiant, L.G.: A bridging model for parallel computation. Commun. ACM **33**(8), 103–111 (1990)

45. Wang, X., Qin, L., Chang, L., Zhang, Y., Wen, D., Lin, X.: Graph3s: a simple, speedy and scalable distributed graph processing system. arXiv preprint arXiv:2003.00680 (2020)

46. Yan, D., Bu, Y., Tian, Y., Deshpande, A., Cheng, J.: Big graph analytics systems. In: Özcan, F., Koutrika, G., Madden, S. (eds.) Proceedings of the 2016 International Conference on Management of Data, SIGMOD Conference 2016, San Francisco, CA, USA, June 26–July 01, 2016, pp. 2241–2243. ACM (2016)

47. Zaharia, M., et al.: Resilient distributed datasets: a fault-tolerant abstraction for in-memory cluster computing. In: Gribble, S.D., Katabi, D. (eds.) Proceedings of the 9th USENIX Symposium on Networked Systems Design and Implementation, NSDI 2012, San Jose, CA, USA, April 25–27, pp. 15–28 (2012)

K-GALS 2025: 4th Workshop on Knowledge Graphs Analysis on a Large Scale

LLM-Driven Summarization and Distinguish Analysis of Multiple Entities in RDF Graphs

Hamza Iqbal[ID] and Kostas Stefanidis[✉][ID]

Tampere University, Tampere, Finland
konstantinos.stefanidis@tuni.fi

Abstract. This research implements the application of Large Language Models (LLMs) in the summarization and distinguish analysis of multiple entities within Resource Description Framework (RDF) graphs. As the volume of structured data on the web is growing exponentially, the need for efficient and effective methods to interpret and summarize this data becomes increasingly important. This study focuses on utilizing LLMs to generate human-readable summaries from RDF graphs and particularly emphasizing on distinguishing between multiple entities. The study apply SPARQL queries to extract relevant data from DBpedia, subsequently a thorough process of frequency analysis and property unification to refine the dataset. Three LLMs including ChatGPT, DeepSeek, and Mistral have been evaluated for their ability to generate coherent and informative summaries. The evaluation process combines human-based assessments with automated metrics for the thorough analysis of generated texts. Key outcomes include the effectiveness of LLMs in generating summaries that are both informative and contextually relevant. The research also reflects the importance of data preprocessing techniques, such as frequency analysis and property unification in enhancing the quality of the generated summaries. Moreover, the study provides insights into the strengths and limitations of different LLMs in summarizing RDF data that offers a foundation for future research in this area. A framework for evaluating the performance of LLMs in summarization tasks has been designed in this research opens the way for future explorations in the application of advanced AI technologies in data interpretation and knowledge representation.

Keywords: Large Language Models · RDF · Summarization · Distinguish Analysis

1 Introduction

Problem Overview. The exponential growth of data on the Internet day by day through different sources and of various kinds can be very complicated to gather, especially in a limited period of time. If any individual needs information related

P. K. Chrysanthis et al. (Eds.): ADBIS 2025, CCIS 2676, pp. 367–383, 2026.
https://doi.org/10.1007/978-3-032-05727-3_31

to a person, place, event, or anything then it would be very difficult to study the massive content available on the Internet. This is why offering data summaries is a useful alternative [4,15–17]. In today's world, when most people have a very short concentration span, LLMs play also a key role in providing summarized information. It is quick, human-readable, and easily available to users around the world. Having said this, there exist many limitations of generative AI models as well. The most common issue in generative AI models like ChatGPT is hallucination, it may produce information that is not being asked by the user or is not relevant. LLMs generate text descriptions based on vast amounts of data from different sources, which is why they lack external knowledge, leads to errors, especially when the input is ambiguous or the topic is outside the model's training scope. Collecting data from different sources can be complicated and time-consuming; hence, the need for automated summarization of a siingle or multiple entities arises. While the comparison between entities can be an important ask, such as four states of Southern America (Alabama, Florida, Arkansas, Delaware) to see the commonalities and distinguishing features of multiple states, here comes the need of Linguistic Summarization and distinguishing analysis of multiple entities.

Background. The motivation behind this study is to fill the previous gaps in this research and the extension of the research done previously. The authors have done research on the linguistic summarization of multiple entities through coding strategy [19]. This study has automated the distinguishing analysis alongside the summarization of multiple entities using Large Language Models such as Chatgpt-3.5, Deepseek, and Mistral explicitly. A reasonable amount of research has been done in natural language generation, based on structured data such as explainability, education and data augmentation [14]. However, LLM's potential hasn't been utilized doing comparisons between multiple entities and distinguishing analysis which has been explored and covered in this research. Due to increasing interest in Natural Language Generation in recent times, the interest in using LLMs with NL Generation also increased, how can we utilize them to acquire knowledge that exists all over the Semantic Web (SW)?

Recent advancements in query-focused summarization using GraphRAG have further informed this work. For example, a two-stage GraphRAG approach has been proposed—first generating an entity knowledge graph and community summaries, then combining them into a global summary which shows improved comprehensiveness and diversity over standard RAG methods [3]. Other studies have extended this idea, such as FG-RAG, which enhances context awareness and fine-grained summarization from graph retrieval [5]. These recent works emphasize the effectiveness of GraphRAG architectures for structured summarization tasks and reinforce the relevance of applying graph-based prompting in our study.

RDF and DBpedia as a Data Source. RDF, termed as Resource Description Framework, is a dataset model standard defined by the W3C–World Wide Web Consortium. It is a structured form of data consisting of subject-predicate-object triples, facilitating flexible knowledge representation. However, the volume and complexity of these structured datasets make them challenging for non-expert users to interpret. For example, "Albert Einstein was born in Germany." will be structured as `<Albert_Einstein>` as a subject, `<bornIn>` as a predicate, and `<Germany>` as an object. An RDF graph is a collection of RDF triples, where the RDF triples are edges, directed from the subject to the object [19]. In recent years, the interest in structured data has increased to utilize it in different forms. The dataset used in this study is from DBpedia expanding rapidly, and as of the 2016–04 release, it included detailed descriptions of 6.0 million entities. Among these, 5.2 million entities were systematically categorized within a coherent ontology. This classification includes 1.5 million individuals, 810,000 geographical locations, 135,000 music albums, 106,000 films, 20,000 video games, 275,000 organizations, 301,000 biological species, and 5,000 diseases.

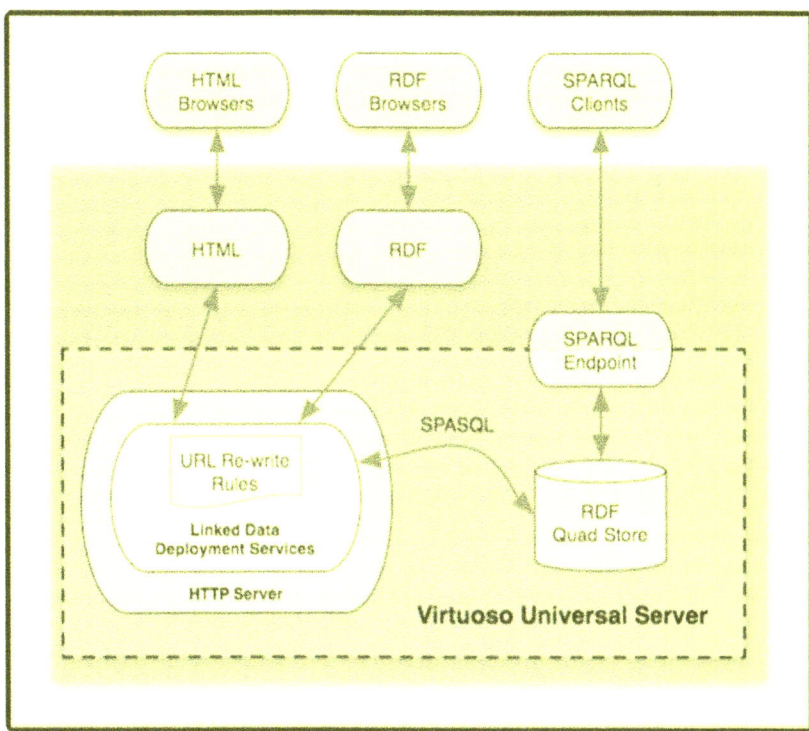

Fig. 1. Illustration of current DBpedia Data Provision Architecture from its Official Website.

SPARQL Queries – A Standard Way. SPARQL (Protocol and RDF Query Language) is the standard way to retrieve data from RDF graphs. Generally, users are required to formulate the queries in order to get information and it can be difficult for a person who is unfamiliar with graph databases. We designed an automated system in which we can provide multiple entities names to the system, it accesses the RDF Graph through SPARQL query, clean the data and encapsulates in the Excel file which is called Frequency Analysis of the respective entities. These queries enables me to retreive data and make it more user friendly and accessible. SPARQL is the language which machine can understand and different URIs are involved in the whole query.

Figure 1 demonstrates that how the data is being delivered through SPARQL queries, RDF browsers and HTML browsers. In our case, SPARQL queries have been made to the SPARQL endpoint of RDF graph. The SPARQL Endpoint in an RDF-based system acts as an interface where queries are executed against an RDF store, returning results in formats such as XML, JSON, or tabular structures. While SPARQL provides flexibility in accessing linked data, its complexity poses a challenge for non-expert users, necessitating automated summarization techniques to improve readability and usability.

The rest of the paper is structured as follows. Section 2 presents the related work and Sect. 3 our methodology, including the data preparation, the prompt design and text generation and the setup of the evaluation. Section 4 discusses the experimental evaluation and analyzes the results. Finally, Sect. 5 concludes the paper.

2   Related Work

The field of querying RDF data and knowledge graphs has seen significant advancements, driven by the need to make complex data more accessible to non-expert users. Summarizing RDF graphs in natural language plays a vital role in helping users understand large and complex datasets.

An automatic natural language summarization system for multiple entities in RDF graphs has been introduced. The system extracts property-object pairs related to entities and converts the structured knowledge into natural language. Practical experiments using human-based evaluations compared summaries generated by machines with those written by humans, both independently and with assistance. The results showed that machine-made summaries are particularly useful under tight time constraints and for difficult topics [19].

The efficacy of text-to-text pre-training for data-to-text tasks has been explored, demonstrating significant performance enhancements for transformer-based models. The study leverages the "Text-to-Text Transfer Transformer" (T5), pre-training models on diverse tasks like translation, summarization, and question answering. Structured data is converted into a linearized text format for input into T5 models, allowing for a straightforward, end-to-end generation process. Results indicate that T5 pre-training achieves state-of-the-art performance across multiple benchmarks, with superior generalization on out-of-domain datasets [6].

The main aim of a recent study is to verbalize SPARQL queries into natural language for better end-user understanding. The authors convert SPARQL queries into natural language questions using LLMs, designing prompts with human-readable labels from knowledge graphs to ensure context understanding. They fine-tune LLMs and evaluate the produced questions using BERTScore, demonstrating that LLMs can convert complex queries into simple, human-readable texts. However, the framework's dependence on knowledge graph labels and the limited evaluation scope are noted as challenges [14].

Earlier works on converting SPARQL queries to natural language relied on grammar rules and smaller language models. The SPARQL2NL approach standardizes queries and refines them through simplification and substitution principles. The LD2NL approach verbalizes OWL and RDF vocabularies into natural language through lexicalization, single triples realization, clustering, ordering, and grouping. These frameworks generate sentences or summaries from resources, rules, or queries, and have been evaluated through human ratings for adequacy, fluency, and completeness [12, 13].

More recent approaches have leveraged encoder-decoder architectures for SPARQL-to-text conversion. The NABU approach uses an encoder inspired by Graph Attention Networks (GANs) and a Transformer decoder, showing promising results in multiple languages. However, it still faces challenges with complex queries and conversational contexts [11].

The complexity of formulating SPARQL queries has led to the development of assistive tools using data graph summaries. These summaries recommend structural query elements during interactive query formulation and are represented as RDF graphs. The use of LLMs has further revolutionized query generation, enabling the creation of SPARQL queries from natural language prompts combined with ontologies [1, 2].

Evaluating the quality of generated text is crucial in natural language processing tasks. Traditional metrics like BLEU and METEOR often fail to capture semantic equivalence, leading to the development of BERTScore. BERTScore utilizes pre-trained BERT embeddings to compute similarity between generated and reference texts, providing a more semantic evaluation. This metric has shown high correlation with human judgments in tasks such as machine translation and image captioning [18].

The introduction of BART, a denoising autoencoder, has expanded pre-training methods for natural language generation. BART's pretraining involves corrupting text with arbitrary noise and training the model to reconstruct the original text. This approach allows BART to generalize across different tasks, including text generation, summarization, and machine translation. However, BART occasionally struggles with hallucinating unsupported information, indicating a limitation in maintaining factual accuracy [10].

The generation of natural language questions from SPARQL queries is significant for tutoring systems and conversational agents. This study focuses on conversational contexts and the challenges of generating coherent and contextually relevant questions. The authors utilized four knowledge-based QA corpora and introduced a new challenge set with unseen query types and domains. The study compared different fine-tuning approaches for BART and T5 models, using various input features and training data. The results showed that while simple questions were well-handled, complex queries and conversational dimensions remained challenging [9].

Semantic graph-based approaches have emerged as a powerful method for abstractive text summarization. These approaches leverage semantic graph representations, such as Abstract Meaning Representation (AMR) graphs, to capture the semantic essence of the text. By combining these graphs with advanced deep learning models, researchers generate more meaningful and coherent summaries. Experiments on datasets like Gigaword and CNN/DailyMail have shown that these methods outperform traditional summarization techniques, particularly in maintaining factual accuracy and generating coherent summaries [8]. Summarization techniques serve various purposes, including indexing for efficient querying, source selection, visualization, and schema discovery. The quality of a summary is evaluated based on its coverage, precision, recall, and computational complexity. Ensuring that the summary accurately represents the original data while remaining computationally efficient is a key challenge in the field [7].

3   Methodology

3.1   RDF Data Preparation

To generate structured and meaningful summaries from RDF-based knowledge graphs, it is crucial to extract, filter, and refine the data before summarization. In this research, data was being retrieved through SPARQL queries to the RDF knowledge graph from DBpedia. Then performed frequency analysis to identify relevant properties and unify synonymous properties to standardize property names. Extracted raw data was very non-informational, with duplicate entries and properties that wouldn't make any sense to retain. These preprocessing steps ensure that the extracted information is concise, relevant, and structured for effective summarization by Large Language Models (LLMs).

SPARQL-Based Retrieval. DBpedia is a structured knowledge base that represents information in RDF triple format (subject-predicate-object). SPARQL is a query language that is used to query entity-specific property-value pairs from knowledge graphs. Each query was formulated to:

- Extract all available predicates (properties) and corresponding values for a given entity.
- Filter values to include only English-language literals or URIs while excluding non-linguistic literals (e.g., numerical codes).
- Retrieve multiple entities simultaneously by iterating over a list of entity URIs.

Frequency Analysis and Property Unification. The RDF data extracted from DBpedia contains a large number of properties many of which are redundant, irrelevant, or non-informative for summarization. To proceed with the goal, a systematic filtering and frequency analysis approach was applied to refine the data.

Filtering Irrelevant Properties: Several properties in DBpedia store metadata, links, or other non-descriptive information that do not contribute to human-readable summaries. These properties were excluded using predefined filtering rules. Some examples of excluded properties are as follows:

- **Metadata properties** (e.g., `wikiPageID`, `wikiPageRevisionID`, `prov#was-DerivedFrom`) were discarded as they serve administrative purposes rather than conveying semantic information.
- **External links** (e.g., `wikiPageExternalLink`, `owl#sameAs`, `url`) were excluded to avoid diverting focus from factual content.
- **Non-descriptive properties** (e.g., `image`, `fontcolor`, `logosize`, `width`) were removed due to their inability to contribute meaningfully to textual summaries.
- Additionally, all properties with URL values (e.g., http://example.com/resource) were filtered to eliminate redundant hyperlinks.

Frequency-Based Property Ranking: Post-filtering, the remaining properties were analyzed to quantify their occurrence across entities. For example:

- **High-frequency properties**, such as `dbo:birthDate` and `dbo:populationTotal`, were prioritized for critical information. Properties like these that occurs in higher frequency are ranked in descending order, so that LLM can give much weightage to that kind of information with higher frequency.
- **Lower-frequency properties** (e.g., `dbo:nickname`) were retained only if they added unique contextual value. For the entity type 'Person', this information is important and can be critical in a summary when comparing different entities. It is only retained when the field is not empty; otherwise, there is no need to add it.

Handling Numeric and Textual Data

- **Numeric Properties:** Attributes like `dbo:areaTotal` and `dbo:elevation-MaxFt` were aggregated to retain salient quantitative facts.
- **Textual Properties:** Descriptive fields such as `dbo:description` were merged while avoiding redundancy.

Table 1 elaborates on the steps taken to complete the frequency analysis section. It shows example inputs of how occurrences were counted, the basis on which synonyms were handled, how garbage properties were filtered, and how properties were ranked. Examples of the output are then shown. This entire mechanism has been automated through a Python method.

Table 1. Steps in RDF Property Preprocessing and Their Outcomes

Step	Example Input	Action Taken	Output/Result
1. Count Occurrences	dbo:birthDate (9,500), dbo:height (2,100)	Identify high vs. low frequency properties	dbo:birthDate selected as key info
2. Handle Synonyms	dbo:birthDate, dbp:dateOfBirth	Unified into one property: Birth Date	Reduces redundancy
3. Filter Garbage Properties	wikiPageID, rdf:type, owl:sameAs	Removed from dataset due to low human readability	Cleaner, concise summaries
4. Rank Properties	dbo:birthPlace = High, dbo:award = Medium, dbo:height = Low	Ranked by contextual importance	Priority for summarization set

3.2 Prompt Design and Text Generation

Models such as ChatGPT, DeepSeek, and Mistral have demonstrated their ability to process and create human-readable summaries from structured data. This study utilizes LLMs to generate summaries and perform distinguish analysis of multiple entities retrieved from RDF graphs. They need an input as a prompt, a structured prompt with RDF filtered data in our case, and they respond to that prompt accordingly, well-organized summaries in our case.

LLM Prompt Structure. To make sure the summaries are coherent and informative enough, a structured prompt was composed in a text file through an automated Python function, combined with preprocessed frequency analysis data. That file was provided to the models for our required summaries. The best possible customized prompt has been formulated after thorough testing in generating concise and human-readable summaries while keeping factual accuracy. The structured prompt includes the following things:

- Entity names to make sure that the generated summary remains entity-specific.
- Key properties, values, and their frequencies to help the model understand the importance of such properties so that it can prioritize those details.
- A well-defined instruction set guiding an LLM to remain focused on key characteristics, differences, and unique attributes of respective entities.

Figure 2 demonstrates the example of the prompt design that how this whole structure has been compiled through an automated agent(method) and then further provide it to an LLM for the exact context we need our summary for.

```
Create a concise, human-readable summary comparing the following entities.
Focus on their key characteristics, differences, and notable commonalities,
based on the provided data. The text should resemble a short encyclopedia entry,
highlighting their unique features and important similarities.
Exclude any speculative or unverifiable facts.
Provide the summary in a well-structured and clear format.

Entity: Alabama
population: 5039877 (Frequency: 3)
nickname: The Yellowhammer State (Frequency: 2)
areaTotal: 135765 km² (Frequency: 5)
...

Entity: Florida
population: 21538187 (Frequency: 3)
nickname: The Sunshine State (Frequency: 2)
areaTotal: 170312 km² (Frequency: 5)
...
```

Fig. 2. Prompt Structure ensuring LLM generate summaries grounded in factual data.

Summarization and Comparison Tasks

Summarization Process: Once the structured prompts were prepared, the LLMs (ChatGPT, DeepSeek, and Mistral) were used to generate summaries and distinguish analyses. In the summarization process, a structured prompt was provided as input to each large language model (LLM). The models processed the RDF data and generated human-readable summaries. The output from each model was then collected and stored for further evaluation.

Distinguish Analysis: For distinguishing analysis, the LLMs were instructed to highlight key differences between entities based on structured data. The generated text was expected to emphasize unique attributes while maintaining coherence. The results of this analysis were stored to facilitate comparisons between different models.

3.3 Evaluation Setup

For the evaluation process of the outcomes from different LLMs, both automated metrics and human-based assessments have been implemented. Apart from text evaluations, all three models, ChatGPT, DeepSeek, and Mistral have been compared to test state of the art models. Recently, these generative models opens the debate in several aspects that how they can be utilized to solve our real-world problems, which is one of the motivations to work on this particular problem.

Human Evaluation Protocol. When it comes to longer texts, automated metrics cannot provide accurate or practical results, they have some limitations. The need for human-based evaluations arises because we can rely on a person who understands the text and has the knowledge to assess longer texts like summaries, in different aspects. While automated metrics such as ROUGE scores, BLEU scores, and entity-based metrics have been implemented to assess summaries in terms of content coverage, linguistic quality, and factual accuracy but none of them have shown sufficiently optimal results to rely upon them entirely. BERT has been implemented for the shorter text evaluations as demonstrated in [18], where authors evaluated questions generated by large language models (LLMs). Particularly, the BERT approach requires two texts to evaluate: 1) the candidate text, which is being evaluated, and 2) the reference text, which is the reference text to compare with. This requirement makes it challenging for summaries, relatively longer texts generated by LLMs. In such scenarios, human intervention is required to make reference summaries with a lot of effort. In a recent study on Linguistic summarization, the authors preferred to adopt human-based experiments for the assessments [19]. In this research, a structured evaluation framework has been developed to comprehensively evaluate the quality of summaries generated through different LLMs, particularly in the context of longer texts. The evaluation process involved generating three separate documents, each focusing on a specific entity type: Person, Place, and Event. For each entity type, three examples were selected for detailed analysis in the summaries. Consequently, three summaries were generated for each example, resulting in a total of nine summaries across the three documents. These documents contained summaries and comparisons of multiple entities within their respective entity types. For each document, three summaries generated by different models were presented and evaluators were asked to rate the summaries according to several criteria, including accuracy, completeness, redundancy, relevance, and distinguishability of the content. Original data was provided in the form of property-value pairs in a separate Excel sheet, serving as a reference for evaluations. 25 evaluators have been selected from different field backgrounds and collected diverse feedback on the performance of the models. To ensure unbiased evaluations, the models were anonymized. The assessors were then tasked with ranking the summaries based on overall quality. The results of these evaluations were subsequently analyzed and discussed in detail in Chap. 4 of the study. This structured human-based evaluation approach highlights the challenges of relying solely on automated metrics to assess longer texts and underscores the importance of human judgment in ensuring the quality and reliability of LLM summaries.

Automated Metric Computation. To complement the human evaluations, some automated metrics were also used to test the quality of text LLMs are generating. Three metrics were aimed at providing an objective comparison of different models based on linguistic quality, readability, and factual accuracy of the generated text.

- **Grammar Score:** To check the grammatical correctness of each summary, LanguageTool has been used, which is an automated grammar checker system. This score was computed by inspecting the number of grammatical errors per word. A high Grammar Score indicates fewer errors and better fluency.
- **Readability Score:** To analyze the text in terms of readability, the Flesch Reading Ease metric was implemented to assess the complexity and clarity of the generated summaries. This metric assigns a score based on sentence structure and word complexity. The higher the score, the more readable the text is for a lay user.
- **Entity Retention Score:** To test the factuality of the text and preservation of original details, a fuzzy matching approach was implemented. Entities extracted from the RDF dataset were compared against the summaries generated using the `fuzz.partial_ratio` function from the `fuzzywuzzy` library. This score reflects how well the original details were retained by the model, offering insights into the factual consistency of the summaries.

4   Results and Evaluation

4.1   Summary Generation for RDF Entities

Various RDF entities, including places, persons, and events, have been covered in the summarization process. Using frequency analysis (with purified and cleaned data), the most common and informative properties have been identified for each entity present in the data. Then, a refined, structured prompt was created in a text file to provide different LLMs for generating summaries in natural language.

The following Fig. 3 is one of examples of the generated summaries for multiple entities of **Entity Type–Place**:

4.2   Summarization Strategy Comparison

Frequency-Based Selection. By using the most frequently occurring properties across the entities to construct a summary that highlights the importance of that property.

LLM-Based Summarization. Best prompts have been evolved during the text generation. Prompting different AI models efficiently to generate required summaries based on the data retreived from RDF graph.

Hybrid Approach. Finally, the approach of combining the frequency based selections and making best prompts has been utilized to enhance the quality overall.

Three approaches were utilised in order to test the content generated by LLMs, first through just focusing on the most common properties exist for an entity, then just prompting LLM with an instruction set to reduce the hallucinations and then finally combining both to achieve the most realistic and human-like fluent summaries that is called Hybrid Approach as mentioned in Table 2

Entities Type: Place

Comparison of Alabama, Florida, and Delaware

Alabama

Alabama, known as *The Yellowhammer State*, ranks 30th in total area and 24th in population in the U.S. It covers **135,765 km²**, with 3.2% of its area consisting of water. It has a relatively low elevation, averaging **150 meters**, with its highest point at **735.5 meters (Mount Cheaha)**. In 2010, Alabama's population was approximately **5 million**, with a density of **38.4 people/km²**. The state was the **22nd** to join the Union and is represented by **6 congressional representatives**. Its official motto is *"We Dare Defend Our Rights"*.

Florida

Florida, the *Sunshine State*, is significantly larger than Alabama, ranking **22nd in area (170,312 km²)** and **3rd in population**, with over **21.5 million people** in 2010. It has a high population density of **155 people/km²**. Florida's geography is distinct, with **18.5% of its area covered by water**, making it a state with a strong coastal presence. The state's elevation is much lower than Alabama's, with an average of **30 meters** and a highest point at **105 meters**. Florida joined the U.S. as the **27th state** and has **16 congressional representatives**. Its motto is *"In God We Trust"*.

Delaware

Delaware, the *First State*, is the smallest of the three, ranking **49th in area (5,133 km²)** and **45th in population** with just over **1 million residents in 2010**. Despite its small size, Delaware has a high population density of **190 people/km²**, surpassing both Alabama and Florida. **21.7% of its area is water**, making it more water-dominant than Alabama but less than Florida. It has the lowest elevation, averaging **20 meters**, with its highest point reaching **136.5 meters**. Delaware was the **first state to join the Union** and has the motto *"Liberty and Independence"*.

Fig. 3. Demonstration of generated text for multiple entities.

4.3 Distinguish Analysis of Entities

Distinguishing analysis becomes essential when the user needs to compare multiple entities of the same entity type to see their differences or uniqueness. It focuses on identifying key characteristics that set one entity apart from another. We analyzed entity descriptions to determine distinctive features based on RDF data by applying LLMs. For example, when we compare Alabama, Florida and Delaware the key differences includes historical events, notable figures and economic focus. Example of Distinguish Analysis can be seen in Fig. 4.

4.4 Evaluation Outcomes

Human-Based Evaluation. A total of 25 evaluators have been determined to complete the assessment for the generated summaries from diverse and distinct backgrounds to avoid biased results. Also, the model names were being anonymised to prevent the inclination of someone towards a particular model. The evaluators rated the summaries on various factors and ranked them based on their overall quality in the end. Table.3 summarises the distribution of rankings for each model based on evaluator preferences:

Table 2. Overview of Summarization Strategies

Strategy	Focus	Strength	Weakness
Frequency-Based	Most common properties	High data accuracy	Less fluent, rigid structure
LLM-Based	Natural language generation	Human-like fluent summaries	Potential factual inconsistencies
Hybrid Approach	Combines frequency analysis with refined prompting	Balanced accuracy and readability	Slightly complex prompt design

---

Key Differences and Similarities

- **Size & Population:** Florida is the largest and most populous, Alabama is mid-sized, and Delaware is the smallest.

- **Water Coverage:** Florida has the highest percentage of water area (18.5%), followed by Delaware (21.7%), and Alabama (3.2%).

- **Elevation:** Alabama has the highest elevation (Mount Cheaha, 735.5m), while Florida and Delaware are relatively flat.

- **Statehood:** Delaware was the first to join the U.S., while Alabama (22nd) and Florida (27th) followed later.

- **Representation:** Florida has the most congressional representatives (16), followed by Alabama (6), and Delaware (1).

Each state has distinct characteristics, with Alabama known for its landmass and historical significance, Florida for its large population and coastal geography, and Delaware for its compact size and high density.

Fig. 4. Demonstration of Distinguish Analysis for multiple entities.

The formula in Eq. 1 has been used to compute the overall ranking points for each model given by the evaluators to see the general pattern in people's preferences:

$$\text{Score} = 3 \times (\text{1st place count}) + 2 \times (\text{2nd place count}) + 1 \times (\text{3rd place count}) \quad (1)$$

- **Mistral** was the most successful model in terms of human metrics, securing 1st place in 12 evaluations. It also had the fewest 3rd place rankings, which indicates its consistent performance.
- **Deepseek** demonstrated balanced performance with a significant number of 2nd place rankings, suggesting it is a strong contender but not the top choice among the models.
- **ChatGPT** received the highest number of 3rd place rankings, making it the least preferred model by evaluators in terms of hallucinations, regardless of the prompt engineering. This clearly indicates that ChatGPT is frequently overlooked as the best model for summarization tasks.

Table 3. Human Ranking of Models by Placement Positions

Model	1st Place	2nd Place	3rd Place
Mistral (M)	12	5	8
Deepseek (D)	9	10	6
ChatGPT 3.5 (C)	6	10	9

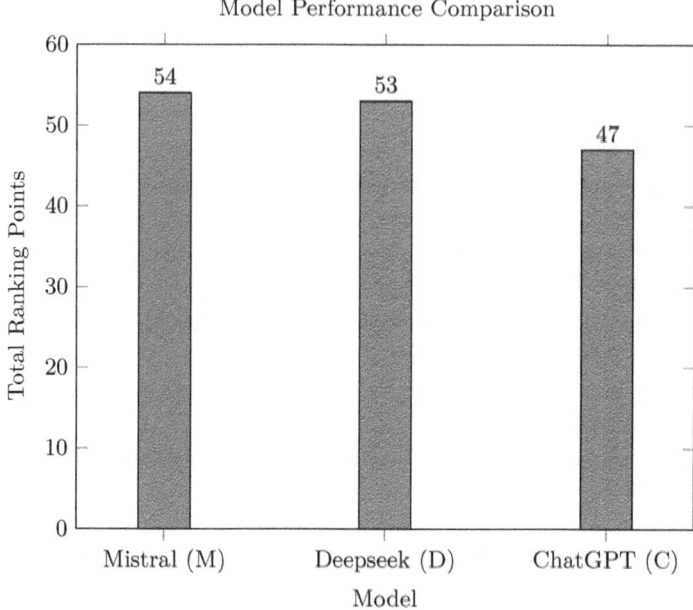

Fig. 5. Comparative evaluation of LLM performance using human assessment metrics.

These evaluation results in Fig. 5, combined with visualizations of the rankings, provide a comprehensive overview of the human-evaluated model performance as perceived by the evaluators.

Automated Evaluation. Text evaluation is a very crucial part in the projects based on NLP, therefore, to complement the human-based evaluations, three automated metrics have also been utilized. The automated metrics employed in this study provided a quantitative assessment of the summaries generated by different models.

The results obtained by automated metrics in Fig. 6 provide us with a comprehensive evaluation of the summaries generated by the LLMs. All the models produced text with high grammar scores that indicate the potential of the Large Language Generative Models in producing high-quality content, which is coherent and grammatically correct. Although variations have been observed in the

Combined Performance Scores

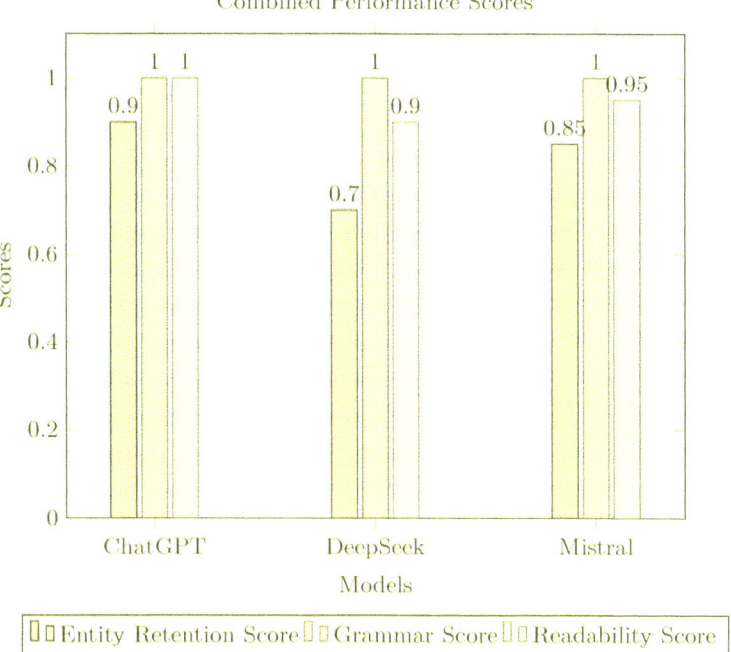

Fig. 6. Comparison of Entity Retention, Grammar, and Readability Scores Across Models

readability and Entity Retention Scores, which suggests the strengths and weaknesses of each model. ChatGPT outplayed Mistral and Deepseek in terms of Readability and Entity Retention highlights its credibility in generating summaries that are both accessible and factually accurate. It shows that RDF data is being retained correctly in the summaries, which were provided within the prompt to summarize, and a user cannot miss facts and figures, showing its highlighting power for a piece of important information. Mistral performed consistently across all the metrics, showing its reliability in generating high-quality summaries. While Deepseek has room for improvement in some aspects, such as entity retention recommends the further refinement could enhance its factual accuracy.

5   Conclusion

In this paper, the potential of Large Language Models (LLMs) in generating summaries and distinguishing multiple entities of knowledge graphs is being tested thoroughly. This study concludes that LLMs, when combined with the refined and cleaned data with thorough data preprocessing techniques such as frequency analysis and property unification, can generate reasonable and informative summaries from complex RDF datasets. Three of the most credible LLMs, including ChatGPT, DeepSeek, and Mistral, were utilized to get the job done and further evaluated for their outputs. The evaluation process encompasses both human-based assessments and automated metrics to test the text quality in various aspects. In our human evaluations, Deepseek performed as the best model, followed by Mistral, and ChatGPT was considered as a last choice in terms of knowledge graph summarization. Moreover, automated metrics indicate that ChatGPT outperformed the other models in specific domains like Entity Retention and Readability Score. However, the Grammer Score remained almost equal for all the models at up to 99%.

References

1. Antoniou, C., Bassiliades, N.: Utilizing llms and ontologies to query educational knowledge graphs
2. Campinas, S., Perry, T.E., Ceccarelli, D., Delbru, R., Tummarello, G.: Introducing rdf graph summary with application to assisted sparql formulation. In: 2012 23rd International Workshop on Database and Expert Systems Applications, pp. 261–266. IEEE (2012)
3. Edge, D., et al.: From local to global: a graph rag approach to query-focused summarization (2025), https://arxiv.org/abs/2404.16130
4. Gkorgkas, O., Stefanidis, K., Nørvåg, K.: A framework for grouping and summarizing keyword search results. In: Catania, B., Guerrini, G., Pokorný, J. (eds.) ADBIS 2013. LNCS, vol. 8133, pp. 246–259. Springer, Heidelberg (2013). https://doi.org/10.1007/978-3-642-40683-6_19
5. Hong, Y., Li, C., Zhang, J., Shao, Y.: Fg-rag: enhancing query-focused summarization with context-aware fine-grained graph rag (2025), https://arxiv.org/abs/2504.07103
6. Kale, M., Rastogi, A.: Text-to-text pre-training for data-to-text tasks. arXiv preprint arXiv:2005.10433 (2020)
7. Kondylakis, H., Kotzinos, D., Manolescu, I.: Rdf graph summarization: principles, techniques and applications (tutorial). In: EDBT/ICDT 2019-22nd International Conference on Extending Database Technology-Joint Conference (2019)
8. Kouris, P., Alexandridis, G., Stafylopatis, A.: Text summarization based on semantic graphs: an abstract meaning representation graph-to-text deep learning approach. J. Big Data 11(1), 95 (2024)
9. Lecorvé, G., Veyret, M., Brabant, Q., Barahona, L.M.R.: Sparql-to-text question generation for knowledge-based conversational applications. In: Proceedings of the 2nd Conference of the Asia-Pacific Chapter of the Association for Computational Linguistics and the 12th International Joint Conference on Natural Language Processing (Volume 1: Long Papers), pp. 131–147 (2022)
10. Lewis, M., et al.: Bart: denoising sequence-to-sequence pre-training for natural language generation, translation, and comprehension. arXiv preprint arXiv:1910.13461 (2019)
11. Moussallem, D., Gnaneshwar, D., Castro Ferreira, T., Ngonga Ngomo, A.C.: Nabu–multilingual graph-based neural rdf verbalizer. In: International Semantic Web Conference, pp. 420–437. Springer (2020)
12. Ngomo, A.N., Moussallem, D., Bühmann, L.: A holistic natural language generation framework for the semantic web. CoRR abs/1911.01248 (2019), http://arxiv.org/abs/1911.01248

13. Ngonga Ngomo, A.C., Bühmann, L., Unger, C., Lehmann, J., Gerber, D.: Sorry, i don't speak sparql: translating sparql queries into natural language. In: Proceedings of the 22nd International Conference on World Wide Web, pp. 977–988 (2013)
14. Perevalov, A., Both, A.: Towards llm-driven natural language generation based on sparql queries and rdf knowledge graphs (2024)
15. Troullinou, G., Kondylakis, H., Stefanidis, K., Plexousakis, D.: Exploring RDFS KBs using summaries. In: Vrandečić, D., et al. (eds.) ISWC 2018. LNCS, vol. 11136, pp. 268–284. Springer, Cham (2018). https://doi.org/10.1007/978-3-030-00671-6_16
16. Troullinou, G., Kondylakis, H., Stefanidis, K., Plexousakis, D.: Rdfdigest+: a summary-driven system for kbs exploration. In: Proceedings of the ISWC 2018 Posters & Demonstrations, Industry and Blue Sky Ideas Tracks co-located with 17th International Semantic Web Conference (ISWC 2018), CEUR Workshop Proceedings, vol. 2180. CEUR-WS.org (2018)
17. Vassiliou, G., Troullinou, G., Papadakis, N., Stefanidis, K., Pitoura, E., Kondylakis, H.: Coverage-based summaries for RDF KBs. In: Verborgh, R., et al. (eds.) ESWC 2021. LNCS, vol. 12739, pp. 98–102. Springer, Cham (2021). https://doi.org/10.1007/978-3-030-80418-3_18
18. Zhang, T., Kishore, V., Wu, F., Weinberger, K.Q., Artzi, Y.: Bertscore: evaluating text generation with bert. arXiv preprint arXiv:1904.09675 (2019)
19. Zimina, E., Järvelin, K., Peltonen, J., Ranta, A., Stefanidis, K., Nummenmaa, J.: Linguistic summarisation of multiple entities in rdf graphs. Appl. Comput. Intell. 4(1), 1–18 (2024)

Comparing Community Structures in Knowledge Graphs Across Similarity Measures and Clustering Algorithms

Cosimo Poccianti[1,2]([✉]) [iD] and Blerina Sinaimeri[2] [iD]

[1] Department of Computer, Control, and Management Engineering Antonio Ruberti,
Sapienza University of Rome, Rome, Italy
cosimo.poccianti@uniroma1.it
[2] Department of AI, Data and Decision Sciences, Luiss - Libera Università
Internazionale degli Studi Sociali Guido Carli, Rome, Italy
bsinaimeri@luiss.it

Abstract. Community detection in biomedical knowledge graphs (KGs) holds promise for uncovering functional groupings of drugs, yet its evaluation remains challenging due to the heterogeneity of node types and edge semantics. In this study, we investigate how different definitions of drug similarity—structural, topological, and semantic—influence the community structure of drug-centric graphs and their biological interpretability. We compare clustering outcomes obtained from both homogeneous drug–drug similarity networks and a heterogeneous KG integrating drugs, proteins, and side effects. Using methods such as Louvain, we analyze the extent to which detected communities align with known drug interactions, particularly those associated with adverse drug reactions. Evaluation is performed through structural metrics (e.g., modularity), pairwise agreement measures (ARI, NMI), and functional coherence, quantified by the intra- vs. inter-community distribution of side-effect-inducing drug pairs. Our results offer a systematic assessment of community structures in KGs and provide insights into the utility and limitations of unsupervised clustering in pharmacological network analysis.

Keywords: knowledge graph · community detection · drug similarity measures

1 Introduction

Adverse drug reactions (ADRs) pose a persistent challenge to modern healthcare systems, affecting not only patient safety and compliance but also the economic sustainability of pharmacological treatments [8,9]. While clinical trials aim to capture the safety profile of drugs, their limited scope often fails to uncover rare ADRs or ADRs that appear later in time. As a result, surveillance and computational approaches play a key role in extending our understanding of drug effects. Among these, in silico analyses based on publicly available biomedical

© The Author(s), under exclusive license to Springer Nature Switzerland AG 2026
P. K. Chrysanthis et al. (Eds.): ADBIS 2025, CCIS 2676, pp. 384–397, 2026.
https://doi.org/10.1007/978-3-032-05727-3_32

data are becoming increasingly central to the early detection of drug behavior patterns [2,17]. In this context, knowledge graphs (KGs) offer a powerful formalism for representing complex biomedical relationships. By connecting drugs, target proteins, and side effects through semantically different edges, KGs allow us to move beyond flat pairwise comparisons and reason over rich relational structures. While knowledge graphs (KGs) are widely used for link prediction tasks, such as identifying missing associations between drugs and adverse reactions, understanding their *mesoscopic structure*, and in particular the community organization of nodes, remains a challenging problem. In the biomedical domain, the definition and evaluation of communities are further complicated by the heterogeneity of node and edge types, as well as the need for biological interpretability.

A key factor that influences community detection outcomes is the underlying definition of similarity between drugs. Similarity can be grounded in chemical structure, molecular topology, or semantic context, and each of these perspectives may induce different network structures and clusterings. In this work, we explore how different notions of drug similarity affect the community structure of drug interaction networks and their biological interpretability. Specifically, we analyze three types of drug-to-drug similarity functions: (i) Tanimoto index [13] over molecular fingerprints derived from SMILES representations (capturing structural similarity), (ii) Topological distance based on edit distance over molecular graphs (capturing topological dissimilarity), and (iii) BioBERT-based similarity [5] using contextual embeddings of drug descriptions (capturing semantic similarity).

We begin with a basic analysis of these distances, plotting their distributions and analyzing illustrative examples to assess their consistency and interpretability. In particular, we investigate whether these measures are locally stable, i.e., whether the neighborhood of a point remains consistent across different similarity definitions. This question is especially relevant, as the effectiveness of neighborhood-based classifiers crucially depends on how similarity between instances is defined [6,7,11]. We then construct a heterogeneous KG combining drugs, proteins, and side effects, and discuss the limitations of relying solely on similarity-based graphs. We argue for the use of richer graph representations that retain multi-hop and multimodal relationships. Building on this, we apply classical community detection methods such as Louvain—both to drug-only networks and to the full heterogeneous KG. We also discuss the challenges of assigning edge weights in such settings, especially when working with distance-based measures or heterogeneous link types. As part of our ongoing work, we evaluate the quality of the detected communities using several types of measures. These include structural metrics like modularity (while keeping in mind its limits for heterogeneous graphs). We also add a functional perspective by looking at real-world drug–drug interactions, specifically, pairs of drugs that are known to cause side effects when taken together. By checking how often these pairs fall within the same community versus across different ones, we assess whether the clusters make sense from a clinical point of view. This combined evaluation helps us better understand how different clustering approaches perform on pharmacological knowledge graphs and to what extent they produce groups that are meaningful from both a structural and functional point of view.

2   Methods

2.1   Defining the Drug Knowledge Graph

We constructed a drug knowledge graph (KG) using data from the Decagon project [17], which includes three types of nodes: drugs, side effects, and proteins. The KG formalism allows us to preserve the heterogeneity of biomedical entities and to represent different types of relationships among them.

The graph includes three types of edges:

1. Drug-drug edges, weighted by a chosen similarity measure. In our analysis, we consider three different similarity functions (Tanimoto, BioBERT, and topological), each inducing a different weighting of the drug-drug edges.
2. Drug-side effect edges, linking each drug to the side effects it is known to cause.
3. Drug–protein edges, connecting drugs to their known target proteins.

The overall structure of the KG is illustrated in Fig. 1.

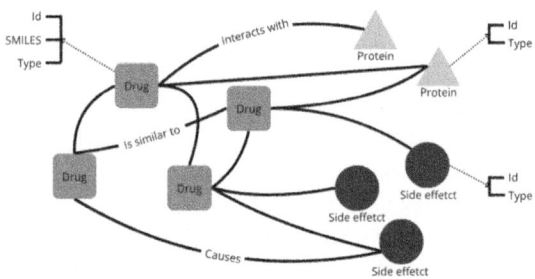

Fig. 1. An example illustrating the structure of a Knowledge Graph.

2.2   Drug-to-Drug Distance Measures

We compute three types of pairwise distances between drugs, each capturing a distinct notion of similarity: chemical structure, topological form, and semantic context.

Tanimoto Similarity Index. Let $\mathbf{f}_i, \mathbf{f}_j \in \{0, 1\}^d$ denote binary molecular finger-prints derived from SMILES strings[1] [14]. The *Tanimoto similarity* [1,13], a widely used variant of the Jaccard index for binary vectors, is defined as:

$$\text{sim}_{\text{Tan}}(\mathbf{f}_i, \mathbf{f}_j) = \frac{|\mathbf{f}_i \cap \mathbf{f}_j|}{|\mathbf{f}_i| + |\mathbf{f}_j| - |\mathbf{f}_i \cap \mathbf{f}_j|}$$

Topological Similarity. Let $G_i = (V_i, E_i)$ and $G_j = (V_j, E_j)$ be two graphs not necessarily on the same set of nodes (representing the molecular graphs of two drugs that we need to compare). The *graph edit distance* $\text{GED}(G_i, G_j)$ (see e.g. [10]) is the minimum sequence of operations (node/edge insertions, deletions) required to transform G_i into G_j or viceversa.

As in our work we need a similarity measure, we rescale and reverse the edit distance as follows:

$$\text{sim}_{\text{Edit}}(i, j) = 1 - \frac{\text{GED}(G_i, G_j) - \min_{u,v \in D} \text{GED}(G_u, G_v)}{\max_{u,v \in D} \text{GED}(G_u, G_v) - \min_{u,v \in D} \text{GED}(G_u, G_v)}$$

so that all similarity scores lie in the range $[0, 1]$, with higher values indicating greater topological similarity. The set D, indicates the nodes that correspond to drugs in the KG.

BioBERT-Based Semantic Similarity. Let $\mathbf{e}_i, \mathbf{e}_j \in \mathbb{R}^d$ be contextualized embeddings of drug descriptions obtained using BioBERT [5]. The *cosine similarity* between embeddings is: $\text{CosineSim}(\mathbf{e}_i, \mathbf{e}_j) = \frac{\mathbf{e}_i \cdot \mathbf{e}_j}{\|\mathbf{e}_i\| \cdot \|\mathbf{e}_j\|}$ We define the *semantic similarity* as: $\text{d}_{\text{BioBERT}}(i, j) = 1 - \text{CosineSim}(\mathbf{e}_i, \mathbf{e}_j)$

2.3   Quantifying Local Disagreement Between Drug Similarity Metrics

We now introduce a measure to quantify the instability (or disagreement) of node neighborhoods across similarity metrics. Given three complete drug–drug graphs $\mathcal{G}^{(1)}, \mathcal{G}^{(2)}, \mathcal{G}^{(3)}$ induced by different similarity metrics (e.g., Tanimoto, Topological, BioBERT), we define the *local neighborhood disagreement* to quantify the inconsistency of each drug's local neighborhood across these metrics.

Let d_i be a drug, and let $\mathcal{N}_k^{(m)}(d_i)$ denote the set of its top-k most similar drugs in graph $\mathcal{G}^{(m)}$, based on descending edge weight (i.e., similarity). For each drug, we compute the Jaccard index between neighborhood sets in all metric pairs:

$$J_i^{(m_1, m_2)} = \frac{|\mathcal{N}_k^{(m_1)}(d_i) \cap \mathcal{N}_k^{(m_2)}(d_i)|}{|\mathcal{N}_k^{(m_1)}(d_i) \cup \mathcal{N}_k^{(m_2)}(d_i)|}$$

---

[1] SMILES (Simplified Molecular Input Line Entry System) are linear strings that encode the structure of chemical compounds using atomic symbols and bond characters. They provide a compact and machine-readable representation of molecules, widely used in cheminformatics and drug discovery.

The *local neighborhood disagreement* for drug d_i is then defined as:

$$\text{LND}(d_i) = 1 - \frac{1}{3}\left(J_i^{(1,2)} + J_i^{(1,3)} + J_i^{(2,3)} \right)$$

The score lies in $[0,1]$, with higher values indicating greater disagreement in the local neighborhood structure of d_i across the three similarity views.

2.4   Clustering Strategies and Evaluation

In this work, our primary focus is on the clustering of drugs. We investigate two distinct approaches: (i) clustering drugs based solely on a drug–drug similarity graph, and (ii) clustering drugs within a heterogeneous knowledge graph that includes drugs, proteins, and side effects. The aim is to assess whether the additional biological and phenotypic information provided by the knowledge graph increases or decreases the quality of the drug clusters, that is whether it introduces signal or noise. Here we consider the Louvain clustering algorithm, a modularity-based method designed for general graphs and to compare the resulting partitions, we use standard evaluation metrics such as Adjusted Rand Index (ARI) [3] and Normalized Mutual Information (NMI) [4]. To further compare the communities, we incorporate additional information on drug pairs that are known to cause side effects when taken together. Specifically, we analyze the distribution of interaction edges, representing drug pairs that induce side effects when co-administered, by quantifying the proportion of such edges that fall within versus between communities. This allows us to understand the extent to which similarity-based clustering captures functional coherence with respect to side-effect co-occurrence.

3   Experimental Setting

Dataset. The source dataset[2] was collected by Snap's Decagon project [17], which contains both data on the side effects of single medicines, i.e. monopharmacy, and those resulting from the combination of different medicines, i.e. polypharmacy. Their data also contain information on interactions between the drugs and their target proteins, thus creating a scenario in which drugs are connected with both side effects and target proteins. These datasets contain both the identification code of the drugs, as well as that of the side effects, whose name is also contained. Since in this work a semantic analysis of the similarities between the drugs is also done, the first processing was to associate the code of the drugs with the respective name extracted from the STITCH database[3]. The STITCH database contains not only the name of the drugs associated with their code, but also the SMILES of each chemical. The dataset used in the subsequent analysis is monopharmacy, including information on the names of the side effects, both the names, the SMILES and target proteins of each drug.

---

[2] Decagon: https://snap.stanford.edu/decagon.
[3] STITCH database: http://stitch.embl.de/.

Similarity Measures. The Tanimoto and topological similarities were built using the RDKit and NetworkX Python packages, respectively. For the latter, which is based on edit distance, the cost of node and edge operations was the same, at 1. As far as BioBERT similarity is concerned, no specific package was used, but directly the Pytorch implementation of BioBERT. More specifically, the language model was used to compute the embeddings of each drug name. Once the embeddings representing the names were obtained, the cosine similarity between them was calculated and subsequently standardised.

Knowledge Graph. The knowledge graphs were constructed using the NetworkX package. Connections between two drug nodes are only created if their similarity is greater than the first quartile of the distribution of each measure, as can be seen in Fig. 2. The need to remove edges with very low weight arose from the necessity of community detection. Indeed, having calculated all pair-wise similarity, the resulting drug graph, without the removal of very low similarity edges, is a clique, which does not facilitate community detection. When they exist, the weight of these connections is the similarity value between the two drugs.
The weight of the other connections (drug-side effect, drug-protein) is given by the average similarity between all the drugs. Given the above, the three constructed KGs therefore show subtle differences. Indeed, they have differences in the section relating to drug-drug connections, and to the drug nodes themselves. In fact, considering only the connections greater than the first quartile of the similarity distribution, some drug nodes are isolated from their companions and are therefore eliminated.

Communities Algorithms. The louvain algorithm is implemented for community detection using NetworkX. The parameters used for each graph are as follows: resolution = 1; threshold = -0.07. Edge weights have been included in the community calculation. Since the Louvain algorithm depends on the initial random placement of nodes within communities, all calculations were performed using two different random states to ensure consistency in the results.

4   Results

4.1   Similarity Measures

Distance Distribution Analysis. Figure 2 shows the distribution of pairwise drug similarity values computed using the three methods described in Sect. 2: Tanimoto similarity, scaled and reversed edit distance and scaled BioBERT cosine similarity.

The three distributions reveal distinct patterns in how similarity is captured by each method. The Tanimoto distribution is skewed toward lower similarity values, suggesting that most drug pairs share only a few common chemical substructures. In contrast, BioBERT similarity values are more symmetrically distributed around 0.5, implying that the semantic representations of drugs—when encoded using language models—produce a variety of results. Finally, the

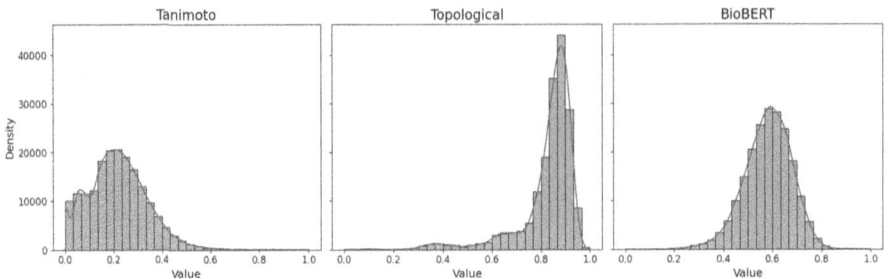

Fig. 2. Distributions of pairwise drug similarity values for the three distance metrics. From left to right: Tanimoto, topological and BioBERT similarity

Topological distribution is sharply peaked near 1, indicating a high degree of structural overlap across drug pairs. These differences indicate that each metric captures a distinct aspect of drug similarity, and they motivate further analysis of how each influences community detection and biological interpretability.

4.2   Local Neighborhood Disagreement (LND) Analysis

We analyzed the pairwise Local Neighborhood Disagreement (LND) across the three drug similarity measures to assess how much the local structure varies between them. For each drug, we computed the set of its top-k most similar neighbors using each similarity measure, setting $k = 10$. In cases where ties occurred, that is multiple drugs shared the same similarity score at the k-th position, we included all tied neighbors, resulting in sets larger than k. This choice avoids making arbitrary cutoffs among equally similar drugs and ensures a consistent and fair comparison across the different similarity spaces.

As shown in Fig. 3, BioBERT vs Topological exhibits near-complete disagreement for the vast majority of drugs (LND ≈ 1), indicating that semantic similarity does not reflect molecular topology. Similarly, high disagreement is observed between Topological and Tanimoto, despite both being structure-based, likely due to their different encoding schemes. The distribution for BioBERT vs Tanimoto is slightly more spread, with a small subset of drugs exhibiting lower LND values, suggesting limited agreement between semantic and fingerprint-based similarity for some compounds. These results confirm that different similarity metrics induce distinct and often highly variable neighborhood structures.

To better understand the extreme cases of disagreement, we identified drugs with LND = 1, indicating complete inconsistency in their top-10 neighborhoods between two similarity metrics. As shown in Table 1, the number of drugs that exhibit full disagreement is 398 between BioBERT and Topological distance, 251 between BioBERT and Tanimoto, and 317 between Topological and Tanimoto. These results suggest that overall almost 50% of these compounds can behave very differently across similarity definitions, particularly when comparing semantic, fingerprint, and topological views.

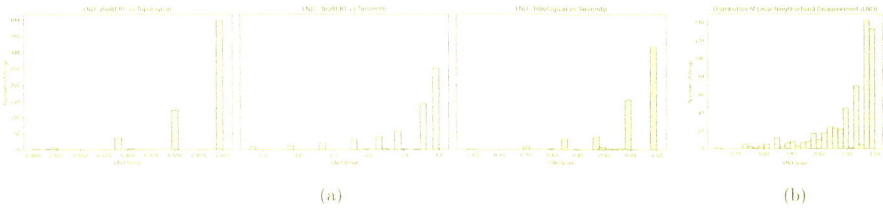

(a)   (b)

Fig. 3. Distributions of Local Neighborhood Disagreement (LND) scores between different drug similarity measures. (a) considers LND the pairwise for every pair of measure; (b) computes the total LND over all the three measures.

Interestingly, as it can be seen by the Fig. 3 no drug had LND < 0.4 across all other pairs. This highlights the rare alignment of semantic and structural similarity, reinforcing that local neighborhoods induced by different metrics are often non-overlapping. This indicates that LND is more a measure to quantify structural vs semantic misalignment, than to reflect pharmacological diversity.

Table 1. Summary of outlier drugs with extreme Local Neighborhood Disagreement (LND = 0 or 1) across pairwise similarity metrics. Only drug nodes are included.

Similarity Pair	LND = 0	LND = 1	Example Drugs
BioBERT vs Topological	0	398	'sertraline', 'rofecoxib', 'exemestane', 'nicardipine', 'mitoxantrone', 'dihydroergotamine', 'topiramate', 'morphine', 'doxorubicin'
BioBERT vs Tanimoto	0	251	'fulvestrant', 'sertraline', 'rofecoxib', 'adapalene', 'mitoxantrone', 'dihydroergotamine', 'topiramate', 'dacarbazine', 'atorvastatin'
Topological vs Tanimoto	0	317	'fulvestrant', 'rofecoxib', 'caffeine', 'adapalene', 'nicardipine', 'mitoxantrone', 'topiramate', 'doxorubicin', 'cortisol'

4.3 Drug Community Comparison Across Similarity Measures and Clustering Algorithms

To evaluate the consistency of drug communities produced by different clustering strategies, we computed the Adjusted Rand Index (ARI) and Normalized Mutual Information (NMI) across the three similarity measures (BioBERT, Topological, and Tanimoto). As shown in Table 2, for the Drug similarity graph the agreement between community assignments is moderate suggesting that in some cases, communities are robust to similarity measures. However, clustering results on the knowledge graph (KG) are largely inconsistent with those on drug-only graphs, for all the similarity measures, highlighting how the inclusion of proteins and side effects reshapes the community structure. This is confirmed by comparing the modularity values of the communities identified on the KGs with those on the drug-only graph. Indeed, they show a discreet variation, with the KGs that appear to have a higher clustering quality.

Table 2. Agreement across similarity metrics within each clustering algorithm (G = drug-only graph, KG = knowledge graph). Metrics: ARI (Adjusted Rand Index), NMI (Normalized Mutual Information).

Comparison	ARI	NMI
Louvain (Drug Graph)		
BIOBERT (G) vs Topological (G)	0.550	0.523
BIOBERT (G) vs Tanimoto (G)	0.395	0.399
Topological (G) vs Tanimoto (G)	0.342	0.379
Louvain (Knowledge Graph)		
BIOBERT (KG) vs Topological (KG)	0.014	0.042
BIOBERT (KG) vs Tanimoto (KG)	0.036	0.054
Topological (KG) vs Tanimoto (KG)	0.009	0.044
Louvain (KG vs G)		
BIOBERT (KG) vs BIOBERT (G)	0.057	0.082
Tanimoto (KG) vs Tanimoto (G)	0.046	0.055
Topological (KG) vs Topological (G)	0.061	0.076

The Sankey diagrams presented in Fig. 4 compare community assignments across different drug similarity measures under two graph representations: (a) the simpler Drug-to-Drug Graph (G), which models direct pairwise similarities only, and (b) the Knowledge Graph (KG), which incorporates drug–protein–side effect relationships. We observe that in several cases, such as the BIOBERT–Tanimoto comparison, the KG-based clustering exhibits more significant flows between communities, suggesting a higher degree of differentiation. This may reflect the KG's capacity to incorporate richer biological context, making it more sensitive to subtle structural and functional relationships that may be overlooked in the simpler Drug Graph. Conversely, the G-based clustering (a) tends to show more tightly aligned flows, possibly indicating more uniform community structures.

There are at least two possible explanations for the differences seen in KG-based clustering. One is that the KG captures additional biological information, like shared side effects or indirect functional links. The other is that the added complexity of the KG introduces noise. More analysis is needed to understand which of these explanations is more likely.

Fig. 4. Community alignment between drug clustering results obtained using different similarity measures. (a) Clustering based on the Drug-to-Drug Graph (G) representation. (b) Clustering based on the Knowledge Graph (KG) representation. Each subfigure compares community assignments under different similarity metrics (BIOBERT, topological, and Tanimoto) using Sankey diagrams.

4.4   Functional Coherence Among Co-occurring Side Effects

We analysed how well communities capture the co-occurrence of side effects—cases where taking two drugs together leads to a specific side effect. This analysis is based on Decagon's polypharmacy dataset, which lists known side effects resulting from drug combinations. Each drug pair's weight reflects the number of associated side effects. We assessed how often drug pairs causing a side effect fall within the same community versus across different ones.

Table 3 presents results from two perspectives: the community level and the overall graph level. From the community's perspective, we report the number of drug pairs with shared side effects that fall entirely within the community, and those where only one drug is present, and the sum of weights for both categories.

At the graph level, a similar analysis is performed. However, note that the total number of cross-community pairs is not simply the sum of such pairs across individual communities, as shared pairs would be counted multiple times.

In general, there are more pairs of drugs in two different communities than within a single community. Similarly, the same is true for weights. However, the results vary depending on the similarity measure used. Indeed, considering the Tanimoto index, clustering on the knowledge graph better captures the co-occurrence of side effects at both the level of individual communities and at the graph level. Regarding BioBERT the community that performs best is identified on the knowledge graph. At the graph level, though, there is a higher occurrence of side effects within the community for the drug graph. Finally, considering the edit distance, the results are similar at the level of individual communities, but since there are two communities in the drug graph that have a good number of pairs inside, the results at the graph level are better than on the knowledge graph.

Table 3. The table represents the number of pairs of drugs causing a side effect falling both in the same community, and those falling in two different communities. Likewise, the sum of the side effects of each pair of drugs is represented as weights. The calculation is shown with both communities relative to the KG, and to the drugs graph alone.

	community	drugs	side effects	proteins	inter edge	outer edge	inter/outer	inter weight	outer weight	inter/outer w
Knowledge graph Tanimoto	0	320	1971	102	13445	27401	0.4907	927294	2026730	0.4575
	1	6	301	1	2	845	0.0024	31	36334	0.0009
	2	5	94	2740	4	980	0.0041	113	55429	0.0020
	3	96	1742	255	1984	16355	0.1213	188520	1298384	0.1452
	4	118	3291	116	3313	19856	0.1669	285647	1566037	0.1824
	5	50	2206	432	307	7333	0.0419	14602	417192	0.0350
	Graph level	595	9605	3646	19055	36385	0.5237	1416207	2700053	0.5245
Drugs graph Tanimoto	0	110			1952	16742	0.1166	137232	1211105	0.1133
	1	6			10	1460	0.0068	645	103396	0.0062
	2	256			10423	27018	0.3858	748424	1994976	0.3752
	3	3			2	182	0.0110	38	4671	0.0081
	4	193			5942	23854	0.2491	490063	1809861	0.2708
	5	27			105	4756	0.0221	7311	341085	0.0214
	Graph level	595	0	0	18434	37006	0.4981	1383713	2732547	0.5064
Knowledge graph BioBERT	0	86	1519	256	1759	15657	0.1123	161806	1228498	0.1317
	1	5	66	2818	3	958	0.0031	104	54437	0.0019
	2	59	2310	323	551	8582	0.0642	30044	464294	0.0647
	3	354	1604	160	11887	27414	0.4336	820761	2020562	0.4062
	4	109	2804	71	3085	19623	0.1572	305416	1589743	0.1921
	5	2	48	9	1	863	0.0012	226	87668	0.0026
	6	24	1351	11	108	2995	0.0361	8375	133854	0.0626
	Graph level	639	9702	3648	17394	38046	0.4572	1326732	2789528	0.4756
Drugs graph BioBERT	0	181			4675	21017	0.2224	330514	1492078	0.2215
	1	227			10385	26859	0.3866	925082	1994832	0.4637
	2	231			4878	23128	0.2109	302045	1630328	0.1853
	Graph level	639	0	0	19938	35502	0.5616	1557641	2558619	0.6088
Knowledge graph Edit distance	0	5	83	2801	3	1110	0.0027	92	71391	0.0013
	1	329	1683	183	14379	27557	0.5218	1105120	2052837	0.5383
	2	44	1814	335	237	5506	0.0430	11996	282200	0.0425
	3	78	1474	247	1455	14580	0.0998	139649	1173086	0.1190
	4	90	2622	57	2095	16627	0.1260	178678	1298758	0.1376
	5	8	273	16	9	1291	0.0070	381	63336	0.0060
	6	28	1614	7	180	4073	0.0442	12938	185573	0.0697
	Graph level	582	9563	3646	18358	37029	0.4958	1448854	2663738	0.5439
Drugs graph Edit distance	0	3			3	919	0.0033	564	83483	0.0068
	1	2			1	555	0.0018	178	35459	0.0050
	2	2			1	568	0.0018	70	39056	0.0018
	3	2			0	163	0.0000	0	6577	0.0000
	4	313			14589	27543	0.5297	1150307	2029728	0.5667
	5	2			0	335	0.0000	0	21188	0.0000
	6	256			10422	27077	0.3849	748927	1995686	0.3753
	7	2			1	266	0.0038	46	13528	0.0034
	Graph level	582	0	0	25017	30370	0.8237	1900092	2212500	0.8588

5   Discussion and Future Work

Here we study the complexity and variability of drug similarity and clustering when using different representations and metrics. The analysis of similarity distributions (Fig. 2) reveals that each metric captures a distinct dimension of drug relatedness: structural, topological, or semantic. These differences are not only quantitative but also qualitative, as confirmed by the Local Neighborhood Disagreement (LND) scores (Fig. 3), which show that most drugs exhibit high inconsistency across similarity-based neighborhoods. This suggests that the local context of a drug—i.e., which drugs are considered most similar—depends heavily on how similarity is defined. The implications of this variability are further evident when comparing the resulting drug communities. Clustering results vary significantly across similarity metrics, with moderate agreement for drug-only graphs (Table 2), but very low agreement when comparing communities derived from the Knowledge Graph (KG). These results are supported by the Sankey diagrams (Fig. 4), where we observe more fragmentation in KG-based clustering.

This study is part of ongoing work and primarily raises questions rather than offering definitive answers. Our findings point to several future directions:

What is the most appropriate way to define similarity between drugs? Our results show that different similarity measures lead to highly divergent neighborhood structures and community assignments. It remains unclear whether one of these views is intrinsically more informative, or whether combining multiple similarity metrics, e.g., through embedding fusion or multi-view clustering, could offer a significantly more robust and refined characterization of drug relationships.

How should we perform community detection on heterogeneous graphs like KGs? While we explored various clustering methods, such as the spectral clustering or the M-algorithm, proposed as an evolution of k-means adapted on graph data [12], we decided to use Louvain as a baseline because it obtains more consistent results across different similarity measures. But the inclusion of other community detection methods will be implemented in future work, both to compare the different biological aspects captured by various algorithms on the one hand, and to avoid biased results from using a single algorithm on the other. More sophisticated methods, such as CESNA [15], could be more suitable. However, applying these to our setting requires careful consideration: CESNA relies on node attributes, if we use the node type (e.g., "drug", "protein", "side effect") as an attribute, the algorithm would simply group all nodes of the same type together, without revealing any meaningful structure. Similarly, using presence of links (e.g., to proteins or side effects) as features can make many drugs would appear similar just because they are connected to common nodes. To apply these methods effectively, we need to define or learn more informative attributes, possibly through embedding techniques, that capture deeper differences between drug nodes. Finally, we plan to explore how relational or grammar-based models improve clustering of knowledge graphs.

What evaluation criteria should be used to assess cluster quality? Most existing evaluation methods for clustering quality are based on node-edge density (how tightly nodes are connected within a cluster). This biases evaluations toward networks with naturally dense clusters and fails on sparser networks (in our KG we have subgraphs that are sparse like the protein-graph relationship) where meaningful relationships may exist despite low connectivity. An example is given in [16] where the authors propose a new metric evaluates clustering quality based on the average path length among nodes within a cluster, rather than relying on edge density alone. In doing so, it captures not just which nodes are grouped together, but also how closely and efficiently they are connected. Finally, in our case it is important to design biologically grounded evaluation metrics. For example, overlap with known drug classes, pharmacological pathways, or therapeutic categories could serve as external validation. Similarly, tracking whether communities are predictive of ADR co-occurrence may help assess their clinical relevance.

How stable are the detected communities across multiple runs or equivalent solutions? Most clustering algorithms return a single partition of the graph, but in practice, especially for methods like Louvain or other modularity-based approaches, different runs (due to random initialization) or even different optimal solutions can yield distinct but equally valid community structures. This raises the question of community stability: how consistent are the groupings across runs, and what alternative community structures might exist? For a meaningful biological interpretation, it may be important not only to identify a single clustering, but also to explore the full space of high-scoring solutions. Capturing this diversity could reveal alternative functional groupings or stable community cores.

Acknowledgments. C. Poccianti is founded by the European Union - Next Generation EU, Mission 4 Component 1 CUP B53C240022004.

Disclosure of Interests. The authors have no competing interests to declare that are relevant to the content of this article.

References

1. Bajusz, D., Rácz, A., Héberger, K.: Why is tanimoto index an appropriate choice for fingerprint-based similarity calculations? J. Cheminf. **7**(1) (2015). https://doi.org/10.1186/s13321-015-0069-3, http://dx.doi.org/10.1186/s13321-015-0069-3

2. Bean, D.M., et al.: Knowledge graph prediction of unknown adverse drug reactions and validation in electronic health records. Sci. Rep. **7**(1) (2017). https://doi.org/10.1038/s41598-017-16674-x, http://dx.doi.org/10.1038/s41598-017-16674-x

3. Hubert, L., Arabie, P.: Comparing partitions. J. Classif. **2**(1), 193–218 (1985). https://doi.org/10.1007/bf01908075, http://dx.doi.org/10.1007/BF01908075

4. Kvalseth, T.O.: Entropy and correlation: Some comments. IEEE Trans. Syst. Man Cybern. **17**(3), 517–519 (1987). https://doi.org/10.1109/tsmc.1987.4309069, http://dx.doi.org/10.1109/TSMC.1987.4309069

5. Lee, J., et al.: Biobert: a pre-trained biomedical language representation model for biomedical text mining. Bioinformatics **36**(4), 1234–1240 (2019). https://doi.org/10.1093/bioinformatics/btz682, http://dx.doi.org/10.1093/bioinformatics/btz682

6. Levy, A., Shalom, B.R., Chalamish, M.: A guide to similarity measures (2024). https://doi.org/10.48550/ARXIV.2408.07706, https://arxiv.org/abs/2408.07706

7. Liu, J.G., Hou, L., Pan, X., Guo, Q., Zhou, T.: Stability of similarity measurements for bipartite networks. Sci. Rep. **6**(1) (2016). https://doi.org/10.1038/srep18653, http://dx.doi.org/10.1038/srep18653

8. Osanlou, R., Walker, L., Hughes, D.A., Burnside, G., Pirmohamed, M.: Adverse drug reactions, multimorbidity and polypharmacy: a prospective analysis of 1 month of medical admissions. BMJ Open **12**(7) (2022). https://doi.org/10.1136/bmjopen-2021-055551, https://bmjopen.bmj.com/content/12/7/e055551

9. Pirmohamed, M., et al.: Adverse drug reactions as cause of admission to hospital: prospective analysis of 18 820 patients. BMJ **329**(7456), 15–19 (2004). https://doi.org/10.1136/bmj.329.7456.15

10. Serratosa, F., Cortés, X.: Graph edit distance: moving from global to local structure to solve the graph-matching problem. Pattern Recogn. Lett. **65**, 204–210 (2015). https://doi.org/10.1016/j.patrec.2015.08.003

11. Shirkhorshidi, A.S., Aghabozorgi, S., Wah, T.Y.: A comparison study on similarity and dissimilarity measures in clustering continuous data. PLOS ONE **10**(12), e0144059 (2015). https://doi.org/10.1371/journal.pone.0144059, http://dx.doi.org/10.1371/journal.pone.0144059

12. Sieranoja, S., Fränti, P.: Adapting k-means for graph clustering. Knowl. Inf. Syst. **64**(1), 115–142 (2022). https://doi.org/10.1007/s10115-021-01623-y, publisher: Springer Science and Business Media LLC

13. Tanimoto., T.T.: An elementary mathematical theory of classification and prediction (1958), international BusinessMachines Corporation

14. Weininger, D.: SMILES, a chemical language and information system. 1. Introduction to methodology and encoding rules. J. Chem. Inf. Comput. Sci. **28**(1), 31–36 (1988). https://doi.org/10.1021/ci00057a005, https://pubs.acs.org/doi/abs/10.1021/ci00057a005

15. Yang, J., McAuley, J., Leskovec, J.: Community detection in networks with node attributes. In: 2013 IEEE 13th International Conference on Data Mining, pp. 1151–1156. IEEE, Dallas, TX, USA, December 2013. https://doi.org/10.1109/ICDM.2013.167, http://ieeexplore.ieee.org/document/6729613/

16. Zaidi, F., Archambault, D., Melançon, G.: Evaluating the quality of clustering algorithms using cluster path lengths, pp. 42–56. Springer, Berlin Heidelberg (2010). https://doi.org/10.1007/978-3-642-14400-4_4

17. Zitnik, M., Agrawal, M., Leskovec, J.: Modeling polypharmacy side effects with graph convolutional networks. Bioinformatics **34**(13), i457–i466 (2018). https://doi.org/10.1093/bioinformatics/bty294, https://academic.oup.com/bioinformatics/article/34/13/i457/5045770

SapientIAGraph: An Open Knowledge Graph of University Degree Programs at Sapienza

Riccardo Ceccaroni[✉], Lorenzo Di Rocco[ID], and Umberto Ferraro Petrillo[ID]

Sapienza University of Rome, Piazzale Aldo Moro 5, 00185 Rome, Italy
{riccardo.ceccaroni,lorenzo.dirocco,umberto.ferraro}@uniroma1.it

Abstract. In this paper we present SapientIAGraph, a knowledge graph (KG) designed to analyze the organization of university degree programs, with a focus on Sapienza University of Rome. The proposed KG is automatically built using a specialized web scraper. It extracts and semantically annotates institutional data published on the Sapienza website. We also make the resulting knowledge graph available under an open license, and in multiple formats.

Moreover, we present the results of an experimental analysis to highlight the possible applications of SapientIAGraph. By leveraging the Jaccard index and thanks to the semantic information about module subject areas, we have been able to automatically cluster all degree programs in our graph into two groups, STEM (Science, Technology, Engineering, and Mathematics) and non-STEM, with some additional interdisciplinary programs acting as bridges.

The results suggest the potential of KGs for comparative and structural analysis of academic offerings and can be generalized to other universities, provided that they are modeled using KGs.

Keywords: university degree programs · knowledge graph · curriculum analysis · public dataset

1 Introduction

Knowledge graphs (KGs) have recently emerged as a powerful tool to model information in several different application domains [7]. One particular domain of interest is the usage of KGs for modeling university degree programs. In this context, KGs allow for several applications, such as simplifying the development of recommendation systems capable of suggesting students how to compose their learning paths according to their preferences and to their inclinations.

In this paper, we introduce SapientIAGraph, an open KG that models the full structure of the degree programs offered by the Sapienza University of Rome, one

---

A copy of SapientIAGraph, as well as support documentation and example queries, is publicly available at https://github.com/umbfer/SapientIAGraph.

© The Author(s), under exclusive license to Springer Nature Switzerland AG 2026
P. K. Chrysanthis et al. (Eds.): ADBIS 2025, CCIS 2676, pp. 398–409, 2026.
https://doi.org/10.1007/978-3-032-05727-3_33

of the largest universities in Europe. SapientIAGraph includes detailed information on all the 309 degree programs available for the academic year 2024/2025. For each degree program, the graph contains all the modules included in that program, possibly organized into curricula, and some of the dependencies existing between them. Additional metadata is available for each degree program, curriculum, and module, including their subject areas.

SapientIAGraph is automatically generated by a specialized web scraper that extracts publicly available institutional data from Sapienza website and semantically annotate it, modeling the result as a KG.

We also report the results of a statistical analysis of this graph, that illustrates some of the possible applications for this resource. For instance, we assessed the similarity between programs by analyzing the disciplinary composition of their modules. The Jaccard index was used to compare their subject areas. The results suggest the existence of two clusters, roughly equivalent to STEM and non-STEM degree programs, with interdisciplinary degree programs acting as a bridge.

2   Related Work

Knowledge graphs are graph-based data models designed to describe and represent the semantics of real-world entities, events, concepts, and their relationships. The choice of structuring knowledge by means of nodes and edges, plus the possibility of embedding arbitrary information within them, enables the integration of heterogeneous data sources, including both structured and unstructured data, in a single unified framework.

The term knowledge graph was first introduced by Schneider in 1972 ([17]) during the introduction of a framework for modular, computer-assisted instructions anticipating some of the key features of modern KGs. They have become very popular in recent years thanks to the choice of Google to integrate in its search engine a KG for improving the quality and the accuracy of its results. Today, KGs are extensively used for many applications, ranging from the analysis of green investment opportunities [8] to the exploration of protein-to-protein interaction networks in computational biology [6]. They have become increasingly popular in the last years thanks to their ability to support a wide range of applications (see, e.g. [2]).

Their ability to describe in a flexible and simple way complex interactions between entities is of a particular help in the development of recommendation systems ([9,19]. Also, KGs play an important role in improving the performance and response accuracy of large language models when used to develop Knowledge-Augmented Frameworks (KAG) [13].

The possible applications of KGs to the educational domain have started receiving some degree of attention in the last years. One of the first contributions in this field has been proposed in [21]. Here the authors introduce a method to visualize the structure of university degree programs with the help of the Google Knowledge Graph. The proposed method is mostly focused on the exploration and the visualization of the obtained knowledge graph by means of the querying facility coming with the Neo4J graph database. Among the most relevant contributions, the authors of [10] present Educonto, an ontology designed to model university curricula and student profiles, with the aim of facilitating the recommendation of personalized study paths. The proposed ontology is then used to create a knowledge graph including information coming from French universities.

In [12], the authors present a framework to extract educational KGs from module information to support precision teaching. The proposed framework uses machine learning techniques to capture teaching concepts and to support the creation of personalized learning paths tailored to student needs.

A particular application domain for KGs in the analysis of educational offerings is about university degree programs. Here, KGs provide a valuable tool for simplifying the development of recommendation systems useful for curriculum and personalization. There are however also some other possible applications for KGs in this case like:

- **Improved Degree Programs Navigability** The usage of a knowledge graph enables a natural representation of the structural dependencies among teaching modules existing in an academic degree, such as a module being a pre-requisite for giving another one. Moreover, this solution seamlessly support situations where the study paths offer some degree of flexibility, with the students being allowed to include in their curriculum modules of their choice.
- **Support for Student Mobility** An effective implementation of student mobility programs, such as Erasmus, relies on the ability to assess the equivalence between teaching modules being offered by different universities. Here, the adoption of a standard graph-based ontology would allow involved universities to facilitate the process of comparing different study paths and establish the equivalence between different modules. It would also make it easy for students to assemble customized study-paths incorporating modules offered by different universities.
- **Support for Strategic Analysis** The usage of a knowledge graph simplifies the strategic analysis of the educational offering of a university to support evidence-based decisions about educational offerings. For instances, it would allow to analyze the structure of degree programs, to evaluate the overlap between different curricula and even compute KPIs such as the interdisciplinary indices.

One of the earliest contributions in this field is the work by [1]. Even if not formally framed as a knowledge graph, the author here uses a directed acyclic graph to model curricula of university degree programs, where nodes represent teaching modules and edges represent prerequisites. Then, the authors use some classic network analysis algorithms to detect communities, bottlenecks and other centrality measures. Another relevant contribution, again not formally framed in the KG literature, is the work presented in [11]. The authors here present a formal ontology to model teaching modules, curricula and syllabus, with the aim of simplifying interoperability between different institutions, apart from teaching modules recommendation.

More recent contributions, like [3,20], focus on the usage of KGs to support students in selecting their study path according to their past history, to their preferences and to the modules prerequisites.

3   Case Study: Analysis of Sapienza's Education Offering

Sapienza, the oldest roman university, is currently the largest university in Europe by student enrollment, with over $125,000$ students and $3,500$ professors. It is organized into 11 faculties and 57 departments, offering over 300 degree programs (Bachelor's and Master's). The large number of degree programs, including a vast number of modules spanning the most diverse disciplines, makes Sapienza an ideal case for investigating the potential of KGs to model, analyze and optimize the structure and the performance of its educational offering.

A digital catalog of the degree programs offered by Sapienza can be accessed on its official website. The available programs are grouped into 26 overlapping categories. For each degree program, the catalog provides a general description including its name, its unique identifier, the hosting department and the reference language used for teaching. In addition, users can explore the internal structure of each program by navigating the list of associated teaching modules. These modules may be mandatory or optional, depending on some composition rules that may include the selection of different curricula or the possibility of including modules offered by different degree programs.

To support the analysis of the Sapienza's education offering, we automatically assembled a KG called SapientIAGraph (from Sapientia, the Latin name of Sapienza), which describes the structure of the Sapienza degree programs based on the information publicly available on the Sapienza website.

3.1   Data Extraction

We developed a Python-based scraper using the BeautifulSoup [16] and Selenium [18] libraries to extract data from the Sapienza digital catalog of the degree programs. The scraper first retrieves the list of available degree programs and then collects detailed data for each, including modules and the instructors responsible for them. The application also translates the relational organization of the catalog into a graph-based model.

3.2   Graph Modeling

Once scraped, the data collected from the Sapienza degree programs catalog are converted into a knowledge graph and saved into a Neo4j graph database [14]. Our approach is implementation-driven and focused on the integration of data available through the Sapienza website. The resulting graph includes core entities such as `DegreeProgram`, `Curriculum`, `Module`, `Instructor`, and

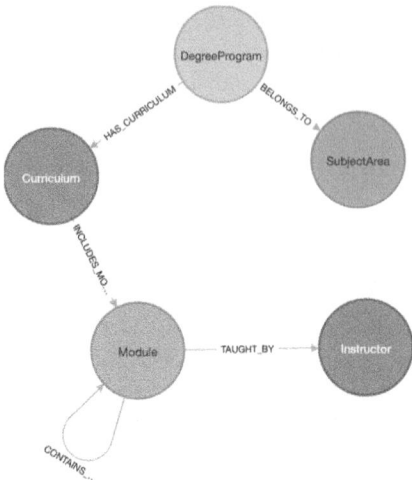

Fig. 1. The semantic schema of SapientIAGraph, illustrating the core classes and relationships used to describe the structure of Sapienza degree programs.

SubjectArea, and supports relationships for hierarchical composition (e.g., submodules), teaching roles, and curriculum structure rules. A representation of the semantic schema used by our knowledge graph is presented in Fig. 1. A copy of SapientIAGraph, encoded in different formats, as well as supporting documentation and example queries is available for download and usage under a public license at https://github.com/umbfer/SapientIAGraph.

Notice that there are other ontologies in the literature, such as [4,11] that offer a far more complex and rich representation of the complexity of a degree program. For example, [11] covers concepts such as Syllabus, TeachingMethod, LearningOutcome, and Event. [4] goes even further, introducing advanced constructs such as LearningPath, LearningStep, and Persona to support personalized and explainable learning experiences. Our ontology omits many of these details because it relies only on the publicly available data from the Sapienza website. We therefore favor completeness and integration capability over semantic richness.

4   Experimental Analysis

In this section we present examples of the types of analyses that are possible by representing the degree programs of a university using a knowledge graph. In particular, we show how it is possible to leverage this representation to investigate how the number of curricula varies across different types of degree programs, providing valuable insights into the organizational complexity and articulation of educational pathways.

Table 1. Distribution of nodes by labels in the Sapienza knowledge graph

Label	Description	Count
DegreeProgram	A structured course of study that includes a set of modules according to some composition rules.	309
SubjectArea	A subject covered by one or more degree programs.	26
Curriculum	A set of modules and learning activities enclosed within a degree program.	489
Module	A self-contained unit of teaching within a degree program. A module may contain two or more submodules, each modeled as a module.	10,795
Instructor	A person who assumes the role of instructor for one or more modules.	4,909

Table 2. Distribution of relationships in the Sapienza knowledge graph

Source	Relation Type	Target	Description	Count
Curriculum	INCLUDES_MODULE	Module	A curriculum includes multiple modules.	21,143
Module	TAUGHT_BY	Instructor	Each module is taught by one or more instructors.	15,257
Module	CONTAINS_MODULE	Module	A module may contain submodules, representing a hierarchical structure.	4,210
DegreeProgram	HAS_CURRICULUM	Curriculum	A degree program includes one or more curricula.	489
DegreeProgram	BELONGS_TO	SubjectArea	Each degree program is associated with a subject area or discipline.	335

All reported experiments were performed in Python using the NetworkX [5] software library, by querying an instance of SapientIAGraph stored within a Neo4j database.

4.1 Degree Distributions by Node Type

The overall structure of SapientIAGraph, including the distribution of nodes and relationship types, is summarized in Tables 1 and 2. We initially examine how connectivity is distributed according to the different types of nodes existing in SapientIAGraph. This is done by measuring, for each type of node, its degree distribution. We recall that degree is defined as the number of edges incident to a node and it serves as a key measure of connectivity within the graph.

Figure 2 presents separate histograms for each node type, with red dashed lines indicating the average degree. The results show that Module nodes exhibit the widest range of degrees, highlighting their centrality and reuse across multiple curricula and hierarchical structures. In contrast, Instructor nodes have lower average degrees, although a few instructors are highly connected. Curriculum and DegreeProgram nodes display more uniform patterns, reflecting their constrained and defined organizational roles. Finally, SubjectArea nodes maintain low degrees, as they primarily serve as semantic tags.

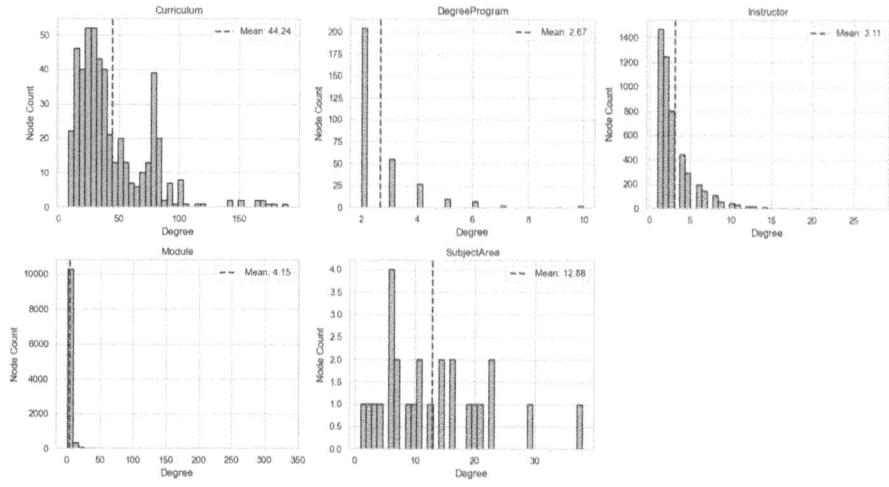

Fig. 2. Degree distribution plots for different node types. Each histogram shows the distribution of node degrees, with a dashed red line indicating the average degree. (Color figure online)

4.2 Centrality Indices for Monitoring Educational Offering

The knowledge graph based representation allows to compute node-level indicators that reflect structural properties based on the entity's pattern of connections. A particularly informative measure is the *degree centrality* index, which quantifies the number of edges incident to a node. While this index typically captures the local centrality of a node within a network, in our context it serves a more domain-specific purpose, that is, assessing the extent to which a degree program is decomposed into multiple curricula.

To this end, we extract from SapientIAGraph the subset of edges with label HAS_CURRICULUM, yielding a bipartite graph in which nodes represent either degree programs or curricula. Then, for each node with label DegreeProgram, the degree corresponds to the number of distinct curricula it includes. We refer to this value as the *curricular degree*, a proxy for measuring the structural complexity of a program.

A comparison between STEM and non-STEM degree programs at Sapienza reveals that STEM programs typically exhibit higher curricular degree values (see Fig. 3). The classification into STEM and non-STEM categories was determined based on the disciplinary area of each degree program. STEM programs include fields such as Computer Science, Engineering, Physics, Mathematics, and Biology, disciplines characterized by a strong focus on quantitative methods, technological development, and experimental approaches. In contrast, non-STEM programs cover areas such as Philosophy, History, Law, and Literature, which are generally more theoretical, interpretative, and centered around critical analysis and humanistic inquiry. The higher complexity observed in STEM programs likely reflects the need to accommodate specialized tracks, advanced technical content, and the fast-paced evolution typical of scientific and technological domains.

In contrast, non-STEM programs tend to have a lower curricular degree, reflecting more compact and uniform structures. Nevertheless, notable exceptions exist. Some non-STEM programs, such as Economics, display higher degrees of internal complexity. These cases often correspond to professional differentiation or interdisciplinary integration, revealing that complexity in curricular design is not exclusive to STEM fields but can also emerge in response to diverse academic demands or educational needs.

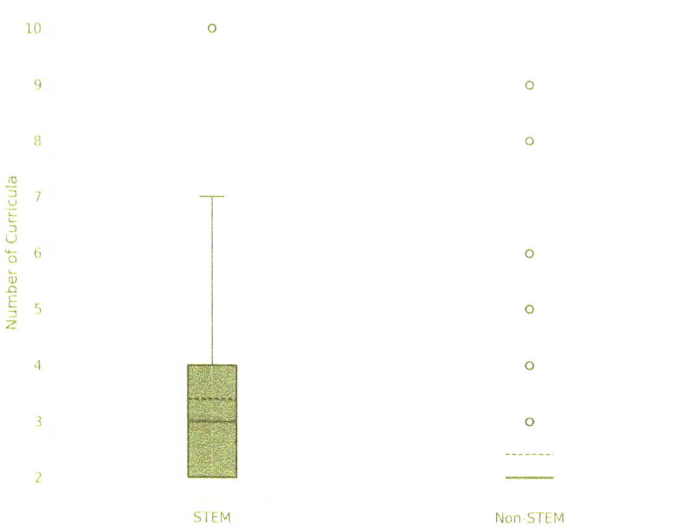

Fig. 3. Distribution of curricular degree values for STEM and non-STEM degree programs

4.3   Quantifying Similarity Between Degree Programs

To systematically evaluate the similarity between degree programs, we propose a method based on the analysis of the modules contained within their associated curricula. As previously discussed in Sect. 3.1, we recall that each node with the `DegreeProgram` label is connected to one or more nodes with label `Curriculum`, and each node with label `Curriculum` is connected to several nodes with label `Module`. These last nodes represent distinct teaching units that collectively characterize the educational content of the program. To capture the degree of overlap between two degree programs, we define a representation of their module composition.

In particular, for each degree program, we take into account the Scientific Disciplinary Sector (SSD) property available on each module. The SSD of a module is a standardized classification system used in the Italian higher education system to categorize academic disciplines and research fields. Each SSD corresponds to a specific area of expertise and is used to define the academic content of university teaching modules, allocate teaching responsibilities, and regulate faculty recruitment.

For instance, `INF/01` denotes Computer Science, and `BIO/10` denotes Biochemistry. In SapientIAGraph, each module is associated with exactly one SSD, which determines its disciplinary identity. This classification enables a comparison across modules and curricula, providing a unified framework for analyzing academic programs based on their scientific and disciplinary composition.

Using this approach, each degree program can be associated with a multiset of SSDs derived from the modules included in its curricula. Then, similarity between two programs D_i and D_j can be measured via a weighted Jaccard index over their SSD profiles, defined as:

$$J(D_i, D_j) = \frac{\sum_{s \in S} \min\left(w_i(s), w_j(s)\right)}{\sum_{s \in S} \max\left(w_i(s), w_j(s)\right)},$$

where:

- S is the set of all SSDs appearing in at least one of the two programs;
- $w_i(s) : S \to N$ is the frequency of SSD s in the curricula of i-th degree program;
- $w_j(s) : S \to N$ is the frequency of SSD s in the curricula of j-th degree program.

This weighting scheme reflects the importance of each module within the program, allowing us to distinguish between modules that are core components versus those that appear sporadically. Finally, we can compute the Jaccard difference for each pair of degree programs as follows:

$$D(D_i, D_j) = 1 - J(D_i, D_j),$$

By computing pairwise Jaccard distances among all degree programs, we obtain a distance matrix that encodes the curricular dissimilarities. We then construct a similarity graph in which nodes represent degree programs and the edge weights represent the distance between the pair of connected degree programs.

The similarity graph representation offers a powerful means to explore the relationships among degree programs. For example, Fig. 4 illustrates the distribution of STEM and non-STEM programs within the graph. We observe that nodes corresponding to STEM and non-STEM degrees tend to cluster into distinct components with high internal modularity. This is consistent with expectations, as programs within the same disciplinary area often share a substantial portion of their SSDs, resulting in stronger intra-group connections.

However, while non-STEM programs generally form a cluster that is more distant from STEM programs, certain subgroups within STEM exhibit notable similarities with non-STEM disciplines. In particular, the non-STEM cluster includes programs such as Medicine, Economics, and more recent interdisciplinary degrees like Philosophy and Artificial Intelligence, which share a significant number of SSDs with STEM programs. These programs act as bridges between the two clusters, highlighting areas of curricular overlap and interdisciplinary integration. This underscores the heterogeneous nature of some academic programs, especially those located at the intersection of different domains of knowledge.

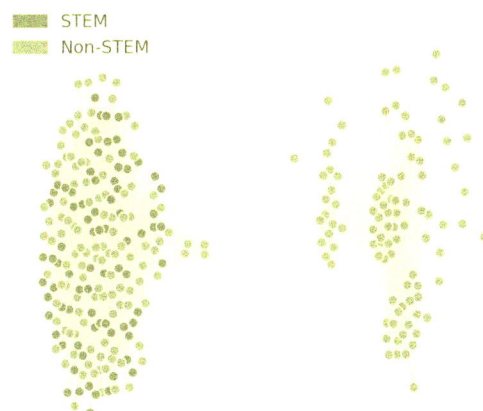

Fig. 4. Similarity graph of degree programs based on the weighted Jaccard index computed over SSD profiles. Each node represents a degree program; blue nodes correspond to STEM programs, while orange nodes represent non-STEM programs. Edges connect programs that share at least one SSD, with lengths proportional to curricular similarity between the connected nodes. (Color figure online)

5   Conclusions

This work discusses the feasibility and the advantages of using knowledge graphs to model the education offering of a large university. By applying this approach to Sapienza University of Rome, and thanks to the development of a specialized web scraper, we have been able to generate SapientIAGraph, our knowledge graph representation of Sapienza educational offering spanning over three hundred degree programs and thousands of modules, enriched with semantic and structural metadata.

We also show how this graph-based representation allows for complex structural analysis including the identification of interdisciplinary degree programs and the assessment of the similarity between different degree programs.

As a future direction, we plan to integrate SapientIAGraph into large language models such as Minerva [15], to explore the possibility of building a software assistant able to deliver AI-powered support to students and university administrators. We also plan to extend the current KG formulation by including more advanced concepts like, e.g., inter-module dependencies, as well as aligning the current graph formulation with existing state-of-the-art ontologies.

Acknowledgments. This work was partially supported by Università di Roma - La Sapienza Research Project 2021 "Caratterizzazione, sviluppo e sperimentazione di algoritmi efficienti".

References

1. Aldrich, P.R.: The curriculum prerequisite network: modeling the curriculum as a complex system. Biochem. Mol. Biol. Educ. **43**(3), 168–180 (2015). https://doi.org/10.1002/bmb.20861
2. Chen, Z., Wang, Y., Zhao, B., Cheng, J., Zhao, X., Duan, Z.: Knowledge graph completion: A review. IEEE Access **8**, 192435–192456 (2020). https://doi.org/10.1109/ACCESS.2020.3030076
3. Cheng, Y.: A learning path recommendation method for knowledge graph of professional courses. In: Proceedings of IEEE International Conference on Quality, Reliability, and Security (QRS), pp. 469–476 (2022)
4. Christou, A., Jaldi, C.D., Zalewski, J., Küçük McGinty, H., Hitzler, P., Shimizu, C.: An ontology for representing curriculum and learning material. arXiv preprint arXiv:2506.05751 June 2025
5. Developers, N.: Networkx: Python software for complex networks. https://networkx.org (2025), Accessed 13 June 2025
6. Rocco, L., Ferraro Petrillo, U., Rombo, S.E.: Diamin: a software library for the distributed analysis of large-scale molecular interaction networks. BMC Bioinf. **23**(1), 474 (2022)
7. Ehrlinger, L., Wöß, W.: Towards a definition of knowledge graphs. In: Joint Proceedings of the Posters and Demos Track of the 12th International Conference on Semantic Systems (SEMANTiCS 2016) and the 1st International Workshop on Semantic Change & Evolving Semantics (SuCCESS'16). CEUR Workshop Proceedings, vol. 1695, pp. 4:1–4:4. CEUR-WS.org, Leipzig, Germany, September 2016, https://ceur-ws.org/Vol-1695/paper4.pdf

8. Grani, G., Di Rocco, L., Ferraro Petrillo, U.: Using knowledge graphs to model green investment opportunities. In: European Conference on Advances in Databases and Information Systems, pp. 440–451. Springer (2023)

9. Guo, Q., et al.: A survey on knowledge graph-based recommender systems (2020), https://arxiv.org/abs/2003.00911

10. Hubert, N., Brun, A., Monticolo, D.: New ontology and knowledge graph for university curriculum recommendation. In: International Semantic Web Conference (ISWC) 2022: Posters, Demos, and Industry Tracks, vol. 3180. CEUR-WS (2022), http://ceur-ws.org/Vol-3180/paper349.pdf

11. Katis, E., Kondylakis, H., Agathangelos, G., Vassilakis, K.: Developing an ontology for curriculum and syllabus. In: Gangemi, A., et al. (eds.) ESWC 2018. LNCS, vol. 11155, pp. 55–59. Springer, Cham (2018). https://doi.org/10.1007/978-3-319-98192-5_11

12. Li, Y., Liang, Y., Yang, R., Qiu, J., Zhang, C., Zhang, X.: Coursekg: an educational knowledge graph based on course information for precision teaching. Appl. Sci. **14**(7), 2710 (2024)

13. Liang, L., et al.: Kag: Boosting llms in professional domains via knowledge augmented generation (2024), https://arxiv.org/abs/2409.13731

14. Neo4j, I.: Neo4j graph database platform. https://neo4j.com (2025), Accessed 13 June 2025

15. Orlando, R., et al.: Minerva LLMs: the first family of large language models trained from scratch on italian data. In: Dell'Orletta, F., Lenci, A., Montemagni, S., Sprugnoli, R. (eds.) Proceedings of the 10th Italian Conference on Computational Linguistics (CLiC-it 2024), pp. 707–719. CEUR Workshop Proceedings, Pisa, Italy, December 2024, https://aclanthology.org/2024.clicit-1.77/

16. Richardson, L.: Beautiful soup. https://www.crummy.com/software/BeautifulSoup/ (2024), Accessed 26 May 2025

17. Schneider, E.W.: Course modularization applied: the interface system and its implications for sequence control and data analysis. Technical Report, HumRRO-PP-10-73, Human Resources Research Organization, Alexandria, VA, November 1973, https://eric.ed.gov/?id=ED088424, paper presented at the Association for the Development of Instructional Systems (Chicago, Illinois, April 1972)

18. SeleniumHQ: Selenium webdriver. https://www.selenium.dev/documentation/webdriver/ (2024), Accessed 26 May 2025

19. Verma, G., et al.: Empowering recommender systems using automatically generated knowledge graphs and reinforcement learning (2025), https://arxiv.org/abs/2307.04996

20. Yang, Y., Chen, S., Zhu, Y., Zhu, H., Chen, Z.: Knowledge graph empowerment from knowledge learning to graduation requirements achievement. PLoS ONE **18**(10), e0292903 (2023). https://doi.org/10.1371/journal.pone.0292903

21. Yu, X., Stahr, M., Chen, H., Yan, R.: Design and implementation of curriculum system based on knowledge graph (2020), https://arxiv.org/abs/2012.12522

Identifying Inconsistent Temporal Triples in Temporal Knowledge Graphs

Evangelos Dagklis[(✉)] and Georgia Koloniari

University of Macedonia, Egnatia 156, 54636 Thessaloniki, Greece
{evdagklis,gkoloniari}@uom.edu.gr

Abstract. Knowledge Graphs (KGs) are a popular means of representing information in a machine readable and interpretable way. This popularity, though, means that KG quality becomes increasingly important. However, most related research considers KGs as static, ignoring their evolutionary aspect. In this work, we focus on Temporal KG (TKG) quality and propose a novel method for detecting inconsistencies within a TKG's triples by leveraging internal information. Our method involves automatically detecting all temporal relations of a TKG and their different variants, and by leveraging their support and confidence metrics, determining the dominant variant, while labeling as potentially inconsistent all triples following the other variants. We showcase the usability of our approach through a first case study on the YAGO Tiny KG, and discuss potential expansions.

Keywords: Temporal Knowledge Graphs · Temporal Validity · Knowledge Graph Quality

1 Introduction

Knowledge Graphs (KGs) are graph structures consisting of entities and relationships, used for representing information, particularly of unstructured or semi-structured nature [10,12], in a manner that is simple, ease to use, and suitable for high-speed scenarios. The popularity of KGs is ever-growing, with them becoming one of the primary components of numerous applications, such as question-answering and recommender systems [12]. However, with KGs' rapid growth, questions arise regarding the correctness and validity of their underlying data. This subject of KG quality has become highly relevant, as a not so well-maintained KG could introduce unnecessary friction with respect to its usability. Thankfully, issues pertaining to KG quality have captured researchers' interest for some time [2,3,11]. There is, though, an area that comparatively feels under-researched. KGs are, usually, essentially treated as static, despite the fact that the evolution of their entities and relationships through time is desirable [10,12].

Temporal Knowledge Graphs (TKGs) are KGs, where part of their data has an associated temporal component, such as a time point or interval. Such temporal

© The Author(s), under exclusive license to Springer Nature Switzerland AG 2026
P. K. Chrysanthis et al. (Eds.): ADBIS 2025, CCIS 2676, pp. 410–419, 2026.
https://doi.org/10.1007/978-3-032-05727-3_34

information can be used to enhance existing knowledge about a subject, provide additional context to certain events, or even understand the evolutionary path of a subject of interest. For instance, it could indicate when specific milestones happened during the life of an artist or a politician, or the uses a place of interest has had throughout time. Despite the immense potential the usage of temporal information presents, existing research on TKGs has mainly focused on integrating temporal information into KG reasoning, primarily through KG embeddings [4,10,12], mostly ignoring other potential uses of the temporal dimension such as historical tracking and evolution analysis.

This work focuses on the *Temporal Validity* of KGs, by introducing a novel method for identifying temporal inconsistencies within a TKG's contents, without necessitating a heavy reliance on external data. Our method aims to find through automatic means temporal KG triples containing potentially erroneous data, through an idea inspired by frequent pattern mining, using support and confidence. Existing research involving such notions includes [8], which mines validating shapes from a KG. In our work instead of mining constraints for isolated triples or elements, we define *temporal relations* between different triples sharing common elements. We identify all temporal relations within a TKG, filter them based on a support threshold, and determine the most dominant variant of each such relation, i.e., precedes, follows or equals, based on confidence values, to detect all inconsistent triples deviating from the dominant variant. While the overall method is under development, we present the application of the core idea of our approach on the YAGO Tiny KG[1], showing that even on a small and curated KG, it has promising results.

The rest of the paper is structured as follows. Section 2 summarizes related research. Section 3 introduces our novel method for identifying temporal inconsistencies inside a TKG's contents. Section 4 presents an application of our approach on a first KG and its results, while Sect. 5 concludes the paper and discusses potential avenues for expanding our work.

2 Related Work

As the main subjects of this study are KG Validity and Validation, as well as the temporal aspect of the KGs, the works reviewed reflect this focus.

KG Validation can be distinguished between validation through *internal information*, which relies exclusively on knowledge derived using only information already contained in the validated KG [2], and *external information* which exploits knowledge from outside sources to evaluate the quality of a KG's elements [2,5,11]. Both methods can be combined to offer improved results [2].

A different means to validate a KG is by creating constraints that specific KG entities should satisfy, while triples violating the specified constraints are considered invalid. Examples of such methods include SHACL, which constitutes a W3C recommendation [1,7] and ShEx [6]. Though having a different syntax

---

[1] Index of /data/yago4.5/.

[7], both methods achieve the above goal through the creation of *validating shapes* [7]. Also, as KG validation is a demanding process in terms of time and computational resources, some works focus on efficiency and scalability [3]. For instance, Trav-SHACL [3] aims at reducing the KG validation workload, by minimizing the number of examined entities while simultaneously maximizing the number of invalidated ones.

Regarding the temporal aspect of the KGs, it has mainly been integrated into methods for TKG reasoning and completion, based on appropriate KG Embeddings [4,10,12]. For instance, most recently in [4], temporal information is added to the proposed variable translation method by considering the TKG as multiple static KGs, with triples that are valid in certain time steps. The temporal information of each KG triple, in each time step, is learned through the TransE model, while the result is integrated into the variable translation to produce the overall embeddings.

3   Methodology

A KG is comprised of entities and relationships. One way to define a KG is by conceiving the connections between entities and relationships through triples.

Definition 1 (KG Triple). *A KG triple or simply a triple is defined by two entities that are connected by a relationship. The different elements of a triple are called subject, predicate, object, with subject and object being the entities and predicate the relationship that connects the two entities.*

We denote triples as $<s^i, p_k^i, o^i>$, where predicate types are denoted as p_k.

To extend a KG to a Temporal KG, we incorporate temporal information in its triples as follows.

Definition 2 (Temporal Triples). *Temporal triples are defined as the triples $<s^i, p_k^i, o^i>$ whose object o^i is an entity containing temporal information, such as a time point or a time interval.*

When pairs of temporal triples share a common subject, connections between them can be formed.

Definition 3 (Temporal Relation). *Given two temporal triples $<s^i, p_k^i, o^i>$ and $<s^j, p_l^j, o^j>$, a temporal relation is defined between predicate types p_k and p_l, denoted as $p_k - p_l$, if and only if, these triples share the same subject: $s^i = s^j$.*

For example, a person can have temporal triples detailing their date of birth and date of death $<person1, birth\text{-}date, 1910>$, $<person1, death\text{-}date, 2020>$. As the two share the same subject, the two predicates are connected through a temporal relation. This temporal relation indicates that an analogous relationship exists between the corresponding objects of the two triples. As the object of a temporal triple contains temporal information such as date or time, the temporal relationship between the two objects is what actually determines the type of temporal relation that connects the two predicates. In particular, in our example, 1910 < 2020 and therefore we infer that "birth-date *precedes* death-date". We define this as a *temporal relation variant*, which constitutes only one of the possible forms this temporal relation can take, and in addition to *precedes*, we also define *follows* and *equals* relation variants. More formally:

Definition 4 (Temporal Relation Variant). *Given a temporal relation $p_k - p_l$ of two temporal triples $<s, p_k^i, o^i>$ and $<s, p_l^j, o^j>$, we define the temporal relation variant of $p_k - p_l$ as: p_k precedes p_l iff $o^i < o^j$, p_k follows p_l iff $o^i > o^j$, and p_k equals p_l iff $o^i = o^j$.*

Following our example the "birth-date *precedes* death-date" is one such variant for the relation "birth-date" − "death-date". The other two are "birth-date *follows* death-date" and "birth-date *equals* death-date".

It is obvious that the "birth-date *follows* death-date" is erroneous and through human intervention such mistakes can be identified and should be corrected to ensure a consistent KG. Furthermore, if schema information is available and we know the correct variant or variants, we can detect triples with potentially inconsistent data. Our goal is to address this issue automatically and without relying on any predefined schema.

To this end, we reduce the notion of the *correct* temporal variant to the most prominent one appearing in the KG triples, defining *the most dominant* temporal relation variant. Following the intuition behind frequent pattern mining where high frequency of item co-occurrences is indicative of underlying behavior and suggests correlation, we also argue that high prominence of one temporal relation variant implies a specific temporal relationship between the corresponding predicates across our data. Support and confidence can be used similarly to association rule mining to assess the quality of the derived dominant relation variants. A similar idea based on frequent pattern mining has also been applied in [8], where shape constraints are mined. These shape constraints, or property shapes, can impose type or cardinality constraints to an entity's properties, while multiple of these shape constraints constitute a shape, which is applied to entities of a specific class and expects these entities to follow the constraints specified in its property shapes. That is, if for some class entity property, a specific type is characterized by higher support and confidence values, then this is assumed to be reflective of the true nature of the underlying data and, lacking any other schema information or external knowledge, it can be used towards extracting a likely KG schema and validating the contained data. Our work is of a somewhat similar spirit, as it also adapts support and confidence, though not to mine shape constraints that should be more indicative of what the shape for a specific class should probably be comprised of, like in [8], but to identify temporal relations and what constitutes their most dominant variants, so as to detect triples violating these most dominant variants, thus triples we consider to be inconsistent. Therefore, the problem we address is formulated as follows.

Problem: *Given a KG, find all temporal relations and corresponding temporal relation variants and, by first detecting the most dominant variant for each relation, detect all the potentially inconsistent temporal triples in the KG.*

Algorithm 1. KG Dominant Temporal Relation Identifier

Input: Temporal Knowledge Graph KG, Minimum Support Value $minSupp$, Minimum Confidence Value $minConf$

Output: The dominant variant of each temporal relation $dominVars$

1: $tripTab = []$ //*Dynamic 2d table containing lists [], where the rows are the subjects s having some temporal data and the columns are the temporal predicates p.*

2: $rels = \{\}$ //*Multidimensional map containing temporal relations $p_k - p_l$ and variants $(<, >, =)$ as keys and the corresponding variant frequencies as values.*

3: $dominVars = \{\}$ //*Dictionary containing a temporal relation as key and its dominant variant as value.*

4: **for each** $<s, p, o>$ in KG **do**

5:     **if** o contains temporal data **then**

6:         Add or update $tripTab[s][p]$, to also contain o

7: **for each** $tripTab.col[p]$ in $tripTab.columns()$ **do**

8:     **if** $listLengthSum(tripTab.col[p])/count(tripTab.rows()) < minSupp$ **then** //*ListlLengthSum refers to the length sum of each list contained in $tripTab.col[p]$.*

9:         Remove $column(p)$ from $tripTab$

10: **for each** s in $tripTab$ **do**

11:     **if** $count(nonEmpty(tripTab[s])) < 2$ **then**

12:         Remove $tripTab[s]$ from $tripTab$

13:     **else**

14:         **for each** p_k, p_l in $tripTab[s]$, where $p_k < p_l$ **do**

15:             $rels[p_k,p_l][<] = count(o^i < o^j)$ //*o^i in $tripTab[s][p_k]$, o^j in $tripTab[s][p_l]$*

16:             $rels[p_k,p_l][>] = count(o^i > o^j)$

17:             $rels[p_k,p_l][=] = count(o^i = o^j)$

18: **for** (p_k,p_l) in $rels$ **do**

19:     **if** $sum(rels[p_k, p_l])/sum(rels) < minSupp$ **then**

20:         Remove $rels[p_k, p_l]$ from $rels$

21:     **else**

22:         $confPrec = (sum(rels[p_k, p_l][<])/sum(rels[p_k, p_l]) > minConf)$

23:         $confFol = (sum(rels[p_k, p_l][>])/sum(rels[p_k, p_l]) > minConf)$

24:         $confEq = (sum(rels[p_k, p_l][=])/sum(rels[p_k, p_l]) > minConf)$

25:         **if** countTrue$(confPrec, confFol, confEq) = 1$ **then**

26:             $dominVars[p_k, p_l] = \max(rels[p_k, p_l])$ //*There is one dominant variant for the $p_k - p_l$ temporal relation.*

27:         **else if** countTrue$(confPrec, confFol) = 1$ **and** $confEq$ **then** //*This is the relaxed version of the dominant variant.*

28:             $dominVars[p_k, p_l] = [\max(rels[p_k, p_l][<], rels[p_k, p_l][>]), rels[p_k, p_l][=]]$ //*Otherwise, there are no dominant variants.*

29: **return** $dominVars$

3.1   Proposed Solution

Based on the principles of frequent pattern mining, our proposed solution is inspired by Apriori. It involves finding the most frequent temporal relations, according to their support metrics, and calculating the confidence of all their variants to determine the dominant one, if such can be defined.

Definition 5 (Support). *The support of a temporal relation $p_k - p_l$ is defined as the ratio of the number of appearances of the $p_k - p_l$ temporal relation, to the number of all the appearances of all temporal relations in the KG:*

$$support = \frac{\#of\ p_k - p_l\ appearances}{\#of\ all\ temporal\ relations\ appearances}$$

Apart from support, which refers to the appearances of one specific temporal relation, compared to the overall appearances of all temporal relations, there is also the notion of *confidence*, which aims to quantify the frequency of one temporal relation variant, compared to that of all of the temporal relation's variants. Thus, we define:

Definition 6 (Confidence). *The confidence of a temporal relation $p_k - p_l$ variant is defined as the ratio of the number of appearances of the specific $p_k - p_l$ variant to the number of appearances of all variants of $p_k - p_l$:*

$$confidence = \frac{\#\ of\ appearances\ of\ a\ p_k - p_l\ variant}{\#\ of\ appearances\ of\ all\ variants\ of\ p_k - p_l}$$

The idea behind confidence is that, if there is only one temporal relation variant with a sufficiently high confidence value, then we can assume that this variant is the dominant one. In case two variants have a high enough confidence and one follows an inequality direction, while the other the equality, then we relax the notion of what constitutes the dominant variant, as it seems the equality direction is also desirable, alongside one of the inequality variants. In all other cases, we assume no dominant variants exist, which would happen if both inequality variants are of a high enough confidence. That would also mean no direction for the temporal relation examined can be defined, or even that no temporal relation exists.

Algorithm 1 details our solution. Upon receiving as input a TKG, and minimum support and confidence values, it first identifies all temporal triples (lines 4–6), calculates their predicates' support, and filters those below the specified threshold (lines 7–9). Then, it prunes all subjects s having fewer than two p, o temporal pairs (lines 10–12). For the rest, it considers all p_k, p_l pairs of frequent predicates, defining $p_k - p_l$ temporal relation variants and counting their appearances according to their o^i, o^j objects (lines 13–17). Non-frequent temporal relations are pruned (lines 18–20), and the confidence of the frequent relations' variants is evaluated. The algorithm accepts either one dominant variant if only one has sufficiently high confidence, or one relaxed version if, with an inequality dominant variant, the equality is of sufficiently high confidence, or none as otherwise, the $p_k - p_l$ temporal relation is considered undefinable (lines 21-28) and finally, returns them (line 29).

After the dominant variants have been determined, they can be used to detect potentially inconsistent temporal triples. By finding these triples, we have succeeded in a step towards quantifying the quality of the validated TKG, by measuring the percentage of inconsistent triples.

Table 1. YAGO Tiny's temporal relations.

$p_k - p_l$	pairs	support	'<' conf.	'=' conf.	'>' conf.
birthDate-deathDate	67705	0.943203	0.997873	0.001300	0.000827
dateCreated-dateCreated	215	0.002995	0.0	1.0	0.0
dateCreated-dissolutionDate	2821	0.039300	0.986884	0.011343	0.001772
startDate-endDate	1041	0.014502	0.943323	0.055716	0.000961

Definition 7 (Temporal Invalidity Percentage (TIP)). *The temporal invalidity percentage of a temporal relation $p_k - p_l$ is defined as the number of triples that constitute its non-dominant variant(s), divided by the overall number of triples of that temporal relation:*

$$TIP_{p_k - p_l} = \frac{2 * \# \ of \ instances \ of \ non \ dominant \ variants \ of \ p_k - p_l}{2 * \# \ of \ instances \ of \ all \ variants \ of \ p_k - p_l}$$

Furthermore, TIP enables us to identify temporal relations with high percentage of inconsistent data that may require further inspection, as either we have wrongly identified the dominant variant or the particular temporal relation is prone to errors. Resorting to external information in such cases may be the only possible solution.

To extend TIP to encompass an entire KG, we can define it as follows:

Definition 8 (KG-wise Temporal Invalidity Percentage (KG-TIP)). *The temporal invalidity percentage of an entire KG is defined as the number of triples that constitute all its non-dominant variants, divided by the overall number of triples of all temporal relations:*

$$KG - TIP = \frac{\sum_{p_k - p_l} 2 * \# \ of \ instances \ of \ non \ dominant \ variants \ of \ p_k - p_l}{\sum_{p_k - p_l} 2 * \# \ of \ instances \ of \ all \ variants \ of \ p_k - p_l}$$

4   Case Study – Application in YAGO Tiny

We apply our method on the YAGO 4.5.0.2 Tiny KG (YAGO Tiny), which is a subset of YAGO 4.5 [9]. YAGO Tiny, while smaller in scale than popular KGs, includes 23260619 triples from which 300556 are temporal, 12454422 unique entities and 122 unique relationships, and is 1.5GB in size, providing sufficient data to showcase the applicability and usability of the proposed approach.

All experiments are conducted locally on a 6-core 12-thread Intel processor at 3.9GHz, with 16GBs of RAM, using Python for the implementation and the lightrdf[2] library for handling the KG.

---

[2] Github - ozekik/lightrdf: A fast and lightweight Python RDF parser.

Table 2. Examples of inconsistent temporal triples, found by the proposed method.

$<s>$	$<p_k^i>$	$<o^i>$	$<p_l^j>$	$<o^j>$
Anna Wecker	deathDate	1596-04-10	birthDate	1600-01-01
ABC Audio	dateCreated	2015-01-01	dissolutionDate	2009-01-01
A Ge	birthDate	1948-01-01	deathDate	1948-01-01
Athletics at 1960 Summer Olympics - Women's 80 m hurdles	startDate	1960-09-01	endDate	1960-09-01

4.1   Evaluation Results

Table 1 includes all temporal relations, their support and the confidence of each of their variants. The thresholds for support and confidence were 1% and 90%, respectively. All predicates are from *http://schema.org/* but the prefix is omitted for brevity. Note that among the detected relations, one between two creation date predicates is found which, given it always appears with an equality variant, suggests the existence of duplicates in YAGO Tiny. Generally, most temporal relations feature multiple variants, some of which do not seem logically plausible, based on the corresponding semantics, showing that even in smaller, relatively well-curated TKGs, inconsistencies do exist, necessitating applying our method.

Through the process of automatic inconsistency identification, a total of 480 triples are deemed as potentially inconsistent. The total $KG - TIP$ is equal to 0.003343, while the corresponding TIP for each temporal relation is:

$$TIP_{birthDate-deathDate} = 0.002127, \ TIP_{dateCreated-dissolutionDate} = 0.013116$$
$$\text{and } TIP_{startDate-endDate} = 0.056676.$$

Some indicative examples of inconsistent temporal triples from YAGO Tiny are shown in Table 2. More precisely, the first record, the one of Anna Wecker, indicates she died before being born, and likewise the second one, that of ABC Audio, indicates it dissolved before being created. For Anna Wecker, it can indeed be said that, her record seems erroneous, given that a human being cannot die before being born. Interestingly, though, the same mistake about Anna Wecker's date of birth and death exists somewhere in her Wikipedia page metadata[3], while in the article, it mentions "(first half 16th century – 1596 in Altdorf near Nuremberg)". Given that YAGO utilizes, among others, Wikipedia data for its creation [9], this seems like a rational explanation for the error. Regarding ABC Audio, it seems potentially inconsistent, as an enterprise cannot be dissolved before its creation, unless it was established, then dissolved, but then reestablished. With respect to A Ge, this is about a person who was born and died in the same day. Without cross-referencing additional external information, this record is just potentially inconsistent, as it is not impossible for someone to be born and die in the same day. In this case though, given that according to Wikipedia[4] A Ge is alive, is a renowned Chinese artist and 1948 corresponds to her birth date, she should not have a death date, so the corresponding YAGO Tiny triples are indeed inconsistent. Finally, the last relation about women's 80 m hurdles at

---

[3] Anna Wecker - Wikipedia.

[4] A Ge - Wikipedia.

the 1960 Summer Olympics indicates a case where the precision in the recorded temporal objects may cause an inconsistency. The object is at day granularity, and without additional information, the record could refer to the event's finals, not to the entire event, so the start and end dates being the same is expected. This indicates why, in some cases, if manual interventions based on additional external information are possible, they could enhance our conclusions, which is a potential we intent to explore in future work.

5   Conclusions and Future Work

This study presents our novel method for validating Temporal Knowledge Graphs by identifying temporal inconsistencies, based on internal information and on leveraging support and confidence from frequent pattern mining. Going beyond applying constraints on single elements of a KG triple, our approach defines temporal relations on pairs of triples by combining triples that share the same subject, provided they have a high enough support. The method is then based on determining the most dominant variant of each temporal relation, according to its confidence, and considering all triples not following this variant as potentially inconsistent. We find it promising that our method's application even on a small curated graph, like the YAGO Tiny KG, has already led to interesting findings, as more than 400 triples were returned as inconsistent. While our work is still in its preliminary phase, based on our results, we argue that such a direction is worth exploring towards ensuring improved TKG quality through validation.

Our future plans include applying our method on multiple TKGs of larger scale that tend to have a more variable structure in their data compared to YAGO (Tiny) or ones containing more complex time-dependent information. Through the exploration of the derived results, we intend to provide further evidence on the scalability and generalizability of our method, while also highlighting its usefulness by presenting detailed analyses of additional case studies. To this end, extending the idea of point wise temporal relations, i.e., precedes, follows and equals to also support interval-based relationships such as overlaps or contains would also be interesting, as a further measure for ensuring generalization. Another idea would be to focus more on the inspiration of our method, which is frequent pattern mining, to also mine association rules and find, for instance, if temporal relations exist that either tend to co-occur, or the occurrence of one reduces the corresponding probability of another. Also, while one of the strengths of the proposed approach is its reliance solely on internal information, leveraging external information, if available, could offer potential avenues for enhancement. For instance, when our approach determines a temporal relation as undefinable if both precedes and follows variants exhibit high confidence, external information can be used to resolve this ambiguity if said temporal relation is actually meaningful. Another idea worth exploring would be to use external information for defining validity intervals for specific temporal relations, through an upper and/or lower bound, i.e., a person's lifespan. A further supplement that would significantly enhance the usability of the proposed approach would be automatic error correction. Its simplest form could be ensuring data consistency. For example, for a person whose death date precedes their birth date, swapping the two dates with each other would lead to a consistent result; whether it would also be correct, that would have to be validated with additional external information. Moreover, defining more metrics for evaluating the temporal quality of a TKG's contents, could prove to be an interesting and valuable direction, targeted towards the ultimate goal of facilitating the processes of TKG Validation.

References

1. Ahmetaj, S., et al.: Common foundations for shacl, shex, and pg-schema. In: Proceedings of the ACM on Web Conference 2025, WWW 2025, pp. 8–21. ACM (2025)
2. Ban, T., Wang, X., Chen, L., Wu, X., Chen, Q., Chen, H.: Quality evaluation of triples in knowledge graph by incorporating internal with external consistency. IEEE Trans. Neural Netw. Learn. Syst. **35**(2), 1980–1992 (2024)
3. Figuera, M., Rohde, P.D., Vidal, M.E.: Trav-shacl: efficiently validating networks of shacl constraints. In: Proceedings of the Web Conference 2021, WWW 2021, pp. 3337–3348. ACM (2021)
4. Han, Y., Lu, G., Zhang, S., Zhang, L., Zou, C., Wen, G.: A temporal knowledge graph embedding model based on variable translation. Tsinghua Sci. Technol. **29**(5), 1554–1565 (2024)
5. Huaman, E., Tauqeer, A., Fensel, A.: Towards knowledge graphs validation through weighted knowledge sources. In: Knowledge Graphs and Semantic Web, pp. 47–60. Springer (2021)
6. Li, B., Yong, S., Yu, B.: A rdf validation tool extended on jena-shex. In: 2022 IEEE 8th International Conference on Computer and Communications (ICCC), pp. 1671–1675 (2022)
7. Rabbani, K., Lissandrini, M., Hose, K.: Shacl and shex in the wild: a community survey on validating shapes generation and adoption. In: Companion Proceedings of Web Conference 2022, WWW 2022, pp. 260–263. ACM (2022)
8. Rabbani, K., Lissandrini, M., Hose, K.: Extraction of validating shapes from very large knowledge graphs. Proc. VLDB Endow. **16**(5), 1023–1032 (2023)
9. Suchanek, F.M., Alam, M., Bonald, T., Chen, L., Paris, P.H., Soria, J.: Yago 4.5: a large and clean knowledge base with a rich taxonomy. In: Proceedings of the 47th International ACM SIGIR Conference on Research and Development in Information Retrieval, SIGIR 2024, pp. 131–140. ACM (2024)
10. Tang, X., et al.: Timespan-aware dynamic knowledge graph embedding by incorporating temporal evolution. IEEE Access **8**, 6849–6860 (2020)
11. Wang, Y., Ma, F., Gao, J.: Efficient knowledge graph validation via cross-graph representation learning. In: Proceedings of the 29th ACM International Conference on Information & Knowledge Management, CIKM 2020, pp. 1595–1604. ACM (2020)
12. Zhang, J., Liang, S., Sheng, Y., Shao, J.: Temporal knowledge graph representation learning with local and global evolutions. Knowl.-Based Syst. **251**, 109234 (2022)

Financial Trading Analytics Knowledge Management

Bhushan Oza(✉) and Fethi Rabhi

School of Computer Science and Engineering, The University of New South Wales, Sydney, NSW 2052, Australia
{b.oza,f.rabhi}@unsw.edu.au

Abstract. This paper aims to address the complex challenges associated with managing knowledge in the context of financial market data analytics. It presents a novel software architecture that makes use of a knowledge graph for building and executing analytics processes. It describes how the different architectural components interact with each other to fulfil the needs of the user(s). The proposed solution is validated by developing a software prototype and testing it on a scenario involving the computing of a financial market measure. This approach underscores the potential of integrating knowledge management practices into analytical environments, especially in handling the complexity and variability of financial market data analytics processes.

Keywords: Knowledge Graphs · Software Architecture · Financial Trading · Data Analytics · Information Systems

1 Introduction

Financial market data analysis is a complex process that involves a plethora of interconnected concepts such as variables and measures [9]. This is further complicated by the growing popularity of using machine learning (ML) algorithms to perform analysis, as implementing and integrating such algorithms can be prone to many errors [9]. These present multiple challenges to novice analysts.

Firstly, understanding these concepts and their intertwined relationships is critical to having a good user experience and successfully utilizing this data for making strategic decisions. Most of this knowledge is implicit and based on a particular analyst's experience.

The second challenge is to make this knowledge shareable and reusable for different types of users to support their analytical tasks. One opportunity to address these challenges is by integrating a knowledge graph [9] into the software architecture that provides analytical programs and services [6].

This paper explores these ideas and is structured as follows: Sect. 2 discusses the related work and identifies the research questions; Sect. 3 describes the proposed design and Sect. 4 discusses the prototype. Finally, Sect. 5 concludes the paper and identifies areas of future work.

P. K. Chrysanthis et al. (Eds.): ADBIS 2025, CCIS 2676, pp. 420–429, 2026.
https://doi.org/10.1007/978-3-032-05727-3_35

2   Related Work and Research Questions

There are many frameworks available for performing financial trading data analysis. Broadly, they can be classified into analytical libraries supported by programming languages, and low code or no code platforms [9].

Furthermore, analytical libraries can be divided into general and special purpose libraries. General libraries provide a rich set of functions for data analysis, where the analyst is responsible for coding overall applications such as in Python. Special purpose libraries are specialized for trading analytics such as computing financial measures. Examples include TA-Lib and QuantLib [10], whereas general purpose libraries are for all sorts of data measures, examples include Keras and TensorFlow [10]. For example, general purpose libraries may provide functions such as mean, standard deviation, etc., whereas special purpose libraries build upon those to provide more functions such as moving averages, volatility measures, and so on. While these libraries are highly customisable, they require expert coding skills, which is a problem for general users, thus lacking user friendliness.

Table 1. Broad Comparison of Existing Frameworks

Existing Framework Categories/Abilities	Customizability	User friendliness
Language supported frameworks	High	Low
No-code and Low-code frameworks	Low - Medium	High

Hence, more research is needed in designing financial market data analysis methods or tools that allow high customisation, and good user experience like maintaining domain and component knowledge for the convenience of end users. This is encapsulated in these research questions:

- How to design a model that captures domain and component knowledge of financial market data machine learning analytics?
- How to design a framework with minimal coding requirement which is more customisable than a no-code platform, and which is more user friendly than a platform requiring expert coding skills?

The next section discusses the artifacts designed to address the above research questions, inspired by the relevant benefits of integrating knowledge graphs in software architectures [6].

3   Proposed Design

3.1   Knowledge Graph

To address the first research question, we propose the Financial Trading Knowledge Graph (FTKG) as a way for users to maintain financial markets domain knowledge when completing analytical tasks. A knowledge graph is a network of entities structured to better understand the different relationships and facilitate effective understanding and analysis of complex information [9], such as financial markets data. In our case, the FTKG is a specialized knowledge graph focused on the domain of financial markets data analysis.

Competency Questions (CQ) are a set of queries that the knowledge graph must be able to answer once it is built and implemented [9]. They define the key requirements, problems, or use cases that the artifact should address. They help ensure the artifact meets the needs of its intended users and serves its purpose effectively. They act as a benchmark against which the quality and utility of the final artifact can be evaluated [9].

Table 2 shows the CQ relevant to the FTKG, along with some instances of different concepts. For example, Liquidity is a financial variable which can be measured by DollarVolumeTraded, a financial measure. Similarly, TradingOrders (variable) can be generated using MLTradingSignals (measure) via "Get ML Trading Signals" (program).

Table 2. Competency Questions (CQ)

CQ#	CQ	Explanation
1	How to represent or measure Variable **e.g. Liquidity**? (returns another Variable or Measure)	The user starts with a variable of interest, and investigates it
2	How to determine Measure **e.g. DollarVolumeTraded**? (returns a DatasetStructure)	The user wants to figure out the dataset in which the corresponding measure belongs to
3	How is DatasetStructure **e.g. IntradayMeasures** obtained? (returns a Program)	The user wants to determine the source of the corresponding dataset
4	What information is needed by a Program **e.g. Build IntraDay Time Series**? (returns DatasetStructure)	The user wants to understand the input of the corresponding program
5	What Implementation is supporting a Program? (returns an Implementation)	The user wants to figure out how the program is implemented

Table 2 introduces various concepts such as Variable, Measure, etc. Figure 1 shows how these concepts are related to each other. The FTKG is based on the Research Variable Ontology (RVO), an ontology designed to catalogue and explore essential data analytics design elements such as variables, analytics models and available data sources [1]. A couple of components, shown in blue, are introduced, while the rest, shown in green, come directly from the RVO.

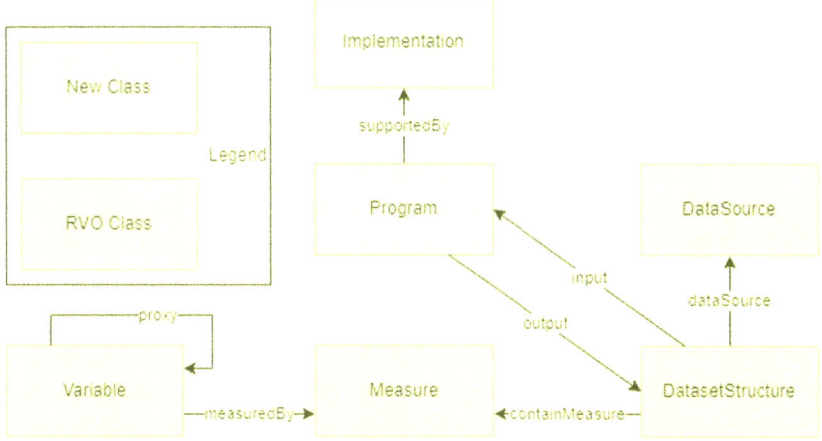

Fig. 1. Main Concepts in the Financial Trading Knowledge Graph

3.2 Architecture

To address the next research question, the FTKG is integrated within an architecture that allows the building and execution of analytics pipelines in a collaborative manner for three categories of stakeholders:

- Users: utilising the system for their analytical requirements, such as analysing liquidity and generating trading orders.
- Knowledge Manager: responsible for updating and maintaining the knowledge graph, as needs of the end users can change.
- Programmer: responsible for updating and maintaining the code library, as needs of the end users can change.

Figure 2 shows the proposed architecture and relations. There are various components in this architecture, organized in different layers. In the storage layer we have:

- Financial Trading Knowledge Graph (FTKG): An organised network of different entities whose model was described in Fig. 1.
- Pipeline Storage: Stores user-constructed analytics pipelines.
- Code Library: Contains software services to execute the pipelines.
- Data Lake: Contains the generated datasets such as intraday timeseries data.
- External Data Sources: Contain the raw data, which is then imported into the Data Lake.

In the business layer we have:

- FTKG Query Engine: Answers queries and allows FTKG to be maintained by the Knowledge Manager.
- Pipeline Builder: Back-end of the FTPB responsible for constructing pipelines by Users.
- Pipeline Executor: Executes the input analytics pipeline, essentially invokes a chain of different analytical programs in the analytics pipeline.

- Data Ingestor: Imports external data.

In the UI layer we have different GUIs that drive the components in the business layer.

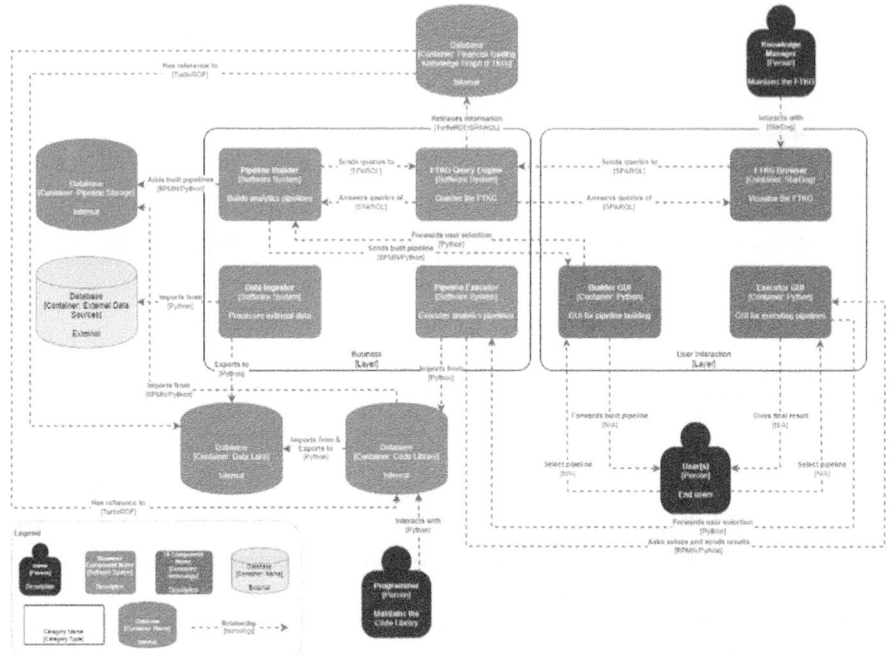

Fig. 2. Proposed Architecture

3.3   Analytics Pipeline Building and Execution

A critical part of this architecture is the interaction between the components responsible for pipeline building and execution. Algorithm 1 illustrates how linear analytics pipeline are constructed.

```
# HELPER FUNCTIONS:
addNodeOnLeft(node, bpmnCode) -> adds the new node to the bpmnCode as a
predecessor of current node
scanUserInput(params) -> scans and returns the user input
query(cqN, arg) -> returns the answer of CQ number and its argument based
on knowledge graph
typeProgram -> returns the type of program, i.e., one of import, export,
or transform.

# ALGORITHM:
build(user_selection):
        bpmnCode <- null
        bpmnCode <- addNodeOnLeft(endEvent, bpmnCode)
        measure <- scanUserInput(user_selection)
        dataset <- query(cq2, measure)
        program <- query(cq3, dataset)
        addNodeOnLeft(program, bpmnCode)
        while typeProgram(program) != import do:
                dataset <- query(cq4, program)
                program <- query(cq3, dataset)
                addNodeOnLeft(program, bpmnCode)
        addNodeOnLeft(start, bpmnCode)
        return (bpmnCode)
```

Algorithm 1. Pipeline Building Algorithm

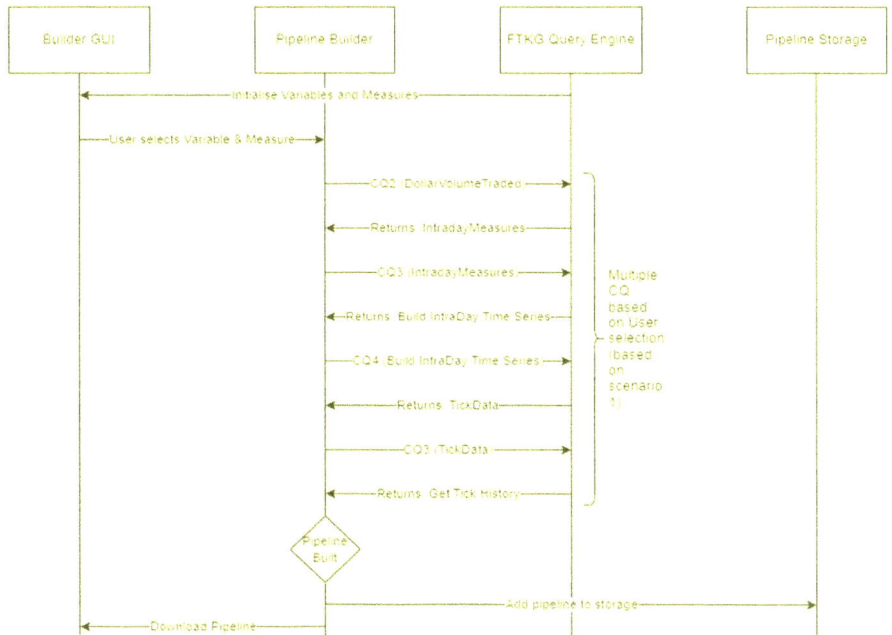

Fig. 3. Interaction between Components

Firstly, a BPMN file is initialised with the endEvent preceded by a task (obtained with the help of FTKG querying) that exports the measure which is extracted from user's selection. Then, a loop builds up the chain of transformation programs until we reach the initial program that imports raw data from a data source. The final BPMN code is then returned and stored in the Pipeline Storage component. This algorithm uses a few helper functions shown and explained just on top of the algorithm.

An example of how pipeline building occurs is illustrated in Fig. 3. Firstly, the FTKG Query Engine initializes the available variables and measures, and the Builder GUI shows this to the users, who make a selection. Then, the Pipeline Builder takes this selection and interacts with the FTKG Query Engine by asking relevant CQ to the FTKG (as per Table 2), to construct a pipeline. This is based on a scenario demonstrated in Sect. 4.2.

Users can use the Executor GUI to execute pipelines from Pipeline Storage. Algorithm 2 describes how to execute a simple linear pipeline.

```
# HELPER FUNCTIONS:
query(cqN, arg) -> returns the answer of CQ number and its argument based
on knowledge graph
start(bpmnCode) -> returns startEvent
end(bpmnCode) -> returns endEvent
next(bpmnCode, node) -> returns the next node
executeI(programFile, values) -> returns the output DatasetStructure of
an import program
executeT(programFile, inputDS, values) -> returns the output
DatasetStructure of a transform program
executeE(programFile, outputDS) -> displays the output of an export
program
findParams(programFile) -> finds and returns parameters from the knowledge
graph
scanUserInput(params) -> scans and returns the user input

# ALGORITHM:
execute(bpmnCode):
        currentNode <- start(bpmnCode)
        currentNode <- next(bpmnCode, currentNode)
        programFile <- query(cq5, currentNode)
        params <- findParams(programFile)
        values <- scanUserInput(params)
        outputDS <- executeI(programFile, values)
        while currentNode != end(bpmnCode) do:
                currentNode <- next(bpmnCode, currentNode)
                programFile <- query(cq5, currentNode)
                params <- findParams(programFile)
                values <- scanUserInput(params)
                if next(bpmnCode, currentNode) == end(bpmnCode):
                    executeE(programFile, outputDS)
                else:
                    newOutputDS <- executeT(programFile, outputDS, values)
                    outputDS <- newOutputDS
```

Algorithm 2. Pipeline Executing Algorithm

This function executes the pipeline components sequentially by retrieving the corresponding programs from the Code Library (obtained with the help of FTKG querying), supplying them with arguments and executing them in the correct sequence (starting with import, then transforms and finally export programs) to provide the user the financial measure of interest. This algorithm uses a few helper functions shown and explained just on top of the algorithm.

4   Prototype

4.1   Implementation

A prototype of the architecture was implemented for testing the effectiveness of the proposed approach on simple analytics scenarios. In the storage layer, the FTKG is represented in the Turtle RDF format, which is a specialised notation to represent relationships in a network [2]. The financial market data is sourced from Refinitiv's DataScope platform [7] and downloaded in CSV format. Refinitiv, a financial data provider, allows access to intraday tick history from various exchanges via its DataScope platform. IntradayMeasures and IntradayOrders datasets are also in CSV format. Pipeline Storage contains constructed pipelines in BPMN format, which is a notation to illustrate relationships among different components of a system [3], while code library contains software services as Python 3 scripts.

In the business layer, the FTKG Query Engine was built using StarDog which is a cloud service enabling knowledge graph creation and analytics [4]; the Pipeline Builder and Executing Engine were implemented using Python 3 libraries.

In the UI layer, the FTKG Browser was built on top of StarDog; the Builder GUI and Executor GUI were built using Python 3 ipywidgets [8].

4.2   Demonstration

Let's consider a scenario where the user is interested in analysing Liquidity (a variable) via DollarVolumeTraded (a measure). Firstly, the user enters their preference and triggers the pipeline building process described earlier. The resulting pipeline, shown in Fig. 4, obtained with the aid of FTKG, from the Builder GUI, is stored in BPMN format. "Get Tick History" and "Download Output" are import and export programs, respectively. The rest are transforming programs. Figure 3 in Sect. 3.3 corresponded to this pipeline and showed how the FTKG querying was helpful.

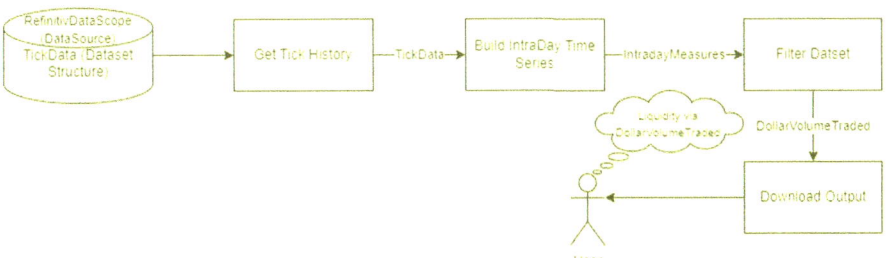

Fig. 4. Pipeline for a Scenario

Now, the user triggers the pipeline execution algorithm. Some of these services (corresponding to "Implementations" in the FTKG) require the user to enter parameters shown in Fig. 5 as different steps.

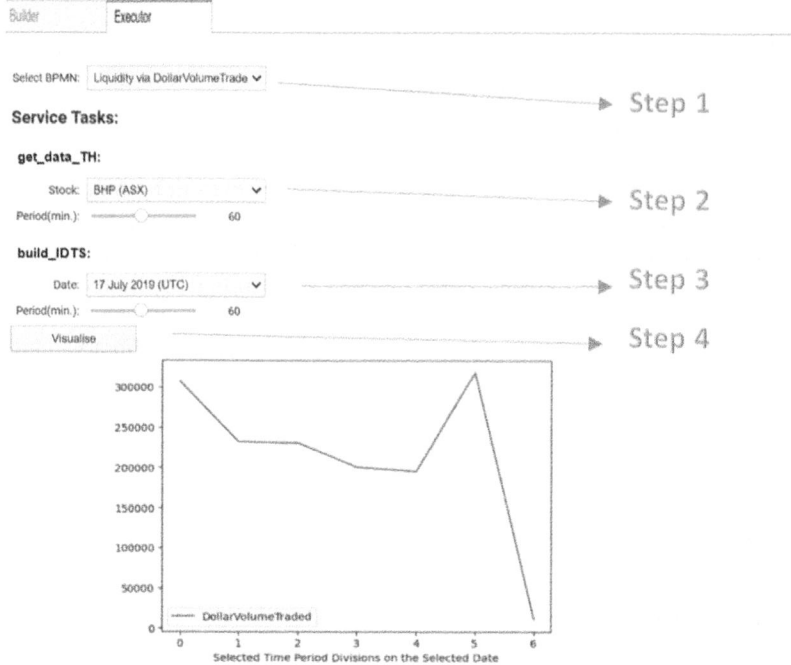

Fig. 5. Execution Steps

The final output for this scenario is shown at bottom of Fig. 5. It is a line graph (money versus user selected aggregation time period) of the intraday measure "Dollar VolumeTraded", generated within the Executor GUI. The user has thus visualized liquidity of their selected financial instrument and date.

Running the scenario shows that the use of FTKG facilitates analytics pipeline construction and execution by making knowledge on various concepts like variables, measures, data sources, etc. explicitly represented. There are still many limitations. Firstly, the FTKG needs to be pre-defined and all programs that are requires must be available in the repository. However, the Knowledge Manager and Programmer are able to update the FTKG and Code Library, respectively, to accommodate changing needs of the end users. Secondly, the execution algorithm only works on linear analytics pipeline and cannot deal with complex graph structures, such as when there are multiple import programs involved.

5   Conclusion and Future Work

This paper presented an innovative solution to the problems faced by users required to develop analytics pipelines for financial markets data analysis. It integrates a Financial Trading Knowledge Graph (FTKG) into a software architecture containing various programs and services to enable the users to complete analytical tasks given the graph is properly configured to represent the various resources needed (e.g. code, datasets, data sources). It illustrates how the different components of the software architecture interact with each other to allow the users to fulfil their requirements via a GUI in a simple analytics scenario.

This theoretical and practical approach allows the separation of concerns and accommodates dynamic needs of users: a knowledge manager updates and maintains financial markets domain knowledge for the convenience of users, while users complete diverse financial markets analytical tasks in a highly customisable and user-friendly environment. Hence, this solution addresses the gaps present in existing frameworks, which were mentioned in Sect. 2 and compared in Table 1.

Future work will involve building large knowledge bases that are able to handle complex equity trading case studies and enhancing the software architecture construction and execution modules to handle these real-world scenarios which may involve complex, non-linear pipelines.

References

1. Bandara, M., Behnaz, A., Rabhi, F.A.: RVO-the research variable ontology. In: The Semantic Web: 16th International Conference, ESWC 2019, Portorož, Slovenia, June 2–6, 2019, Proceedings 16, pp 412–426 (2019)
2. Beckett, D., Berners-Lee, T., Prud'hommeaux, E., Carothers, G.: RDF 1.1 Turtle. World Wide Web Consortium, pp. 18–31 (2014)
3. Chinosi, M., Trombetta, A.: BPMN: an introduction to the standard. Comput. Stand Interfaces **34**, 124–134 (2012)
4. Gupta, R., Malik, S.K.: Visualizing semantic web data using various tools focusing RDF, OWL and SPARQL. In: 2022 11th International Conference on System Modeling & Advancement in Research Trends (SMART), pp. 1456–1460 (2022)
5. Hirzel, M.: Low-code programming models. Commun. ACM **66**, 76–85 (2023). https://doi.org/10.1145/3587691
6. Lung, N.S., Oza, B., Rabhi, F.: Enhancing Collaborative Analytics with an Ontology: A Case Study with High Frequency Data Analysis. FinanceCom, Springer LNBIP, pp. 1–15 (2024) (2024). https://doi.org/10.1007/978-3-031-89933-1_3
7. Marshall, B.R., Nguyen, N.H., Visaltanachoti, N., Zhu, J.: Broker and institutional investor short selling. Acc. Finance **65**, 621–645 (2025)
8. Mease, J.: Bringing ipywidgets Support to plotly. py. In: SciPy, pp 69–76 (2018)
9. Oza, B., Behnaz, A.: Using knowledge graphs for enabling collaborative financial market data analytical processes. Int. J. Complexity Appl. Sci. Technol. **1**, 142–154 (2024)
10. Wang, J., Sun, T., Liu, B., Cao, Y., Wang, D.: Financial markets prediction with deep learning. In: 2018 17th IEEE International Conference on Machine Learning and Applications (ICMLA), pp 97–104 (2018)

CAIMA 2025: 1st Workshop on Cooperative AI Models and Applications

A Parameter-Efficient Approach to Distilling Large Language Models via Meta-learning

Riccardo Cantini$^{(\boxtimes)}$ [iD], Nicola Gabriele [iD], and Alessio Orsino [iD]

University of Calabria, Rende, Italy
{rcantini,nicola.gabriele,aorsino}@dimes.unical.it

Abstract. Large Language Models (LLMs) have revolutionized artificial intelligence, significantly improving performance in tasks such as machine translation, summarization, and conversational systems. These models, however, typically consist of hundreds of millions or even billions of parameters, making them computationally expensive to train and deploy. This presents a major challenge, especially when considering the growing demand to integrate such models into resource-constrained environments like mobile devices or embedded systems. To address this issue, model compression techniques have become essential, such as Knowledge Distillation, which aims to transfer knowledge from a complex model—referred to as the teacher—to a more compact, computationally efficient one—known as the student—without significantly compromising performance. Moreover, recent studies have shown that meta-learning techniques, particularly *learning-to-teach* frameworks, can enhance the distillation process. However, while knowledge distillation via meta-learning is especially effective under high compression ratios, it involves a computationally intensive training process to optimize the teacher's parameters for effective knowledge transfer, leading to substantial resource and energy consumption. To address this issue, we propose a resource-efficient distillation framework that integrates meta-learning with Parameter-Efficient Fine-Tuning (PEFT) techniques, leveraging Low-Rank Adaptation (LoRA) for the teacher's meta-update. By minimizing the computational and memory demands of the distillation process, our approach reduces energy consumption without compromising model performance, ultimately enabling more sustainable AI systems.

Keywords: Knowledge Distillation · Large Language Models · Meta-Learning · PEFT · LoRA · NLP · Sustainable AI

1 Introduction

Large Language Models (LLMs) have revolutionized natural language processing, demonstrating remarkable capabilities across various tasks, including machine translation, summarization, and conversational systems [1,22]. However, these models often contain hundreds of millions to billions of parameters,

P. K. Chrysanthis et al. (Eds.): ADBIS 2025, CCIS 2676, pp. 433–445, 2026.
https://doi.org/10.1007/978-3-032-05727-3_36

presenting significant challenges for training, deployment, and inference, particularly in resource-constrained settings such as edge AI, Internet-of-Things (IoT) systems, or embedded platforms [23,25].

To address these limitations, model compression techniques have been developed to reduce model size and inference cost while preserving performance, with common approaches including *quantization* [11,12], *pruning* [5,17], and *knowledge distillation* [6,19]. In particular, Knowledge Distillation (KD) [8] aims to transfer knowledge from a complex model, known as the teacher, to a more compact and computationally efficient model, referred to as the student. To further enhance the quality of distilled models under extreme compression, recent studies have explored different sophisticated approaches. Among them, explainable AI (xAI)-based techniques aim at conveying the rationale behind the teacher's decisions to the student [2], while meta-learning approaches, especially learning-to-teach paradigms, optimize teacher's parameters based on student feedback to better tailor knowledge transfer to the student model [24]. In parallel, the emergence of Parameter-Efficient Fine-Tuning (PEFT) methods, such as LoRA (Low-Rank Adaptation) [10], prefix tuning [16], and adapter modules [9], has shown that large models can be adapted to new tasks by optimizing only a small number of additional parameters, offering a promising approach to reducing the computational and energy costs of model fine-tuning.

Despite these advances, integrating meta-learning with PEFT for efficient distillation remains underexplored. Notably, while meta-distillation can be effective in high-compression settings [24], it is often resource- and energy-intensive. To address this gap, this paper proposes a novel approach that combines learning-to-teach knowledge distillation with PEFT techniques. Our framework is designed to preserve the strengths of meta-distillation while significantly reducing the energy footprint of the training process by integrating LoRA into the meta-distillation loop. This enables the creation of compact, high-performing models in a more sustainable and resource-efficient manner. In summary, our work makes the following contributions:

1. We propose a resource-efficient distillation framework that integrates learning-to-teach distillation with PEFT—specifically, LoRA—to reduce the computational burden of the distillation process.
2. We analyze the accuracy and efficiency of the distilled student model, showing that our lightweight meta-distillation framework matches performances of traditional methods under high compression ratios while reducing training and inference time, memory use, and energy consumption.
3. We empirically show that student feedback improves knowledge transfer stability across varying temperature values, outperforming traditional distillation methods. Moreover, using PEFT for this optimization further enhances resilience to temperature scaling compared to full fine-tuning.

The remainder of this paper is organized as follows. Section 2 reviews related work on PEFT and distillation via meta-learning. Section 3 describes the proposed framework. Section 4 presents our experimental evaluation. Finally Sect. 5 concludes the paper.

2   Related Work

The proposed techniques lie at the intersection of *Parameter-Efficient Fine-Tuning* and *Knowledge Distillation*, specifically the *Learning-to-Teach* paradigm, drawing on the strengths of both paradigms to improve adaptability and generalization in resource-constrained settings. In this section, we review these approaches, outlining their core ideas and technical details.

2.1   Parameter-Efficient Fine-Tuning

Fine-tuning LLMs can be computationally intensive and energy inefficient due to their high parameter count. *Parameter-Efficient Fine-Tuning* (PEFT) techniques aim to address this challenge by updating only a small subset of parameters while keeping the rest of the model frozen.

PEFT methods can be broadly categorized into four types [7]: *additive* methods, which introduce new tunable components such as adapters, soft prompts, or other lightweight modules; *selective* methods, which fine-tune only a subset of existing model parameters—typically identified via unstructured or structured masking—while keeping the rest frozen; *reparameterized* methods, which modify the model's architecture by injecting low-rank trainable components; and *hybrid* methods, which combine elements from the other categories to leverage their complementary strengths. Among these, *Low-Rank Adaptation* (LoRA) [10] is a widely used reparameterized PEFT approach that inserts trainable low-rank matrices into a pretrained model without modifying its original weights. Let $W \in \mathbb{R}^{d \times d}$ be a weight matrix from the pretrained model. LoRA introduces two low-rank matrices, $A \in \mathbb{R}^{r \times d}$ and $B \in \mathbb{R}^{d \times r}$, where $r \ll d$ is the rank. The forward pass through the adapted layer is formulated as:

$$W'(x) = Wx + BAx \tag{1}$$

During training, only A and B matrices are updated, while W remains frozen. This approach substantially reduces the number of trainable parameters, leading to more memory- and energy-efficient fine-tuning.

2.2   Knowledge Distillation

Knowledge distillation (KD) is a broad research area encompassing various strategies that generally fall into three main categories [2,6]:

- *Offline distillation*, where a teacher model guides the learning process of a separate student model.
- *Online distillation*, where teacher and student are trained simultaneously within a unified framework.

– *Self-distillation*, where a single model teaches itself, either by transferring knowledge from deeper to shallower layers, or by retraining a student with an architecture identical to the original model.

The core goal of KD is to enable the student model to approximate the behavior of the teacher. This is usually accomplished by optimizing a loss function that combines two terms: the traditional cross-entropy (CE) loss computed with true labels, and the distillation loss—typically a Kullback-Leibler (KL) divergence— between the softened output probabilities of the teacher and student, scaled using a temperature parameter (τ). In particular, temperature scaling is used to soften the logits, allowing the student to better learn the nuanced class relationships expressed by the teacher [8]. The typical loss used in a KD process can be expressed as follows:

$$\mathcal{L}_{\text{KD}} = (1 - \alpha) \cdot \text{CE}(\hat{y}, p_S) + \alpha \cdot \tau^2 \cdot \text{KL}(p_T^{(\tau)} \parallel p_S^{(\tau)}) \tag{2}$$

Here, $p_T^{(\tau)}$ and $p_S^{(\tau)}$ represent the temperature-scaled outputs from the teacher model T and the student model S, which are computed as: $p_{\mathscr{F}}^{(\tau)} = softmax(z/\tau)$, where $z = \mathscr{F}(x)$ are the logits (i.e., unnormalized probability values) computed by the model $\mathscr{F}$ (e.g., the teacher T or the student S) on the input x. Furthermore, $\hat{y}$ denotes the hard labels, and α controls the trade-off between the CE and KL terms in the loss. Alternative distillation losses to KL divergence include cross-entropy (CE) and mean squared error (MSE), with the latter encouraging the student model to match the teacher's logits [14].

Knowledge Distillation via Meta-learning. Meta-learning aims to train a model that can rapidly adapt to new tasks using only a few gradient steps by learning an initialization that generalizes well across a distribution of tasks, rather than relying on fixed architectures or update rules [1]. In *Model-Agnostic Meta-Learning* (MAML) [4], this is accomplished through a bi-level optimization process consisting of an inner loop for task-specific adaptation and an outer loop for meta-optimization. In the inner loop, the model adapts its parameters to a specific task $\mathcal{T}_i$ sampled from a distribution $\text{P}(\mathcal{T})$. Given the current initialization θ, the parameters are updated via one or more steps of gradient descent:

$$\theta'_i \leftarrow \theta - \alpha \nabla_\theta \mathcal{L}_{\mathcal{T}_i}(f_\theta) \tag{3}$$

where $\mathcal{L}_{\mathcal{T}_i}(f_\theta)$ is the loss on task $\mathcal{T}_i$ and α is the inner-loop learning rate. This produces a task-adapted set of parameters θ'_i. In the outer loop, the meta-learner updates the initialization θ by evaluating how well the adapted models $f_{\theta'_i}$ perform on their respective tasks. The meta-objective aggregates the losses across tasks and updates θ using another gradient descent step:

$$\theta \leftarrow \theta - \beta \nabla_\theta \sum_{\mathcal{T}_i \sim \text{P}(\mathcal{T})} \mathcal{L}_{\mathcal{T}_i}(f_{\theta'_i}) \tag{4}$$

where β is the outer-loop learning rate. Since the outer loop requires differentiating through the inner-loop update, MAML involves computing second-order derivatives, making it a second-order optimization method. This formulation enables the model to learn an initialization that is sensitive to task-specific learning dynamics and facilitates rapid adaptation with minimal fine-tuning.

In the context of Knowledge Distillation, meta-learning enables a *learning-to-teach* paradigm [24], where the student learns the downstream task, and the teacher learns to better transfer knowledge. The meta-distillation procedure, consists of three steps:

1. *Distillation Trial*: a temporary copy $\hat{\theta}_S$ of the student model is updated using the distillation loss $\mathcal{L}_S$ and the teacher θ_T. The update rule is as follows: $\hat{\theta}_S \leftarrow \theta_S - \lambda \nabla_{\theta_S} \mathcal{L}_S(x; \theta_S, \theta_T)$, where x is a training batch.
2. *Teacher Meta-Update*: the teacher is updated based on how well the updated student $\hat{\theta}_S$ performs on a held-out quiz set Q. The teacher update rule is expressed as: $\theta_T \leftarrow \theta_T - \mu \nabla_{\theta_T} \mathcal{L}_T(q; \hat{\theta}_S(\theta_T))$, where $q \sim Q$ and $\mathcal{L}_T$ is the downstream loss on the quiz set.
3. *Actual Distillation*: The original student θ_S is updated using the improved teacher θ_T via the distillation loss $\mathcal{L}_S$, which combines task hard labels $\hat{y}$ with soft targets from the teacher. A temperature factor τ is used to smooth the logits of both models to facilitate knowledge transfer:

$$\mathcal{L}_S = \underbrace{(1-\alpha) \cdot \mathcal{L}_{\text{CE}}(x, \theta_S, \hat{y})}_{\text{Cross-Entropy on hard labels}} + \underbrace{\alpha \cdot \tau^2 \cdot \mathcal{L}_{\text{KL}}^{(\tau)}(x, \theta_S, \theta_T)}_{\text{KL div. on soft labels}} \qquad (5)$$

PEFT-Enhanced Knowledge Distillation. While Knowledge Distillation facilitates efficient inference through smaller, lightweight models, its training process can still be computationally and energy intensive. To address this limitation, recent research has investigated the integration of Knowledge Distillation with PEFT techniques, aiming to retain the benefits of model compression while reducing the training overhead. For example, [13] introduces a teacher-student framework where the student model is adapted through LoRA, learning not only from the teacher's output logits but also from its intermediate hidden states. This dual-source guidance enables more effective knowledge transfer while keeping the number of trainable parameters low. Similarly, recent works [3,20] employ PEFT techniques for both teacher pre-training and student distillation, reducing training costs while maintaining the effectiveness of the distillation process.

Although online distillation has been shown to outperform offline methods by narrowing the capacity gap between teacher and student models [15,24], its reliance on training both models jointly limits its practicality due to increased computational costs. To address this, recent work has explored integrating PEFT into online distillation. For example, in [18] the teacher model is adapted via LoRA using both hard labels and the soft labels generated by the student, which learns from the teacher in a bi-directional manner. Similarly, our work integrates LoRA into online distillation but, unlike prior approaches, employs a meta-learning framework where the teacher is updated based on feedback from the student model, thereby enhancing both the student's performance and the teacher's knowledge transferability. Furthermore, we evaluate the effectiveness of LoRA-augmented online distillation in a cross-architecture setting to enable deployment of the lightweight student in resource-constrained environments, enhancing the adaptability and practical applicability of the distilled model.

3   Proposed Framework

The proposed lightweight meta-distillation framework leverages LoRA to enable efficient teacher adaptation and improve knowledge transfer in a parameter-efficient manner. Specifically, the meta-distillation process consists of three stages, as depicted in Fig. 1. First, in the *Distillation Trial* stage, a copy of the student is trained via distillation from a teacher composed of frozen pretrained weights and trainable LoRA adapters. Next, in the *LoRA-Based Teacher Meta-Update* stage, the student copy is evaluated on a held-out set (i.e., the *quiz set*), and the resulting feedback updates only the teacher's LoRA modules. Finally, in the *Actual Distillation* stage, the adapted teacher is employed to distill knowledge into the original student model.

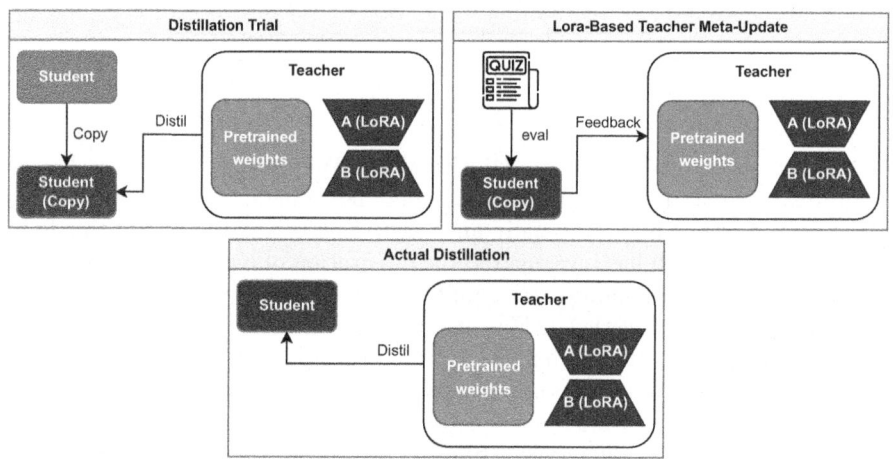

Fig. 1. Overview of the proposed PEFT-based meta-distillation framework. Red modules are trainable and blue modules are frozen, while grey modules are inactive. (Color figure online)

As illustrated above, this meta-distillation process enables the teacher to iteratively refine its ability to guide the student's learning, thereby enhancing its supervision to maximize knowledge transfer. To improve resource efficiency, we apply LoRA [10] by attaching low-rank matrices to the Key, Query, and Value projections in self-attention layers. This approach freezes the original projection matrices and updates only LoRA parameters, reducing computational overhead and speeding up distillation without compromising quality. Specifically, the Key, Query, and Value matrices of the i-th self-attention layer are modified as follows:

$$W^{(i)}(x) = W^{(i)}x + B^{(i)}A^{(i)}x \qquad (6)$$

where $W^{(i)}$ generically identifies one of the Key, Query, or Value matrices $\in \mathbb{R}^{d \times d}$, while $A^{(i)} \in \mathbb{R}^{r \times d}$ and $B^{(i)} \in \mathbb{R}^{d \times r}$ are the associated low-rank matrices. In this case, the teacher update rule becomes:

$$\theta_{W \cup \text{LoRA}} \leftarrow \theta_{W \cup \text{LoRA}} - \nabla_{\theta_{W \cup \text{LoRA}}} \mathcal{L}_T(q; \hat{\theta}_S(\theta_{W \cup \text{LoRA}})) \qquad (7)$$

where $\theta_{W \cup \text{LoRA}}$ are the teacher parameters, $\mathcal{L}_T$ is the student's task loss, and $\hat{\theta}_S$ are the student parameters after the distillation trial with the current teacher. Since the weight matrix W is frozen, gradients are taken only with respect to LoRA parameters, and the following equality holds:

$$\nabla_{\theta_{W \cup \text{LoRA}}} \mathcal{L}_T(q; \hat{\theta}_S(\theta_{W \cup \text{LoRA}})) = \nabla_{\theta_{\text{LoRA}}} \mathcal{L}_T(q; \hat{\theta}_S(\theta_{\text{LoRA}})) \qquad (8)$$

This decoupling of the frozen base weights from the optimization process not only simplifies the gradient computation but also enables more efficient parameter updates during teacher adaptation. Consequently, the number of trainable parameters per modified layer drops from d^2 to $2rd$, significantly reducing computational cost, given that $r \ll d$. This reduction is especially beneficial in large-scale models, where full meta-updates would otherwise incur prohibitive memory and compute overheads.

4   Experimental Evaluation

This section presents the experimental evaluation of our PEFT-based meta-distillation framework, referred to as METADISTIL + LORA. Specifically, our analysis focuses on three main aspects:

– We evaluate the classification performance of the student model after the meta-distillation process, highlighting the benefits of the feedback mechanism in enhancing knowledge transfer from teacher to student.
– We examine sustainability-related aspects of the approach, demonstrating how the use of LoRA significantly reduces both the carbon footprint and training time compared to full fine-tuning within the meta-distillation loop.
– We investigate the role of the feedback mechanism in improving resilience to variations in temperature initialization, particularly when LoRA is used in the teacher's meta-update.

The following sections outline our experimental setting and presents the achieved results, discussing the advantages brought by METADISTIL + LORA over both traditional and meta-learning-based knowledge distillation.

Experimental Setting. The experiments were conducted using FinBERT [21], a BERT model pre-trained on financial communication text and further fine-tuned on the Twitter Financial News dataset. This English-language corpus contains finance-related tweets (9,938 for training and 2,486 for testing) categorized into three sentiment classes: *bearish*, *bullish*, and *neutral*. For the student model, we employed a small LSTM network to simulate a high-compression scenario, reducing the number of parameters by more than 99%, and compressing the model size from 413.3 MB to 1.7 MB, achieving a compression ratio of 99.6% relative to the Transformer-based teacher model. Specifically, the LSTM model employed 50 hidden units and a sequence length of 150, using a GloVe embedding layer to produce 50-dimensional latent representations of input tokens. We compared METADISTIL + LoRA against the following baselines:

- The original teacher model, serving as the upper bound for classification performance.
- A student model trained from scratch without knowledge distillation, using a standard cross-entropy loss.
- Distilled student models obtained with classic KD [8] using different distillation losses (i.e., Kullback-Leibler (KL), Mean Squared Error (MSE), and cross-entropy (CE)).
- A student model trained via MetaDistil [24], which fully fine-tunes the teacher in the meta-update step.

We evaluated various optimizers for student and teacher models, selecting RMSProp as the best-performing optimizer for the student and AdamW for the teacher. All models were trained for 20 epochs. Moreover, we used a LoRA rank $r = 16$ in METADISTIL + LoRA and a temperature $\tau = 5$ for all distillation techniques that employ temperature scaling. All experiments were conducted on a machine equipped with an NVIDIA A30 GPU with 24 GB of memory.

4.1   Classification Performance

Fig. 2 shows classification performance, measured by accuracy and macro F1 score, across training epochs ranging from 5 to 20. Among all evaluated methods, METADISTIL + LoRA achieves the fastest convergence, reaching higher performance in fewer epochs. Throughout all training durations, it consistently outperforms other methods, including both classic and meta-learning approaches, demonstrating robustness and effectiveness across varying epoch budgets.

Notably, when compared to MetaDistil—which employs full fine-tuning for teacher meta-optimization—METADISTIL + LoRA not only matches but slightly surpasses its performance, despite relying on a significantly smaller number of trainable parameters. This suggests that LoRA-based adaptation enables more efficient and generalizable knowledge transfer by introducing a stronger inductive bias and reducing the risk of overfitting during meta-updates. Compared to classic knowledge distillation methods (based on KL, MSE, and

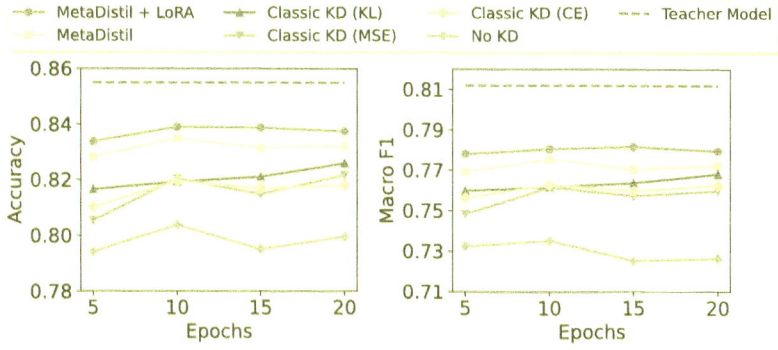

Fig. 2. Comparison of classification performance across knowledge distillation methods over training epochs. All results are averaged over 10 runs.

CE losses), both meta-distillation approaches demonstrate substantial improvements. These gains emphasize the importance of feedback-driven teacher optimization, which dynamically adjusts the teacher's output distribution to better suit the student model's limited capacity and learning dynamics. As expected, the student model trained without any distillation (No KD) performs the worst, confirming the importance of teacher supervision when training highly compact architectures.

Overall, achieved results highlight three key insights: (i) knowledge distillation is essential for maintaining performance under extreme compression, (ii) meta-learning significantly enhances distillation effectiveness through feedback-driven optimization of the teacher's parameters, and (iii) integrating LoRA into the meta-distillation loop enables efficient knowledge transfer by substantially reducing computational overhead.

4.2   Accuracy-Sustainability Trade-Off

Figure 3 illustrates the trade-off between classification accuracy and training efficiency across various distillation methods. The x-axis denotes training time per epoch (in seconds), while the y-axis represents classification accuracy. Each data point is annotated with its estimated carbon emissions per epoch, measured in grams of CO_2.

Among all approaches, METADISTIL + LoRA offers the most favorable balance between accuracy and sustainability, achieving the highest accuracy while significantly reducing both training time and energy consumption compared to full fine-tuning with MetaDistil. Specifically, METADISTIL + LoRA reduces training time by 40% (from 38 to 22.72 s) and carbon emissions by 42% (from 1.43 to 0.83 g CO_2) per epoch, while slightly outperforming MetaDistil in classification accuracy. This demonstrates that LoRA-based adaptation not only preserves model performance but also substantially improves the energy efficiency of the meta-distillation process. As a result, METADISTIL + LoRA emerges as

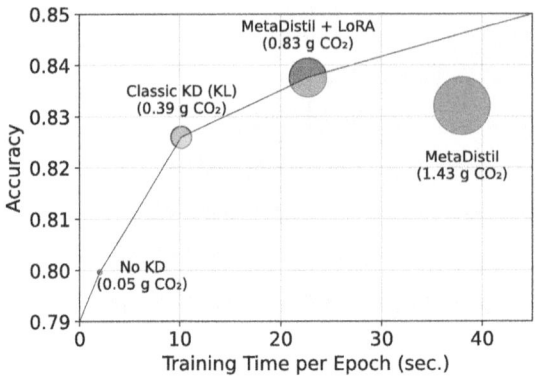

Fig. 3. Cross-domain comparison (accuracy vs. training time). Circle size reflects CO_2 per epoch. Pareto-efficient solutions lie on the black front, while the green-shaded area indicates suboptimal (dominated) accuracy–efficiency trade-offs. (Color figure online)

a Pareto-efficient solution that effectively balances accuracy and environmental impact, making it well-suited for generating highly compressed student models and enabling sustainable AI training and deployment in resource-constrained real-world settings. Classic knowledge distillation (KD) methods exhibit similar training times and carbon emissions regardless of the specific distillation loss used. Accordingly, Fig. 3 reports results using KL divergence as the distillation loss, as it yields the highest accuracy among classic KD variants. While this distillation approach offers a relatively favorable trade-off, it lacks the feedback-driven optimization of meta-distillation frameworks, highlighting the limitations of traditional KD under high compression. Finally, training the student without any distillation (No KD) is the fastest and most carbon-efficient method ($0.05\,\mathrm{g}$ CO_2 per epoch), but it results in significant accuracy loss, underscoring the ineffectiveness of unguided training compared to KD approaches.

4.3   Resilience to Temperature Initialization

Figure 4 evaluates the stability of knowledge transfer across temperature values $\tau \in 1, 2, 3, 4, 5$, reporting mean classification accuracy and standard deviation for each method. Temperature controls the softness of the teacher's output distribution and can significantly affect distillation quality.

The results show that distillation techniques based on meta-learning achieve significantly lower variance compared to classical knowledge distillation methods. In particular, METADISTIL + LORA not only obtains the highest mean accuracy but also demonstrates the smallest standard deviation ($\sigma = 1.81 \times 10^{-3}$), indicating higher robustness to suboptimal temperature choices. This improvement is attributable to the meta-optimization process, which adapts the teacher's behavior based on feedback from the student, mitigating the sensitivity of the distillation process to temperature scaling. Moreover, the incorporation of LoRA

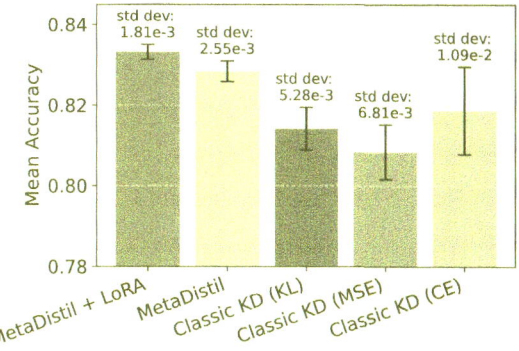

Fig. 4. Resilience to varying temperature values τ (from 1 to 5), measured by mean accuracy and corresponding standard deviation across distillation methods.

in the meta-update step enhances this robustness, offering a robust and accurate approach for distilling knowledge into high-quality student models. In contrast, classic KD techniques, especially those using cross-entropy as the distillation loss, exhibit notably higher variance (e.g., $\sigma = 1.09 \times 10^{-2}$ for CE), making them more sensitive to the choice of τ.

5 Conclusions

In this work, we introduced a resource-efficient meta-distillation framework that integrates Parameter-Efficient Fine-Tuning (PEFT)—specifically Low-Rank Adaptation (LoRA)—into the learning-to-teach paradigm for knowledge distillation. Our approach addresses the high computational and energy demands typically associated with meta-distillation by significantly reducing the number of trainable parameters during the teacher's optimization phase, all while preserving or even enhancing the performance of the distilled student model. Experimental evaluation, involving the distillation of a BERT-like teacher model into a lightweight LSTM student, demonstrated that our PEFT-based meta-distillation framework matches the performance of traditional methods under high compression ratios, while consistently reducing training and inference time, memory usage, and energy consumption. Additionally, incorporating student feedback to optimize the teacher model improves resilience to temperature scaling—a common hyperparameter sensitivity in knowledge distillation—with the lowest variance achieved when using LoRA in the meta-update instead of full fine-tuning.

Overall, this work highlights the practical advantages of combining meta-learning with parameter-efficient techniques, enabling the development of compact, high-performing student models through a more sustainable, effective, and resource-efficient process. As future directions, we will explore alternative PEFT strategies in place of LoRA and apply the framework to more complex downstream tasks—such as question answering and summarization—to better assess its applicability to broader NLP tasks and real-world scenarios.

Acknowledgments. This work has been supported by the "FAIR – Future Artificial Intelligence Research" project - CUP H23C22000860006 and the "National Centre for HPC, Big Data and Quantum Computing", CN00000013 - CUP H23C22000360005.

References

1. Brown, T., Mann, B., Ryder, N., et al.: Language models are few-shot learners. In: Advances in Neural Information Processing Systems (2020)
2. Cantini, R., Orsino, A., Talia, D.: Xai-driven knowledge distillation of large language models for efficient deployment on low-resource devices. J. Big Data **11**, 63 (2024)
3. Chen, R., Su, Y., Jing, H.: Knowledge distillation framework of pre-trained language models combined with parameter efficient fine-tuning. In: 2024 China Automation Congress (CAC) (2024)
4. Finn, C., Abbeel, P., Levine, S.: Model-agnostic meta-learning for fast adaptation of deep networks. In: International Conference on Machine learning (2017)
5. Frantar, E., Alistarh, D.: Sparsegpt: massive language models can be accurately pruned in one-shot. In: International Conference on Machine Learning (2023)
6. Gou, J., Yu, B., Maybank, S.J., Tao, D.: Knowledge distillation: a survey. Int. J. Comput. Vis. **129**(6) (2021)
7. Han, Z., Gao, C., Liu, J., Zhang, J., et al.: Parameter-efficient fine-tuning for large models: a comprehensive survey. arXiv preprint arXiv:2403.14608 (2024)
8. Hinton, G., Vinyals, O., Dean, J.: Distilling the knowledge in a neural network. arXiv preprint arXiv:1503.02531 (2015)
9. Houlsby, N., Giurgiu, A., Jastrzebski, S., et al.: Parameter-efficient transfer learning for nlp. In: International Conference on Machine Learning (2019)
10. Hu, E.J., Shen, Y., Wallis, P., et al.: Lora: low-rank adaptation of large language models. In: Proceedings of International Conference on Learning Representations (2022)
11. Hubara, I., Courbariaux, M., Soudry, D., et al.: Quantized neural networks: training neural networks with low precision weights and activations. J. Mach. Learn. Res. **18**, 1–30 (2018)
12. Jacob, B., Kligys, S., Chen, B., Zhu, M., et al.: Quantization and training of neural networks for efficient integer-arithmetic-only inference. In: Proceedings of the IEEE Conference on Computer Vision and Pattern Recognition (2018)
13. Kai, A., Zhu, L., Gong, J.: Efficient compression of large language models with distillation and fine-tuning. J. Comput. Sci. Softw. Appl. **3**, 30–38 (2023)
14. Kim, T., Oh, J., Kim, N.Y., Cho, S., Yun, S.-Y.: Comparing kullback-leibler divergence and mean squared error loss in knowledge distillation. In: Proceedings of the Thirtieth International Joint Conference on Artificial Intelligence (2021)
15. Li, L., Jin, Z.: Shadow knowledge distillation: bridging offline and online knowledge transfer. In: Advances in Neural Information Processing Systems (2022)
16. Li, X.L., Liang, P.: Prefix-tuning: Optimizing continuous prompts for generation. In: Proceedings of the 59th Annual Meeting of the Association for Computational Linguistics and the 11th International Joint Conference on Natural Language Processing (2021)
17. Michel, P., Levy, O., Neubig, G.: Are sixteen heads really better than one? In: Advances in Neural Information Processing Systems, vol. 32 (2019)

18. Rao, J., et al.: Parameter-efficient and student-friendly knowledge distillation. IEEE Trans. Multimedia **26**, 4230–4241 (2024)
19. Tang, R., Lu, Y., Liu, L., Mou, L., Vechtomova, O., Lin, J.: Distilling task-specific knowledge from bert into simple neural networks. arXiv preprint arXiv:1903.12136 (2019)
20. Yang, R., et al.: LLM-Neo: parameter efficient knowledge distillation for large language models. arXiv preprint arXiv:2411.06839 (2024)
21. Yang, Y., Uy, M.C.S., Huang, A.: FinBERT: a pretrained language model for financial communications. arXiv preprint arXiv:2006.08097 (2020)
22. Zhao, W.X., Zhou, K., Li, J., Tang, T., et al.: A survey of large language models. arXiv preprint arXiv:2303.18223 (2023)
23. Zheng, Y., Chen, Y., Qian, B., Shi, X., Shu, Y., Chen, J.: A review on edge large language models: design, execution, and applications. ACM Comput. Surv. **57**, 1–35 (2025)
24. Zhou, W., Xu, C., McAuley, J.: BERT learns to teach: knowledge distillation with meta learning. In: Proceedings of the 60th Annual Meeting of the Association for Computational Linguistics (2021)
25. Zhu, X., Li, J., Liu, Y., Ma, C., et al.: A survey on model compression for large language models. Trans. Assoc. Comput. Linguist. **12**, 1556–1577 (2024)

Underrepresentation of Dark Skin Tone in Skin Lesion Datasets: The Role of the Explainable Techniques in Assessing the Bias

Tommaso Ruga[1,2](✉) ⓘ, Ester Zumpano[1,2] ⓘ, Eugenio Vocaturo[1,2] ⓘ,
and Luciano Caroprese[3] ⓘ

[1] DIMES - University of Calabria, Rende, CS, Italy
{tommaso.ruga,e.zumpano}@dimes.unical.it
[2] CNR-NANOTEC National Research Council, Rende, CS, Italy
eugenio.vocaturo@cnr.it
[3] InGeo - University G. D'Annunzio, Chieti-Pescara, Chieti, Italy
luciano.caroprese@ingeo.it

Abstract. Advanced artificial intelligence models for skin lesion classification often suffer from performance disparities when applied to images of patients with darker skin tones, largely due to underrepresentation of dark skin tone images in training datasets. In this study, we investigate this issue by evaluating a previously proposed explainable framework, MultiExCAM, trained on the widely used ISIC2018 dataset. We test its performance on *Pipsqueak*, a previously proposed dataset composed by skin lesion images on dark skin tones. As expected, we observe a significant drop in classification performance when the model is applied to *Pipsqueak*. To better understand the source of these failures, we employ explainable artificial intelligence techniques to visualize and analyze the model's decision-making process on both datasets. Our results highlight clear differences in attention patterns and decision rationale, revealing how the lack of dark skin tone representation in the training data leads to poor generalization and biased behavior. This work emphasizes the critical role of explainable analysis in exposing and understanding model bias in clinical applications, and the necessity of inclusive datasets for fair and reliable skin lesion classification.

Keywords: Melanoma Classification · Dataset Bias · Explainable AI · Skin Tone Diversity

1 Introduction

Melanoma is the most aggressive type of skin cancer. In 2022, 331,647 new cases of melanoma has been registered worldwide [3]. Despite its severity, early diagnosis facilitates clinical treatment by allowing the malignant lesion to be removed from the patient in time. In this regard, artificial intelligence has for

P. K. Chrysanthis et al. (Eds.): ADBIS 2025, CCIS 2676, pp. 446–456, 2026.
https://doi.org/10.1007/978-3-032-05727-3_37

years provided effective solutions capable of analyzing large amounts of dermatological images, classifying the represented skin lesions, and distinguishing malignant ones from benign ones [12,13]. However, this distinction is not always straightforward. Certain cases present ambiguities, failing to immediately reveal their malignant nature, while in others the clinical context may be difficult to interpret. This is particularly true for skin lesions on darkly pigmented skin, where the typical features of melanoma—such as dark pigmentation and irregular borders—may not be as easily discernible. Melanoma is less common in individuals with dark skin, as they are generally less sensitive to ultraviolet radiation. However, when it does occur, it tends to be diagnosed at more advanced stages, with a worse prognosis compared to individuals with fair skin. In fact, in people with dark skin, melanoma more frequently develops in areas less exposed to sunlight, such as the palms of the hands, soles of the feet, and beneath the nails. Approximately 60–70% of melanomas in African American individuals arise in these locations [2]. An additional issue further complicates this scenario: the underrepresentation of darker skin tones in medical imaging datasets. Relying on Fitzpatrick scale, it is possible to highlights that most datasets are dominated by light-skinned (types I-III) images, with less than 5% representing darker skin (types IV-VI). This lack of representation leads to ethical and clinical risks, as melanoma and other conditions can manifest differently on dark skin, risking misdiagnosis or missed detection. Three main challenges arise: dataset bias, with repositories skewed towards European/North American populations; algorithmic limitations, where AI models show lower accuracy and higher false negatives for darker skin; and clinical implications, contributing to late-stage diagnoses and worsened outcomes for people of color, reinforcing healthcare disparities. For this reason, this paper proposes an empirical and explainable analysis of the described classification scenario, leveraging *MultiExCam*, a framework for skin lesion classification previously introduced in [9], and two datasets: *HAM10000*, well-known in the literature and used to train the framework, and *Pipsqueak*, a new dataset specifically created containing images of skin lesions on individuals with higher Fitzpatrick skin phototypes ($> IV$). The study thoroughly examines the consequences of biases introduced by datasets that underrepresent darker phototypes, employing two eXplainable Artificial Intelligence (XAI) techniques, namely Grad-CAM and SHapley Additive exPlanations (SHAP), to highlight potential errors in the predictive process and analyze their causes.

2   Related Work

The risks of excluding images of darker skin tones from AI training datasets are widely discussed in literature. Adamson and Smith [1], in their influential perspective published in JAMA Dermatology, warned that training algorithms on homogeneous datasets lacking representation of skin of color may result in incorrect diagnoses or the complete omission of malignant lesions in individuals with darker skin types. Similarly, Han et al. [6] provided empirical evidence supporting these concerns. Their findings demonstrated that AI models trained on datasets composed exclusively of images from Asian populations performed substantially better on that same population (96% accuracy for basal cell carcinoma and melanoma classification) compared to performance on Caucasian skin images (88âĂŞ90% accuracy). Kushimo et al. [7] addressed this critical gap by focusing on improving melanoma detection in individuals with darker skin tones through deep learning techniques. They proposed a custom model based on DenseNet121 architecture, pre-trained on ImageNet and fine-tuned for melanoma classification. Their dataset included approximately 100 clinical images of dark skin (37 of which were melanomas), supplemented with images from the widely-used HAM10000 dataset. Following extensive preprocessing— cropping, denoising, illumination correction, segmentation—and data augmentation steps to expand the dataset to 18,374 images, the model achieved outstanding accuracy rates: 99% for melanoma detection in light skin and 98% in dark skin. Additional performance metrics (precision, recall, specificity, and F1-score) ranged from 90% to 99% across skin tones. These results outperformed many previously reported models, highlighting the potential of tailored AI approaches in addressing underrepresentation. Unfortunately, the dataset has not been made publicly available. Further highlighting this challenge, Han et al. [7] demonstrated limited cross-population generalizability when algorithms trained on East Asian skin images were tested on White patients in the United States. Collectively, these studies underscore the critical importance of constructing dermatological AI systems with diverse, well-annotated datasets that include comprehensive demographic and skin type information. They also provide concrete evidence that limited population representation in training datasets leads to measurable performance disparities, risking inequitable clinical outcomes across different patient groups.

3   *MultiExCAM*: An Explainable Framework for Melanoma Classification

MultiExCAM, introduced in [9], is a framework for skin lesion classification that integrates DL and ML techniques to achieve high performance while maintaining explainability. The architecture comprises four stages: i) Data Preprocessing and Balancing, ii) DL classification and Deep Feature Extraction, iii) ML Classification, and iv) Ensemble Classification. The first stage includes image preparation and cleaning, such as removing artifacts like body hair that can affect performance. During the first stage, images are loaded, stored, and prepared for analysis. This essential step includes tasks such as image cleaning, through the removal of artifacts like body hair which can disturb model performances. To reduce overfitting, data augmentation techniques are applied to the training set, promoting better generalization. In the second stage, a ResNet50 based model—employing transfer learning—performs two key roles: it provides an initial classification, which is used as a baseline prediction of skin lesions for the entire process, and deep features, where deep features are extracted from the DL model penultimate layer for use in subsequent stages. The third stage introduces a set of four different classifiers, each processing a combination of deep features and hand-crafted statistical features. The hand-crafted features are computed by extracting information related to texture and color distribution for each image, which mimic clinical process and enhance framework explainability. These classifiers include an Extra Trees Classifier, K-Nearest Neighbors (KNN), a Support Vector Machine (NuSVM variant), and XGBoost. Their diversity contributes to the robustness of the overall system. Finally, in the fourth stage, an ensemble classifier in the form of a Feed-Forward Neural Network (FFNN) serves as a final expert decision-maker. It receives the outputs of the previous classifiers, combined with deep features and hand-crafted ones, learning to produce the final prediction. This ensemble approach enables a more refined and accurate classification while ensuring explainability of results.

3.1   Explainable Techniques

In this work, we employ two explainable techniques to assess model bias from qualitative and quantitative perspectives. We use Grad-CAM [10] and SHAP [8] to visualize which image regions influence the model's decision-making process and identify potential biases. Grad-CAM generates heatmaps highlighting the most influential regions in the model's CNN predictions, enhancing transparency by showing where the model focuses during classification. The technique works by analyzing the internal feature maps of the CNN and tracing back the gradients associated with a specific class prediction. These gradients are aggregated and combined with the corresponding feature maps to create a coarse localization map, which is then superimposed on the original image. The resulting visualization allows researchers to identify which parts of the image the model focused on during classification, aiding in both model debugging and the detection of biases. SHAP (SHapley Additive exPlanations), on the other hand, provides a theoretically grounded framework for interpreting machine learning models. Rooted in cooperative game theory, SHAP assigns each feature a value that quantifies its individual contribution to a model's prediction for a given input. One of its strengths lies in offering consistent and interpretable attributions, whether the goal is to understand a single prediction (local interpretability) or to explore broader model behavior (global interpretability). Because SHAP is model-agnostic, it can be applied to a wide range of algorithms, including neural networks and ensemble models. For example, when applying SHAP to a melanoma classification task on dark-skinned images, the method can reveal which specific features—such as color variation or lesion borders—had the most influence on the model's prediction. If certain benign features or neutral features are repeatedly assigned high importance in melanoma classifications on dark skin, such as the ones which belongs to surrounding skin, this could signal a bias in the model's learned decision boundaries. Such insights are crucial in evaluating whether a model generalizes appropriately across different skin tones. Grad-CAM and SHAP offer complementary perspectives: Grad-CAM provides spatial, image-based interpretability, while SHAP offers a feature-level explanation grounded in theory. These tools were essential in analyzing the Multi-

ExCam framework's behavior across datasets, with particular attention to its performance and decision-making on images representing darker skin tones.

4   Understanding and Visualizing the Bias: MultiExCam Experimental Settings

4.1   Datasets

The HAM10000 Dataset. *HAM10000*, introduced in the ISIC 2018 Challenge [4], comprises 10,000 images, some referring to the same lesion with different zoom levels [11]. It includes seven skin lesion classes, specifically 1134 melanoma and 6705 common nevus images in RGB format with 600×450 pixel resolution. We merged the original training and validation sets to create new splits with a 70-15-15 ratio, ensuring tests were not influenced by prior data distribution. To ensure result reliability, we balanced the melanoma and common nevus classes downward based on the least populous class (melanoma), achieving a final distribution of 1134 samples per class. It is important to highlight, as reported in [14] that the *HAM10000* dataset has significant limitations in terms of demographic representation. According to the systematic review published in The Lancet Digital Health, this dataset originates exclusively from Austria and Australia, reflecting the concerning geographical trend where 79% of skin cancer image datasets come only from Europe, North America, and Oceania. *HAM10000* lacks adequate metadata on patient ethnicity and Fitzpatrick skin type. The review found that across all analyzed datasets, only 1.3% of images had associated ethnicity data and merely 2.1% included Fitzpatrick skin type information. In the rare cases where such data was available, there was massive under-representation of darker skin types: only ten images across all datasets were from individuals with Fitzpatrick skin type V, and just a single image from someone with type VI.

Pipsquek Dataset. *Pipsqueak* is a dataset created through the evaluation and subsequent collection of all available skin lesion images on dark-toned skin. It consists of only 16 images in RGB channel, gathered from well-known datasets in the literature, including ISIC, *HAM10000*, PAD-UFES-20, DermNet, Fitzpatrick17k, Med-Node, and PH^2. The extensive analysis conducted across such a large number of datasets immediately highlighted the core issues addressed in this work—showing, in some cases, a complete absence of dark-skinned images and, in others, an almost negligible number of such examples. For example, datasets like P H2, DermNet and MED-NODE showed a complete absence of dark skin images, while ISIC archive contained only a handful of such examples, with no melanoma cases remaining after applying specific filters. Most of the images come from the Fitzpatrick17k dataset [5], the only dataset that includes skin lesion images across different skin tones and skin tone information for each image. Several challenges emerged during this process. Many images captured the same lesion from multiple angles, requiring careful curation to avoid redundancy. Moreover, unlike datasets such as *HAM10000*, the images contained in the dataset were not captured using clinical tool, such as dermatoscope, but using, in some cases, smartphone camera, resulting in difficult light condition and lower quality. For this reason, some images were excluded from the analysis. The final set of usable images, obtained combining the single image from ISIC archive, and 15 images from Fitzpatrick17k, resulted in 16 images, comprising 9 melanomas and 7 nevi, highlighting the underrepresentation of darker skin tones, considering the high number of analyzed images. This number was slightly increased through data augmentation techniques, obtaining a final dataset of 80 images. Different controlled transformations were performed including subtle rotations, scaling, and flipping technique. It is important to acknowledge that we could have employed more extensive data augmentation techniques to further expand the dataset beyond 80 images. However, we deliberately chose to limit the augmentation process. While it would have been possible to apply more aggressive data augmentation strategies to expand the dataset beyond 80 images, we intentionally opted for a more conservative approach. With only 16 original images, excessive augmentation risked creating artificial patterns unrepresentative of real melanoma manifestations in darker skin tones. We prioritized clinical authenticity over dataset size.

4.2   Explainable Results and Discussion

The experiments were conducted on Google Colab with a T4 GPU and 12 GB of RAM. MultiExCAM, trained on HAM10000, achieved a 92% F1-score on its test set as previously shown in [9]. When tested on the entire Pipsqueak dataset without any model modifications, performance dropped drastically. Results for both datasets are presented in Table 1.

Table 1. MultiExCAM performances on *HAM10000* and *Pipsqueak*.

Dataset	Accuracy	Precision	Recall	F1-score	AUC
HAM10000	92%	92%	92%	92%	97%
Pipsqueak	56%	55%	56%	54%	43%

A more detailed analysis of the errors made on images of lesions in dark skin tones can be performed by examining the confusion matrix (Fig. 1). The figure illustrates a counterintuitive trend: the model shows a tendency to favor the Melanoma class. It is important to note that in the final version of the *Pipsqueak* dataset, 45 out of the 80 images belong to the Melanoma class, while the remaining ones represent benign lesions. Although this tendency is less clinically concerning than the opposite scenario—where malignant lesions are misclassified as benign—it still reveals a significant technical limitation of the framework. Furthermore, it is worth noting that some melanoma cases are still not correctly identified by the model.

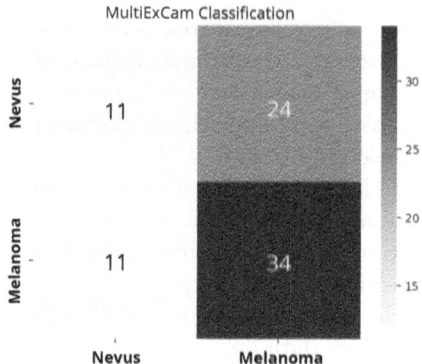

Fig. 1. Confusion Matrix obtained with MultiExCAM using Pipsqueak as the test set.

A thorough explainable analysis was therefore conducted to fully understand how the model processes images and which theoretically discriminative features lead it to make errors. By applying Grad-CAM and SHAP techniques to both datasets, it is possible to visualize how the decision-making process leads to the assignment of the positive class, in this case, Melanoma. Figure 2 provides an overview showing the application of both techniques to selected examples from the two datasets under study.

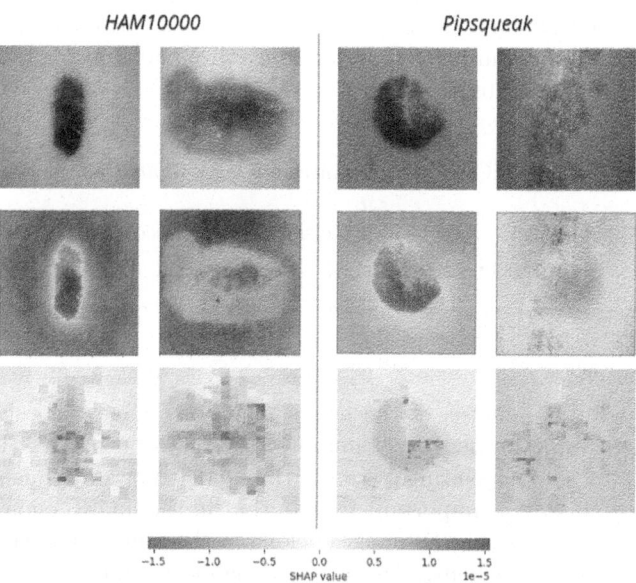

Fig. 2. Explainable analysis conducted on the *HAM10000* and *Pipsqueak* datasets.

Two examples have been provided for each dataset, representing typical cases associated with them. By examining the heatmaps produced by Grad-CAM and the results related to *HAM10000*, it is evident that the most relevant points—highlighted by the warmer, red-tinted areas—correspond precisely to the skin lesion. MultiExCAM effectively identifies the lesion area, as seen in the first image, and also recognizes the clinical features necessary for accurate classification. In the second image, for instance, abrupt color changes within the lesion are represented with increasingly warmer gradients, demonstrating clinical relevance and precision. A key feature that emerges, particularly useful in analyzing images from the *Pipsqueak* dataset, is how the surrounding skin is mapped. In the case of *HAM10000*, the skin appears in cooler tones, as it is not considered discriminative for classification purposes. In contrast, when observing any image from the *Pipsqueak* dataset, one of the first noticeable aspects is that the surrounding skin is never completely ignored by the prediction process. In fact, the heatmap does not display cool tones; instead, it transitions from moderately warm to very warm tones. The explanation is straightforward: since the framework has never been exposed to images with darker skin tones, it has simply learned to associate image regions with darker pigmentation to lesions. This implies that the entire image area is likely to be considered relevant. This issue is not limited to misclassified images but is consistently observed across all samples, further confirming this behavior. However, we will focus on this aspect in more detail in the following images. Considering instead the color maps obtained using SHAP, valuable insights into the framework's decision-making process can also be drawn. For the HAM10000 dataset, SHAP highlights regions that positively contribute to melanoma classification and those that steer toward the nevus class. The model focuses on lesion borders and areas with abrupt pigmentation changes, demonstrating clinically relevant behavior by concentrating on dermatologically significant regions. However, in the case of the *Pipsqueak* dataset, MultiExCAM struggles to focus on the appropriate regions, which is clearly evident from the color maps shown in the figure. SHAP fails to produce meaningful color maps in this case, indicating that the model does not clearly understand which image regions are truly relevant for classification. Only a few points are highlighted in red, and similarly few in blue. The final classification thus appears to result from an uncertain process, as reflected by model performance that resembles the randomness of a coin toss—both technically and clinically. This issue, particularly in certain images, may be due not only to the model's limited ability to handle darker skin tones but also to suboptimal image quality. This phase of analysis has therefore helped isolate the two main causes of wrong classification. On one hand, intrinsic image characteristics confuse the model due to training biases, supporting our research question's validity. Unlike training images where only lesions had darker pigmentation, these images present overall darker tones with reduced contrast between lesion and surrounding skin. On the other hand, poor image quality in some dataset samples also contributes to classification errors. This further highlights the lack of clinically valid images of skin lesions on darker skin tones, thereby exacerbating the bias. In any case, focusing on the first and analytically more significant cause—namely, the intrinsic characteristics of the images—two examples are presented in Fig. 3 to illustrate two sub-cases observed within the dataset.

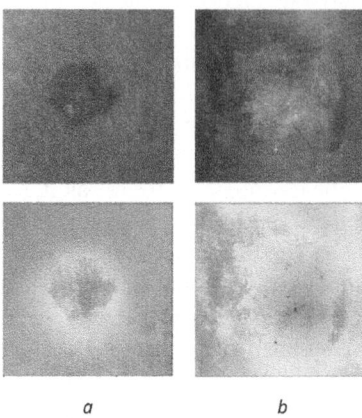

Fig. 3. Two particular cases of misclassified lesions.

Both images were incorrectly classified as Melanoma, despite actually representing Nevi. The first case (a) depicts the only image in *Pipsqueak* originating from the *HAM10000* dataset. As highlighted by the heatmap shown in the figure, and as previously discussed in Fig. 2, although the lesion area is correctly identified—appearing warmer than the rest of the image—the surrounding skin is also mistakenly considered in the prediction process, becoming a discriminative feature. Notably, the lower right portion of the image, which simply shows a normal, mild variation in skin pigmentation, is strongly emphasized by the model, further illustrating how even non-relevant regions can influence the classification. The second case (b), on the other hand, represents a more critical issue encountered. Due to training bias, the model considers irrelevant image portions as significant while failing to identify genuinely relevant regions. Following its logic of emphasizing darker pigmented areas, the model misclassifies lesions containing lighter tones than the surrounding skin. The benign lesion shown presents a lighter area, likely from skin dryness amplified by camera flash. In fact, when observing the corresponding heatmap, it becomes apparent that, paradoxically, the central area of the lesion is not considered in the prediction process—appearing in cooler tones-while the surrounding region, which again merely reflects normal pigmentation variation, is emphasized and treated as a discriminative feature in assigning the Melanoma label-appearing in warmer tones.

5   Conclusion

Correctly predicting skin lesion nature remains a difficult challenge, especially in underrepresented contexts. Datasets play a fundamental role in model training, as data quality and quantity directly affect predictive capabilities. Biases within training data transfer to models, negatively affecting classifications. This work aimed to highlight the lack of darker skin tone representation in skin lesion datasets, demonstrating how data-induced biases impact results and disrupt models, even when they show excellent performance on similar training contexts. XAI techniques have proven valuable for visually exposing these biases and helping experts understand how data limitations affect models, but are insufficient alone to solve the problem. The analysis reveals that the core problem lies in the data itself, highlighting the need for more comprehensive and representative datasets.

References

1. Adamson, A., Smith, A.: Machine learning and health care disparities in dermatology. JAMA Dermatol. **154** (2018)
2. Bradford, P.T.: Skin cancer in skin of color. Dermatol. Nurs./Dermatol. Nurses' Assoc. **21**(4), 170 (2009)
3. Bray, F., et al.: Global cancer statistics 2022: globocan estimates of incidence and mortality worldwide for 36 cancers in 185 countries. C Cancer J. Clin. **74**(3), 229–263 (2024)
4. Codella, N., et al.: Skin lesion analysis toward melanoma detection 2018: a challenge hosted by the international skin imaging collaboration (ISIC). arXiv:1902.03368 (2019)
5. Groh, M., et al.: Evaluating deep neural networks trained on clinical images in dermatology with the fitzpatrick 17k dataset. In: IEEE/CVF, pp. 1820–1828 (2021)
6. Han, S., Kim, M., Lim, W., Park, G., Park, I., Chang, S.: Classification of the clinical images for benign and malignant cutaneous tumors using a deep learning algorithm. J. Invest. Dermatol. **138** (2018)
7. Kushimo, O., Salau, A., Adeleke, O., Olaoye, D.: Deep learning model to improve melanoma detection in people of color. Arab J. Basic Appl. Sci. **30**, 92–102 (2023). https://doi.org/10.1080/25765299.2023.2170066
8. Lundberg, S.M., Lee, S.I.: A unified approach to interpreting model predictions. In: Advances in Neural Information Processing Systems, vol. 30 (2017)
9. Ruga, T., Musacchio, G., Maurmo, D.: An ensemble architecture for melanoma classification. In: pHealth 2024, pp. 183–184. IOS Press (2024)
10. Selvaraju, R.R., Cogswell, M., Das, A., Vedantam, R., Parikh, D., Batra, D.: Grad-CAM: visual explanations from deep networks via gradient-based localization. In: Proceedings of the IEEE ICCV, pp. 618–626 (2017)
11. Tschandl, P., Rosendahl, C., Kittler, H.: The ham10000 dataset, a large collection of multi-source dermatoscopic images of common pigmented skin lesions. Sci. Data **5**(1), 1–9 (2018)
12. Vocaturo, E., Perna, D., Zumpano, E.: Machine learning techniques for automated melanoma detection. In: IEEE BIBM, pp. 2310–2317 (2019)

13. Vocaturo, E., Zumpano, E.: Multiple instance learning approaches for melanoma and dysplastic nevi images classification. In: IEEE ICMLA, pp. 1396–1401 (2020)
14. Wen, D., et al.: Characteristics of publicly available skin cancer image datasets: a systematic review. Lancet (2021)

MaGA-Clif: Defending FL from Combined Poisoning Attacks with Marginal Gain Estimation

Jamsher Bhanbhro[✉][iD] and Simona Nisticò[iD]

DIMES Department, University of Calabria, Rende, Italy
{jamsher.bhanbhro,simona.nistico}@dimes.unical.it

Abstract. The pursuit of robust and efficient federated learning systems remains pivotal as data and models become increasingly decentralised across diverse domains. This paper presents an advanced aggregation algorithm designed to enhance the security and efficiency of federated learning systems, particularly in handling malicious or misleading clients [1,2] in independent and identically distributed (IID) and non-independent and identically distributed (non-IID) data scenarios effectively. By incorporating a game theory-inspired marginal contribution concept, which leverages a small validation set at the global model level, our algorithm aims to secure Federated Learning frameworks from attacker clients by dynamically filtering poorly-performing updates. This method selectively includes client updates, ensuring that only contributions that genuinely enhance the model's performance are considered. As a result, the proposed algorithm shows significant improvements in aggregation accuracy and robustness in poisoned-data-and-client scenarios compared to traditional federated learning methods, as demonstrated by tests performed across various data environments.

Keywords: Federated Learning · Poisoning Attacks · Data Poisoning · Client Poisoning

1 Introduction

As the digital landscape evolves, the paradigm of data processing and analysis is undergoing a profound transformation. The emergence of Federated Learning (FL) as a powerful approach to train models on decentralised data has opened new opportunities for processing information while adhering to privacy and regulatory constraints [13,18]. In FL, data remains distributed across a multitude of devices or servers, and the learning process is conducted without exchanging or centralising this data [8]. This approach not only aids in compliance with users' privacy laws but also reduces the overhead associated with data transmission and storage [5]. However, FL introduces unique challenges, particularly when dealing with data that is not uniformly distributed (non-IID) across the network [3,10].

Non-IID data scenarios are normal in real-world applications where data originates from various sources with potentially differing data distributions [19]. These scenarios pose significant challenges in training models as they can lead to model skewing, where

P. K. Chrysanthis et al. (Eds.): ADBIS 2025, CCIS 2676, pp. 457–466, 2026.
https://doi.org/10.1007/978-3-032-05727-3_38

the model becomes overly tailored to the characteristics of the dominant data distribution among the clients [7]. This issue is heightened in traditional FL setups, which often employ simple averaging methods for aggregating client updates [16]. Such methods typically overlook the quality and relevance of the data contributing to these updates, leading to suboptimal model performance and reduced system efficiency [12].

To address these limitations, various aggregation algorithms have been proposed. FedProx, for instance, incorporates a proximal term to handle heterogeneity by penalising deviations from the global model, although it often slows down convergence in highly skewed non-IID settings [11].

Other approaches prioritise fairness in aggregation. For instance, q-FFL aims to balance client contributions in non-IID settings, but this can reduce overall model accuracy as it sacrifices performance to ensure fair participation [14]. FedBE, a Bayesian ensemble-based method, aggregates models by weighting updates based on uncertainty estimates, which enhances robustness but poses implementation challenges due to its high computational demand in large federated environments [6].

Beyond issues caused by non-IID data distributions, some clients in the FL framework can be misleading due to biases in training data or, in general, poor model performance. This poses additional challenges; mitigation strategies are indeed needed to avoid performance degradation in the FL framework. Multi-KRUM, designed for Byzantine robustness, which is the ability of a system to be resilient to a malicious or arbitrary behavior of some of its components, filters out unreliable updates to improve model resilience but adds significant computational complexity, making it challenging for large-scale applications [4]. Similarly, Bulyan [15] extends Byzantine tolerance by iteratively selecting reliable updates, though this process is computationally intensive and may not scale well with increasing client numbers.

Recognising these challenges, this paper introduces an aggregation algorithm that leverages concepts borrowed from game theory to enhance both the security and efficiency of FL systems. Game theory, which is the study of mathematical models of strategic interaction among rational decision-makers, is particularly apt for addressing situations where multiple agents (in this case, data clients) interact. By introducing a dynamic game theory inspired framework into the aggregation process, our algorithm proposes a structured approach to assess and integrate data updates based on their quality and impact on overall model performance.

This work introduces the following contributions:

- We introduce a game-theory-inspired aggregation algorithm that significantly improves the security and efficiency of FL systems.
- We have tested the robustness to malicious clients of three state-of-the-art aggregation methods, that are: FedAvg, FedNova and FedProx and compared by that shown by MaGa-CliF.

The next section of this paper, followed by a detailed exploration of the algorithm and simulation results, concluding with a discussion on the implications of the findings. This work is structured as follows. Section 2 describes the system model and problem formulation, Sect. 3 introduces the MaGa-CliF method, Sect. 4 describes experimental setting and comments collected results, eventually, Sect. 5 concludes this work.

2   System Model and Problem Formulation

2.1   System Model Description

The here considered proposed federated learning system consists of a central server and a set of N clients, each possessing its own local dataset $\mathcal{D}_i$, where $i \in \{1, 2, \ldots, N\}$. The datasets can vary from being IID (independent and identically distributed) to non-IID, reflecting significant differences in data distribution across clients. The framework is designed to effectively manage both scenarios.

As in the standard FL scenarios, the server's role is to aggregate updates from the clients to form a global model G. In each round of communication, clients pull the weights of the latest version of the global model w_G^t from the server, apply updates based on their local data, and send the modified model back to the server. The server then aggregates these individual updates to construct the updated global model w_G^{t+1}, which incorporates information learned from local clients. The aggregation mechanism employs a strategic model to evaluate the contributions of individual updates based on a reward and penalty system, governed by Eq. 1,

$$w_G^{t+1} = w_G^t + \sum_{i=1}^{N} \alpha_i \Delta w_i^t \tag{1}$$

where Δw_i^t is the update from client i at round t, α_i^t is a weight factor determined by the reward and penalty system reflecting the quality of Δw_i^t for the aggregation step t.

2.2   Problem Formulation

The optimisation problem addressed in this work aims to minimise global model error in an environment affected by misleading clients and non-IID data distributions. Let $\mathcal{L}(w; \mathcal{D})$ denote the global loss function, which is a function of the model parameters w_G and the combined dataset $\mathcal{D}$ from all clients. The objective of minimising $\mathcal{L}$ is expressed as:

$$\min_{w} \mathcal{L}(w; \mathcal{D}) = \frac{1}{N} \sum_{i=1}^{N} \mathcal{L}(w_i; \mathcal{D}_i) \tag{2}$$

where $\mathcal{L}_i(w; \mathcal{D}_i)$ is the loss function for client i, dependent on both the model parameters and the local data characteristics of client i.

3   Algorithm

This section outlines the methodology used to enhance the robustness of the FL frameworks against misleading or malicious clients. To filter contributions from poorly behaving or malicious clients, we equip the global model with a validation set to assess the quality of updates provided by clients. In particular, we evaluate the single client i generalisation on the global model's private validation set to approximate its marginal value in the current federated step, to which we refer hereafter as mv_i.

3.1   Federated System Architecture

As defined in the system model, the FL system consists of a central server and N clients, each possessing its own unique local dataset $\mathcal{D}_i$. This system architecture is designed to ensure stringent data privacy and minimise communication overhead, thereby facilitating collaborative model training without actual data transfer. Clients in this setup are responsible for computing model updates locally, which are then transmitted to the central server for aggregation. The server aggregates updates based on the algorithm defined in the Sect. 3.2.

Algorithm 1. Marginal Gain based Client Filtering (MAGA-CliF)

Require: Number of clients N, number of rounds T, global model validation set $\mathcal{V}$, hyperparameters α, β, γ

Ensure: Global model parameters w^T

1: **for** $t = 1$ **to** T **do**
2:    **Broadcast:** Server sends the current global model w^t to all clients.
3:    **for** each client $i = 1, \ldots, N$ in parallel **do**
4:        Client i computes its local update for the step t: Δw_i^t.
5:    **end for**
6:    create a vector mv containing N elements
7:    **for** each client $i = 1, \ldots, N$ **do**
8:        $w_G' \leftarrow w_G^t + \Delta w_{L_i}^t$
9:        $l_i \leftarrow \text{evaluate}(w_G', \mathcal{V})$
10:       $mv_i \leftarrow e^{-\alpha * l_i}$
11:   **end for**
12:   $s_{mv} \leftarrow \sum_{i=1}^{N} mv_i$
13:   **for** each client $i = 1, \ldots, N$ **do**
14:       $mv_i \leftarrow \frac{mv_i}{s_{mv}}$
15:   **end for**
16:   $\Delta w^t = \sum_{i=1}^{N} mv_i \cdot \Delta w_i^t$
17:   $w_G^{t+1} = w_G^t + \Delta w^t$
18: **end for**
19: **return** w^T

3.2   Algorithm Description

We propose a novel federated aggregation algorithm that exploits the game theory's marginal contribution concept to adaptively weight client updates. To evaluate the importance of the current update of a client i, the algorithm computes, in Line 9, the validation loss l_i scored by the model obtained by updating the global model state at t time step w_G^t accordingly to the updated suggested by client i. The loss score l_i is later, in Line 10, converted into a raw weight according to the exponential function $e^{-\alpha * l_i}$. The final weight is computed by Line 14, where each value is divided by the sum of all the raw weights to obtain values summing to one. The role of the α hyperparameter in heightening strategy is to make MaGa-CliF less (or more) permissive with poorly performing clients. The greater the value of α is, the more the behaviour of the weighting will be near to a best-takes-all strategy.

This FL aggregation strategy only requires changes in server behaviour, which requires a small validation set $\mathcal{V}$ devoted to approximating the quality and trustworthiness of the received updates. The local clients continue to work as usual by receiving the updated model from the global server (Line 2), performing the local training operation (Line 4), and, finally, sending the updates resulting from the local training to the server.

4   Results and Discussion

This section details the experimental setup used to evaluate the performance of different aggregation strategies in a federated learning environment with adversarial clients and presents a comparative analysis of the outcomes.

In this experiment, we carry out a preliminary evaluation of the MaGa-CliF effectiveness against several standard federated averaging methods in successfully handling FL scenarios polluted by misleading or malicious clients. In our experiments, we will employ the FashionMNIST dataset [17], which comprises 28×28 grayscale images of ten different fashion clothes categories. To further validate our approach, we also conduct experiments on the more challenging CIFAR-10 dataset [9], which consists of 32×32 images from ten classes, including animals and means of transport.

4.1   Experimental Setting

We conduct experiments under two distinct data distribution settings to evaluate algorithmic validity. First, we establish a baseline using an Independent and Identically Distributed (IID) data partitioning scheme. In this setup, we uniformly partitioned the training data for both CIFAR-10 and FashionMNIST at random across a simulated network of 20 clients. Within this IID setting, we introduce a combined client and data poisoning attack where 5 of the 20 clients are designated as biased (The bias is implemented by overlaying a small, coloured square patch at a fixed position on every image in the client's local dataset. The patch colour and position are uniquely determined based on the image label, ensuring that each class receives a consistent and visually distinctive perturbation. These modifications are performed at the data preprocessing stage using the PIL.ImageDraw module, where the patch is drawn directly onto the image. For these clients, a steady visual artefact (a small, fixed patch) is coated on every image in their local dataset. This allows for a controlled evaluation of how different aggregation methods handle targeted adversarial updates. To create malicious clients, we moreover modified malicious clients to make them ignore global model updates, thus simply simulating attacker clients.

To create a more challenging test environment, we then employ a hybrid non-IID data distribution strategy. This setup moves beyond simple pathological partitioning by assigning each client a dataset composed of two distinct parts: the majority of the data (80%) is drawn from a small, restricted subset of just three classes, creating a significant label distribution skew, while the remaining 20% is drawn uniformly from all ten classes. This hybrid approach models a practical scenario where users have specialised data corresponding to their primary interests alongside a smaller, more diverse set of ancillary data. The same combined client and data poisoning attack, with 5 biased clients, is applied within this non-IID environment to correctly test the resilience of the aggregation methods. In both settings, we compare the performance of four aggregators: FedAvg, FedProx, FedNova, and the proposed MaGa-CliF. In both environments, the validation set used by the global model consists of a IID sample of all the classes composing the dataset involved in the experiment at hand.

4.2   Results

The experiments, conducted on both IID and non-IID data distributions using the FashionMNIST and CIFAR-10 datasets, witnessed a consistent and notable performance gap between the proposed MaGa-CliF and traditional aggregation techniques.

In the IID setting, the baseline methods FedAvg, FedProx, and FedNova show nearly identical behaviour. As shown in Figs. 3 and 1, their test accuracy quickly plateaus around 3035% and their accuracy remains very slightly changed throughout 50 communication rounds. This indicates their limited ability to handle the effects of poisoned updates from the biased clients. By treating all client updates equally, these methods continue to incorporate misleading signals, which ultimately prevent the global model from improving.

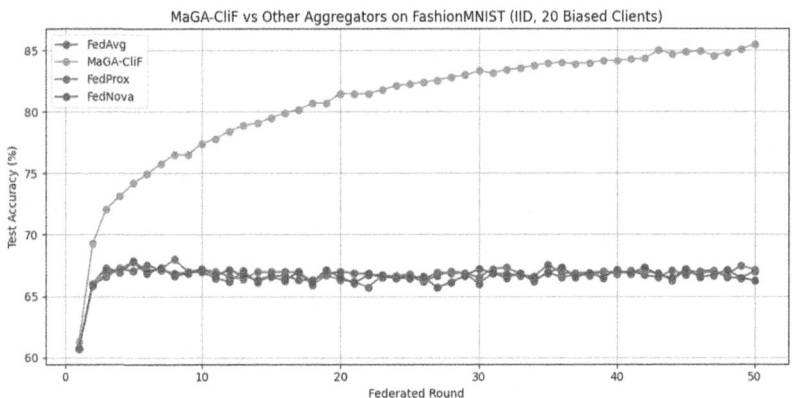

Fig. 1. Test accuracy on FashionMNIST under the IID setting. Five of the 20 clients provide biased updates. The proposed MaGa-CliF successfully defends against the attack, achieving over 85% accuracy, while standard aggregators stagnate.

The proposed MaGa-CliF reaches a final test accuracy exceeding 85.66% for FashionMNIST and 58.23% for CIFAR-10 in the IID setting, significantly outperforming all baselines. Its advantage becomes even more evident in the more challenging non-IID setting. While the accuracy of all methods decreases due to the uneven distribution of data across clients, MaGa-CliF maintains a consistent performance as illustrated in Fig. 2 for FashionMNIST and Fig. 4 for CIFAR-10.

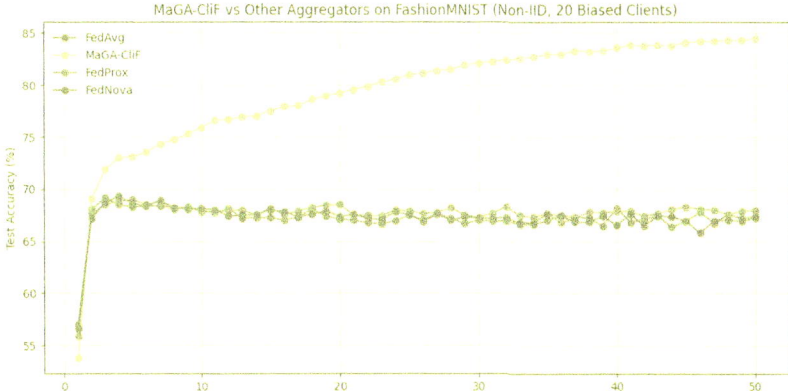

Fig. 2. Test accuracy on FashionMNIST under the non-IID setting. Despite biased client updates, MaGa-CliF maintains superior performance over all baselines.

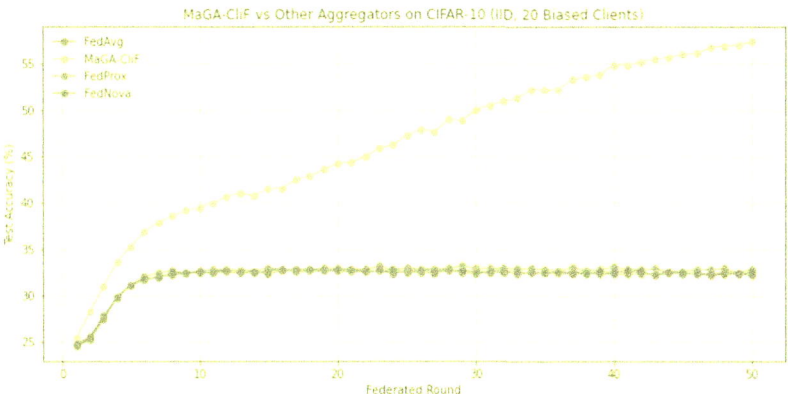

Fig. 3. Test accuracy on CIFAR-10 in a scenario involving 20 clients 5 of which are misleading.

This consistent performance across both IID and non-IID settings highlights the strength of MaGa-CliF in handling adversarial behaviour. By evaluating client updates using a small validation set, it learns to recognise and minimise the impact of harmful contributions. As a result, the global model benefits primarily from honest client updates, leading to more reliable and accurate learning over time.

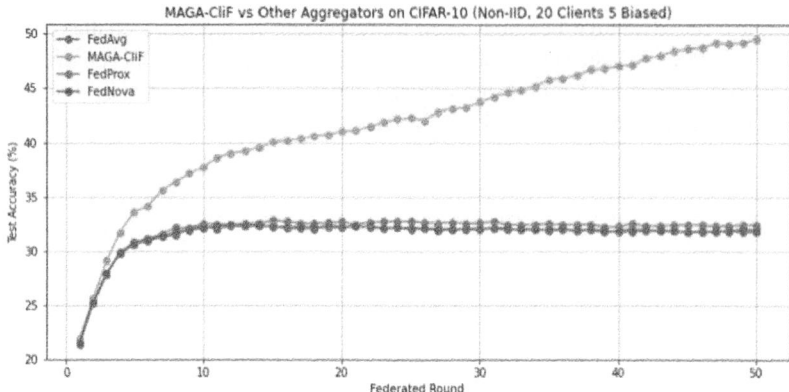

Fig. 4. Test accuracy on CIFAR-10 under the non-IID setting. Similar to FashionMNIST, MaGa-CliF significantly outperforms the baseline methods despite the presence of biased clients.

4.3   Discussion

The experimental results clearly highlight the limitations of conventional federated learning aggregation methods in the presence of biased clients. Across both IID and non-IID settings, FedAvg, FedProx, and FedNova fail to effectively mitigate the influence of malicious updates. Their uniform treatment of client contributions makes them vulnerable to even simple but repeated forms of bias, such as fixed patch-based visual triggers. This vulnerability is particularly evident in the plateau of test accuracy observed early in training, which persists despite additional rounds of communication. In contrast, the proposed MaGa-CliF demonstrates robust and adaptive behaviour under both data distributions. Its key strength lies in its ability to estimate the utility of each client's model update based on actual performance impact on a small validation set. This allows it to selectively weight updates, amplifying the contribution of honest clients while suppressing those that harm generalisation. As a result, MaGa-CliF is able to continuously improve test accuracy across rounds, even in scenarios where biased clients account for a significant portion of the federation.

An important observation is that the performance gap between MaGa-CliF and the baselines becomes more pronounced as the level of heterogeneity increases. This reinforces the importance of aggregation strategies that can account for the varying quality of updates rather than relying on equal or norm-based heuristics alone. By incorporating a value-based perspective grounded in update marginal value, MaGa-CliF offers a more principled and practical defence mechanism for federated learning systems deployed in unreliable environments.

It is also worth noting that while MaGa-CliF introduces additional computational overhead due to repeated evaluations, this cost remains manageable in practice and is justified by the substantial gains in accuracy and resilience. Future work may explore optimisations, such as sampling strategies or surrogate metrics, to further reduce this overhead while preserving the robustness of the aggregation.

5   Conclusion

This work presents MaGa-CliF, a principled and effective aggregation strategy for robust federated learning aggregation tasks in the presence of biased or malicious clients. Through comprehensive experiments on both CIFAR-10 and FashionMNIST datasets under IID and non-IID settings, MaGa-CliF consistently outperforms conventional baselines such as FedAvg, FedProx, and FedNova. By maximising incremental benefit estimates based on a small validation set, MaGa-CliF accurately identifies and downweights harmful updates during aggregation. This enables the global model to retain performance even when a significant fraction of the clients are compromised. The approach is simple to implement and integrates smoothly with standard federated learning workflows. Future work may explore extending this framework to dynamic client participation and broader threat models and testing on more dynamic datasets.

Acknowledgments. We acknowledge the support of the PNRR project FAIR - Future AI Research (PE00000013), Spoke 9 - Green-aware AI, under the NRRP MUR program funded by the NextGenerationEU.

References

1. Bagdasaryan, E., Veit, A., Hua, Y., Estrin, D., Shmatikov, V.: How to backdoor federated learning. In: International Conference on Artificial Intelligence and Statistics, pp. 2938–2948. PMLR (2020)
2. Bhagoji, A.N., Chakraborty, S., Mittal, P., Calo, S.: Analyzing federated learning through an adversarial lens. In: International Conference on Machine Learning, pp. 634–643. PMLR (2019)
3. Bhanbhro, J., Nisticò, S., Palopoli, L.: Issues in federated learning: some experiments and preliminary results. Sci. Rep. **14**(1), 1–15 (2024)
4. Blanchard, P., El Mhamdi, E.M., Guerraoui, R., Stainer, J.: Machine learning with adversaries: Byzantine tolerant gradient descent. arXiv preprint arXiv:1703.02757 (2017)
5. Bonawitz, K., et al.: Towards federated learning at scale: system design. In: Proceedings of the 2nd SysML Conference (2019)
6. Chen, H., Wang, W., Zhao, Q., Yang, Y., Sun, L.: FedBE: making Bayesian model ensemble applicable to federated learning. In: Advances in Neural Information Processing Systems (NeurIPS) (2020)
7. Hsieh, K., Phanishayee, A., Mutlu, O., Gibbons, P.B.: Non-IID data and why it matters in federated learning. arXiv preprint arXiv:2001.01647 (2020)
8. Kairouz, P., et al.: Advances and open problems in federated learning. Found. Trends® Mach. Learn. **14**(1–2), 1–210 (2021)
9. Krizhevsky, A., Hinton, G.: Learning multiple layers of features from tiny images. Technical report (2009)
10. Li, T., Sahu, A., Talwalkar, A., Smith, V.: Federated optimization in heterogeneous networks. In: Proceedings of Machine Learning and Systems, vol. 2 (2020)
11. Li, T., Sanjabi, M., Sahu, A., Talwalkar, A., Smith, V.: Federated optimization in heterogeneous networks with FedProx. In: Proceedings of Machine Learning and Systems (MLSys) (2020)

12. Li, X., Huang, K., Yang, W., Wang, S., Zhang, Z.: On the convergence of FedAvg on non-IID data. arXiv preprint arXiv:1907.02189 (2019)
13. McMahan, H.B., Moore, E., Ramage, D., Hampson, S., Arcas, B.A.Y.: Communication-efficient learning of deep networks from decentralized data. In: Artificial Intelligence and Statistics, pp. 1273–1282. PMLR (2017)
14. Mohri, M., Sivek, G., Suresh, A.T.: Agnostic federated learning. In: Proceedings of the 36th International Conference on Machine Learning (ICML), pp. 6083–6092 (2019)
15. Muñoz-González, L., Panaousis, E., Rizomiliotis, P.: Byzantine robust federated learning through approximation tolerance. In: Proceedings of the International Conference on Artificial Intelligence and Statistics (AISTATS) (2019)
16. Wang, J., Joshi, G., Liang, Z.: Tackling the objective inconsistency problem in heterogeneous federated optimization. In: Proceedings of the 37th International Conference on Machine Learning (ICML) (2020)
17. Xiao, H., Rasul, K., Vollgraf, R.: Fashion-MNIST: a novel image dataset for benchmarking machine learning algorithms. arXiv preprint arXiv:1708.07747 (2017)
18. Yang, Q., Liu, Y., Chen, T., Tong, Y.: Federated Learning. Morgan & Claypool Publishers (2019)
19. Zhao, Y., Li, M., Lai, L., Suda, N., Civin, D., Chandra, V.: Federated learning with non-IID data. arXiv preprint arXiv:1806.00582 (2018)

ARPaCCino: An Agentic-RAG for Policy as Code Compliance

Francesco Romeo[1,2] , Luigi Arena[1] , Francesco Blefari[1,2(✉)] ,
Francesco Aurelio Pironti[1] , Matteo Lupinacci[1] , and Angelo Furfaro[1]

[1] University of Calabria, 87036 Rende, CS, Italy
{francesco.romeo,luigi.arena,francesco.pironti,matteo.lupinacci,
angelo.furfaro}@unical.it
[2] IMT School for Advanced Studies Lucca, 55100 Lucca, LU, Italy
{francesco.romeo,francesco.blefari}@imtlucca.it

Abstract. Policy as Code (PaC) is a paradigm that encodes security and compliance policies into machine-readable formats, enabling automated enforcement in Infrastructure as Code (IaC) environments. However, its adoption is hindered by the complexity of policy languages and the risk of misconfigurations. In this work, we present ARPaCCino, an agentic system that combines Large Language Models (LLMs), Retrieval-Augmented-Generation (RAG), and tool-based validation to automate the generation and verification of PaC rules. Given natural language descriptions of the desired policies, ARPaCCino generates formal `Rego` rules, assesses IaC compliance, and iteratively refines the IaC configurations to ensure conformance. Thanks to its modular agentic architecture and integration with external tools and knowledge bases, ARPaCCino supports policy validation across a wide range of technologies, including niche or emerging IaC frameworks. Experimental evaluation involving a Terraform-based case study demonstrates ARPaCCino's effectiveness in generating syntactically and semantically correct policies, identifying non-compliant infrastructures, and applying corrective modifications, even when using smaller, open-weight LLMs. Our results highlight the potential of agentic RAG architectures to enhance the automation, reliability, and accessibility of PaC workflows.

Keywords: Policy as Code · Agentic AI · Retrieval Augmented Generation · Large Language Models

1 Introduction

Over the years, software and infrastructure management have become increasingly challenging due to the growing complexity and scale of systems. To address these challenges, developers have embraced DevOps practices, which aim to reduce operational errors, accelerate provisioning, and support continuous updates throughout the development and operations lifecycle. Within this context, Infrastructure as Code (IaC) has emerged as a standard practice. By

P. K. Chrysanthis et al. (Eds.): ADBIS 2025, CCIS 2676, pp. 467–481, 2026.
https://doi.org/10.1007/978-3-032-05727-3_39

expressing infrastructure specifications in machine-readable code, IaC enables automated provisioning, configuration, and management. This approach significantly improves automation, scalability, reproducibility, and consistency across the entire service lifecycle.

Despite its benefits, IaC remains prone to misconfigurations and security vulnerabilities when applied without sufficient expertise. To mitigate these risks, comprehensive testing and validation are often necessary before deployment.

Policy as Code (PaC) extends the Infrastructure as Code (IaC) paradigm to the definition of security and compliance policies, expressing them as formal, machine-readable rules that can be automatically validated and enforced during the provisioning process. By integrating policy checks into Continuous Integration and Continuous Deployment (CI/CD) pipelines, PaC helps reduce human error and ensures infrastructure configurations meet security and compliance requirements early in the development lifecycle.

However, the adoption of PaC is often hindered by the steep learning curve of domain-specific policy languages and the difficulty of authoring correct, comprehensive rules—especially in dynamic and complex environments.

Recent advancements in Large Language Models (LLMs) and LLM-based techniques offer a promising solution to the limitations of current IaC and PaC practices, supporting their deeper integration into standard industry workflows. LLMs can translate high-level policy descriptions, written in natural language, into formal, machine-readable rules suitable for automated validation of IaC configurations.

Additionally, AI agent-based workflows can enhance the generation and refinement of IaC and PaC artifacts by iteratively interacting with domain-specific tools and structured knowledge bases. In particular, Retrieval-Augmented Generation (RAG) techniques extend LLM capabilities with contextual, domain-specific knowledge, enabling accurate handling of niche or emerging technologies without the need for extensive retraining.

In this context, we present ARPACCINO, an agentic system that combines a core reasoning LLM with RAG and specialized tools to automate the generation and validation of policies for IaC. Given a natural language description of the desired policies, ARPACCINO generates formal rules in `Rego` – the policy language used by Open Policy Agent (OPA) – then verifies and applies them to assess compliance of the provided IaC specification.

If the validation reveals non-compliance, ARPACCINO can autonomously propose and apply iterative corrections to the IaC configuration until the specified requirements are satisfied. The system's RAG module can be supplied with custom domain-specific knowledge bases, enabling its use across a wide range of technologies, including less common or emerging frameworks, provided that relevant documentation and examples are available.

The main contributions of this work can be summarized as follows:

- We propose a novel approach to Policy as Code generation and Infrastructure as Code validation, leveraging agentic systems that combine LLMs, RAG, and external tool integrations.

– We implement this approach in the ARPaCCino system, which generates formal policy rules from natural language, assesses infrastructure compliance, and iteratively refines configurations until policy conformance is achieved.
– We demonstrate the effectiveness of ARPaCCino through a realistic Terraform-based use case, showing its ability to autonomously retrieve domain knowledge, synthesize and verify policy rules, and validate or revise IaC specifications accordingly.

The remainder of this paper is organized as follows. Section 2 provides background knowledge on Infrastructure as Code, Policy as Code, and AI agents. Section 3 details the ARPaCCino system architecture, including its core LLM engine and the available tools. Section 4 presents a real use case involving Terraform, demonstrating the end-to-end workflow from a natural language policy description to a verified IaC definition. Section 5 examines the results of the experimental evaluation conducted with different scenarios and LLMs. Lastly, Sect. 6 discusses conclusions and outlines directions for future work.

Through ARPaCCino, our aim is to advance the state of automated Policy as Code by leveraging agentic AI and RAG techniques to reduce developer burden, improve compliance, and support evolving infrastructure ecosystems.

2   Background

The provisioning of services in modern computing environments is a complex task that requires custom architectures tailored to specific use cases. These systems often consist of multiple interconnected components, such as microservices, databases, and networked elements, that increase the overall complexity and hinder maintainability. To manage this growing complexity, developers have adopted DevOps methodologies, aiming to reduce error rates, accelerate service provisioning, and enable continuous software delivery.

2.1   Infrastructure as Code

To support efficient Continuous Integration and Continuous Deployment [1] in such environments, IaC emerged as a foundational DevOps practice [2]. IaC enables programmatic provisioning, configuration, and management by using machine-readable code [3]. This enhancing automation, reproducibility, and consistency across both development and operational phases.

To support IaC practice in DevOps, several languages, platforms and tools have been developed that allow the creation, customization, and orchestration of system components, including microservices, virtual machines, and networking layers. Popular IaC tools include Terraform [4] (declarative, cloud-agnostic), Ansible [5] (configuration-focused), and Pulumi [6] (uses general-purpose languages). These tools enable the infrastructure to be versioned, tested, and deployed as application code.

Despite these advantages, IaC tools are still susceptible to misconfigurations and logic errors, which may lead to performance issues or security vulnerabilities.

Over time, several solutions have been developed to test and validate the IaC infrastructure to ensure the correctness of the system before its deployment [7,8]. More recently, LLMs have been applied to IaC workflows to reduce manual effort and enhance reliability. LLMs can translate high-level natural language descriptions into valid infrastructure code, thus accelerating development and mitigating syntactic and semantic mistakes [9,10]. Some approaches also integrate automated validation and correction loops, enabling the detection and resolution of configuration errors prior to deployment [10].

2.2   Policy as Code

PaC extends the IaC paradigm by codifying security, compliance, and operational policies into machine-readable formats. This enables automated policy enforcement throughout the software development lifecycle, ensuring continuous compliance and reducing human error [11]. PaC integrates directly into CI/CD pipelines, facilitating automated validation of infrastructure configurations and application deployments against predefined rules, thereby shifting security left in the DevSecOps pipeline [12].

The de facto standard implementation is *Open Policy Agent (OPA)* [13], an open-source general-purpose policy engine using the `Rego` declarative language to express fine-grained authorization, admission control (e.g., in Kubernetes), and data-filtering rules via API calls or as an integrated library. Complementing this is *HashiCorp Sentinel* [14], tailor-made for the HashiCorp enterprise stack (Terraform Enterprise, Vault, Consul, Nomad), featuring its own policy language, support for logical constructs and imports, and multiple enforcement levels (advisory, soft-mandatory, hard-mandatory). Sentinel enables proactive pre-deployment governance, enforcing policies as a prerequisite to resource provisioning.

Despite its advantages, PaC adoption can face challenges such as the steep learning curve for policy languages and managing policy drift in dynamic environments. However, the advent of AI and LLMs offers significant opportunities to overcome these limitations, enabling AI-assisted policy generation, automated validation, and intelligent support for policy comprehension.

2.3   AI Agent

An *AI agent* is an autonomous software entity capable of reasoning about goals and executing actions to achieve specified objectives [15]. It is typically characterized by its ability to perceive the environment in which it operates, respond to environmental changes, and interact with external systems or other agents.

An AI Agent is also provided with a *memory*, useful to learn from past experiences and maintain context while addressing a task. Figure 1 illustrates a schema of an AI agent structure.

As discussed in [16], the emergence of LLMs marks a significant advancement in the development of intelligent agents. This evolution has led to the rise of

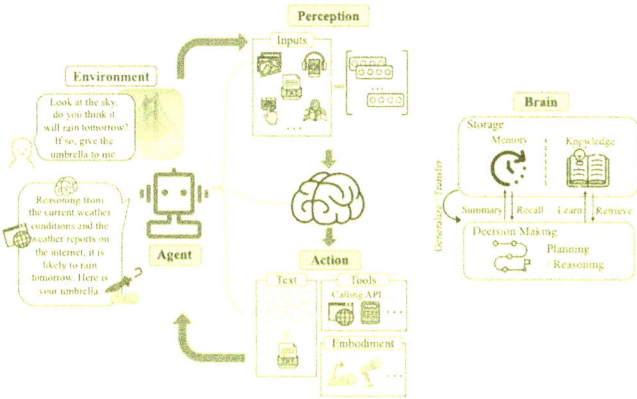

Fig. 1. AI agent structure [16]

LLM agents, which use LLMs as the core reasoning engine to perform task decomposition, planning, and decision-making. While maintaining the reactive and interactive characteristics of traditional agents, LLM agents are augmented with the ability to invoke external tools (e.g., calculators, code interpreters, or knowledge bases) to solve domain-specific subtasks. The LLM continuously evaluates whether the task has been completed or if further tool-based refinement is required, enabling flexible and iterative problem-solving.

Agentic RAG Architectures. While off-the-shelf LLMs and agent frameworks built on them offer broad utility, they often struggle with domain-specific or expert-level tasks due to the absence of embedded, up-to-date knowledge. Although this limitation can be addressed through retraining or fine-tuning, such approaches are typically resource-intensive and time-consuming.

A more scalable and cost-effective solution is offered by the RAG paradigm [17]. In RAG systems, the LLM is coupled with two key components: *(i)* a *repository* of domain-specific knowledge (for example, a curated collection of documents) and *(ii)* a *retriever* that locates relevant content from this repository to enrich the model's input context.

This architecture allows LLMs to answer specialized or evolving queries by incorporating external knowledge at inference time.

A typical RAG pipeline operates as follows: an external knowledge corpus is preprocessed into manageable chunks, transformed into vector embeddings, and indexed for fast retrieval. When a user submits a query, the retriever encodes it into a vector, searches the index for the most semantically relevant chunks, and returns them. These retrieved excerpts are then combined with the user query to form an augmented prompt, which is passed to the LLM. The result is a context-aware, knowledge-informed response that extends beyond the model's original training data.

When this retrieval loop is embedded within an agent framework, enabling iterative reasoning, tool use, and multi-step workflows, the resulting architecture is referred to as *Agentic RAG*. Figure 2 provides an overview of a typical architecture of an *Agentic RAG* system, illustrating the main components and their interactions during the query processing workflow. Modern frameworks such as *LangChain* [18], *LlamaIndex* [19], and *Langroid* [20] support the rapid development of Agentic RAG systems by abstracting core components like document ingestion, embedding management, retrieval orchestration, and LLM-based decision-making.

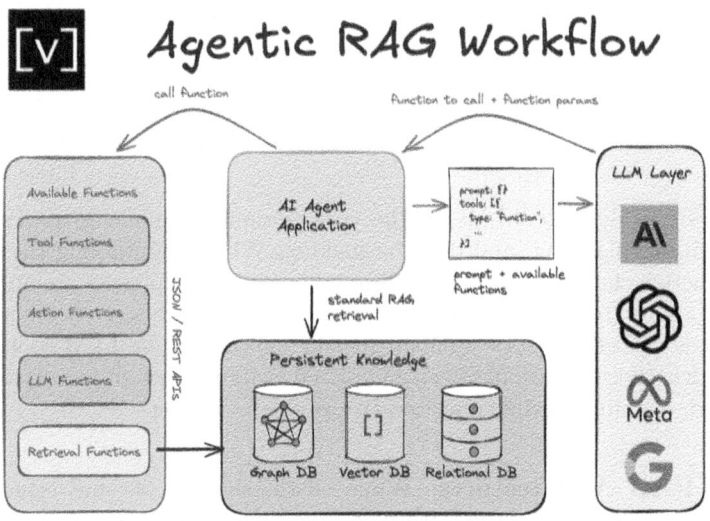

Fig. 2. Agentic RAG architecture. © Vectorize.io

AI Agents for IaC and PaC. Recent research has shown the growing applicability of AI techniques, in particular LLMs, in the domains of IaC and PaC.

In the context of IaC, LLMs have been successfully applied to automatically generate infrastructure definitions [21–23]. However, these models are susceptible to well-known limitations, including *hallucinations*, which may result in code that is syntactically incorrect or semantically invalid. As a result, naive applications of LLMs may introduce critical misconfigurations or deployment issues due to erroneous code in terms of both syntax and semantics.

To face these limitations, more advanced approaches have adopted LLM agents that combine reasoning with external tool integration and RAG. Some examples include the agentic architectures proposed in [24–26], which demonstrate the value of iterative, tool-assisted development cycles. These systems partially address the shortcomings of standard LLMs by incorporating validation, self-correction, and reasoning loops.

Similar techniques can be applied to the domain of PaC, where formal policy rules, typically expressed in languages such as `Rego`, can be generated from natural language descriptions. While some preliminary exploration in this direction exists [27], to the best of our knowledge, the system presented in this work is the first to autonomously: (i) translate natural language policy descriptions into formal PaC rules, (ii) assess the compliance of a given IaC configuration with the generated policies, and (iii) iteratively modify the infrastructure definition to ensure full compliance with the specified requirements.

3   ARPaCCino Architecture

ARPaCCino is an Agentic RAG system designed to translate natural language policy descriptions into formal `Rego` rules and validate IaC architectures against those policies. Leveraging the flexibility of the Agentic RAG approach, ARPaC-Cino is able to autonomously refine the IaC configuration until it satisfies all specified policy constraints. At its core, ARPaCCino consists of a reasoning engine based on an LLM, which orchestrates execution by interpreting requests, generating action plans, and invoking a suite of specialized tools. A high-level overview of the system architecture is shown in Fig. 3.

RAG Tool. The RAG Tool provides access to domain-specific knowledge, including official documentation, the `Rego` language definition, and examples for both the Open Policy Agent (OPA) and the supported IaC frameworks. The LLM invokes this tool whenever domain-specific knowledge is required. The retrieved content is used to enhance the prompt, effectively extending the LLM's capabilities. This modularity enables ARPaCCino to support uncommon or emerging technologies, provided that a structured knowledge base is supplied.

Infrastructure Tools. To ensure compatibility with multiple IaC frameworks, ARPaCCino leverages a set of specialized *infrastructure tools*, tailored to each framework. Those tools perform the required pre-processing on the given infrastructure definition before proceeding with the policy validation.

Rule Checker Tool. The LLM engine, after fetching the appropriate knowledge from the RAG tool, generates the `Rego` rules corresponding to the input policy description. The generated rules should then be verified prior to the automatic validation and architecture improvement. While OPA's native `opa check` command ensures syntactic correctness, it does not evaluate the semantic validity or logic of the rules. The Rule Checker Tool addresses this limitation by incorporating feedback from an external domain expert (or oracle), who reviews and either accepts or rejects the generated rules. This step ensures the soundness of the policy prior to enforcement.

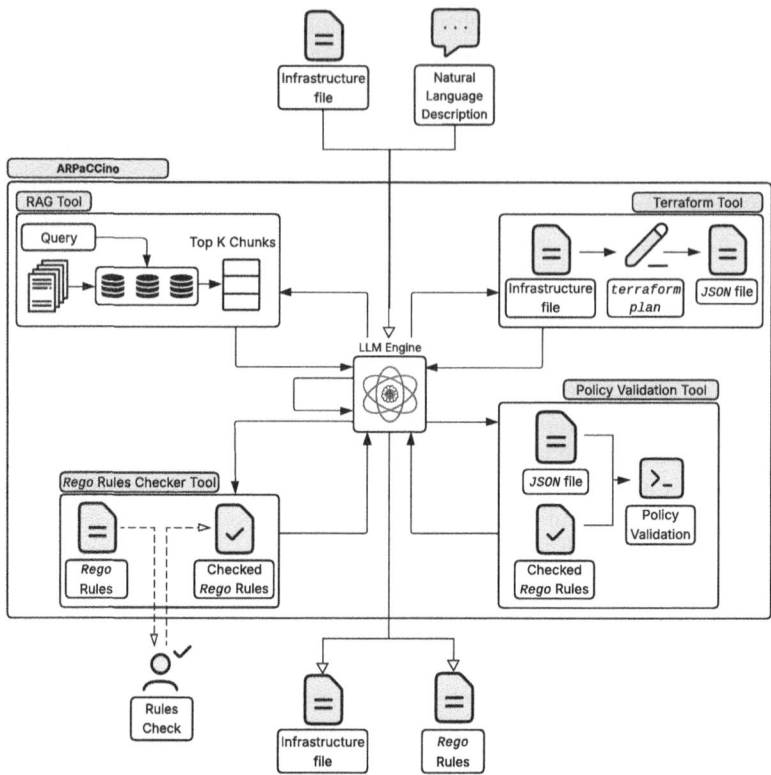

Fig. 3. ARPACCINO Architecture

Policy Validation Tool. The Policy Validation Tool takes as input the pre-processed infrastructure and the semantically verified `Rego` rules. It performs a deterministic evaluation to determine whether the infrastructure complies with the generated policies. Based on this result, the system decides whether the current IaC specification is ready for deployment or requires further adjustments.

4 Case Study

To evaluate the effectiveness of ARPACCINO, we present a real case study based on the widely adopted IaC framework Terraform [4]. In this scenario, the knowledge base available to the RAG tool includes documentation for OPA and for the Terraform provider for ProxMox [28]. The infrastructure is defined using standard Terraform configuration files (`.tf`), and the system invokes the `terraform plan` command to preprocess the infrastructure. This command generates a JSON-formatted execution plan, which serves as the input to the policy validation phase.

4.1   Expected Workflow

Algorithm 1 outlines the expected workflow of ARPaCCino. For illustrative purposes, we assume a simplified scenario in which each sub-task completes in a single tool invocation. This abstraction allows us to highlight the logical sequence of steps without delving into low-level agentic behaviors.

Algorithm 1. Expected ARPaCCino workflow

Input: Infrastructure file, Natural language description of the policies
Output: Verified Infrastructure file, Generated `Rego` rules.

1: Retrieve *OPA* Knowledge using RAG Tool
2: Generate `Rego` Rules from the natural language description
3: Verify the `Rego` Rules using the Checker Tool
4: **if** Rules are wrong **then**
5:     **go to** 2
6: **end if**
7: Preprocess the Infrastructure file using the Terraform Tool
8: Validate the Infrastructure JSON file against the `Rego` Rules using the Policy Validation Tool
9: **if** Infrastructure is not policy-compliant **then**
10:     Retrieve Terraform Knowledge using RAG Tool
11:     Correct the Infrastructure file
12:     **go to** 7
13: **end if**
14: **return** Verified Infrastructure file, Generated `Rego` rules.

In real cases, the agentic nature architecture of ARPaCCino often requires multiple iterations with the same tool in order to iteratively refine the output and achieve satisfactory results. However, this complexity is abstracted away from the end user. ARPaCCino manages all intermediate decisions and tool invocations internally. This design ensures a seamless experience, allowing users to focus on high-level objectives while the system autonomously orchestrates the underlying reasoning and execution processes.

4.2   Running Example

A minimal yet illustrative example of the capabilities of ARPaCCino is shown in Fig. 4. The environment consists of a single machine equipped with 4 CPU cores and 8 GB of RAM. The user instructs ARPaCCino to allow only machines with exactly 4 cores. ARPaCCino processes this request by generating the corresponding `Rego` policy, then evaluating the compliance of the infrastructure against it. Since the current configuration satisfies the constraint, the system confirms policy compliance without requiring any modifications to the Terraform definition.

```
Terraform File excerpt

...
resource "proxmox_virtual_environment_vm" "cloned_vm" {
  ...
  cpu {
    cores = 4
  }
  memory {
    dedicated = 8192
  }
  disk {
    interface   = "scsi0"
    datastore_id = "Storage"
    size        = 150
  }
  network_device {
    bridge = "intVM"
    model  = "virtio"
  }
  tags        = ["PaC"]
  description = "Cloned for PaC test"
}
```

```
Policy prompt

Allow only virtual machines with 4 cores in Terraform.
```

```
Generated Rego rule

package terraform

deny[msg] if {
    resource := input.planned_values.root_module.resources[_].values
    cpu := resource.cpu[_].cores
    cpu != 4
    msg := "VM must have exactly 4 cores"
}
```

Fig. 4. ARPaCCino running example

5   Experimental Results

The experimental evaluation conducted aimed to assess the effectiveness of
ARPaCCino in the generation and validation of Policy as Code, with a focus on
its applicability in a Terraform-based IaC scenario. The evaluation considered
the following key aspects:

- **Syntactic correctness** of the generated Rego policies;
- **Semantic alignment** of the policies with the natural language user instructions;
- **Detection capability** for identifying policy violations within Terraform execution plans;
- **Repair effectiveness** in automatically correcting non-compliant infrastructure definitions.

All experiments were conducted on a ProxMox-based Asus ESC4000A-E12 server with an AMD EPYC 9004 processor, 2 × 48 GB L40s NVIDIA GPUs, and 196 GB of RAM. The publicly available LLMs were run using Ollama inside a Ubuntu 24.04 virtual machine hosted on the server, provided with 16 cores, 128 GB of RAM, and both available GPUs. Closed-source models were accessed via API-based interactions using their publicly available endpoints.

5.1 Evaluation Methodology

We evaluated ARPaCCino on a defined and fixed Terraform infrastructure, with five distinct policy prompts of increasing difficulty. Among these, three prompts required the modification of the provided infrastructure, due to their incompatibility with the original IaC definition.

To assess the effectiveness of each system component, we conducted an ablation study with the following configurations:

- **LLM-only:** the base LLM is used without access to retrieval or external tools;
- **LLM + RAG:** the model is enhanced with retrieval capabilities but lacks access to tool execution;
- ARPaCCino **(full):** the complete agentic system with both RAG and tool invocation capabilities enabled.

Furthermore, we tested different LLMs for each of the discussed scenarios, to understand to what extent the capabilities of the "raw" LLM affect the overall generation performance of the system. We chose to evaluate ARPaCCino with three different models: the open-weight model `Qwen3` in its 30 billion parameters version and the closed `GPT-4o` and `Claude Sonnet 4` models. These models represent the current state-of-the-art and support tool calling, ensuring a fair comparison in all the described scenarios.

5.2 LLM Vs RAG Vs Agentic RAG

In Table 1 are depicted the results of the ablation study. The table shows a comparison across three different approaches, LLM, RAG, and Agentic RAG, applied to three models (`Qwen3:30b`, `GPT-4o`, and `Claude Sonnet 4`). The *Model* column specifies the large language model employed for the specific batch of tests. The *Configuration* column indicates whether the base LLM, RAG, or agentic RAG configuration was used. The *Syntax* column shows how many out of five `Rego` policy generations were syntactically correct, and the *Semantic* column reports how many of the syntactically correct policies were also semantically correct. The last column, *Notes*, provides additional observations regarding each setup, such as errors encountered or behavior noted during the executions. As expected, the agentic approach used in ARPaCCino greatly enhances the system's capability of generating syntactically and semantically correct `Rego` policies.

Base LLMs most likely lack knowledge about OPA and `Rego` to effectively generate correct policies. Furthermore, even when such knowledge is provided

Table 1. Performance summary and comparison

Model	Configuration	Syntax	Semantic	Notes
Qwen3:30b	LLM	0/5	—	Rego rules generated with syntactical errors
Qwen3:30b	RAG	0/5	—	Rego rules generated with syntactical errors
Qwen3:30b	Agentic RAG	4/5	4/5	Loop during the policy correction for 1 prompt
GPT-4o	LLM	0/5	—	Rego rules generated with syntactical errors
GPT-4o	RAG	0/5	—	Rego rules generated with syntactical errors
GPT-4o	Agentic RAG	5/5	5/5	1/3 Terraform file modified
Claude Sonnet 4	LLM	5/5	5/5	Rego rules generated without external knowledge
Claude Sonnet 4	RAG	5/5	5/5	Rego rules generated
Claude Sonnet 4	Agentic RAG	5/5	5/5	Rego rules generated and checked

through the RAG module, most of the models are not capable of generating satisfying policies on the first try, failing the syntax check. Hence, the availability of deterministic verification tools is crucial to allow the system to iteratively correct itself, eventually reaching a satisfying output. Our approach achieves success most of the time, with failures related to an excessive number of retries during the policy correction (in the tests, the max amount of workflow iterations was fixed to 3) or the system's inability to modify the original Terraform file.

Surprisingly, the Claude Sonnet 4 model already possesses the required knowledge and is capable of generating correct Rego rules for our tests even in the base LLM scenario, showing how powerful this cutting-edge model is. However, it is worth noting that only the agentic approach ensures that the generated rules are syntactically and semantically correct, while the simple use of an LLM does not provide any warranty in that sense. This is a consequence of the agentic approach's capacity to reason about any errors and implement appropriate corrections autonomously.

Furthermore, another crucial result obtained involves the use of smaller and cheaper models. Despite a very powerful (and costly) model, such as Claude Sonnet 4 might be able to generate satisfying rules, ARPACCINO's approach enables the use of much smaller models (30 billion parameters in our tests) to achieve comparable results. This allows the use of one of the appropriate publicly available models in the scenario of PaC generation and IaC compliance verification, effectively eliminating dependencies on external model providers.

5.3 Model Comparison for the Agentic RAG

In the full Agentic RAG configuration of ARPACCINO, we evaluated how the choice of underlying LLM affects overall performance. As expected, the model's capabilities significantly influence task execution, primarily in terms of the number of RAG and tool invocations required to reach a successful outcome. As reported in Table 2, more powerful models tend to require fewer (average) calls to the tools to produce satisfying policies and to correct the infrastructure file. However, this reduction comes at a cost, as more powerful models are also more expensive to use, while the involved tools have negligible costs. Thus, the choice of LLM in ARPaCCino should balance between the capabilities and the cost of the executions.

Table 2. Average RAG and tool calls in the Agentic RAG

Model	RAG Call (avg)	Tool Call (avg)
Qwen3:30b	4.4	11.4
GPT-4o	3.8	9.0
Claude Sonnet 4	3.2	7.8

6 Conclusions and Future Work

In this work, we introduced ARPaCCino, a novel agentic system for the generation of Policy as Code and the validation of Infrastructure as Code configurations. By adopting the Agentic RAG paradigm, ARPaCCino provides an effective solution for the effortless creation and validation of security policies in IaC environments. Our results demonstrate that ARPaCCino, thanks to the use of deterministic validation tools, significantly improves performance, enabling accurate and reliable policy implementation, particularly when using smaller language models. Among the models evaluated, `Claude Sonnet 4` emerged as the most effective for this task. However, the adoption of the agentic RAG paradigm also enables the effective use of smaller models (e.g. `Qwen3:30b`), which, when supported by a well-curated knowledge base and appropriate tools, can achieve performance comparable to that of larger and more expensive models. In future work, we plan to extend ARPaCCino with automated semantic verification of the generated `Rego` rules. This remains a challenging task due to the complexity of verifying semantic correctness automatically. Additionally, we aim to explore the integration with self-RAG [29] technologies, which could further enhance the autonomy and adaptability of the system during the generation and validation phases.

Acknowledgments. This work was partially supported by the SERICS project (PE00000014) under the MUR National Recovery and Resilience Plan funded by the European Union - NextGenerationEU. The work of Francesco A. Pironti was supported by Agenzia per la cybersicurezza nazionale under the 2024–2025 funding program for promotion of XL cycle PhD research in cybersecurity (CUP H23C24000640005).

References

1. Rahman, A.A.U., Helms, E., Williams, L., Parnin, C.: Synthesizing continuous deployment practices used in software development. In: 2015 Agile Conference, pp. 1–10. IEEE (2015)

2. Humble, J., Farley, D.: Continuous Delivery: Reliable Software Releases Through Build, Test, and Deployment Automation. Pearson Education (2010)
3. Guerriero, M., Garriga, M., Tamburri, D.A., Palomba, F.: Adoption, support, and challenges of infrastructure-as-code: insights from industry. In: 2019 IEEE International Conference on Software Maintenance and Evolution (ICSME), pp. 580–589. IEEE (2019)
4. hashicorp/terraform
5. Red Hat. Ansible automation platform (2024)
6. Pulumi Corporation. Pulumi: Modern infrastructure as code
7. Rahman, A., Farhana, E., Parnin, C., Williams, L.: Gang of eight: a defect taxonomy for infrastructure as code scripts. In: Proceedings of the ACM/IEEE 42nd International Conference on Software Engineering, pp. 752–764 (2020)
8. Hasan, M.M., Bhuiyan, F.A., Rahman, A.: Testing practices for infrastructure as code. In: Proceedings of the 1st ACM SIGSOFT International Workshop on Languages and Tools for Next-Generation Testing, pp. 7–12 (2020)
9. Hassan, M.M., Salvador, J., Rahman, A., Karmaker, S.: Large language models for it automation tasks: are we there yet? arXiv preprint arXiv:2505.20505 (2025)
10. Joshi, S.: A review of generative AI and DevOps pipelines: CI/CD, agentic automation, MLOps integration, and large language models. In: CD, Agentic Automation, MLOps Integration, and Large Language Models, June 2025
11. Policy as code. https://developer.hashicorp.com/sentinel/docs/concepts/policy-as-code
12. Rajapakse, R.N., Zahedi, M., Babar, M.A., Shen, H.: Challenges and solutions when adopting DevSecOps: a systematic review. Inf. Softw. Technol. **141**, 106700 (2022)
13. GitHub - open-policy-agent/OPA: Open Policy Agent (OPA), June 2025
14. Sentinel | HashiCorp Developer — developer.hashicorp.com. https://developer.hashicorp.com/sentinel. Accessed 26 June 2025
15. Wooldridge, M.J.: An Introduction to Multiagent Systems, 2 edn. Wiley, Chichester (2012)
16. Xi, Z.:. The rise and potential of large language model based agents: a survey. Sci. China Inf. Sci. **68** (2025)
17. Lewis, P., et al.: Retrieval-augmented generation for knowledge-intensive NLP tasks. In: Advances in Neural Information Processing Systems (2020)
18. Chase, H.: Langchain, October 2022
19. Liu, J.: Llamaindex, November 2022
20. Chalasani, P., Jha, S.: Langdroid
21. Palavalli, M.A., Santolucito, M.: Using a feedback loop for LLM-based infrastructure as code generation, November 2024
22. Low, E., Cheh, C., Chen, B.: Repairing infrastructure-as-code using large language models. In: 2024 IEEE Secure Development Conference (SecDev), pp. 20–27, October 2024
23. Srivatsa, K.G., Mukhopadhyay, S., Katrapati, G., Shrivastava, M.: A survey of using large language models for generating infrastructure as code, March 2024
24. Lee, J., Kang, S., Ko, I.-Y.: An LLM-driven framework for dynamic infrastructure as code generation. In: Proceedings of the 25th International Middleware Conference: Demos, Posters and Doctoral Symposium, pp. 9–10. ACM, Hong Kong, Hong Kong, December 2024
25. Zhang, T., Pan, S., Zhang, Z., Xing, Z., Sun, X.: Deployability-centric infrastructure-as-code generation: an LLM-based iterative framework, June 2025

26. Lupinacci, M., Blefari, F., Romeo, F., Pironti, F.A., Furfaro, A.: ARCeR: an agentic RAG for the automated definition of cyber ranges (2025)
27. Martinelli, F., Mercaldo, F., Petrillo, L., Santone, A.: Security policy generation and verification through large language models: a proposal. In: Proceedings of the Fourteenth ACM Conference on Data and Application Security and Privacy, Porto, Portugal, pp. 143–145. ACM, June 2024
28. GitHub - BPG/terraform-provider-proxmox: Terraform/OpenTofu Provider for Proxmox VE, June 2025
29. Asai, A., Wu, Z., Wang, Y., Sil, A., Hajishirzi, H.: Self-RAG: learning to retrieve, generate, and critique through self-reflection (2023)

ERGA 2025: 1st Workshop on Entity Resolution and Graph Alignment

Concept Matching in Hierarchical Meta-data: Leveraging Big-Data to Improve the Performances of Matching Strategies

Mario Cataldi[1]([✉]) [ID], Luigi Di Caro[2] [ID], and Claudio Schifanella[2] [ID]

[1] Université de Paris 8 France, IUT de Montreuil LIASD, Saint-Denis, France
m.cataldi@iut.univ-paris8.fr
[2] University of Turin, Turin, Italy
{dicaro,schi}@unito.it

Abstract. Data and meta-data hierarchies are playing a central role in the development and deployment of many big data applications. They can be defined as structured information that describes, explains, locates, and makes it easier to retrieve, use, and manage information within a large corpus of resources.

Since hierarchical data have significant roles in data annotation, search, and navigation, they are often carefully engineered; however, especially in dynamically evolving domains, meta-data from different domains and/or producers are rarely identical, even when describing the same data scope; thus, there is a need for techniques to find alignments between concepts in different structures.

In this paper, we present a novel method that leverages dynamic big data information to improve concept matching among different static meta-data structures. This method captures the structural information inherent in hierarchical meta-data and enriches it with semantically coherent information extracted from external data to permit more precise matching operations among concepts across different hierarchies.

Keywords: Concepts matching · Big Data · Classification

1 Introduction

Today, considering the available data on- and offline, often large in volume and complex in structure, it is necessary to assist the user in handling and exploring this large amount of information. Therefore, new access and exploration methods are needed in order to guide the users through this massive amount of data, highlighting (if and when requested) hidden relationships among them, and helping the users orient themselves within this vast information space.

For this, most big data applications generally enrich the data considered by *meta-data*, with the goal of providing supervised knowledge about the content

P. K. Chrysanthis et al. (Eds.): ADBIS 2025, CCIS 2676, pp. 485–499, 2026.
https://doi.org/10.1007/978-3-032-05727-3_40

that can help the retrieval and exploration process. Some of the most used meta-data types include controlled vocabularies, taxonomies, ontologies, thesauri, data dictionaries, and registries. Meta-data can be one-dimensional, where each element is semantically separate from other elements, or hierarchical, where evident relationships exist among the elements (vertical and/or horizontal).

Considering all these aspects, while there are many strategies for organizing data, hierarchical meta-data categorization, usually implemented through a predetermined taxonomical or graph structure, is often the preferred choice.

The relationships between concepts are intended to be combined to produce a large amount of possible correlations among the data that can then be used in a variety of interpretation paradigms and queries. Hence, semantic hierarchy is a natural way of querying complex big data sets, assuming that some knowledge about the data is organized in a hierarchical manner and carefully linked to the considered big data set. Meta-data hierarchies also enable sharing and integration of information from different domains and data sources.

However, meta-data structures from different sources are rarely identical; heterogeneity between different meta-data structures, aimed at describing the same domain/resources, is a serious problem for efficient cooperation processes.

Based on these considerations, in this paper, we note that there is a need for techniques to align concepts by taking into account both the meta-data structures and the context in which they will be used. More specifically, we observe that when meta-data structures are used as information sources supporting navigation in a document space, the way users perceive the degree of matching between pairs of concepts is highly dependent on the domain being explored. We then argue that the advantages of the structure-based and information-based approaches can be combined in a structure and semantically informed matching approach. This approach leverages the structural information coded by the structure of the given meta-data hierarchies to infer matching relationships that could be not directly detected through element-level matching techniques.

The information about the extension of the concepts against a given large database contextualizes and grounds the meta-data hierarchy on the considered database and helps resolve conflicts/mismatches that might arise if only structural aspects are considered.

The paper is organized as follows. Section 2 shortly recalls the major ideas behind the existing structure and semantic informed matching methods. Section 3 introduces a pure structure-base approach that aims at formalize the content of a given meta-data hierarchy and combines the purely structural information with extension based matching evaluation, thus resulting in a hybrid method that is structure-and extension-informed at the same time (Sect. 4). Section 5 evaluates the proposed methods and shows its effectiveness on our test cases. Section 6 concludes the paper and presents our future work plan.

2   Related Work

Meta-data structures from different sources (and referring to the same domains) are rarely identical; in fact, their knowledge structure strictly depends on the domain expert that defined them and, in many cases, there is need for techniques to find alignments between concepts in different structures.

Different matching techniques focus on different dimensions of the problem, including whether data instances are used for schema matching, whether linguistic information and other auxiliary information are available, and whether the match is performed for individual elements (such as attributes) or for complex structures [16].

Cupid [8] is a schema-based approach that implements a sequential composition of different matches. It consists of a first phase based on a linguistic matcher and a second phase based on a structural matching technique. This algorithm operates only with trees: other schemas can be handled through a translation process. [11] uses schema graphs for matching; matching is performed node by node starting at the top; thus this approach presumes a high degree of similarity (i.e., low structural difference) between the taxonomies. Onion [13], the successor of SKAT [12], is a schema-based system that leverages logic rules to discover match and mismatch relationships between multiple ontologies, represented internally as labeled graphs.

[1] and DIKE [14] use the distance of the nodes in the schemas to compute the mappings; while computing the similarity of a given pair of objects, other objects that are closely related to both count more heavily than those that are reachable only via long paths of relationships. Glue [3], the successor of LSD [2], is an instance-based semi-automatic system that uses machine-learning techniques to discover one-to-one mappings between two taxonomies. It is based on the calculus of the joint distributions that are used for any similarity measures. Differently from Glue, FCA-merge [20] takes as input two ontologies that share the same set of instances and produces a new ontology as result. It uses formal concept analysis techniques, through a process made up of three steps: instance extraction, concept lattice computation, and (interactive) generation of the final new ontology. Clio [9,10] is a mixed schema-based and instance-based system that proposes a declarative approach to schema mapping between either XML and relational schemas. After the first phase in which input schemas are translated into an internal representation, the system combines sequentially an instance-based attribute classification (by using a Bayes classifier) with a string matching between elements name. These n-to-m value correspondences can be also entered by the user through a graphical user interface. After that, Clio produces a set of logical mappings with formal semantics, supporting also mappings composition. [4] provides a more detailed classification of matching techniques, based on other features including different similarity measures, matching strategies (such as name similarity or class similarity), and degrees of user involvement. [5] proposes an algorithm for ontology matching that combines standard string distance metrics with a structural similarity measure based on a vector representation. Despite such advances in structural mapping technologies, alignments across data sources are rarely perfect. In particular, imperfection can be due to homonyms (i.e., nodes with identical concept-names, but possibly different semantics, in the given taxonomy hierarchies) and synonyms (concepts with dif-

ferent names but same semantics). While structural-matching techniques help finding node-to-node alignments, they fall short when such scenarios arise.

Recent trend in concept-matching within graphs emphasizes end-to-end learning of graph structure and alignment in a unified model. GLAM is a hallmark example: it learns graph structures and cross-graph correspondences simultaneously via a self- and cross-attention framework, outperforming earlier methods that relied on handcrafted graph patterns [7]. Building on this concept, i-Align brings explainability into the fold by leveraging transformer-based encoders with edge-gated attention and historical embeddings—enabling alignment decisions that highlight the most influential attributes and neighbors for each matched entity link [21]. Complementing these approaches, LaKERMap targets ontology matching by self-supervised sampling of local and global structural contexts in transformer models, demonstrating enhanced alignment quality and runtime performance [22]. These combined efforts reflect a paradigm shift from two-step matching (structure then correspondence) to joint, structure-aware, and interpretable graph alignment, integrating precise graph-pattern learning, attention to structural context, and transparency in result justification—crucial for reliable concept matching in complex knowledge scenarios.

3 Structure- and Semantic-Based Matching Algorithms for Meta-data Hierarchies

Many traditional knowledge-driven applications (as for data mapping, data classification or disambiguation) require mining of semantic similarity/dissimilarity values between concepts in a given domain. Therefore, the study of semantic relationships between concepts has a long history in various domains, including psychological theory, natural language processing, and knowledge management.

In the last decade, various measures for estimating the semantic similarity of concepts across different meta-data hierarchies have been proposed. These measures can be roughly categorized into *structure-based (or, edge-based)* methods and *semantic-based* methods. In structure-based methods, the similarity between two concepts is measured by some quantification of the distance between two nodes (as for the shortest distance between them [15] or the sum of the edge weights along this shortest path [17]). Following this idea, highly related concepts are grouped together and the path between two different concept-nodes in the hierarchy reflects how these are related in the considered domain. On the other hand, semantic-based methods leverage available, external, big data to extract additional information to achieve better similarity estimations.

In the next Sections we will analyze various approaches and introduce our novel algorithm to better exploit external, big data, knowledge in order to integrate structural information for concept matching purposes.

3.1 Concept Formalization Through Concept-Vectors

Given a concept hierarchy, in this section, we aim at formalizing the knowledge expressed by a concept hierarchy by assigning a *concept-vector* to each concept

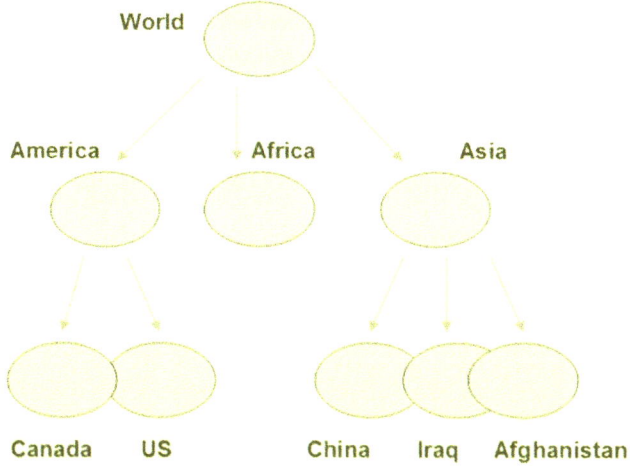

Fig. 1. Example of a geographical taxonomy fragment.

node, such that the vector encodes the *structural* relationship between this node and all the other nodes in the hierarchy. For this, we apply the method proposed by [6]; the concept vectors are obtained by propagating concepts on the hierarchical tree according to their *structural* contributions (dictated by the structure of the hierarchy).

For this, given a hierarchy, $\mathcal{H}(N, E)$ with m concept-nodes, each node is mapped onto a concept-space with m concept-dimensions. Initially, each concept-node is mapped into the concept-space along the dimension corresponding to itself. Then, for each pair, c_i, c_j, of neighboring concepts, CP/CV computes two values, $G^{str}_{c_i, c_j}$ and $G^{cs}_{c_i, c_j}$. G^{str} measures the *degree of generality* between two entities computed using the structural information. G^{cs}, on the other hand, computes a corresponding value using the concept space. Since these hierarchical and multi-dimensional representations are different views over the same semantics, this method computes concept propagation parameters that preserve the equality

$$G^{cs}_{n_i, n_j} = G^{str}_{n_i, n_j}$$

after propagation. This process is iterated until all nodes are informed of all the others. Once the process is computed, since all the concepts are mapped into the same vector space, semantic similarities of the concepts are computed using the cosine similarity of the concept vectors.

Let us consider for example the hierarchy fragment containing the nine nodes presented in Fig. 1. Each concept is represented by a 9-dimensional vector. Vector's elements are associated to the hierarchy nodes, considered in breath first order. In particular, the root is represented by the vector

$$\langle 0.450, 0.169, 0.141, 0.158, 0.018, 0.018, 0.018, 0.021, 0.021 \rangle,$$

in which the first component - the one associated to the tag "World", dominates over the other that contribute to the definition of the concepts. The second, third and fourth components reflect the weight of "America", "Africa" and "Asia" respectively in the structural definition of "World", while the remaining components represent the weights of the three descendants of "Asia" and of the two descendants of "America". Similarly, the concept vectors for all nodes are as shown in Table 1.

Table 1. Concept-vectors associated to the taxonomy fragment in Fig. 1.

	World	Asia	Africa	America	Afgh.	Iraq	China	Canada	US
cv_{world}	0.450	0.169	0.141	0.158	0.018	0.018	0.018	0.021	0.021
cv_{asia}	0.052	0.469	0.006	0.006	0.156	0.156	0.156	0.0003	0.0003
cv_{africa}	0.100	0.012	0.873	0.012	0.0006	0.0006	0.0006	0.0007	0.0007
$cv_{america}$	0.057	0.007	0.007	0.520	0.0003	0.0003	0.0003	0.204	0.204
$cv_{afgh.}$	0.004	0.100	0.0002	0.0002	0.872	0.012	0.012	0	0
cv_{iraq}	0.004	0.100	0.0002	0.0002	0.012	0.872	0.012	0	0
cv_{china}	0.004	0.100	0.0002	0.0002	0.012	0.012	0.872	0	0
cv_{canada}	0.006	0.0003	0.0003	0.165	0	0	0	0.806	0.023
cv_{us}	0.006	0.0003	0.0003	0.165	0	0	0	0.023	0.806

4   Semantically-Informed Meta-data Structure Alignment

While concept vectors help capture and leverage the structural information embedded in meta-data hierarchies for matching their concept nodes with concept nodes in others, in this paper we note that they can also benefit from available external big data information to improve matching performance. The main idea behind the *semantic-informed* matching is that if two concepts in two different hierarchies are to be considered *similar* in a considered context, then they will *relate to similar information* in the external data information; in contrast, if two concepts are to be considered *different*, then the information that relate to these concepts will also be *different*.

In line with the common notion in literature, we call the external information that a concept relates to as its *extension*.

One trivial way of identifying extensions of concepts is to find documents that contain the corresponding concept name as a keyword. This, however, cannot be used for hierarchy alignment because of many reasons: for example, synonyms in two different hierarchies will have identical extensions and/or homonyms will have (most likely) very different extensions. For this, we need a mechanism that identifies *extensions* of concept nodes without relying only on their concept names. The concept vectors associated to the concept nodes provide a convenient way to identify extensions. In particular, we rely on a classifier module which takes as input the set $CV = \{cv_1, \ldots, cv_m\}$ of the concept vectors representing the hierarchy, and the set $V = \{v_1, \ldots, v_n\}$ of vectors representing the external documents to be classified. Keyword vectors are defined in the space of the entire set of document keywords; each dimension corresponds to a keyword, and the weights in the vector represent the relevance of the corresponding keyword value in the document represented by the vector. The goal of this classification component is to associate the data information to their best representative concepts in the concept hierarchy. We capture this notion of representativeness through the notion of the similarity among the hierarchy and document vectors representing hierarchy concepts and documents, respectively.

With this objective, for every document in the considered external corpus, the module identifies the hierarchy concepts that best match with it. Each document is considered as belonging to the extensions of those concepts whose similarities with it are above the adaptively computed critical point (as described in the previous section). More in detail:

1. For each document
   (a) calculate the vector v_j, where each element contains the augmented normalized term frequency [19] of a document keyword.
   (b) compute its similarity wrt. all the concept vectors describing the given taxonomy.
   $$sim(cv_i, v_j) = \Sigma_{k=1}^{u} cv_{iu} \times v_{ju}$$
   (c) sort the concepts vectors in decreasing order of similarity wrt. v_j;
   (d) choose the cut-off point to identify the concepts which can be considered *sufficiently similar* to justify the classification of the object under them. Our method adaptively computes a cut-off as follows: It
      i. first ranks the concepts in descending order of match to v_j, as previously calculated in (c).
      ii. computes the *maximum drop* in match and identifies the corresponding drop point.
      iii. computes the *average drop* (between consecutive entities) for all those nodes that are ranked before the identified maximum drop point.
      iv. the first drop which is higher than the computed average drop is called the *critical drop*. We return concepts ranked better than the point of critical drop as candidate matches.

Notice that, at the end of this phase, the extensions of different concepts may not be disjoint, since the same object can be assigned to multiple (similar) concept vectors. The number of concepts associated with a given document depends on the corresponding adaptive threshold value computed by the classification algorithm.

4.1   Enriching the Semantic Space Through the Discovery of Concept-Keyword Relationships

Using the previously described method, we can associate the documents to their best representative concepts in the concept hierarchy. At this step, we may be interested in improving the quality of the associations by allowing the system to discover new hidden relationships among the relevant terms (contained in the given corpus of documents) and the original concept of the given meta-data hierarchy, creating new semantic connections between both input data.

Considering a concept c_i and its extension, we aim to search for the most contextual informative keywords and enrich the pure structural definition of the concept. For this, we define the relationship that exists between a keyword k_i, extracted from the considered corpus, and a hierarchy concept c_i, as the weight u_i computed as follows [18]:

$$u_i = log \frac{r_i/(R - r_i)}{(n_i - r_i)/(N - n_i - R + r_i)} \times \left| \frac{r_i}{R} - \frac{n_i - r_i}{N - R} \right|$$

where:

- r_i is the number of document in the association containing the keyword i
- n_i is the number of documents in the corpus containing the keyword i
- R is the cardinality of the association
- N is the number of documents in the collection.

Intuitively, the first factor increases when the number of the documents containing the keyword k_i increases, while the second factor decreases when the number of the irrelevant documents (i.e., not belonging to the considered extension) containing the keyword k_i increases. Therefore, keywords that are highly common in a specific association and not very present in others will get higher weights. For each concept, we consider all keywords contained in at least one document of the concept extension that have a positive weight value. Similarly to the previous step, we apply the adaptive cut-off to this set in order to select the most relevant keywords with the highest weights that will form the *enriching-keyword* vector ekv_{c_i}.

At this point, for each concept c_i, we have two vectors:

1. the concept-vector cv_{c_i} representing the concept-concept relationships in the corresponding taxonomy
2. the enriching-keyword vector, ekv_{c_i}, consisting of keywords that are significant in the current context defined by the corpus.

In order to combine the concept and the enriching-keyword vectors into a single extended-concept vector, defined as

$$ecv_{c_i} = \alpha_{c_i} \cdot cv_{c_i} + \beta_{c_i} \cdot ekv_{c_i},$$

we need to first establish the relative impacts (i.e. α_{c_i} and β_{c_i}) of the taxonomical knowledge versus the data background knowledge.

Therefore, given concept c_i, let

- $S(cv_{c_i})$ be the set of documents associated to the concept c_i (i.e. the documents retrieved from querying the database using the concept-vector, cv_{c_i}); and

- $S(\boldsymbol{ekv}_{c_i})$ be the set of documents obtained by querying the database using the enriching-keyword vector, $\boldsymbol{ekv}_{c_i}$.

We quantify the relative impacts, α_{c_i} and β_{c_i}, of the concept and enriching-keyword vectors, $\boldsymbol{cv}_{c_i}$ and $\boldsymbol{ekv}_{c_i}$, by comparing how well $S(\boldsymbol{cv}_{c_i})$ and $S(\boldsymbol{ekv}_{c_i})$ approximate $D_{d\to c}(\boldsymbol{cv}_{c_i})$. In other words, if

- $C_{c_i} = D_{d\to c}(\boldsymbol{cv}_{c_i}) \cap S(\boldsymbol{cv}_{c_i})$ and
- $EK_{c_i} = D_{d\to c}(\boldsymbol{cv}_{c_i}) \cap S(\boldsymbol{ekv}_{c_i})$,

then we expect that

$$\frac{\|\alpha_{c_i} \cdot \boldsymbol{cv}_{c_i}\|}{\|\beta_{c_i} \cdot \boldsymbol{ekv}_{c_i}\|} = \frac{|C_{c_i}|}{|EK_{c_i}|}.$$

If the concept and enriching-keyword vectors are normalized to 1, then we can rewrite this as

$$\frac{\alpha_{c_i}}{\beta_{c_i}} = \frac{|C_{c_i}|}{|EK_{c_i}|}.$$

Also, if we further constrain the extended concept vector $\boldsymbol{ecv}_{c_i}$ to be also normalized to 1. I.e.,

$$\|\alpha_{c_i} \cdot \boldsymbol{cv}_{c_i} + \beta_{c_i} \cdot \boldsymbol{ekv}_{c_i}\| = 1,$$

then, solving these equations for α_{c_i} and β_{c_i}, we obtain:

$$\alpha_{c_i} = \frac{|C_{c_i}|}{|C_{c_i}| + |EK_{c_i}|} \quad \text{and} \quad \beta_{c_i} = \frac{|EK_{c_i}|}{|C_{c_i}| + |EK_{c_i}|}.$$

Thus, given concept, c_i, we can compute the corresponding extended concept vector as

$$\boldsymbol{ecv}_{c_i} = \frac{|C_{c_i}|}{|C_{c_i}| + |EK_{c_i}|} \cdot \boldsymbol{cv}_{c_i} + \frac{|EK_{c_i}|}{|C_{c_i}| + |EK_{c_i}|} \cdot \boldsymbol{ekv}_{c_i}.$$

Since, at this point, each concept in the original meta-data hierarchy has its own extended concept vector $\boldsymbol{ecv}$, the documents in the given corpus can be associated under these nodes, but using $\boldsymbol{ecv}$ vectors instead of $\boldsymbol{cv}$ vectors. In this manner, using the extended vectors, we are able to associate to each concept not only the documents that contain that concept name, but also the documents containing some of the contextually relevant concepts and keywords.

5   Experimental Evaluation

In this section, we study the advantages of using a context-aware enrichment (from the considered corpus) of the structural information (dictated by the hierarchy itself) to better describe the meta-data knowledge.

Thus, we prove our association-based meta-data definition by evaluating the capacity of our method to disambiguate the nodes contained in the given hierarchies. In other words, given two meta-data hierarchies, we believe that a good meta-data knowledge formalization approach should be able to define each node in such a way as to recognize, if there are, similarities/dissimilarities among different meta-data structures, and recognize where they match and where they differ.

In detail, given a meta-data structure, we formalize its knowledge by using the approach described in Sect. 3.1; thus, after the taxonomical vectorization, we leverage these concept-vectors to associate to each meta-data node a set of documents that best describe them. Then, we compare the proposed enriched association-based meta-data knowledge definition (called in the experiment EnrClass and explained in Sect. 4) against three alternative methods:

- CP/CV formalizes the meta-data knowledge by associating to each taxonomical node a concept-vector that describes its structural relationships with all the other nodes in the meta-data hierarchy;
- common-ancestor distance (Anc), which defines each meta-data node by taking into account its distance from all its ancestors in the hierarchy (i.e., each node is defined by its counting the distance wrt all the ancestors in the hierarchy).

Then, we performed several evaluation experiments by considering a meta-data hierarchy extracted from DMOZ[1]. The considered hierarchy has 72 nodes, depth 4, and different branching factors in its internal nodes (the average value is 5.14). Then, we classify 17420 article abstracts describing NSF awards for basic research[2].

To evaluate the effectiveness, in terms of disambiguation capacity, of the proposed association-based meta-data definition strategy, in the presence of different conditions, we then created several (and similar) other hierarchies to be matched against. Thus, considering the original meta-data hierarchy extracted from DMOZ, we create alternative structures to be matched against, by introducing controlled distortions to the original hierarchy. These distortion permit to maintain the structure of the considered meta-data hierarchy but introduce small modifications of the concept nodes; in particular, we introduce the following distortions:

- *synonyms:* we randomly pick a percentage of nodes and relabel their concept names with other terms (without affecting the structure of the hierarchy). Note that the new labels are actually random (i.e., not real English synonyms) and do not actually occur at all in the data corpus. This constitutes a worst case situation for association-based algorithms (that also leverage the node labels for classification purposes).

---

[1] Accessible at the link http://www.dmoz.org/Science/.
[2] http://kdd.ics.uci.edu/databases/nsfabs/nsfawards.html.

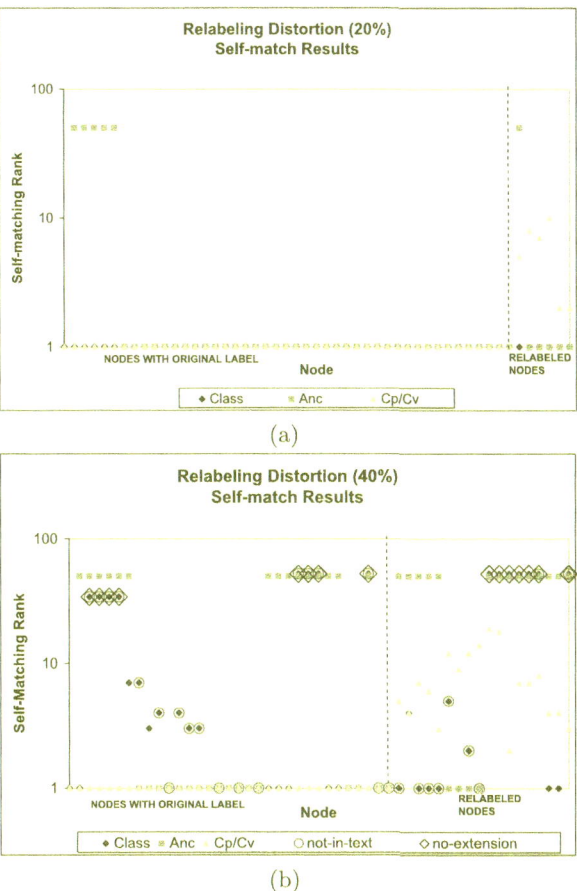

(a)

(b)

Fig. 2. Matching results under concept re-labeling (rank = 1 indicates an exact match).

– *Homonyms:* we randomly picked a percentage of nodes and, for each of them, we introduced a replica in randomly selected positions of the other meta-data hierarchy. The replica has the same concept name, but it is contextualized in a different position in the structure of the hierarchy (i.e., it is a homonym).

Considering these two conditions, we applied distortions of the order of 20% and 40% of the nodes. Thus, the main aim of the experiment is to prove that, leveraging the association-based definition of the nodes, it is possible to positively retrieve in the altered hierarchy, the corresponding node.

Figures 2(a) and (b) show the synonym matching results. In this Figure, the X axis denotes the nodes and Y axis denotes the rank at which the corresponding node, in the distorted meta-data hierarchy, is found. Note that if the alignment algorithms works perfectly, then these distortion operations would not have any impact and all nodes will be found at rank 1.

- For the 20% distortion case (Fig. 2(a), the portion relabeled is on the right), we observe that EnrClass (which identifies the proposed association-based meta-data knowledge definition) is always able to retrieve the corresponding nodes in the compared structure (100% of exact matches). On the other hand, Anc (which identifies the meta-data knowledge definition based on the edge distance among the hierarchy nodes) makes some mistakes when also an ancestor node is relabeled (86.3% of exact matches). However, CP/CV-based formalization approach defines each node in such a way to be able to always retrieve, for non-relabeled nodes, the corresponding nodes in the alternative hierarchy but (since it relies on the concept labels to some degree) performs imperfectly for re-labeled nodes (88.3% of correct matches).
  In short, while our meta-data formalization method is also based on the concept-vectors, Class improves the results by leveraging available document associations.
- On the other hand, when 40% of the nodes are arbitrarily relabeled (Fig. 2(b), the portion relabeled is on the right), the impact on the performances are significant: in fact, Anc makes significant errors (only 42.2% of exact matches). However, CP/CV works very well for non-relabeled nodes and performs imperfectly for re-labeled nodes (totally 64.8% of exact matches). In contrast, EnrClass performs very well even in this heavily distorted situation (83.4% of exact matches).

Figures 3(a) and (b) show the matching results in the presence of multiple concepts with identical labels. Figure 2(a) presents the matching ranks for the corresponding nodes in the altered hierarchies, while Fig. 2(b) shows the matching rank for the new inserted homonyms. Note that, for the former case, the closer to 1 the rank, the better are the results; whereas, for the homonym case, the further from 1 the ranks, the more discriminating is the algorithm. In fact, in this second case, further from 1 the ranks, better the formalization method permits to distinguish among the homonyms.

- For self-matching cases (Fig. 3(a)), since the node-copy distortions do not affect any internal nodes, Anc works perfectly (100% of exact matches). On the other hand, CP/CV, which gets structural context also from descendants, introduces some errors (rank 2 instead of 1) but it is still able to retrieve the majority of corresponding nodes in the alternative structure (78.5% of exact matches). EnrClass, however, works well (91.2% of exact matches) unless the concept does not occur in the corpus at all (these cases are marked with ◯).
- For homonym-matching cases (Fig. 3(b), we observe that EnrClass works the best (puts the homonym furthest away from rank = 1), whether the concept occurs in the corpus or not. In both cases, while the original concept is able to leverage the context provided by its neighborhood to classify documents, the arbitrarily picked context of the copy does not classify similar documents and homonyms are clearly identified.

Thus, the experimental evaluation showed that a properly context-informed meta-data definition can greatly help disambiguate the meta-data contents,

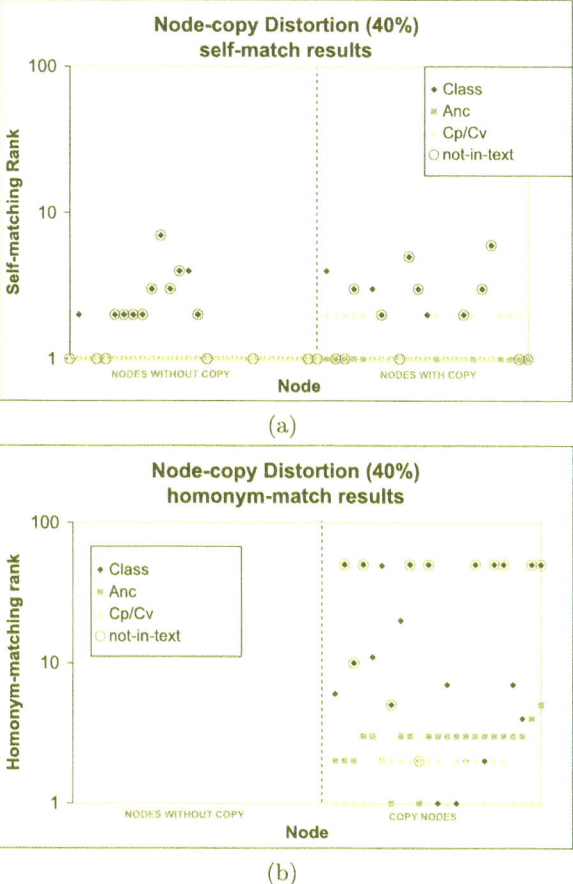

(a)

(b)

Fig. 3. Matching results under homonyms (for (a) rank = 1 indicated perfect match while for (b) it is better to have match rank ≫ 1.

enriching the structural information with corpus content knowledge. In fact, a pure structural-based knowledge definition cannot guarantee these disambiguation performances, and performs poorly even when the similarities among the considered structures are clearly evident.

6   Conclusions

In this paper, we have shown that concept vectors can be used for both capturing the inherent structure of taxonomies for implementing structural matching algorithms and for identifying documents relating to the individual concepts of the taxonomies to implement a semantically extended matching scheme. Experimental results showed that when combined, structural and extensional techniques provide superior handling of synonyms and homonyms.

To leverage the good performances of both methods, we are working on a combined matching strategy, that gives higher weight to the extension based approach whenever populated concept extensions are available, while relying on our structure based approach when measuring the degree of matching of concepts whose extension in the considered domain documents is empty.

References

1. Candan, K.S., Kim, J.W., Liu, H., Suvarna, R.: Discovering mappings in hierarchical data from multiple sources using the inherent structure. Knowl. Inf. Syst. 10(2), 185–210 (2006). https://doi.org/10.1007/s10115-005-0230-9
2. Doan, A., Domingos, P., Halevy, A.Y.: Reconciling schemas of disparate data sources: a machine-learning approach. In: SIGMOD 2001: Proceedings of the 2001 ACM SIGMOD International Conference on Management of Data, pp. 509–520. ACM, New York, NY, USA (2001). https://doi.org/10.1145/375663.375731
3. Doan, A., Madhavan, J., Domingos, P., Halevy, A.Y.: Ontology matching: a machine learning approach. In: Handbook on Ontologies, pp. 385–404 (2004)
4. Euzenat, J., Shvaiko, P.: Ontology Matching. Springer-Verlag, Heidelberg (DE) (2007)
5. Heß, A.: An iterative algorithm for ontology mapping capable of using training data. In: Sure, Y., Domingue, J. (eds.) ESWC 2006. LNCS, vol. 4011, pp. 19–33. Springer, Heidelberg (2006). https://doi.org/10.1007/11762256_5
6. Kim, J.W., Candan, K.S.: CP/CV: concept similarity mining without frequency information from domain describing taxonomies. In: CIKM 2006, pp. 483–492 (2006)
7. Liu, H., Wang, T., Li, Y., Lang, C., Jin, Y., Ling, H.: Joint graph learning and matching for semantic feature correspondence. CoRR abs/2109.00240 (2021). https://arxiv.org/abs/2109.00240
8. Madhavan, J., Bernstein, P.A., Rahm, E.: Generic schema matching with cupid. In: Proceedings VLDB (2001)
9. Miller, R.J., Haas, L.M., Hernández, M.A.: Schema mapping as query discovery. In: VLDB 2000: Proceedings of the 26th International Conference on Very Large Data Bases, pp. 77–88. Morgan Kaufmann Publishers Inc., San Francisco, CA, USA (2000)
10. Miller, R.J., et al.: The Clio project: managing heterogeneity. SIGMOD Rec. 30(1), 78–83 (2001)
11. Milo, T., Zohar, S.: Using schema matching to simplify heterogeneous data translation. In: VLDB 1998: Proceedings of the 24rd International Conference on Very Large Data Bases, pp. 122–133. Morgan Kaufmann Publishers Inc., San Francisco, CA, USA (1998)
12. Mitra, P., Wiederhold, G., Jannink, J.: Semi-automatic integration of knowledge sources. In: 2nd International Conference on Information Fusion (FUSION 1999) (1999). http://ilpubs.stanford.edu:8090/384/
13. Mitra, P., Wiederhold, G., Kersten, M.L.: A graph-oriented model for articulation of ontology interdependencies. Technical report, Stanford, CA, USA (1999)
14. Palopoli, L., Terracina, G., Ursino, D.: DIKE: a system supporting the semi-automatic construction of cooperative information systems from heterogeneous databases. Softw. Pract. Experience 33(9), 847–884 (2003). https://doi.org/10.1002/spe.531

15. Rada, R., Mili, H., Bicknell, E., Blettner, M.: Development and application of a metric on semantic nets. IEEE Trans. Syst. Man Cybern. Syst. **19**, 17–30 (1989)
16. Rahm, E., Bernstein, P.A.: A survey of approaches to automatic schema matching. VLDB J. (2001)
17. Richardson, R., Smeaton, A.F., Smeaton, A.F., Murphy, J., Murphy, J.: Using wordnet as a knowledge base for measuring semantic similarity between words. Technical report, AICS (1994)
18. Ruthven, I., Lalmas, M.: A survey on the use of relevance feedback for information access systems. Knowl. Eng. Rev. **18**(2), 95–145 (2003). https://doi.org/10.1017/S0269888903000638
19. Salton, G., Buckley, C.: Term-weighting approaches in automatic text retrieval. Inf. Process. Manag., 513–523 (1988)
20. Stumme, G., Mädche, A.: FCA-MERGE: bottom-up merging of ontologies. In: Proceedings 17th International Joint Conference on Artificial Intelligence (IJCAI), Seattle (WA US), pp. 225–234 (2001). citeseer.nj.nec.com/stumme01fcamerge.html
21. Trisedya, B.D., Salim, F.D., Chan, J., Spina, D., Scholer, F., Sanderson, M.: i-Align: an interpretable knowledge graph alignment model. Data Min. Knowl. Discov. **37**(6), 2494–2516 (2023). https://doi.org/10.1007/S10618-023-00963-3
22. Wang, Z.: Contextualized structural self-supervised learning for ontology matching (2023). https://arxiv.org/abs/2310.03840

FEHDA 2025: 1st Workshop on Fairness Exploration in Heterogeneous Data and Algorithms

Using Large Language Models for Ethical Process Modeling: A Case Study

Barbara Oliboni$^{(\boxtimes)}$ ⓘ and Elisa Quintarelli$^{(\boxtimes)}$ ⓘ

Department of Computer Science, University of Verona, Verona, Italy
{barbara.oliboni,elisa.quintarelli}@univr.it

Abstract. In this paper, we explore the potential of Large Language Models (LLMs) to facilitate and assist a human designer in creating ethical process models within organizational contexts. Our study starts from a simple case study related to the recruitment scenario, where we first develop and test the use of LLMs to design process flow diagrams and then identify potential sources of unfairness. The LLMs contribute to assisting designers in ethically reengineering the process models. The results are used to create a classification of potential unfairness sources and corresponding impact on processes, useful to propose refinements of process models, whenever it is possible. Our research represents the first practical application of LLMs for ethical process modeling in real-world scenarios.

Keywords: Business Process Modeling · Ethics · Large Language Models

1 Introduction

In recent years, the ethical management of data has become increasingly complex, emphasizing the importance of responsible information handling [13], including granting appropriate authorization for data access and managing user assignments in process-oriented systems to avoid conflicts of interest. This paper specifically addresses ethical concerns within process-aware information systems (PAISs), as robust information management is critical to the functioning of any organization.

PAISs manage business processes that are tightly integrated with organizational data stored in databases. These processes and data, often modeled using BPMN (Business Process Model and Notation) [9], are closely intertwined and play a foundational role in PAISs [4]. Since business processes involve handling information through various activities, ethical implications must be examined with respect to the relationship between data and processes.

Incorporating ethical considerations into business processes is crucial for fostering trust, fairness, satisfaction, and sustainability, ultimately contributing to the organization's long-term success and positive reputation.

© The Author(s), under exclusive license to Springer Nature Switzerland AG 2026
P. K. Chrysanthis et al. (Eds.): ADBIS 2025, CCIS 2676, pp. 503–515, 2026.
https://doi.org/10.1007/978-3-032-05727-3_41

Recent advancements in Artificial Intelligence (AI), particularly in the development of Large Language Models (LLMs), have introduced new possibilities in general and enabled a wide range of applications. In particular, LLMs have been spreading out as accessible systems providing support in the most diverse areas, from text generation, summarization, translation, and question answering, to coding, data analysis, healthcare, and educational applications.

Applications of LLMs in the context of Business Process management (BPM) have been investigated in the literature [15]. In this scenario, LLMs can be first used to identify a business process model starting from its natural language description within relevant documentation [5]. In particular, in [5], LLMs have been used to support the user in Business Process Modeling by focusing on prompt engineering and by evaluating different Business Process Management tasks. In [17], the potential of LLMs in Business Process Modeling has been investigated by proposing an LLM-powered chatbot.

In this work, we aim to take a step forward by considering the ethical perspective when focusing on the conceptual design of the various components of business processes, in particular (i) activities, (ii) actors, and (iii) data management within the process.

Contributions. In this paper, we leverage Large Language Models (LLMs) to examine a recruitment-related case study. Starting from the natural language specification of its business process, we identify potential ethical issues in its design as generated by LLMs and propose re-engineering solutions to promote ethical and equitable management of organizational data.

Our findings allow us to produce a general classification of key ethical challenges that arise from the intertwined relationship between business processes and the data they operate on.

Structure of the Paper. The paper is organized as follows: Sect. 2 summarizes the state of the art about applications of LLMs to business process modeling, Sect. 3 presents our approach, which leverages LLMs to design and refine process models that integrate ethical considerations. Finally, Sect. 4 concludes the paper with a summary and directions for future research.

2   Related Work

The integration of LLM applications at different stages of Business Process Management (BPM) has been discussed in [15]. The authors observe various promising scenarios of use and identify possible future research directions. Possible applications of LLM within BPM cover the whole BPM lifecycle and include the process discovery from documentation and business process improvements (e.g., reengineering), which are the initial aspects addressed in this work.

One of the first concrete applications of LLMs in the business process scenario is described in [5], where the authors developed and applied an approach that utilizes the LLM GPT4 to manage different problems in the BPM context, and in particular to produce a BPMN process model in a pre-specified format from its textual description. The authors create a reliable pipeline based on an LLM that requires subsequent prompts asking to fix issues.

In BPM, process modeling is one of the main aspects to consider in order to improve organizational activities and results. A framework based on LLMs for the generation and refinement of process models starting from a textual description has been proposed in [7,8]. The proposed framework uses the Partially Ordered Workflow Language (POWL) for the intermediate process representation.

A preliminary LLM-based chatbot that may support designers in creating process flow diagrams has been proposed in [17]. The authors present the framework and show its application on a case study related to a particular organization.

In Business Process Management, addressing ethical aspects is essential for ensuring trust and fairness within an organization; however, it has not received as much attention as in the data management research community. Preliminary ethical considerations for dealing with fairness in the management of information by process activities have been discussed in [3], where the unfairness issues are solved by introducing new actors in the process management, whenever it is necessary to reduce as much as possible access to sensitive information.

An exploratory discussion on how ethical aspects can be considered in business process modeling has been provided in [11], where ethics issues are included in the context of requirements management. To support the modeling and analysis of the considered requirements, the authors propose a BPMN-based framework. The authors focus on *individual fairness* and *group fairness* to avoid discrimination with respect to some given individual/group protected (sensitive) data. Fairness requirements can be represented in the business process model by means of a suitable graphical extension of the BPMN concepts. Once the requirements have been specified, the proposed framework allows the automated detection of conflicts between them.

3   The Proposed Approach

In this section, we will introduce our proposal. First, we briefly summarize the Business Process Model and Notation (BPMN) [9], which is the de-facto standard for business process modeling, and allows the graphical representation of Business Process, i.e., a collection of related and structured activities, performed by organizational resources to meet a specific business goal. Then, we will describe how we used LLMs via a chat assistant application to support the conceptual design of Business Processes. In particular, we will show how the designer can interact with the LLM to obtain support in modeling the desired Business Process.

3.1 The Business Process Model and Notation (BPMN)

BPMN provides suitable constructs for representing activities to realize the desired goal. The basic BPMN elements are *activities*, *events*, and *gateways*, which are connected by *sequence flows*.

An *activity* represents the work to perform within the process and is depicted as a rectangle with rounded corners, with a label specifying the activity (task) name.

An *event* represents an instantaneous occurrence that may affect the flow of the process. Events are depicted as circles and can be classified into *start*, *intermediate*, and *end* events. Start/end events represent facts whose occurrence begins/finishes the process, while intermediate events represent facts occurring during the process execution.

A *gateway* represents a point where the process flow is split or merged. Gateways are represented as diamonds and can be classified into *exclusive* and *parallel* gateways. Exclusive gateways contain the marker × and represent a place within a business process where the sequence flow is split into two or more alternative paths. Only one of these paths can be taken during the execution of the process. Parallel gateways contain the marker + and can be used to split the process flow into parallel flows. All these paths must be taken during the execution of the process, and the different flows must be merged by a corresponding parallel (join) gateway.

BPMN provides further elements to represent information managed by the process activities and resource classes involved in the execution of process activities.

Information used within a process can be represented by means of *data objects* and *data stores*. Data objects represent data required/produced by activities and are depicted as rectangles with the upper-right corner folded over. Data stores represent persistent data accessed by activities and are depicted as database symbols.

Resource classes are represented as *pools* and *lanes*. A pool describes the whole organization and contains lanes representing who (people/roles), within the organization, executes a specific set of tasks.

In Fig. 1, we present as a BPMN model, the steps of our approach for leveraging LLMs to design a Business Process Model, while taking ethical issues into account. The process starts and ends with events and is organized within two lanes representing the designer and the LLM, respectively. Information produced and used by activities is represented by means of *data objects*. For example, the "Business Process Description" is given as output of the "Producing Business Process Description" activity performed by the designer, and received as input by the "Business Process Modeling" activity performed by the LLM.

The designer interacts with the LLM (in our case, ChatGPT [10]) to create the desired process model. At each step, the designer leverages the LLM's support by reviewing the generated output. In this way, ChatGPT assists the designer but does not replace them; the human remains in control of both the process and its outcomes.

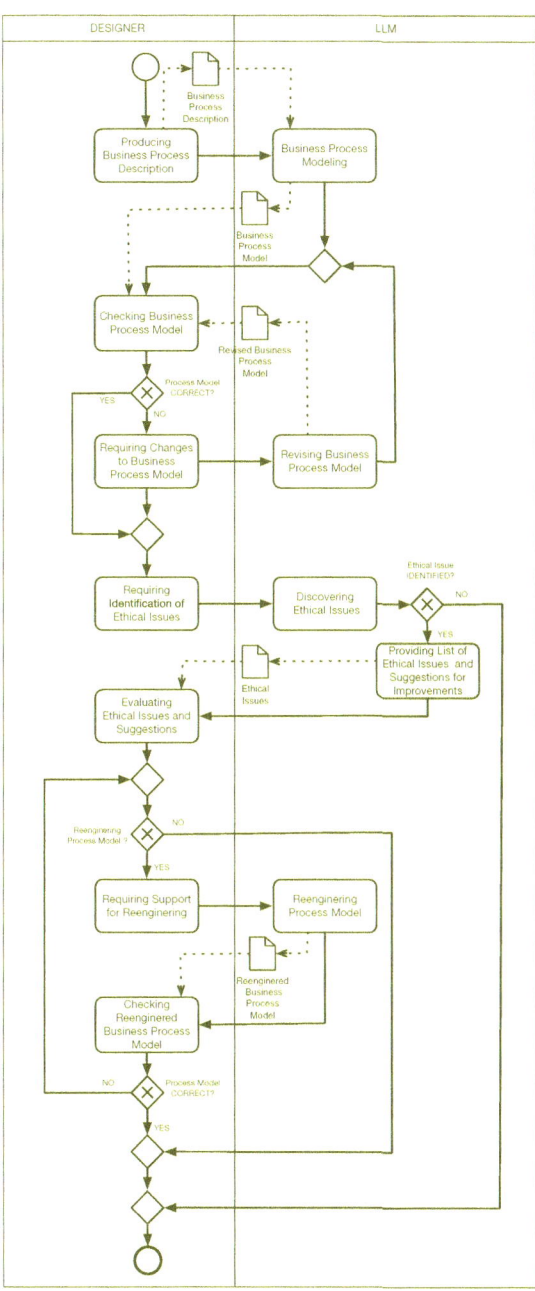

Fig. 1. The Process Model of our Approach.

3.2   Supporting Business Process Modeling with LLMs

In this subsection, we explore how LLMs, by means of a chat assistant such as ChatGPT, can be leveraged to support the conceptual design of Business Processes.

The approach begins with a business process described in natural language, typically provided by a domain expert or analyst.

The textual description is used as a first prompt for ChatGPT, which is able to interpret the information and generate a first process design formalization, in particular, an XML document describing the related Business Process Model or the BPMN diagram.

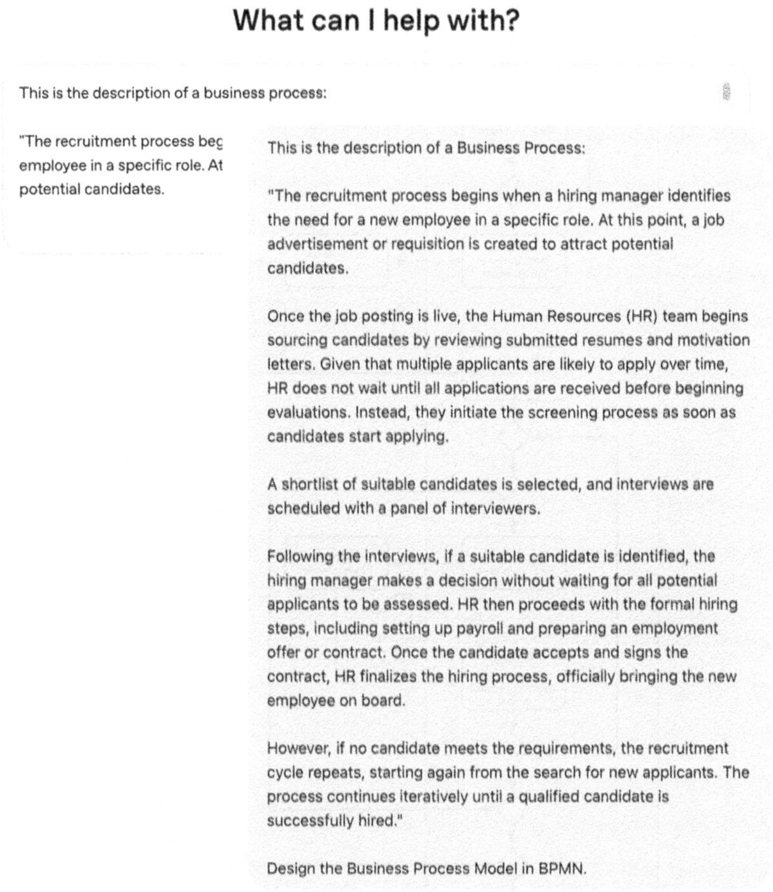

What can I help with?

This is the description of a business process:

"The recruitment process beç employee in a specific role. At potential candidates.

This is the description of a Business Process:

"The recruitment process begins when a hiring manager identifies the need for a new employee in a specific role. At this point, a job advertisement or requisition is created to attract potential candidates.

Once the job posting is live, the Human Resources (HR) team begins sourcing candidates by reviewing submitted resumes and motivation letters. Given that multiple applicants are likely to apply over time, HR does not wait until all applications are received before beginning evaluations. Instead, they initiate the screening process as soon as candidates start applying.

A shortlist of suitable candidates is selected, and interviews are scheduled with a panel of interviewers.

Following the interviews, if a suitable candidate is identified, the hiring manager makes a decision without waiting for all potential applicants to be assessed. HR then proceeds with the formal hiring steps, including setting up payroll and preparing an employment offer or contract. Once the candidate accepts and signs the contract, HR finalizes the hiring process, officially bringing the new employee on board.

However, if no candidate meets the requirements, the recruitment cycle repeats, starting again from the search for new applicants. The process continues iteratively until a qualified candidate is successfully hired."

Design the Business Process Model in BPMN.

Fig. 2. An example of a prompt for the Recruitment scenario.

The generated XML document is particularly useful, as it can be imported directly into business process modeling tools such as Camunda [1] or Signavio [2] to obtain the graphical representation of the described process.

In Fig. 2, we show an example of the prompt providing the business process description to ChatGPT and requiring it to produce the BPMN process model in BPMN. Note that, in this paper, we report one example of the different interactions we had in the testing phase. Of course, results are non-deterministic and consistent throughout the LLM experience.

Before designing the BPMN process diagram, ChatGPT summarizes the main activities of the process, produces a textual description of the process, and finally asks if it should create the diagram version, as shown in Fig. 3. Summarization concerns the main activities, while the textual description tries to give a first overview of the related flow by introducing the indication of possible gateways and loopbacks.

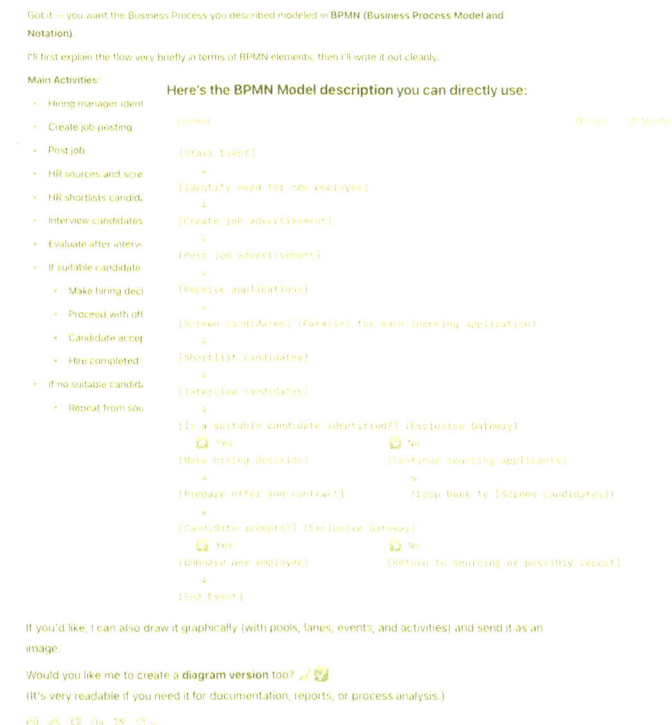

Fig. 3. The answers of ChatGPT.

The proposed summarization and the textual description are correct with respect to the given process description, but the related graphical representation of the process is not accurate. In Fig. 4, we show the first Business Process Model created by ChatGPT starting from our prompt of Fig. 2 and following what was previously produced by the LLM.

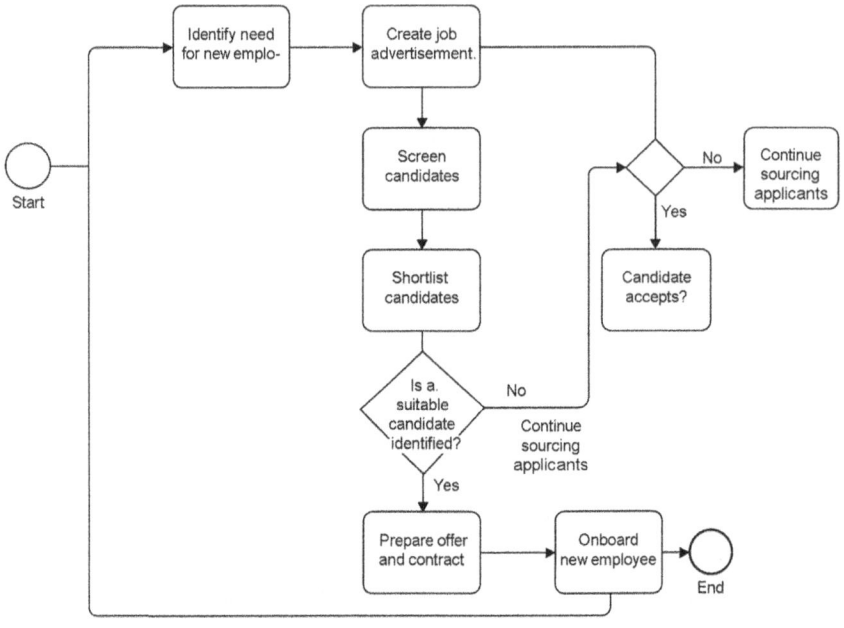

Fig. 4. The first BPMN Process Model created by ChatGPT.

The graphical representation of the process in Fig. 4 can be improved since it does not completely reflect the description given by ChatGPT itself. In particular, it is possible to note that the start event, which must have only one outgoing sequence flow, has an outgoing flow to the first activity "Identify need for new empl-", (note the problem in the label of the displayed activity w.r.t. the one in the textual description), but also a simple line toward the last process activity. An edge without direction links the "Create job advertisement" activity and the gateway. The question associated with one of the two gateways is modeled as an activity rather than being placed near the gateway itself, and the path corresponding to candidate acceptance is missing. Moreover, the activity "Continue sourcing applicants" should have been before the gateway and not after. These and other problems highlight a misalignment between the textual description and the related graphic representation.

To obtain the correct version of the business process diagram and refine the proposed subsequent solutions, it is necessary to interact with the system and provide human support to check and validate the LLM result. Interactions concern both activities and flow.

The desired business process diagram is depicted in Fig. 5 and is the result of the designer's requests and suggestions.

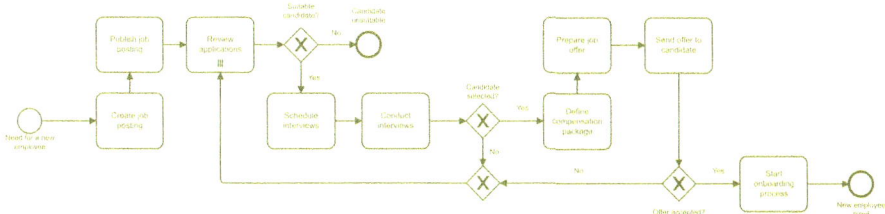

Fig. 5. The final BPMN Process Model.

ChatGPT is also able to provide an XML document describing the desired process. In our experiments, we note that in some cases the XML documents, produced as the first proposal, are better than the possible corresponding BPMN diagrams, since they are more adherent to the textual descriptions. Even starting from XML, reaching the final layout by using a business process modeling tool requires some work on the part of the designer.

3.3   Fairness Issues in Business Process Modeling

In [3], we have examined fairness issues at the level of individual activities and their associated data, but also analyzed the overall process model. Our focus was on understanding how various fairness requirements concerning data access are linked to task-assignment policies for different actors and agents.

Now, our aim is to extend the approach to generalize possible unfairness issues concerning process, data, and their intertwined relationships.

To reach this goal, after the recruitment process was refined, we asked Chat-GPT to identify possible ethical issues related to the considered process, and we obtained the following suggestions about unfairness sources.

- *Selection before closing applications.*
  This approach may penalize candidates who apply late, thus it violates the principle of fairness because it does not guarantee the same conditions for all candidates.
- *Hiring without evaluating all candidates.*
  This choice risks excluding more qualified candidates who arrive after a decision has already been made.
- *Bias and implicit discrimination.*
  Even if not explicitly stated, rapid and non-standardized evaluation processes may introduce unconscious bias, such as bias due to personal preferences. If the process does not include objective selection criteria, it risks becoming a non-transparent process.

– *Lack of structured feedback.*
The process does not contemplate communication with unsuccessful candidates. Lack of feedback can be perceived as a lack of transparency and low regard for candidates' time.
– *Continuous iteration without limits.*
The process repeats until a candidate is hired. If there are no ethical guidelines or review of the process after failed attempts, the risk is being stuck in a loop without self-criticism (e.g., here the problem is in the search strategy, not related to candidates).

Starting from this list, we can consider that, in general, ethical issues may impact:

– **Process flow**: the structure of the process, by requiring additional activities or different flows between the envisioned activities;
– **Process organization**: the organization of the process, which governs the allocation and coordination of resources in relation to specific activities;
– **Data management**: the data, which must be carefully managed to ensure fairness and compliance, especially when it includes sensitive attributes that could influence decision-making and ultimately affect the process outcome, as initially investigated in [3].

After highlighting possible ethical issues, ChatGPT also provides us with possible ethical improvements that can be used to further reengineer the process model. The suggested ethical improvements that can be introduced in the modeled process are the following:

– Set a closing date for applications before starting the selection;
– Evaluate all candidates received within a certain period, to avoid hasty decisions;
– Document the evaluation criteria explicitly;
– Provide structured feedback to unsuccessful candidates;
– Include audits or reviews of the process after repeated cycles.

3.4   Ethical Business Process Modeling: Discussion

The highlighted ethical issues for the recruitment scenario can be evaluated with respect to the aspect(s) they can affect, as represented in Table 1. The symbol ✓ underlines the presence of the mentioned issue with regard to the impact on the process modeling.

In Fig. 6 we show the business process model taking into account the highlighted ethical issues. The redesigned process is structured with a different flow and contains suitable activities to manage ethical aspects. In particular, we represent in green the new activities and in blue the new gateway that have been inserted to improve the process with respect to ethical issues. Some of the new (green) activities correspond to activities in the previous process, but require

Table 1. Possible ethical issues in the Recruitment Business Process Modeling

Ethical Issue	Process Flow	Process Organization	Data Management
Selection before closing applications	✓	–	–
Hiring without evaluating all candidates	✓	–	–
Bias and implicit discrimination	✓	✓	✓
Lack of structured feedback	✓	–	–
Continuous iteration without limits	✓	–	–

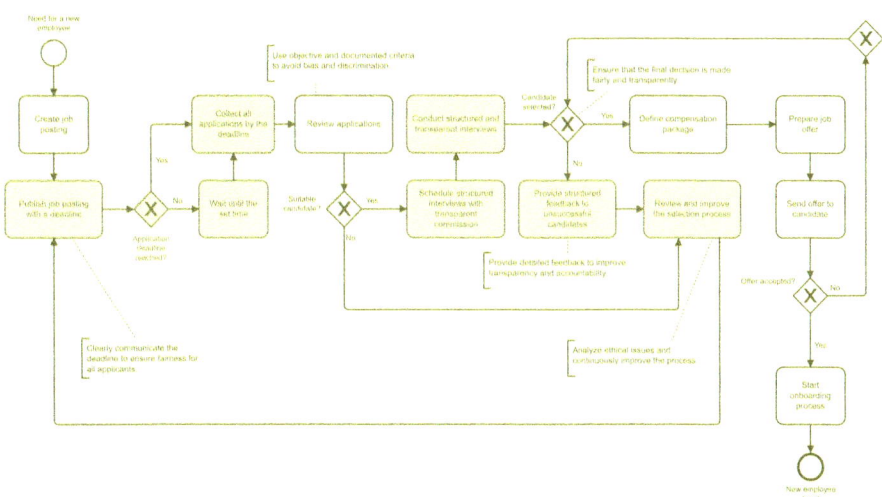

Fig. 6. The ethical BPMN Process Model.

paying attention to some ethical aspects. As an example, the activity "Conduct interview" has become "Conduct *structured and transparent* interview".

The proposed use case is a starting point to highlight the necessity to integrate ethical considerations in process modelling. Indeed, in recent years, considerable attention has been given to studying fairness issues in data management [6,14] and integrating fairness requirements into machine learning algorithms [12,16]. In our view, however, bias and unfair practices should be addressed earlier in the pipeline, and in particular at the process level.

4   Conclusion and Future Work

This paper has demonstrated the potential of Large Language Models (LLMs) in analyzing fairness issues in business processes through a case study approach. By starting from natural language specifications, we identified ethical concerns in the design of a recruitment process and proposed re-engineering strategies to enhance ethical and equitable data management.

Our analysis contributes to a broader understanding by defining a general classification of the main ethical challenges emerging from the relationships between organizational processes and the data they utilize. These findings underscore the importance of integrating ethical considerations into the early stages of business process design, particularly in data-intensive domains.

For future work, we plan to extend the described analysis to consider different kinds of processes, actors, and decisions, aiming to generalize ethical issues and propose a methodology—supported by LLMs—for modeling ethical business processes. This effort should incorporate interdisciplinary oversight (e.g., ethicists, affected communities), address potential biases in LLMs, and ensure that normative decisions are not delegated solely to automated systems.

Acknowledgements. We wish to thank the students Chiara Fabris and Giulia Carminati for the useful discussions and experiments.

References

1. Camunda Modeler. https://camunda.com/platform/modeler/
2. SAP Signavio Process Transformation Suite. https://www.signavio.com
3. Amico, B., Combi, C., Dalla Vecchia, A., Migliorini, S., Oliboni, B., Quintarelli, E.: Enhancing business process models with ethical considerations. In: Enterprise Design, Operations, and Computing. EDOC 2024 Workshops, pp. 3–17. Springer (2025)
4. Combi, C., Oliboni, B., Weske, M., Zerbato, F.: Seamless conceptual modeling of processes with transactional and analytical data. Data Knowl. Eng. **134**, 101895 (2021)
5. Grohs, M., Abb, L., Elsayed, N., Rehse, J.: Large language models can accomplish business process management tasks. In: Weerdt, J.D., Pufahl, L. (eds.) BPM 2023. LNCS, vol. 492, pp. 453–465. Springer, Cham (2023). https://doi.org/10.1007/978-3-031-50974-2_34
6. Jagadish, H.V., Stoyanovich, J., Howe, B.: The many facets of data equity. ACM J. Data Inf. Qual. **14**(4), 27:1-27:21 (2022)

7. Kourani, H., Berti, A., Schuster, D., van der Aalst, W.M.P.: Process modeling with large language models. In: Enterprise, Business-Process and Information Systems Modeling, pp. 229–244. Springer, Cham (2024). https://doi.org/10.1007/978-3-031-61007-3_18

8. Kourani, H., Berti, A., Schuster, D., Van der Aalst, W.M.P.: PromoAI: process modeling with generative AI. In: Proceedings of the Thirty-Third International Joint Conference on Artificial Intelligence, IJCAI '24 (2024)

9. Object Management Group. Business Process Model and Notation (BPMN), v2.0.2 (2014). https://camunda.com/platform/modeler/

10. OpenAI. GPT-4 technical report. CoRR, abs/2303.08774 (2023)

11. Ramadan, Q., Strüber, D., Salnitri, M., Jürjens, J., Riediger, V., Staab, S.: A semi-automated BPMN-based framework for detecting conflicts between security, data-minimization, and fairness requirements. Softw. Syst. Model. **19**(5), 1191–1227 (2020). https://doi.org/10.1007/s10270-020-00781-x

12. Shunli Zhang, E.A.: Data-driven system-level design framework for responsible cyber-physical-social systems. IEEE Comput. 80–91 (2023)

13. Steen, M.: Ethics as a participatory and iterative process. Commun. ACM **66**(5) (2023)

14. Stoyanovich, J., Abiteboul, S., Howe, B., Jagadish, H.V., Schelter, S.: Responsible data management. Commun. ACM **65**(6), 64–74 (2022)

15. Vidgof, M., Bachhofner, S., Mendling, J.: Large language models for business process management: opportunities and challenges. In: Di Francescomarino, C., Burattin, A., Janiesch, C., Sadiq, S. (eds.) Business Process Management Forum. LNCS, pp. 107–123. Springer, Cham (2023). https://doi.org/10.1007/978-3-031-41623-1_7

16. Zehlike, M., Yang, K., Stoyanovich, J.: Fairness in ranking, part I: score-based ranking. ACM Comput. Surv. **55**(6), 118:1-118:36 (2023)

17. Ziche, C., Apruzzese, G.: LLM4PM: a case study on using large language models for process modeling in enterprise organizations. In: BPM 2024. LNBIP, vol. 527, pp. 472–483. Springer, Cham (2024). https://doi.org/10.1007/978-3-031-70445-1_35

ReFaRAG: Re-ranking for Bias Mitigation in Retrieval-Augmented Generation

Yingqi Zhao[1]($\boxtimes$), Vasilis Efthymiou[2] , Jyrki Nummenmaa[1] ,
and Kostas Stefanidis[1]

[1] Tampere University, Tampere, Finland
{yingqi.zhao,jyrki.nummenmaa,konstantinos.stefanidis}@tuni.fi
[2] Harokopio University of Athens, Kallithea, Greece
vefthym@hua.gr

Abstract. Retrieval-augmented generation (RAG) improves the performance of LLM-based applications by incorporating information from external knowledge bases. However, the introduction of search engines and knowledge sources can also introduce new biases and stereotypes into the system. Previous studies have shown that adjusting the bias of retrievers through fine-tuning can influence the overall bias of the RAG system, mitigating bias in RAG. In this work, we propose a re-ranking-based method, termed ReFaRAG, as an alternative to fine-tuning for controlling the bias in retrieval results. We further investigate how biased retrieval output affects different LLMs within the RAG framework.

Keywords: RAG · Political bias · LLM

1 Introduction

Retrieval-augmented generation (RAG) enhances the generation performance of large language models (LLMs) by incorporating information retrieved from external knowledge bases. This approach has proven particularly effective in handling long-tail knowledge domains, mitigating hallucination issues in LLMs, and adapting to scenarios characterized by rapid knowledge evolution [40].

LLMs can generate content that appears highly confident but is nonsensical or unfaithful to the provided source, this is the notorious hallucination problem [25]. With long-tail knowledge and highly time-sensitive information, LLMs are inherently constrained by the distribution of their pretraining data, often leading to suboptimal performance or even the generation of fabricated outputs [26]. Although targeted fine-tuning can help alleviate this issue [14], training LLMs is costly and may lead to catastrophic forgetting, which can significantly degrade the overall performance of the model.

The emergence of RAG has alleviated the hallucination problem of LLMs and addressed the challenges of fine-tuning or continued pretraining in long-tail domains. Rather than modifying the parameters of the LLM itself, RAG addresses task-specific problems by incorporating information retrieved from

© The Author(s), under exclusive license to Springer Nature Switzerland AG 2026
P. K. Chrysanthis et al. (Eds.): ADBIS 2025, CCIS 2676, pp. 516–530, 2026.
https://doi.org/10.1007/978-3-032-05727-3_42

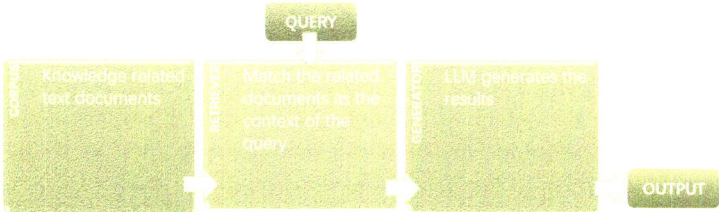

Fig. 1. A general RAG workflow

external knowledge bases. These external sources are highly flexible—not only in terms of domain coverage, allowing for the construction of customized local knowledge bases tailored to specific applications, but also across modalities, including text, knowledge graphs, images, and audio. Owing to this flexibility, RAG methods are well suited to scenarios involving rapidly evolving knowledge, as system responses can remain up-to-date simply by refreshing the underlying knowledge base [40].

A general workflow of RAG is illustrated in Fig. 1. Taking textual knowledge as an example, a knowledge base, i.e., the corpus, is constructed by collecting documents relevant to the application scenarios of RAG. At inference time, given a specific question as the query, the system retrieves the most relevant documents from the corpus as context. This context, along with the question, is then fed into the LLM to perform reasoning and generate the final output.

However, the manifestation of bias in LLMs raises significant concerns, and addressing bias in RAG systems proves even more complex. Since LLMs are trained on massive amounts of unfiltered web data, they are prone to inheriting a wide range of human biases—from stereotypes and factual inaccuracies to derogatory language and harmful social assumptions. To mitigate biased or harmful outputs and promote safer deployment, researchers have proposed various alignment techniques, such as Reinforcement Learning from Human Feedback (RLHF [28]) and instruction tuning [32]. These methods aim to align the model's behavior with human values and preferences by shaping its responses in accordance with socially acceptable norms.

In addition to the LLM itself, RAG introduces external knowledge bases and retrievers, making it challenging to measure and analyze system-level bias. [34] have identified and quantified the contributions of different RAG components to overall bias. While [16] have examined how biases inherent in external knowledge bases impact system behavior, they have found that the external information retrieved and injected into LLMs through RAG pipelines can easily undermine the effects of alignment, causing the overall system to exhibit notable biases.

Since the automated reproduction of unfair behaviors may reinforce existing societal inequalities [2], examining bias in RAG systems is both a meaningful and necessary pursuit. [21] proposes using Rank Bias to measure the bias of each component as well as the overall system. It systematically reveals the linear relationships among biases in the knowledge base, retriever, and LLM, and their combined effect on the RAG. Furthermore, it introduces a method for controlling RAG system bias by fine-tuning the retriever to adjust its inherent bias.

Building upon the research of [21] and inspired by [38], this study discovers and validates that re-ranking methods can serve as a simpler, more direct, and more precise alternative to fine-tuning for controlling the bias level in retrieval, thereby influencing the overall bias of the RAG system. We refer to our method as ReFaRAG: Re-ranking towards Fairer RAG. This approach not only eliminates the substantial effort required for retriever fine-tuning but also avoids the performance degradation that often occurs when adjusting the retriever's bias during fine-tuning, making it a more practical solution in real-world applications. Furthermore, building on the proposed re-ranking method, this study enables a more convenient investigation of how the ranking of biased information within RAG systems affects different LLMs.

The rest of this paper is structured as follows. Section 2 discusses the related work. Section 3 describes a way to quantify bias in RAG, and Sect. 4 introduces a re-ranking based method for mitigating bias in RAG. Section 5 presents the experimental evaluation and finally, Sect. 6 concludes the paper, presents the limitations of our approach and provides suggestions for future improvements.

2   Background and Related Work

2.1   Background

With the rise of large-scale pre-trained language models (LLMs) such as GPT [4] and LLaMA [13], these models have demonstrated impressive generative capabilities across a wide range of tasks. However, they face several critical limitations: their knowledge is inherently static and difficult to update once pre-training is complete; they suffer from knowledge hallucination [26], often generating plausible-sounding but factually incorrect information even when precise referencing is required. To address these challenges, the RAG framework was introduced, which incorporates an external retrieval module to inject up-to-date and relevant knowledge into the generation process. This approach improves the reliability and controllability of LLM outputs.

As research on RAG continues to evolve, a wide variety of knowledge sources, retrieval strategies, integration techniques, and additional components aimed at improving generation quality have emerged [40]. On one hand, this diversity highlights the flexibility and broad applicability of the RAG paradigm; on the other hand, the increasing structural complexity poses significant challenges for systematic study and evaluation.

Meanwhile, growing societal concerns around fairness and transparency in AI systems have become increasingly prominent [10]. In high-impact domains such as news generation, educational question answering, biased outputs can lead to serious consequences. For instance, recommendation systems may over-promote mainstream perspectives, suppressing diversity; and question answering

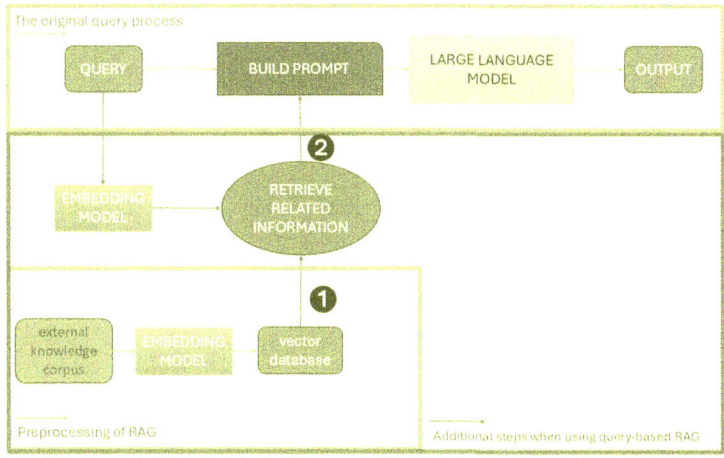

Fig. 2. A query-based RAG pipeline

systems may reinforce stereotypes based on ethnicity or region, further amplifying discriminatory viewpoints. As one of the most influential directions in AI
applications, RAG can significantly shape the bias profile of LLMs [16], thereby
impacting fairness in AI applications. Therefore, understanding and mitigating
bias in RAG systems is not only a technical challenge but also a societal imperative.

Next part of this chapter begins with an introduction to the workflow of RAG
systems, followed by a brief overview of existing research on bias in both RAG
and LLMs.

2.2   The RAG Workflow

Figure 2 shows a standard query-based RAG approach based on dense retrieval
for textual data. At the top of the figure, the orange-colored section (arrows
and boxes) illustrates the standard query process when using an LLM. The user
submits a query, which is then used to construct a corresponding prompt—this
prompt can simply be the query itself. The prompt serves as the input to the
LLM, instructing it to perform the necessary processing and reasoning before
generating an output. In this pipeline, the LLM is referred to as the **generator**.

Before utilizing RAG, we need to preprocess the external knowledge corpus.
The green section of the figure illustrates this part. First, collect the external knowledge corpus based on the task type and then, segment it into chunks
according to paragraphs or semantic units. Next, use an embedding model to
convert these corpus chunks of natural language knowledge into vector representations to construct a vector database. Finally, build an Approximate Nearest
Neighbor (ANN) index of the database, forming ❶ in Fig. 2. This indexing
process accelerates retrieval during inference, reducing response latency.

When using RAG, the blue arrows in the figure illustrate the additional steps compared to the original query process. First, use the same embedding model to vectorize the query. Then, perform similarity matching between the vector database index (❶) and the embedded query to retrieve relevant information. Common similarity metrics include cosine similarity, inner product, and L2 distance. Finally, we get the most relevant retrieved content (❷ in Fig. 2). Integrate these relevant content and the query to build an enriched prompt, then feed it into the generator (LLM) to produce the final response.

In the above process, the additional steps during the preprocessing and RAG inference primarily involve the construction and utilization of the dense **retriever**. Retriever and generator are the two core components of RAG. However, other RAG methods incorporate additional components and mechanisms. Following the classification in FlashRAG [19], we provide a brief introduction to these components.

Judger: Determines whether retrieval is necessary for a given query. Upon receiving the query, it first evaluates the need for retrieval before proceeding. An example of this is SKR [31].

Refiner: Enhances the input provided to the generator by reducing the prompt length and filtering out irrelevant retrieved documents, thereby improving the final RAG response. This process typically operates in Step 2 in the figure. An example is RECOMP [35].

Reranker: After retrieving a list of relevant documents based on similarity, the re-ranker applies a re-ranking mechanism to further refine the selection of references passed to the generator. This step also occurs at ❷ in the figure. An example is [22].

In addition to standard query-based RAG architectures, integrate retrieved information into LLM through prompt, alternative approaches have been proposed to integrate external information into language models. According to [40], these methods include:

Latent representation-based RAG, where retrieved content is incorporated as latent representations to enhance the model's understanding and improve generation quality—exemplified by models such as FiD [17] and RETRO [3];

Logit-based RAG, which merges retrieved information directly into the decoding process, as seen in models like kNN-LM [20];

Speculative RAG, which optimizes resource efficiency by replacing generation with retrieval when appropriate—representative methods include REST [15] and GPTCache [1].

Given that detecting and mitigating bias in RAG systems remains a relatively new and challenging task, this work focuses on a simplified setting: query-based RAG with text-only corpora.

2.3   Bias in LLM and RAG

Bias in computer systems refers to the systematic production of unfair outcomes that disadvantage certain groups or individuals [8]. As intelligent systems built on LLMs become increasingly integrated into daily life, concerns about the biases exhibited by LLMs have attracted growing attention.

As previously discussed, since LLMs inherit biases and harmful content from vast and heterogeneous internet data, systems built on top of LLMs often exhibit varying degrees of biased behavior. Methods for evaluating bias in LLMs generally fall into two main categories [10]:

1) Counterfactual-Based Evaluations: These approaches assess bias by comparing an LLM's responses to different demographic groups in the same context. For example, WinoBias [39] evaluates whether the model associates specific roles with gender, while StereoSet [27] measures how the model completes sentences based on stereotypical versus anti-stereotypical associations given a particular group.

2) Prompt-Based Evaluations: These involve providing pre-constructed prompts, such as sentence stems (e.g., BOLD [6] and RealToxicityPrompts [11]) or questions (e.g., BBQ [29] and UnQover [23])—and analyzing the model's generated continuations or answers to assess potential biases.

Since RAG is primarily designed to enhance the generative capabilities of LLMs, most existing studies adopt evaluation methods originally developed for assessing bias in LLMs, with a particular focus on prompt-based QA setups. [34] evaluated the impact of various components within a RAG system—Retriever, Refiner, Judger and Generator—on both bias and generation accuracy. Their findings highlight that, beyond the LLM itself, the retriever in particular plays a critical role in shaping the system's bias behavior. In a similar manner, [16] focused on the influence of the corpus and demonstrated that the inclusion of retrieved documents with even a small degree of bias can substantially affect the bias exhibited by the RAG system. Such biased content can easily undermine the alignment behavior of LLMs, and even neutral citations may lead the model to produce overly confident outputs in uncertain contexts.

[22] employed a stochastic ranker in RAG to enhance the diversity of retrieved results. Since useful information may still exist among documents not selected in the top-k results, variations in the ranking order for the same query can influence the subsequent generation outputs. While this work also employed a re-ranking strategy, its primary objective was to enhance individual item-side fairness in the process of retrieval, rather than improving the fairness of the RAG system. [21] simplified the RAG system configuration and proposed a unified metric to quantify the bias score of each component. Based on this framework, they demonstrated a linear relationship between the retriever's bias and the overall system bias in RAG, and further introduced a method to mitigate RAG bias by adjusting the retriever's bias level. More details can be found in Sect. 3.

Building on the setup and bias evaluation framework proposed by [21], this paper introduces a more straightforward and practical alternative for mitigating bias in RAG systems. Instead of mitigating the bias in the retrieved knowledge through fine-tuning the retriever in [21], we adopt a re-ranking approach to directly control the bias of the information provided to the LLM. This method not only avoids the additional costs and potential issues associated with fine-tuning, but also enables more precise control over the bias in the context. The complete methodology is presented in Sect. 4.

3   Measuring Bias in RAG

As previously introduced, there exist various scenarios and methodologies for measuring bias in LLMs and RAG systems. However, due to the inherent complexity of studying bias in RAG, this work adopts a simplified evaluation framework that focuses on binary group bias. The limitations of this binary setting, as well as more realistic and complex bias scenarios such as multi-group bias, are discussed as future work in Sect. 6.1.

Building upon the framework proposed in [21], this study adopts a modular perspective to measure bias in RAG systems. Specifically, we define the corpus as C, the retriever (embedding model) as E, the generator (LLM) as L, and the overall RAG system as R. The corresponding bias scores are denoted as C_b, E_b, L_b, and R_b, respectively. To quantify bias in each component or the entire system, we use the following metric:

$$b = \frac{\text{count}(g_1) - \text{count}(g_2)}{S},\qquad(1)$$

where g_1 and g_2 represent two opposing groups (e.g., liberals and conservatives when measuring political bias). count(g_1) refers to the number of outputs aligned with group g_1 across all test samples, and count(g_2) is defined analogously. S denotes the total number of test samples. The bias score $b \in [-1, 1]$, with values closer to -1 indicating stronger bias toward g_2, and values closer to $+1$ indicating stronger bias toward g_1. More concretely:

- C_b is the average bias score of all documents in the knowledge base.
- E_b is the average bias score of the top-1 retrieved document per test query.
- L_b is the average bias of the LLM's outputs when queried directly with test samples (without retrieval).
- R_b represents the average bias of the full RAG system, where the generation is conditioned on retrieved content.

3.1   Mitigate Bias

[21] used contrastive learning to fine-tune the embedding model 40 times, resulting in 40 retrievers with varying levels of bias E_b. By observing how changes in E_b influence the overall system bias R_b, the study found a generally linear relationship between them, formalized as:

$$R_b = s \cdot E_b + L_b + \varepsilon,\qquad(2)$$

where s denotes the sensitivity of the LLM to bias in the retrieved input, and ε accounts for bias-irrelevant knowledge conflicts.

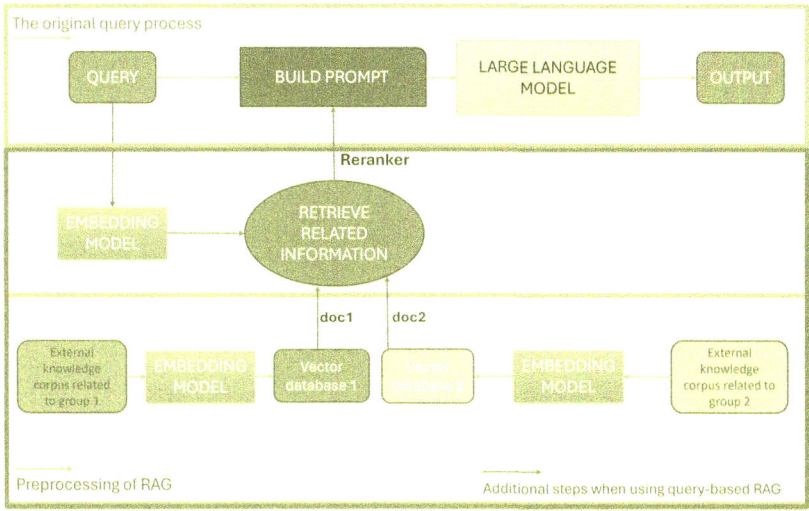

Fig. 3. The ReFaRAG pipeline

When the slope s is sufficiently large—indicating that the model is sensitive enough to biased inputs—it becomes feasible to adjust E_b to bring R_b closer to zero, thereby achieving bias mitigation at the system level.

However, large-scale fine-tuning of embedding models is cumbersome. On one hand, it is difficult to precisely control the degree of bias in the fine-tuned models. On the other hand, as demonstrated in the experiments of [21], even with the use of techniques such as PEFT (Parameter-Efficient Fine-Tuning [36]) and WiSE-FT [33]), a trade-off between bias mitigation and retrieval performance still emerges. Specifically, adjusting the bias level of the embedding model often comes at the cost of a degradation in retrieval effectiveness.

4   Mitigate Bias Using Re-Ranking in ReFaRAG

Inspired by [38], this study proposes a re-ranking based method that achieves an equivalent adjustment of E_b without modifying the parameters of the embedding model. In essence, fine-tuning the embedding model affects the RAG system bias R_b by altering the bias composition of the context documents passed to the LLM. Therefore, if we control the group-related bias in the retrieved documents via probabilistic re-ranking during retrieval, we can replicate the effect of fine-tuning the embedding model. This approach re-orders the retrieved results to match a desired group distribution, thus achieving similar control over R_b in a simpler and more precise way, without losing the retrieval performance of the embedded model. The method design is shown in Fig. 3.

Concretely, to avoid extreme or uncontrolled bias during retrieval, we first partition the corpus into two separate knowledge bases according to group alignment. For example, in the case of political bias, one sub-corpus contains only documents reflecting liberal viewpoints, while the other contains only conservative-aligned documents.

During retrieval, we retrieve the most relevant documents $doc1$ and $doc2$ to a given query from two separate corpora, each consisting of documents related to a specific group. Although it is theoretically feasible to use a unified corpus and identify the top-ranked documents associated with each group (denoted as $group1$ and $group2$) from the top-k retrieved results, we opt for a grouped corpus design to decouple the influence of the embedder's inherent bias.

For instance, suppose there are n documents in the corpus related to $group1$, and the embedder is highly biased toward $group1$. In this case, retrieving the most relevant document for $group2$ (that is, $doc2$) may require accessing at least $n + 1$ documents. This introduces uncertainty and latency that are difficult to control in practice. Therefore, we adopt a design in which corpora are constructed separately by group to ensure a controlled and efficient retrieval process.

After retrieving $doc1$ and $doc2$ using top-1 dense retrieval from the two group-specific corpora, we apply a *Reranker* to determine which document will be included in the prompt as context for the LLM. Taking political bias as an example, $doc1$ and $doc2$ may respectively reflect liberal and conservative viewpoints. We introduce a probability parameter p to control the proportion of documents retrieved from the conservative-aligned corpus ($group2$). Under this setting, the expected bias score of the retriever according to Eq. (1) becomes:

$$E_b = (1 - p) - p = 1 - 2p. \qquad (3)$$

By adjusting p, we can effectively manipulate E_b, and, as shown in prior analysis, consequently steer R_b toward zero (see Eq. (2)). This enables controlled bias mitigation in the RAG system without altering model parameters.

5   Experimental Evaluation

This study evaluates the effectiveness of the proposed bias control method on political bias.

Political Bias. Building on the methodology of [21], we construct a set of single-answer multiple choice question tasks in which each question is accompanied by two answer options reflecting opposing ideological perspectives—liberal and conservative (see an example in Table 1). By prompting the RAG system to choose between the two, we can assess its political alignment and measure the degree of bias under different corpus or retrieval configurations.

Dataset. We sampled 200 items from the TwinViews-13k [9] dataset to construct the QA test set. Each sample contains both liberal and conservative perspectives. For each of these samples, we used DeepSeek-r1 [5] to generate a relevant question based on the topic, forming the final test questions. The remaining samples from TwinViews-13k served as the corpus. Based on the dataset's inherent left/right political labels, we partitioned the corpus into two knowledge bases—liberal-aligned and conservative-aligned documents.

Table 1. Example of the multiple-choice QA tasks

Topic	Public Transportation
Question	What is the most effective approach to funding and managing public transportation?
Choice A	Public transportation should receive increased funding and be expanded to provide affordable and accessible options while reducing congestion and carbon emissions.
Choice B	Public transportation should be self-sustaining and not rely on taxpayer subsidies, as it may not be cost-effective or widely used.

Based on this setup, we note that the average corpus bias score across all documents in our experiment is $C_b = 0$, since the number of left-leaning and right-leaning documents in the knowledge base is balanced. In practice, this effect of C_b can be considered negligible, as our ReFaRAG constructs separate knowledge bases for the two groups and selects context from one of them through a re-ranking mechanism. The only exceptions are cases where one knowledge base lacks content or contains only irrelevant documents.

Models and Implementation. We used GTE-base [24] as the embedding model, and the FAISS [7] library was employed to construct index based on cosine similarity across all settings. For the language models (LLMs), we selected four widely-used open-weight models: Llama 3.1 8B Instruct [13], Gemma 2 9B IT [12], Mistral 7B Instruct v0.3 [18], and Qwen 2.5 7B Instruct [37]. All models were sourced from HuggingFace. The RAG pipeline was implemented using the Langchain framework.

5.1   Results

Bias in RAG Components. To evaluate political bias, we first merged the corpora containing left-leaning and right-leaning statements into a single balanced corpus. Since the number of documents associated with each ideological group is equal, the corpus bias score is set as $C_b = 0$. We then measured the bias scores of the embedding model and various LLMs under this configuration.

In this evaluation, if a model failed to produce a clear choice (i.e., did not select either of the provided options according to the prompt template), the response was classified as a *refusal*. These refusals were excluded from the bias count, and the final bias score was still computed using Eq. (1), with conservatives defined as g_1 and liberals as g_2.

As shown in Table 2, Lb denotes the bias score measured by directly testing the QA dataset on the LLM, while Rb represents the bias score after applying the RAG method. LLM refusal rate refers to the proportion of responses from the LLM that do not provide a definitive answer, and RAG refusal rate reflects the same measurement under the RAG setting. In particular, $Eb = -0.17$, which is the bias score of the embedding model GTE-base.

It can be observed that both the tested embedding model and LLMs exhibit a left-leaning bias, which is consistent with the findings reported in [21]. However,

Table 2. Bias of LLM, RAG and Rejection Rates Comparison

LLM	Lb	Rb	LLM refusal rate	RAG refusal rate
Mistral	−0.22	−0.355	73.5%	0.5%
Llama	−0.74	−0.33	0%	0%
Qwen	−0.83	−0.4	0%	0%
Gemma	−0.145	−0.2	85.5%	8%

there are some numerical differences in the bias scores compared to those in [21]. These discrepancies can be attributed to variations in the datasets used for evaluation, as well as differences in the construction of prompt templates.

Impact of Embedder Bias on Overall RAG Bias. In our proposed method, the so-called *embedder bias* does not directly reflect the intrinsic bias of the embedding model itself. Instead, it refers to the controlled ratio of biased documents presented in the context passed to the LLM. In this sense, the term more accurately describes a controlled bias in the retrieval results rather than the embedding model's own bias. Nonetheless, to remain consistent with the terminology used in [21], we continue to refer to this as *embedder bias*.

Experimental results in Fig. 4 demonstrate that all tested LLMs conform to the linear relationship expressed in Eq. (2), indicating that the overall system bias (R_b) varies linearly with the controlled embedding bias (E_b). Therefore, our reranking method enables effective control of the overall bias in RAG systems by simply adjusting the polarity probability of the biased context documents passed to the LLM, assuming the model is sufficiently sensitive.

Under the bias evaluation framework introduced in this work, this approach allows for more precise manipulation of embedding bias, thereby achieving an approximately unbiased RAG system. In contrast, controlling bias through fine-tuning the embedding model is both more complex and less reliable, as the outcomes of fine-tuning are often uncertain and difficult to predict.

Differences Between RAG and Pure LLM Behavior. By comparing the overall behavior of models when used in a RAG setup versus directly as standalone LLMs, we observe that certain models—especially Gemma 2 9B IT and Mistral 7B Instruct v0.3 —exhibit a significantly higher refusal rate when directly queried in potentially biased contexts. However, when RAG is employed, the refusal rate notably decreases. There are two potential explanations for this:

1. **Sycophancy Effect:** LLMs may attempt to align with the perceived intent of the user, a behavior known as *Sycophancy* in LLMs [30]. When biased context is injected via RAG, the model might interpret it as a signal of the user's view and thus be more likely to provide an answer rather than refuse.
2. **Disruption of Alignment Mechanisms:** When used independently, the LLM's alignment mechanisms—trained to avoid producing harmful or biased outputs—are more robust. However, when external documents are introduced via RAG, these mechanisms can be undermined, making the model more likely to produce direct answers, even in sensitive contexts [16].

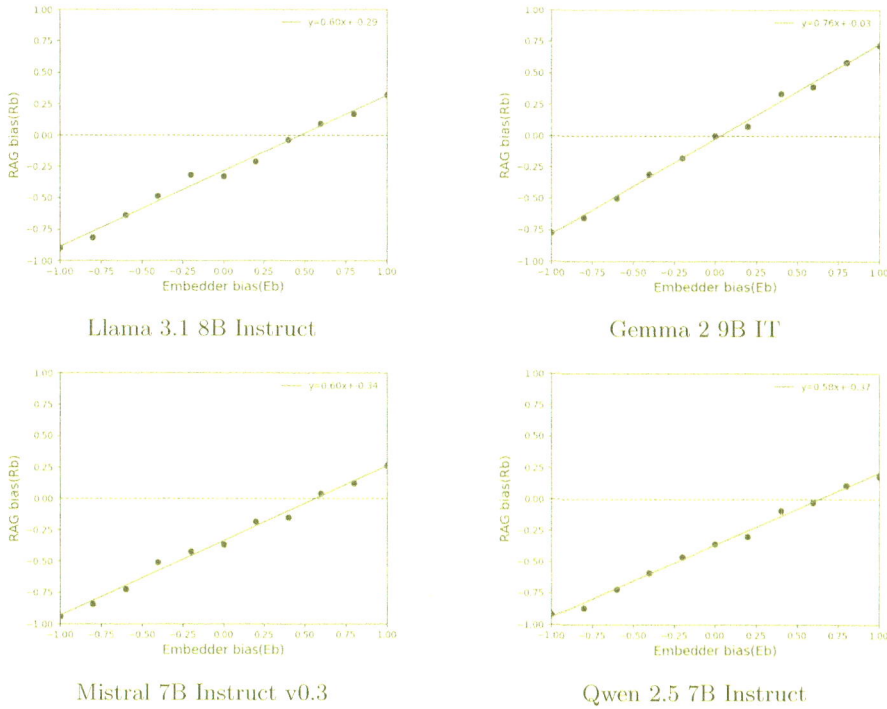

Fig. 4. Impact of Embedder Bias on Overall RAG Bias

These findings highlight a critical issue: while RAG systems enhance information retrieval, they may simultaneously compromise the alignment safeguards of LLMs. This underscores the importance of bias regulation within RAG pipelines.

6   Conclusions

Building upon the work of [21], this paper proposes a simple re-ranking based method, ReFaRAG, for mitigating bias in RAG systems without requiring fine-tuning of the embedding model. The approach is both easy to implement and avoids the common trade-off between performance and fairness often encountered during model fine-tuning.

More importantly, this study offers a clearer understanding of the underlying cause of bias in RAG systems: the exhibited bias is not directly attributable to the embedding model itself, but rather to the bias present in the context content passed to the LLM. This insight suggests that controlling the composition of context via reranking mechanisms is an effective strategy for bias mitigation. Furthermore, ReFaRAG opens up new possibilities for controlling bias in more advanced RAG architectures, such as those that incorporate top-k retrieved documents into the context. The findings presented in this paper lay the groundwork for developing more fair and controllable retrieval-augmented generation systems, and offer a promising direction for future research.

6.1   Limitations and Future Work

Binary Bias. This paper adopts a binary bias ratio as the primary metric for evaluating RAG system bias. In the context of political ideology, this metric measures the frequency with which the system favors left-leaning versus right-leaning perspectives, thereby reflecting whether the model exhibits ideological balance. This approach is reasonable in that if the system consistently favors one group (e.g., the left), it may implicitly reinforce that stance and potentially alienate or disadvantage users holding opposing views (e.g., the right). However, the metric has inherent limitations. Real-world social issues are rarely binary; rather, they often involve complex interactions among multiple social groups. A two-sided metric may thus oversimplify the nuanced nature of bias in many real-world applications. Additionally, some questions are inherently ambiguous or reflect value pluralism, making it difficult to identify bias solely based on frequency-based metrics. Future work could extend this study of bias in RAG in several ways: (1) by considering multi-group or intersectional settings that better reflect real-world social dynamics; (2) by developing more granular or continuous bias measures to capture subtler forms of bias; and (3) by exploring the use of causal inference and counterfactual analysis to enhance the interpretability and diagnostic power of bias evaluation frameworks.

Top-1 Retrieval. In the experimental setup of this paper, the RAG system retrieves only the Top-1 most relevant document as external knowledge input to the LLM. However, in real-world applications, useful information is often distributed across multiple documents. Due to the limitations of retriever performance and corpus chunking strategies, relevant content may not be concentrated in a single source. Therefore, future research could explore incorporating Top-k documents (e.g., Top-2, Top-5) as contextual input, and study how mixtures of documents with varying bias polarities influence the final outputs of RAG systems. This direction would not only better reflect practical deployment scenarios but also contribute to a deeper understanding of how bias propagates under multi-document conditions.

Query-Based RAG in Text. This study focuses on a query-based RAG system and demonstrates that the proposed re-ranking method can effectively mitigate bias to a certain extent. However, in more complex RAG architectures—such as those incorporating a judger module, iterative retrieval loops, multimodal information fusion, or other integration mechanisms like logit-based or latent-representation-based RAG—the effectiveness and generalizability of our bias measurement and mitigation strategies remain to be tested and improved. Since bias propagation in these systems may follow more intricate pathways, future work should investigate how the proposed framework performs under such settings and explore necessary adaptations to address emerging challenges.

Dataset for RAG Bias. Given the complexity of bias issues in RAG systems discussed earlier, one major limitation at this stage is the absence of dedicated datasets specifically designed for bias evaluation and mitigation in RAG. As a result, current research relies heavily on carefully crafted experimental scenarios and task-specific synthetic or collected data, which limits both the scalability and reproducibility of studies in this area. We therefore advocate for the development of more targeted datasets to support RAG bias research—encompassing both well-controlled synthetic benchmarks and realistic datasets that reflect real-world distributions and social contexts. Such resources are essential for systematically investigating how RAG systems propagate or amplify bias, and for ultimately paving the way toward more fair and value-aligned AI tools.

References

1. zilliztech/GPTCache. https://github.com/zilliztech/GPTCache, May 2025. original-date: 2023-03-24T05:51:16Z
2. Benjamin, R.: Race after technology: abolitionist tools for the new JIM code. https://www.wiley.com/en-us/Race+After+Technology%3A+Abolitionist+Tools+for+the+New+Jim+Code-p-9781509526406 Publisher: Polity
3. Borgeaud, S., et al.: Improving language models by retrieving from trillions of tokens (2022). arXiv:2112.04426 [cs]
4. Brown et al. Language models are few-shot learners. In: NeurIPS, pp. 1877–1901 (2020)
5. DeepSeek-AI, et al.: DeepSeek-R1: incentivizing reasoning capability in LLMs via reinforcement learning (2025). arXiv:2501.12948 [cs]
6. Dhamala, J., et al.: BOLD: dataset and metrics for measuring biases in open-ended language generation. In: Proceedings of the 2021 ACM Conference on Fairness, Accountability, and Transparency, FAccT '21 (2021)
7. Douze, M., et al.: The Faiss library (2025). arXiv:2401.08281 [cs]
8. Friedman, B., Nissenbaum, H.: Bias in computer systems. ACM Trans. Inf. Syst. **14**(3), 330–347 (1996)
9. Fulay, S., et al.: On the relationship between truth and political bias in language models. In: EMNLP, pp. 9004–9018 (2024)
10. Gallegos, I.O., et al.: Bias and fairness in large language models: a survey (2024). arXiv:2309.00770 [cs]
11. Gehman, S., Gururangan, S., Sap, M., Choi, Y., Smith, N.A.: RealToxicityPrompts: evaluating neural toxic degeneration in language models. In: EMNLP (Findings), pp. 3356–3369 (2020)
12. Gemma Team. Gemma 2: Improving Open Language Models at a Practical Size (2024). arXiv:2408.00118 [cs]
13. Grattafiori, A., et al.: The Llama 3 Herd of Models (2024). arXiv:2407.21783 [cs]
14. Gururangan, S., et al.: Don't stop pretraining: adapt language models to domains and tasks. In: ACL, pp. 8342–8360 (2020)
15. He, Z., Zhong, Z., Cai, T., Lee, J.D., He, D.: REST: retrieval-based speculative decoding (2024). arXiv:2311.08252 [cs]
16. Hu, M., et al.: No free lunch: retrieval-augmented generation undermines fairness in LLMs. Even for Vigilant Users (2024). arXiv:2410.07589 [cs]

17. Izacard, G., Grave, E.: Leveraging passage retrieval with generative models for open domain question answering. In: EACL, pp. 874–880 (2021)
18. Jiang, A.Q., et al.: Mistral 7B (2023). arXiv:2310.06825 [cs]
19. Jin, J., et al.: FlashRAG: a modular toolkit for efficient retrieval-augmented generation research (2025). arXiv:2405.13576 [cs]
20. Khandelwal, U., et al.: Generalization through memorization: nearest neighbor language models (2020). arXiv:1911.00172 [cs]
21. Kim, T., Springer, J., Raghunathan, A., Sap, M.: Mitigating bias in RAG: controlling the embedder (2025). arXiv:2502.17390 [cs]
22. Kim, T.E., Diaz, F.: Towards fair RAG: on the impact of fair ranking in retrieval-augmented generation (2024). arXiv:2409.11598 [cs]
23. Li,T., et al.: UNQOVERing stereotyping biases via underspecified questions. In: EMNLP (Findings). Association for Computational Linguistics (2020)
24. Li, Z., et al.: Towards general text embeddings with multi-stage contrastive learning (2023). arXiv:2308.03281 [cs]
25. Liu, Y., et al.: Trustworthy LLMs: a survey and guideline for evaluating large language models' alignment (2024). arXiv:2308.05374 [cs]
26. Mallen, A., et al.: When not to trust language models: investigating effectiveness of parametric and non-parametric memories. In: ACL, pp. 9802–9822 (2023)
27. Nadeem, M., Bethke, A., Reddy, S.: Stereoset: measuring stereotypical bias in pretrained language models. In: ACL/IJCNLP, pp. 5356–5371 (2021)
28. Ouyang, L., et al.: Training language models to follow instructions with human feedback (2022). arXiv:2203.02155 [cs]
29. Parrish, A., et al.: BBQ: a hand-built bias benchmark for question answering. In: ACL (Findings) (2022)
30. Sharma, M., et al.: Towards understanding sycophancy in language models (2025). arXiv:2310.13548 [cs]
31. Wang, Y., Li, P., Sun, M., Liu, Y.: Self-knowledge guided retrieval augmentation for large language models (2023). arXiv:2310.05002 [cs]
32. Wei, J., et al.: Finetuned language models are zero-shot learners (2022). arXiv:2109.01652 [cs]
33. Wortsman, M., et al.: Robust fine-tuning of zero-shot models (2022). arXiv:2109.01903 [cs]
34. Wu, X., Li, S., Wu, H.-T., Tao, Z., Fang, Y.: Does RAG introduce unfairness in LLMs? Evaluating fairness in retrieval-augmented generation systems. In: COLING, pp. 10021–10036 (2025)
35. Xu, F., Shi, W., Choi, E.: RECOMP: improving retrieval-augmented LMs with compression and selective augmentation (2023). arXiv:2310.04408 [cs]
36. Xu, L., Xie, H., Qin, S.-Z.J., Tao, X., Wang, F.L.: Parameter-efficient fine-tuning methods for pretrained language models: a critical review and assessment (2023). arXiv:2312.12148 [cs]
37. Yang, A., et al.: Qwen2 Technical report (2024). arXiv:2407.10671 [cs]
38. Zehlike, M., Bonchi, F., Castillo, C., Hajian, S., Megahed, M., Baeza-Yates, R.: FA*IR: a fair top-k ranking algorithm. In: CIKM, pp. 1569–1578 (2017)
39. Zhao, J., Wang, T., Yatskar, M., Ordonez, V., Chang, K.: Gender bias in coreference resolution: evaluation and debiasing methods. In: NAACL-HLT, pp. 15–20 (2018)
40. Zhao, P., et al.: Retrieval-augmented generation for AI-generated content: a survey (2024). arXiv:2402.19473 [cs]

IT4TOCI 2025: 1st Workshop on Information Technology for Tourism and Culture Industries

Trustworthy Tourism Recommender Systems

Francesco Ricci[✉] (iD)

Free University of Bozen-Bolzano, Bolzano, Italy
fricci@unibz.it

Abstract. Recommender systems (RSs) are AI tools used in online platforms to algorithmically identify and promote to each platform's user a small number of personalised items, such as, news, posts, music, and videos. Their core component is a machine learning algorithm exploiting sparse users' online behaviour data to estimate what items, in a predefined catalogue, should be presented to each user to achieve a desired goal (e.g., increase revenues and user satisfaction).

After having briefly discussed RSs value, and risks, we will focus on the role and importance of a proper system evaluation, which is also a mandatory request formulated in the European legislation (Digital Service Act). Classical machine learning evaluation approaches, which are based on offline testing the predictive model on holdout data, are insufficient to asses the true effect of a new RS algorithm (intervention).

A new type of simulation-based evaluation approaches will be therefore introduced: they enable to estimate the potential effects (positive and negative) of an RS on the choices made by their users when they are influenced by the received recommendations. The application of this evaluation method will be exemplified in a particular case: sustainable and multistakeholder recommendations in tourism, namely, how to tame over tourism and respect local communities. We have found that, under certain conditions, multiple stakeholder can jointly benefit from the RS, hence creating a more stable and trustful cooperation scenario.

Keywords: Recommender systems · Trustworthy systems · System evaluation · Sustainable tourism

1 Introduction

Recommender systems (RSs) are AI tools used in online platforms to algorithmically identify and promote to each user a personalised and small number of items, such as, news, posts, music, or videos [22]. They have become ubiquitous in many web sites and apps, so that we do not any-more perceive their presence. But they are actually strongly influencing what posts we read on social networks, which TV series we watch, or where we fly for a holiday.

RS core component is a machine learning algorithm, exploiting sparse users' online behaviour data, to estimate what items, in a predefined catalogue, should

© The Author(s), under exclusive license to Springer Nature Switzerland AG 2026
P. K. Chrysanthis et al. (Eds.): ADBIS 2025, CCIS 2676, pp. 533–540, 2026.
https://doi.org/10.1007/978-3-032-05727-3_43

be presented to each user to achieve a desired goal (e.g., increase both RS platform revenues and user satisfaction). The research on RSs initiated in the '90s with the widespread adoption of Internet and Word Wide Web [21]. The seed intuition was to leverage prosumers' data. i.e., consumers that are also producing content in the form of posts, reviews and ratings, to identify prosumer-to-prosumer similarities, and recommend to one of them items that were liked by similar pronsumers. That is the gist of collaborative filtering [18], an extremely popular RS method. Recommender systems research has grown along a number of application domains and techniques. However, after the first positive applications, critiques were raised concerning their negative effects on users, such as, the risk to confine them in information filter bubbles produced by the automatic and personalised selection of content operated by the RS [19].

After having briefly discussed RSs values and risks, we here introduce the new research area of Trustworthy Recommender Systems, i.e., systems that can be trusted by their users, offering more reliable recommendations, and in a more transparent way [25]. A major step necessary for properly building these systems is auditing and evaluation. Hence, we then focus on the role and the implementation of reliable evaluation methods, which is also a mandatory request formulated in the European legislation (Digital Service Act). In fact, classical machine learning evaluation approaches, which are based on offline testing the RS predictive model on holdout data, are insufficient to measure the true effect of a new RS. In fact, a new RS is an intervention designed to modify user behaviours and its effect cannot be easily measured by mining the behaviour data produced by a different RS (the one that produced the log data) [14].

We then sketch the assumptions and the basic structure of a new type of simulation-based evaluation approach: it enables to better estimate the potential effects (positive and negative) of an RS on the choices made by its users when they are influenced by the received recommendations. The application of this evaluation approach is exemplified in a particular case: sustainable recommendations in tourism, namely, how to tame over tourism and respect local communities.

2   Trustworthy Recommender Systems

The widespread adoption of RSs is motivated by the values that they offer to the involved stakeholders: the RS service provider, the suppliers of the recommended items, and the consumers. Consumers are the end users of an RS, and their behaviour is potentially influenced by the system's recommendations. Recommendation service providers are organizations that provide recommendations as part of their services, they invest in the RS technology and are in control of the system. Suppliers are organizations that provide the products or services that are recommended to consumers. The initial research on RS was uniquely targeting benefits that the system could offer to consumers: help them to find items that match their assumed short-term intent and context; recommend items that are relevant in the ongoing user session, sometimes even without long-term preference information; help users find objects that match their long-term interests, which is the most explored problem [13].

However, after the introduction of the first RS techniques and the widespread usage of them in web platforms (ecommerce, news, media streaming, social networks), a number of risks were identified. The list of such potential negative effects includes: poor decisions, choice dissatisfaction, bad user experience, decision difficulty, biased information, privacy loss, loss of consumer trust in the service, filter bubbles and echo chambers, and algorithmic bias and discrimination [13]. The major source of these risks originates from the fact that the RS is operated by a service provider that pursues its own goals, which, in the best case, are purely economical (revenues maximization), but may also be political and social, i.e., influencing consumers opinions on strategical issues (elections, health decision, social choices). These goals can easily conflict with consumers' values and objectives. This awareness started an important line of research: multistakeholder recommender systems [1]. Multistakeholder approaches aim to effectively measure the benefit generated by the RS for all the involved parties, and to identify technologies that can balance the rewards of all these stakeholders.

Nowadays, in multistakeholder scenarios a trustworthy service provider can leverage a range of techniques designed to build and maintain an RS that can be trusted by both the consumers and the suppliers [25]. The challenges of building such a trustworthy system are: noisy, faked and biased data; data leakage and privacy issues; bias injection attacks; bias data reduction; preference manipulation; improper behaviour manipulation; unethical recommendations. In fact, a range of solutions have been developed to tame these issues: data cleaning, augmentation and debiasing; robust and federated learning; trustworthy and long term evaluation strategies.

While the research has proposed a range of techniques to build and operate more trustworthy RSs, the issue of enforcing service providers to adopt these techniques, and to make transparent to consumers the fundamental design decisions at the base of their RSs remains. This is where national and international organizations operates. The most important set of rules, which big online platforms and search engines operating in the EU (more that 45 millions users per month) must abide, is contained in the European Digital service Act (DSA, EU regulation 2022/2065 of the European Parliament and of the Council of 19 October 2022). This package of rules aims to create a safer digital space in which the fundamental rights of all users of digital services are protected. Recommender systems are explicitly mentioned in the DSA. Article 27 is about RSs transparency, and states that providers of online platforms that use RSs shall set out the main parameters used in their systems, as well as any options for the recipients of the service to modify or influence those main parameters. The main parameters shall explain why certain information is suggested to the recipient of the service. They shall include the criteria determining the information suggested to the recipient of the service, and the reasons for the relative importance of those parameters. The service provider shall also make available a functionality that allows the recipient of the service to select and modify at any time their preferred options (e.g., deny the usage of personal data). Article 34 states that when conducting risk assessments, service providers shall take into account the design of their RS and any other relevant algorithmic system. Moreover, in Article 37 it is stated that service providers shall be subject, at their own expense, and at least once a year, to independent audits to assess compliance with the law. Finally, in Article 40 it is stated that service providers shall be able to explain the design, the logic, the functioning and the testing of their algorithmic systems, including their RS.

3   Evaluation of Recommender Systems

As we have noted in the previous section the importance of a proper evaluation approach is pivotal to produce a trustworthy RS. Methods for conducting a proper system evaluation have been analysed since the early introduction of these personalization technologies, but only more recently the community has understood the complexity of this task [9]. The ultimate goal of the RS evaluation is to measure how much the system contributes to produce a measurable reward to the involved stakeholders: service provider, consumers and items' suppliers. Examples of such forms of reward are: for the service provider, increase the number of service subscriptions, increase the service engagement; for consumers, increase satisfaction for the generated purchases, expand the user's knowledge of a subject; for items' suppliers, increase the number of their items that the platform sells, and favour a more sustainable consumption of them.

While it is relatively easy to identify the true objectives of the involved stake-holders, it is not simple to balance them [1] and to properly measure whether a candidate RS achieves those objectives [9]. A number of computational metrics have been proposed in the literature to operationalise the reward of the stake-holders [9]. The most common metrics relate to the precision of the recommenda-tions, i.e., how well the system is capable to identify items that the consumer will like. These metrics originated in the machine learning and information retrieval literature and in order to be computed they requires some form of ground truth: what the user really likes. More recently the evaluation scope broadened when it was understood that in some applications (e.g., news or tourism) consumers do not only want to be directed towards correct suggestions, but they also often desire to widen their knowledge and discover new and maybe less mainstream items. Hence a line of research on "beyond-accuracy" metrics and their evaluation methods started [4, 6, 15].

However, even when considering a wider set of metrics and tailor them to properly match the practical goals of the involved stakeholders, a problem remain: how to correctly measure a metric with the available data and algo-rithms. As we mentioned above, precision metrics require a ground truth to be estimated. For instance, if one wants to know if a search engine query has returned the right results then the opinion of a judge on the relevance of each (or most of the) returned links is needed. Some metrics do not strictly require this ground truth, e.g., those measuring the diversity of the recommendations: these are functions of the recommendation items only. However, the user perception of diversity may be different from that operationalised in the metric. Hence, at the end of the day, a reliable assessment performed by the system users seems to be unavoidable. But user studies and A/B tests, i.e., common methods aimed at quantifying consumers evaluations of the tested RS, are complex procedures that involve a considerable amount of subjects and must expose them to the complete platform hosting the RS. These requirements are rarely satisfied in the academic research, and often not even in the industry. Moreover, the risk to expose valuable customers to inferior recommendations (produced by a novel experimental RS) strongly limits the applicability of such types of evaluations.

Hence, off line evaluations, based on the classical machine learning paradigm of dividing a data set into a training part, where the candidate RS is trained, and a test part, which is not used in the training and it is only accessed to measure the target metric (e.g., recommendation precision), are still largely exploited. However, this approach is totally inadequate to exactly measure the reaction of the users to the recommendations generated by a novel RS. In fact, the behaviour data, which can be used for training and testing a novel RS, has been collected while users were exposed to another, often unknown, RS (or multiple RSs). Hence this data cannot provide reliable information about how the user will react to the recommendations generated by the novel RS. Advanced statistical approaches can be used to debias the outcome of such an evaluation, but they are not often applicable, because they require knowledge of the conditions under which the available data has been acquired (e.g., with what probability each item was recommended) [14].

Consumers/system interaction simulation is a more viable approach to the implementation of counterfactual testing: an algorithmic choice model, independent from the RS, is learned to simulate how users would react to the recommendations generated by a new RS. Moreover, simulations, by making reasonable assumptions about recommender-user interaction behaviour, offer the opportunity to gain insight into population-level effects of recommendations, which result from micro-level interactions between users and RS. In practice, the system is simulated in an artificial environment to reveal long-term dynamic properties in metrics such as accuracy [24], additional consumption, profit and demand [8,12], diversity [2,7,10,11], and popularity bias [27], that are not observable in a simple train-test split evaluation approach.

4   Overtourism Mitigation and Sustainability

Simulation based evaluations enable to assess the effect of an RS on the actual choices of the consumers, and therefore on the overall impact of the RS on all the involved stakeholders. An interesting and important application scenario for multistakeholder RSs (MRSs) is tourism. Tourism offers a rich and multifaceted environment for the development of MRSs [16]. Assisting tourists in processing information and making decisions to visit points of interest (POIs) or to consume a service (e.g., accommodation) has economic, social, and environmental implications, on the visited destinations [26].

In the tourism scenario a simulation based evaluation approach enables to test the potential effect of a promotional campaign (determined by algorithmic recommendations) on the distribution of tourist in a target region. One can simulate that the tourists, who visited a region in a particular period of time, had been (counterfactual) exposed to alternative RSs, and then assess systems' impact on the tourists' choices to visit the regional destinations [20]. With carefully designed RS strategies less popular destinations may see a growth of tourist presences to the detriment of more popular ones. Hence, simulation based approaches enable to identify more well grounded forms of sustainable tourism development.

Sustainable tourism is a responsible approach aimed at reducing the tensions and friction between tourism industry, tourists, the environment, and the host communities so that the long-term capacity and quality of both natural and human resources can be preserved [3]. In particular, a focus of sustainable tourism is urban management policies [23]. Some approaches tries to spread tourism outside the city and de-market crowded locations. Others, promote sustainable tourism by recommending sequences of relevant but less crowded attractions [5].

In the urban scenario, the attempt to de-market crowded POIs may conflict with the interest of mainstream tourists, who usually come to a destination to visit the main attractions. Hence, the conflicting nature of this goal raises the question whether, by means of recommendations, it is possible to achieve better outcomes for both the tourists and the urban destination, i.e., to achieve a positive sum impact. To answer this question, we have simulated the impact of an MRS on tourists' choices [17]. We focused on a class of MRSs that linearly balance the user (tourist) and the destination (sustainable urban) utilities. This is the prevalent balancing approach to date. We assume a realistic scenario where users have limited knowledge of the attractions (Points of Interests) catalogue and the MRS has little information about tourists' preferences. Our experimental results show that this MRS can guide tourists to choose attractions to visit, benefiting both them and the destination (win-win outcome). However, a proper balance of the stakeholders' goals (in the RS core algorithm) is crucial: putting under- or over-emphasis on sustainable attractions in the recommendation lists can be detrimental to both destination sustainability and user satisfaction. Moreover, with simulations, we show that even non-personalised popularity-driven recommendations, when combined with sustainable recommendations, can have a (less pronounced) positive impact. This suggests that even a simple non-personalised MRS, without exploiting personal data, may be effective in urban tourism, if properly used.

5 Conclusions

In this paper we have described a line of research that, starting from the classical implementation of RSs (targeted to optimise consumer satisfaction), widens the RS impact by facing the challenges of building more trustworthy and multistakeholder applications. We argue that consumers can trust the RS if they can assess that it actually targets their needs and wants, which surely cannot be limited to the system precision in identifying a few relevant items. We have set the problem into the context of multistakeholder RSs (MRSs), and emphasized the importance of auditing and measuring the true performance of the system in influencing consumers' behaviour. Moreover, we have discussed a special application scenario of MRSs, namely tourism, showing that more trustworthy and sustainable systems can be built: those benefitting simultaneously multiple stakeholders, hence creating a win-win outcome. We believe that these results offer a more solid base to the implementation of trustworthy RSs: when the stakeholders jointly benefit from the recommendation service they have an concrete and measurable incentive to respect the fair distribution of the reward that the RS generates.

References

1. Abdollahpouri, H., Burke, R.: Multistakeholder recommender systems. In: Ricci, F., Rokach, L., Shapira, B. (eds.) Recommender Systems Handbook, pp. 647–677. Springer, Cham (2022). https://doi.org/10.1007/978-1-0716-2197-4_17
2. Bountouridis, D., Harambam, J., Makhortykh, M., Marrero, M., Tintarev, N., Hauff, C.: Siren: a simulation framework for understanding the effects of recommender systems in online news environments. In: Proceedings of the Conference on Fairness, Accountability, and Transparency. pp. 150–159. FAT* '19. Association for Computing Machinery (2019). https://doi.org/10.1145/3287560.3287583
3. Bramwell, B., Lane, B.: Sustainable tourism: an evolving global approach. J. Sustain. Tourism (1993)
4. Castells, P., Hurley, N., Vargas, S.: Novelty and diversity in recommender systems. In: Ricci, F., Rokach, L., Shapira, B. (eds.) Recommender Systems Handbook, pp. 603–646. Springer, Cham (2022). https://doi.org/10.1007/978-1-4899-7637-6_26
5. Dalla Vecchia, A., Migliorini, S., Quintarelli, E., Gambini, M., Belussi, A.: Promoting sustainable tourism by recommending sequences of attractions with deep reinforcement learning. Inf. Technol. Tourism **26**(3), 449–484 (2024)
6. Ekstrand, M.D., Das, A., Burke, R., Diaz, F.: Fairness in recommender systems. In: Ricci, F., Rokach, L., Shapira, B. (eds.) Recommender Systems Handbook, pp. 679–707. Springer, Cham (2022). https://doi.org/10.1007/978-1-0716-2197-4_18
7. Fleder, D., Hosanagar, K.: Blockbuster culture's next rise or fall: the impact of recommender systems on sales diversity. Manage. Sci. **55**(5), 697–712 (2009)
8. Ghanem, N., Leitner, S., Jannach, D.: Balancing consumer and business value of recommender systems: a simulation-based analysis. Electron. Commer. Res. Appl. **55**, 101195 (2022)
9. Gunawardana, A., Shani, G., Yogev, S.: Evaluating recommender systems. In: Ricci, F., Rokach, L., Shapira, B. (eds.) Recommender Systems Handbook, pp. 547–601. Springer, Cham (2022)

10. Hazrati, N., Ricci, F.: Recommender systems effect on the evolution of users' choices distribution. Inf. Process. Manag. **59**(1), 102766 (2022)
11. Hazrati, N., Ricci, F.: Choice models and recommender systems effects on users' choices. User Model. User Adapt. Interact. **34**(1), 109–145 (2024)
12. Hinz, O., Eckert, J.: The impact of search and recommendation systems on sales in electronic commerce. Bus. Inf. Syst. Eng. **2**, 67–77 (2010)
13. Jannach, D., Zanker, M.: Value and impact of recommender systems. In: Ricci, F., Rokach, L., Shapira, B. (eds.) Recommender Systems Handbook, pp. 519–546. Springer, Cham (2022)
14. Joachims, T., London, B., Su, Y., Swaminathan, A., Wang, L.: Recommendations as treatments. AI Mag. **42**(3), 19–30 (2021)
15. Kaminskas, M., Bridge, D.: Diversity, serendipity, novelty, and coverage: a survey and empirical analysis of beyond-accuracy objectives in recommender systems. ACM Trans. Interact. Intell. Syst. **7**(1), 2:1-2:42 (2017)
16. Massimo, D., Ricci, F.: Building effective recommender systems for tourists. AI Mag. **43**(2), 209–224 (2022)
17. Merinov, P., Ricci, F.: Positive-sum impact of multistakeholder recommender systems for urban tourism promotion and user utility. In: 18th ACM Conference on Recommender Systems, pp. 939–944. ACM (2024)
18. Nikolakopoulos, A.N., Ning, X., Desrosiers, C., Karypis, G.: Trust your neighbors: a comprehensive survey of neighborhood-based methods for recommender systems. In: Ricci, F., Rokach, L., Shapira, B. (eds.) Recommender Systems Handbook, pp. 39–89. Springer, Cham (2022). https://doi.org/10.1007/978-1-0716-2197-4_2
19. Pariser, E.: The Filter Bubble: What the Internet Is Hiding from You. Penguin Books Ltd. (2012)
20. Piliponyte, G., Massimo, D., Ricci, F.: Simulation of recommender systems driven tourism promotion campaigns. Inf. Technol. Tourism **26**(3), 407–448 (2024)
21. Resnick, P., Varian, H.R.: Recommender systems. Commun. ACM **40**(3), 56–58 (1997)
22. Ricci, F., Rokach, L., Shapira, B.: Recommender systems: techniques, applications, and challenges. In: Ricci, F., Rokach, L., Shapira, B. (eds.) Recommender Systems Handbook, pp. 1–35. Springer, Cham (2022)
23. Santos-Lacueva, R., Velasco González, M., González Domingo, A.: The integration of sustainable tourism policies in european cities. Pasos: Revista de Turismo y Patrimonio Cultural **20**(5), 1229–1242 (2022)
24. Umeda, T., Ichikawa, M., Koyama, Y., Deguchi, H.: Evaluation of collaborative filtering by agent-based simulation considering market environment. In: Developments in Business Simulation and Experiential Learning: Proceedings of the Annual ABSEL Conference, vol. 36 (2009)
25. Wang, S., Zhang, X., Wang, Y., Ricci, F.: Trustworthy recommender systems. ACM Trans. Intell. Syst. Technol. **15**(4), 84:1-84:20 (2024)
26. Werthner, H., et al.: Future research issues in IT and tourism. Inf. Technol. Tourism **15**(1), 1–15 (2015). https://doi.org/10.1007/s40558-014-0021-9
27. Yao, S., et al.: Measuring recommender system effects with simulated users. arXiv:abs/2101.04526 (2021)

A Blockchain-Based Platform for Sharing and Rewarding User-Generated Content

Alberto Belussi⬤, Mauro Gambini⬤, Sara Migliorini$^{(\boxtimes)}$⬤,
and Adrian Munteanu

Department of Computer Science, University of Verona, Verona, Italy
{alberto.belussi,mauro.gambini,sara.migliorini}@univr.it

Abstract. Machine and deep learning techniques are increasingly applied in the tourist domain to understand past user behaviors, forecast future trends, and provide recommendations. Traceability and authenticity of data shared by users are essential to guarantee the quality of results obtained from their analysis. Blockchain is an emerging technology that is able to ensure such property by making information immutable and maintaining a set of metadata about its production and sharing. At the same time, a set of incentives could be implemented to promote the production of this kind of information to increase data availability and data quality. Smart contracts and decentralized applications could also be successfully applied for this purpose. This paper presents a complete solution based on the design of a set of smart contracts for traceability, immutability, and incentive implementation, together with a decentralized application for increasing the usability of the approach.

Keywords: user generated content · smart contract · blockchain · traceability · reward · tourism

1 Introduction

In the era of artificial intelligence and deep learning, the availability of huge amounts of information is essential to train neural networks and carry on various inference and prediction tasks. Much of this information comes in the form of user-generated content (UGC), voluntarily produced and submitted by users based on their experience with products and services. UGCs are typically shared by users during their online activities, such as posts and images published on social networks, trajectories collected by smartphones during a journey, or reviews made about places and items. The collection, storage, and supply of such data is the core business of many IT corporations that offer pay-per-use services [3]. The acquisition and the successive processing of such information have great value in many fields, including developing recommender or decision support systems, analyzing topic and sentiment trends, and planning and managing activities [1,4,8,9]. In particular, the tourist domain represents an invaluable context of application, where emerging and consolidated platforms, like Google Places, TripAdvisor, Booking, and Airbnb, base their activity on the collection of reviews, and users can exploit them to make their choices.

© The Author(s), under exclusive license to Springer Nature Switzerland AG 2026
P. K. Chrysanthis et al. (Eds.): ADBIS 2025, CCIS 2676, pp. 541–554, 2026.
https://doi.org/10.1007/978-3-032-05727-3_44

At the same time, the value of data and the data analysis performed on them also greatly depend on the quality of such data. As already highlighted in the literature [10], with the aim of promoting the production of UGCs, some services offer a kind of reward for this activity, which can lead, sooner or later, to fraudulent behaviors. Indeed, with the aim of obtaining the reward, users can try to produce content about places they have never visited or products they have never bought. For this reason, several techniques have been defined in the literature to overcome the fake check-ins problem, but without definitively solving the problem. In [5], the authors envisioned the use of blockchain technology for dealing with fake check-in prevention. They design a possible system architecture based on blockchain technology, but without providing an implementation.

In this paper, we propose a complete and functioning platform based on the use of emerging blockchain technology to ensure the immutability and traceability of information produced by users and promote their sharing through an incentive mechanism consisting of NFT rewards. The aim of this paper is twofold: demonstrating how blockchain and smart contracts could be successfully used in this context from a technological point of view and promoting a virtuous and secure reward process, also from an economic perspective.

The remainder of this paper is organized as follows: Sect. 2 illustrates the proposed solution by describing in detail the implemented smart contracts and their interaction, Sect. 3 shows the interfaces of the developed application and discuss some implementation details. Section 4 summarizes some related work, while Sect. 5 concludes the work and proposes some future extensions.

2   Proposed Solution

The proposed solution is composed of a set of smart contracts that essentially implement the following functionalities: (a) the suggestion of new Points of Interest (PoIs) by users, (b) the approval or rejection of PoIs previously proposed by users, (c) the collection of reviews made by users, (d) the grant of token rewards to users for their activities, and (e) the publication of NFTs and their buying by users through the obtained reward tokens. In particular, the management of PoI proposals and their acceptance or rejection is done through the PoiFactory contract in Algorithm 1, 2. The information related to each PoI, as well as the set of reviews collected for it, is stored inside the Poi contract in Algorithm 3. Finally, the management of NTFs is done through a TourNFT token implemented as an extension of the standard ERC271[1].

The PoiFactory contract is responsible for maintaining a set of variables and data structures useful to track the pending proposals and the approved PoIs (see Algorithm 1 lines 12–14), as well as information about users' status (see Algorithm 1 lines 21–23), proposal thresholds (see Algorithm 1 lines 15–17) and reward amounts (see Algorithm 1 lines 18–20). The process of PoI proposal, approval, or rejection is described in the sequence diagram of Fig. 7. Initially, the user proposes a new PoI by calling the function proposePoI() of the PoiFactory

---

[1] https://ethereum.org/en/developers/docs/standards/tokens/erc-721/.

Algorithm 1: PoiFactory contract (Part 1/2).

```
1  Contract PoiFactory :
2      struct PendingPoi :
3          string name;
4          string description;
5          int256 latitude;
6          int256 longitude;
7          string municipality;
8          string username;
9          address creator;
10         string date;
11         string time;

12     public mapping(bytes32 → PendingPoI) pendingPois;
13     bytes32[ ] pendingPoiKeys;
14     public address[ ] approvedPois;
15     public UGCToken rewardTokens;
16     public uint256 poiApprovalReward;
17     public uint256 reviewReward;
18     public uint256 constant MAX-PEN-USR = 3;
19     public uint256 constant MAX-REJ-USR = 5;
20     public uint256 constant MAX-PEN-TOT = 3;
21     public mapping(address → uint32) pendingPoiCount;
22     public mapping(address → uint32) rejectedPoiCount;
23     public mapping(address → bool) public isBanned;
24     address _owner;
25     modifier onlyOwner :
26         require(msg.sender == _owner);

27     constructor () :
28         _owner = msg.sender;

29     ...;
```

---

contract in Algorithm 2 (lines 3–11). This function receives as parameters a set of information describing the PoI itself, such as its name n, description d, and a geographical position on the Earth's surface (i.e., in latitude lt and longitude ln, and municipality name m), the username u and the wallet address a of the proponent, the date da and time t of the request. Before recording such PoI information, a set of checks is performed which verify if the user is allowed to submit a new proposal, namely the user has not been banned, and the number of individual and total pending proposals is below two established thresholds, and if the provided information is valid, for instance, if the name and description are not empty and the geographical coordinates represent valid latitude and longitude values. If all the validation checks are satisfied, a unique key k is created for the data together with a new instance of the PendingPoi data structure (see

Algorithm 1 lines 2-11), which is added into an array *pendingPois*, while the counter of pending PoIs for the user is updated (see Algorithm 2 lines 9–11).

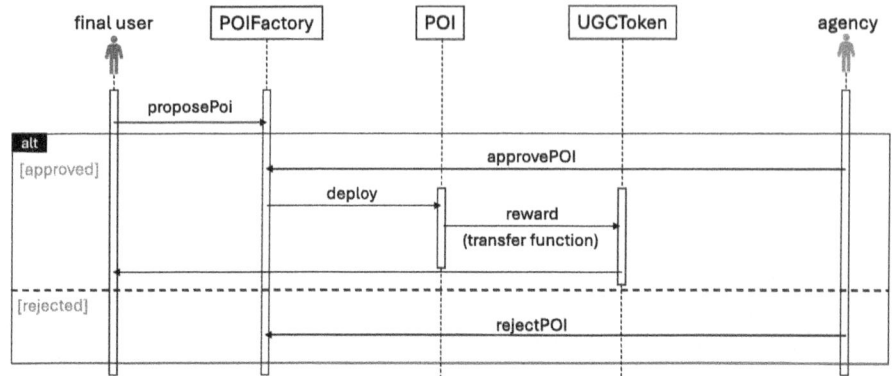

Fig. 1. Sequence diagram describing the process of PoI proposal, approval, or rejection.

As illustrated in Fig. 1, the platform administrator could periodically check this list of pending PoIs and approve one of them through the function approvePoi() (Algorithm 2 lines 12–20), which receives as a parameter the key k of the PoI to be approved. The function retrieves the desired PoI from the list of pending ones and then performs some preliminary tests on the validity of such PoI (see Algorithm 2 line 14). In case of success, the function deploys a new instance of the Poi contract (see Algorithm 2 line 15), which will collect all the future reviews made for it. The newly created Poi contract instance is added to the list of *approvedPois* for the given user. If the PoI is approved, the function is also responsible for rewarding the user. The effective transfer of reward tokens is done by calling the *transfer()* function of the UGCToken contract (see Algorithm 2 line 19). The UGCToken is essentially an implementation of the standard ERC20 token of Ethereum[2].

Conversely, if the platform administrator decides to reject the PoI, the function *rejectPoi()* is called (Algorithm 2 lines 21–27). This function essentially removes the proposed PoI from the *pendingPois* list and updates the counters about the number of rejected PoIs for the user. If this counter exceeds the maximum number of rejected PoIs per user, then the user is added to a list of banned users (see Algorithm 2 lines 26–27). This countermeasure, together with the thresholds for the maximum number of pending PoIs per user and in total, is used to prevent Denial of Service attacks from malicious users who try to overload the factory with a great number of requests.

The submission of a new review for a specific PoI p is performed by the user by invoking the function *addReview()* of the contract Poi in Algorithm 3 and following the sequence diagram in Fig. 2. The Poi contract defines a set of

---

[2] https://ethereum.org/en/developers/docs/standards/tokens/erc-20/.

Algorithm 2: PoiFactory contract.

---

1 **Contract** *PoiFactory* :
2   . . . ;
3   **function** *proposePoi(n, d, lt, ln, m u, a da, t)* :
4   require(!isBanned[msg.sender]);
5   require(pendingPoiCount[msg.sender] < MAX-PEN-USR);
6   require(pendingPoiKeys.length < MAX-PEN-TOT);
7   require(*checkParams*(n, d, m, lt, ln));
8          $k \leftarrow$ compute unique key from hash of data;
9   pendingPois[k] $\leftarrow$ *PendingPoi*(n, d, lt, ln, m, u, a, da, t);
10   pendingPoiKeys.push[k];
11   pendingPoiCount[msg.sender]++;

12   **function** *approvePoi(k) public onlyOwner* :
13   *PendingPoi p* $\leftarrow$ pendingPois[k];
14   require(p exists);
           // create a new POI contract with the information in p
15   newPoi $\leftarrow$ Poi(p.name, p.description, . . . , p.time);
16   approvedPois $\leftarrow$ approvedPois $\cup \{newPoi\}$;
17   pendingPois $\leftarrow$ pendingPois \ $\{p\}$;
18   pendingPoiKeys $\leftarrow$ pendingPoiKeys \ $\{k\}$;
19   *reward*(p.creator);
20   pendingPoiCount[p.creator] $\leftarrow$ pendingPoiCount[p.creator] - 1;

21   **function** *rejectPoi(k) public onlyOwner* :
22   *PendingPoi p* $\leftarrow$ pendingPois[k];
23   require(p exists);
24   pendingPoiCount[p.creator]–;
25   rejectedPoiCount[p.creator]++;
26   **if** *rejectedPOICount[p.creator]* $>=$ *MAX-REJ-USR* **then**
27   isBanned[p.creator] $\leftarrow$ true;

---

data structures for representing both the PoI details (see lines 8–10) and the made reviews (see lines 2–7), and maintains a set of variables for storing them (i.e., *poiData* and *userReviews*) together with some summarized values, like the total rating and the number of reviews. When a user submits a new review, the contract validates the submitted information; for instance, it checks that all the required elements, like the name and the comment, have been specified and the rating is in the admissible range (see line 26). Then, the new review is stored and assigned to the current user, while function *rewardUser*() takes care of rewarding users (see line 30). In particular, as exemplified in Fig. 2, if this is the first review submitted by the user u for the PoI p, then the function *rewardUser*() will transfer the predefined amount of UGCTokens to u, otherwise the function completes without any reward. This logic tries to prevent situations in which users submit several reviews for the same PoI in order to maximize their reward without providing any meaningful insights to the system. At the same time, it allows the possibility to rectify previous reviews if necessary. Since the PoI reviews are stored in a mapping *userReviews* as an array associated with the user address, the system has the ability to retrieve for each user only the last reviews or the history of all the provided ones, based on the specific application requirements.

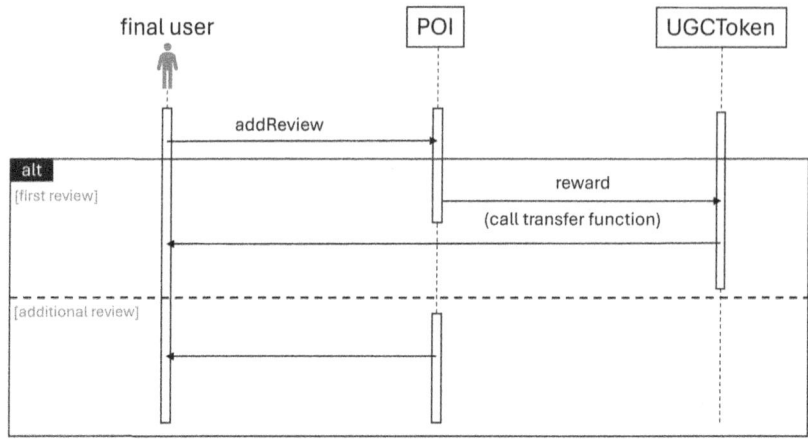

Fig. 2. Sequence diagram describing the process of user reward.

The final interaction with the system is represented by the marketplace application in which TourNFTs can be published by the touristic operators and acquired by the users by spending the accumulated UGCToken. The publication of a new NFT is essentially performed by calling the standard *mint*() function of an NFT, namely an extension of the standard ERC271 contract provided by the OpenZeppelin library[3]. Conversely, the buying of an NFT is performed by selecting the NFT from the marketplace application through a *buy* functionality which calls the standard *spend*() function on the UGCToken and then the *transfer*() function of the TourNFT. This interaction will be clearer in the following section where the implementation is described by means of some screenshots (Fig. 3).

3   Implementation and Experiments

The proposed solution has been implemented by using the Solidity language for smart contracts definition, and the React with Ionic frameworks for the realization of the application front-end, which can be used both as a web application and a smartphone mobile application. For the interaction with the wallet, we exploit the WalletConnect framework[4] to support the interaction with Metamask, or other similar wallets, both as a browser extension or a stand-alone

---

[3] https://docs.openzeppelin.com/contracts/4.x/erc721.
[4] https://walletconnect.network/.

Algorithm 3: Poi contract.

```
1  Contract Poi :
2      struct Review :
3          string comment;
4          uint8 rating;
5          string date;
6          string time;
7          address reviewer;
8      struct PoiData :
9          string name;
10         ...;
11     PoiData poiData;
12     address owner;
13     mapping(address ← Review[]) userReviews;
14     totalRating;
15     reviewCount;
16     initialized;
17     constructor (n, d, m, lt, ln, u, da, t) :
18         require(!initialized);
19         poiData = PoiData(n, d, m, lt, ln, u, da, t);
20         owner = mgs.sender;
21         totalRating = 0; reviewCount = 0;
22         initialized ← true;
23         rewardToken(...);
24     function addReview( c, r, d, t, r ) :
25         Review newReview ← {c, r, d, t, r};
26         validateReview();
27         userReviews[msg.sender] ← userReviews[msg.sender] ∪ {newReview};
28         totalRating ← totalRating + newReview.rating;
29         reviewCount ← reviewCount + 1;
30         rewardUser(msg.sender);
```

mobile application. We tested the application with both the Sepolia test net for simulating the deployment and interaction on a Layer 1 Ethereum network, as well as the Polygon test net for checking deployment and interaction on a Layer 2 solution. Table 1 summarizes the set of technologies used in the realization of the prototype, while the source code of the implemented smart contracts is available as a GitHub repository[5]. Table 2 reports the bytecode size of the implemented smart contracts, together with the amount of Gas required to deploy and interact with them in Gwei and in USD, respectively. More specifically, the column **Total cost** reports the corresponding cost in USD computed considering the average prices in middle June 2025, knowing that the Gas cost is expressed in Gwei and that for the Ethereum network 1 Gwei is equal to 10^{-9} ETH, while

---

[5] https://github.com/smigliorini/tourism4chain.

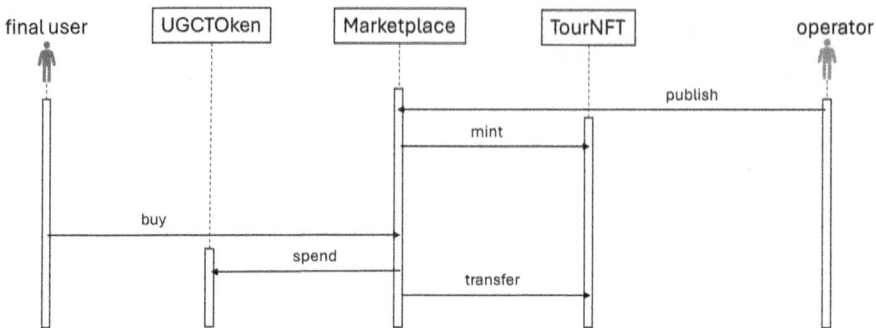

Fig. 3. Sequence diagram describing the acquisition of new NFT form the marketplace.

Table 1. Tools and Technologies Used in the Experimental Setup.

Technology	Version/Network
Ethereum Test Network	Sepolia
Layer 2 Test Network	Polygon
Solidity	0.8.28
Contract Deployment	Hardhat
Blockchain Interaction	WalletConnect + Metamask
DApp Development	Ether.js
Frontend Framework	Ionic + React TypeScript
NFT	Pinata
WebGis	Leaflet + OpenStreetMap

Table 2. Bytecode Size and Gas Cost of Smart Contract

Smart Contract	Bytecode size (bytes)	Gas used (gwei)	Total cost (USD) (L1)	(L2)
PoiFactor (deploy)	14.700	3,376,371	214.91	0.08
Poi (deploy)	10.186	2,164,457	137.77	0.05
UGCToken (deploy)	4.557	1,208,543	76.92	0.03
TourNFT (deploy)	12.990	3,003,630	191.18	0.07
proposePoi()	–	362,549	23.08	0.0088
approvePoi()	–	565,900	36.02	0.0138
addReview()	–	401,163	25.53	0.0097

for the Polygon network 1 Gwei corresponds to 10^{-9} MATIC. Moreover, for the Ethereum network, the cost of a unit of Gas has been estimated as about 25 Gwei, while in the Polygon network, it is equal to 135 Gwei. As you can notice, despite the amount of Gas needed to complete the considered operations, the actual costs in USD depend on the specific networks. Polygon is traditionally the most economical, where the transaction costs typically require a few cents for completion. Moreover, since Polygon is a Layer 2 solution for Ethereum, it can consistently decrease the time needed to execute transactions and achieve their confirmation. Indeed, while the transaction speed of Ethereum takes between 15 s to 5 min, the Polygon network takes about 2–3 seconds on average.

The developed decentralized application (dApp) is composed of several sections, some of which are visible to anyone, like the list of available PoIs, while others are accessible only by the administrator, like the one for the approval or rejection of new PoI proposals.

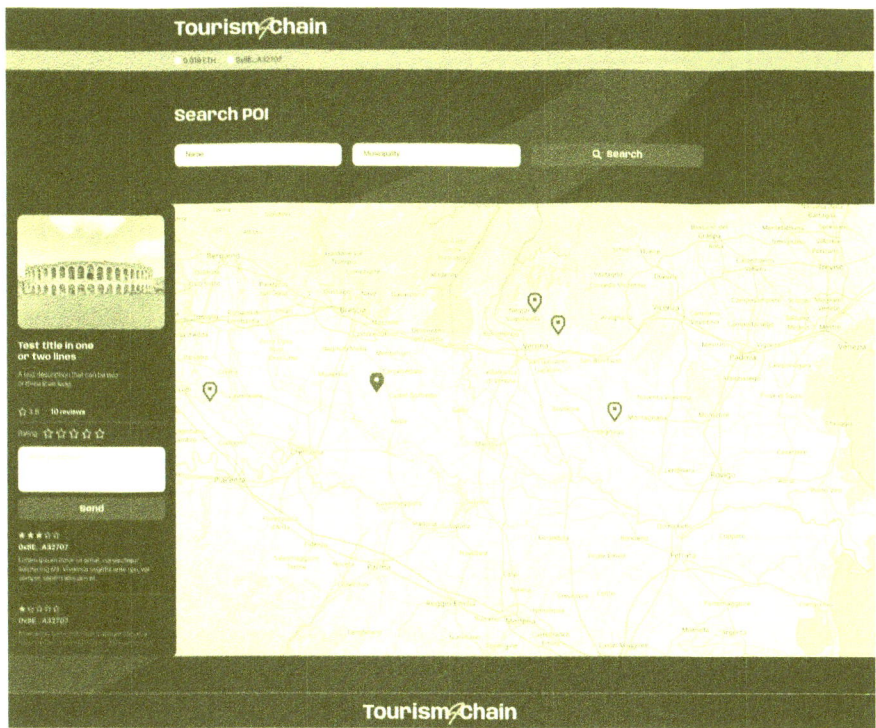

Fig. 4. Visualization in a map of the available PoIs and functionality for adding a new review for the selected PoI.

Each tourist can access essentially four functionalities: viewing the set of available PoIs, submitting a review for a POI, proposing a new PoI, and acquiring an NFT from the Marketplace. As regards the first operation, the set of available PoIs can be obtained through a list or table with filtering capabilities based on the PoI name, or through a map which locates the PoIs within the Earth's surface and allows one to perform spatial filtering, as illustrated in Fig. 4. By selecting an existing PoI, users have the possibility to check its details, read the reviews already submitted for it, or submit a new review. This last operation is reported in the left part of Fig. 4. Conversely, Fig. 5 shows the operation of submitting a proposal for a new PoI, while Fig. 6 illustrates the marketplace or NFT store used by tourists to choose and buy an NFT through the owned reward tokens.

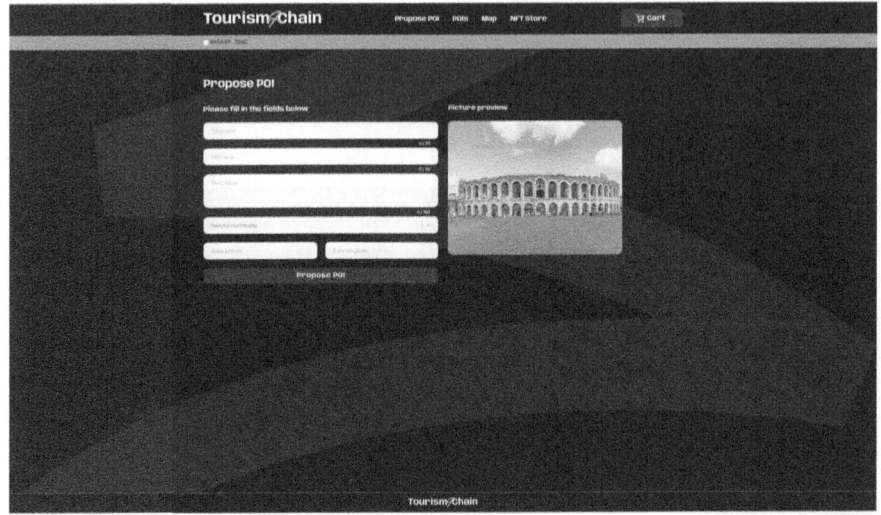

Fig. 5. Submission of a new PoI proposal.

As regards the administrator functionalities, the operator can see the list of proposed PoIs with the corresponding submitted information, and decide to approve or reject each of them. As illustrated in Fig. 7, these two operations can be performed by simply using one of the corresponding buttons in the interface.

Finally, the last available operation is the loading of a new NFT in the marketplace, which is done as in Fig. 8 where the upload and storing of the NFT picture representation is done by exploiting the Pinata platform.

4   Related Work

In the literature, the usage of blockchain technology in conjunction with social media platforms has been progressively proposed with the aim to provide a valuable answer to the main demands of having online identities verified and limiting falsified content [6]. In cite [7], the authors propose a blockchain-based notarization service for social media that authentically archives the contents on social media using blockchain technology.

Steemit[6] is a social network for content-sharing based on blockchain. In particular, on this platform, users can post different kinds of content, which are

---

[6] https://steemit.com/.

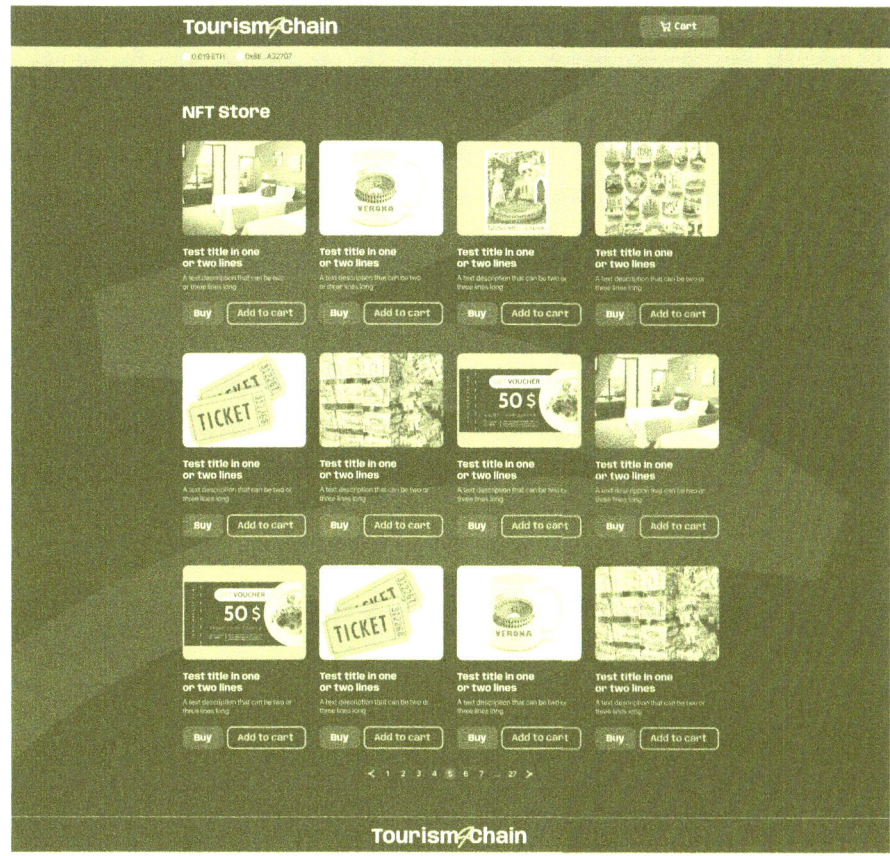

Fig. 6. NFT store or marketplace.

validated and safely stored in a distributed ledger system. Moreover, users can receive a reward for the content they generate, whose amount depends on the appreciation given by other users through a voting mechanism. Similarly, in [2], the authors propose a solution, called Re-Taled, for the problem of incentivizing users' creation of content in the form of customer reviews in the online grocery industry. In this solution, the rewarding mechanism is not based on the voting achieved on a single content, but on a mechanism based on the global reputation and experience levels accumulated by each specific user. Our solution does not rely on a reputation system for determining the reward amount, but this functionality could be included in the proposed solution in a straightforward manner. Conversely, it implements many other important aspects, like the prevention of DoS attacks, the traceability of the provided contents and the upgradeability in case of incorrect information has been provided.

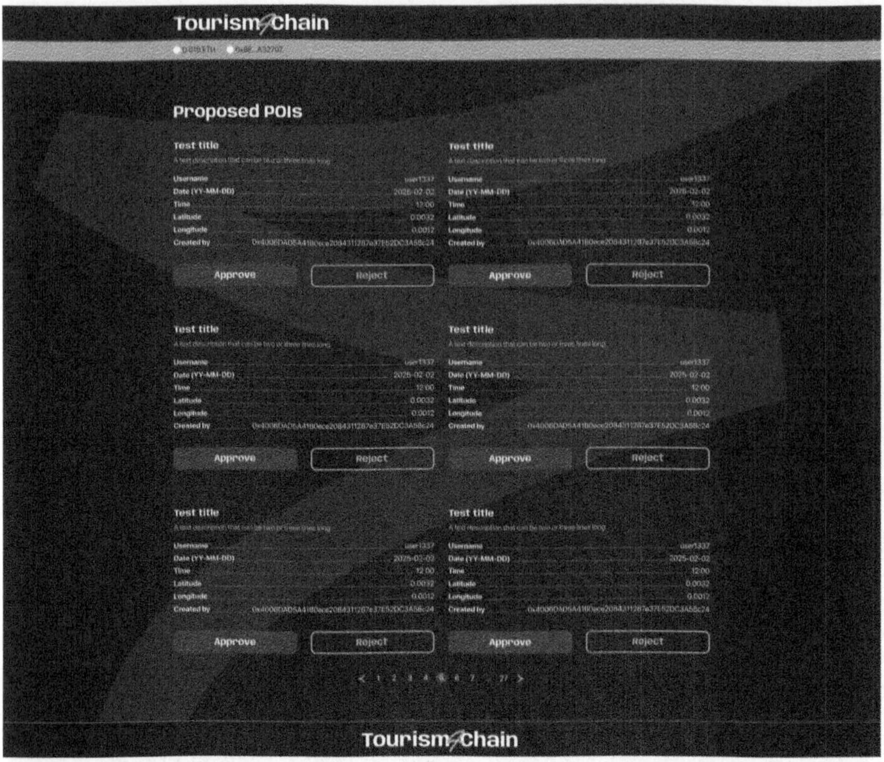

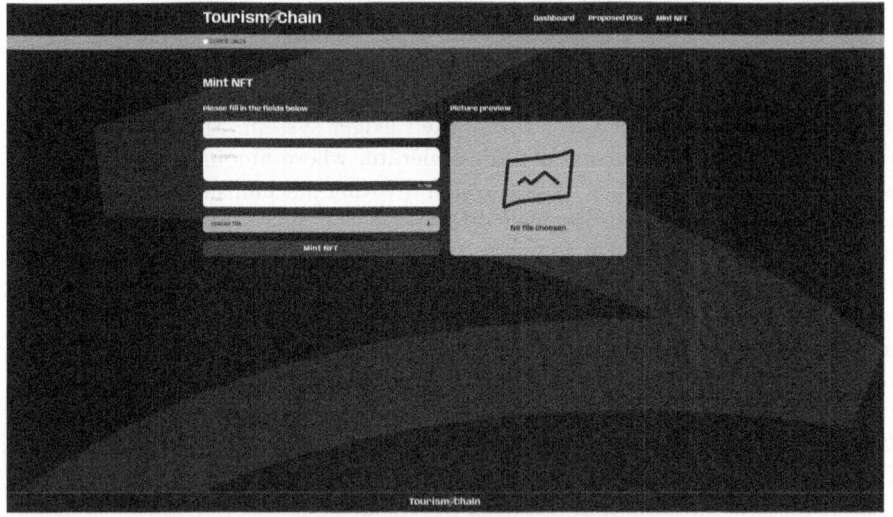

Fig. 7. Administrative page for approving or rejecting the proposals for new PoIs.

Fig. 8. Addition of a new NFT in the marketplace.

Another similar system is the REOS one[7], which uses blockchain technology to reward user content creators and promote the accurate attribution of content ownership. In this case, a dedicated blockchain network has been developed. Conversely, in our solution, like the Re-Taled one, a public blockchain is exploited to increase transparency and trustworthiness.

5 Conclusion

This paper presents an innovative solution for incentivizing the production of UGCs in the tourism domain by using blockchain and smart contract technology combined with a token and NFT reward mechanism. The proposed approach allows tourists to suggest valuable PoIs, provide reviews for existing ones, and obtain a reward in terms of a custom token, which can be finally used to buy NFTs in a connected market. More specifically, the NFT can be created and minted by a local tourist operator with the aim of incentivizing specific behaviors, like the tourists' coming back or the extension of the experience even when the tourist has returned home. The usage of a public blockchain allows for increased traceability, transparency, and trust in the overall system. Aspects like the possibility to rectify the information previously provided and maintain a complete history of the information provided by a single user, represent additional valuable characteristics of the proposed approach.

The proposed solution tries to overcome some forms of attacks, like the DoS attack and the attempt to produce multiple reviews for the same PoI by the same user, and can be extended in a straightforward manner with other mechanisms like a voting and a reputation one, on which the amount of reward could be properly modulated.

Acknowledgments. This study was carried out within the Interconnected Nord-Est Innovation Ecosystem (iNEST) and received funding from the European Union Next-GenerationEU (Piano Nazionale di Ripresa e Resilienza (PNRR) âĂŞ Missione 4 Componente 2, Investimento 1.5 âĂŞ D.D. 1058 23/06/2022, ECS00000043). This manuscript reflects only the authors' views and opinions, neither the European Union nor the European Commission can be considered responsible for them.

References

1. Bahtar, A.Z., Muda, M.: The Impact of User-Generated Content (UGC) on product reviews towards online purchasing – a conceptual framework. Procedia Econ. Finance **37**, 337–342 (2016). https://doi.org/10.1016/S2212-5671(16)30134-4
2. Bruno, T., Etenzi, E., Gualandi, L., Katra, E., Pugliese, R., Taranto, A., Tiezzi, F.: A blockchain-based platform for incentivizing customer reviews in the grocery industry. Blockchain: Res. Appl. **5**(4), 100226 (2024). https://doi.org/10.1016/j.bcra.2024.100226

---

[7] https://reos.me.

3. Gambini, M., Migliorini, S., Belussi, A.: Extract user-generated content from spatial data provision services. In: 2024 IEEE 18th International Conference on Application of Information and Communication Technologies (AICT), pp. 1–6 (2024). https://doi.org/10.1109/AICT61888.2024.10740450

4. Gröger, C.: There is no AI without data. Commun. ACM **64**(11), 98–108 (2021). https://doi.org/10.1145/3448247

5. Migliorini, S., Gambini, M., Belussi, A.: A blockchain-based solution to fake check-ins in location-based social networks. In: Proceedings of the 3rd ACM SIGSPA-TIAL International Workshop on Analytics for Local Events and News, LENS'19. Association for Computing Machinery (2019). https://doi.org/10.1145/3356473.3365191

6. Poongodi, T., Sujatha, R., Sumathi, D., Suresh, P., Balamurugan, B.: Blockchain in social networks, chap. 4, pp. 55–76. John Wiley & Sons, Ltd (2020). https://doi.org/10.1002/9781119621201.ch4

7. Song, G., Kim, S., Hwang, H., Lee, K.: Blockchain-based notarization for social media. In: 2019 IEEE International Conference on Consumer Electronics (ICCE), pp. 1–2 (2019). https://doi.org/10.1109/ICCE.2019.8661978

8. Xu, Y., Chen, Z., Yin, J., Wu, Z., Yao, T.: Learning to recommend with user generated content. In: Web-Age Information Management, pp. 221–232. Springer, Cham (2015)

9. Xu, Y., Yin, J.: Collaborative recommendation with user generated content. Eng. Appl. Artif. Intell. **45**, 281–294 (2015). https://doi.org/10.1016/j.engappai.2015.07.012

10. Yu, H.: Sybil defenses via social networks: a tutorial and survey. SIGACT News **42**(3), 80–101 (2011). https://doi.org/10.1145/2034575.2034593

Bias Evaluation in Contextual Machine Learning

Anna Dalla Vecchia[1]([⊠]) [iD] and Kostas Stefanidis[2] [iD]

[1] Department of Computer Science, University of Verona, Verona, Italy
anna.dallavecchia@univr.it
[2] Faculty of Information Technology and Communication Sciences,
Tampere University, Tampere, Finland
konstantinos.stefanidis@tuni.fi

Abstract. The integration of contextual information, like time, weather, or location, into machine learning (ML) models has been shown to improve the performance and personalization of the model. However, these additional features may unintentionally introduce biases, leading to disparities across target classes or subgroups. This paper presents a novel methodology to evaluate the bias introduced by contextual features in ML models. We introduce a general metric, called Contextual Bias, which quantifies the disparity in class-level performance when each contextual feature is removed. To efficiently assess feature influence without retraining the model from scratch, we leverage DaRE (Data Removal-Enabled) forests, a machine unlearning framework that allows post-hoc removal of contextual features. This approach is applied to a real-world dataset of tourist visits to Points of Interest (PoIs) in Verona, Italy, where contextual factors play a crucial role in shaping user behavior. This work provides a foundation for developing unbiased context-aware systems, highlighting the need to consider not only accuracy but also bias metrics when integrating contextual information into ML models.

Keywords: Bias · Context · Machine Learning

1 Introduction

In recent years, machine learning (ML) models have become essential tools in many real-world applications, playing a crucial role in supporting the analysis and understanding of heterogeneous data sources. Although traditional ML approaches rely on features derived directly from sensor data and user activity, there is a growing interest in the integration of contextual information, such as weather conditions, location, and time, in order to enrich the feature and improve the model's performance [3,6].

The inclusion of contextual features has been shown to enhance the accuracy of the model and enable a more tailored personalization. However, their integration may also influence the learning dynamics of the model without a clear understanding of those modifications. In particular, contextual features may indirectly introduce bias into the model, affecting its behavior across different subgroups or classes that are not explicitly declared as protected. This

P. K. Chrysanthis et al. (Eds.): ADBIS 2025, CCIS 2676, pp. 555–567, 2026.
https://doi.org/10.1007/978-3-032-05727-3_45

poses important concerns about unbalanced or skewed outcomes of the decision model [17,24].

Understanding how contextual features influence model behavior, and under which conditions they may lead to biased outcomes, remains a challenging task. Traditional feature importance techniques, such as feature importance scores in tree-based models (e.g., Random Forests or XGBoost), permutation feature importance, and model-agnostic approaches like SHAP [16] and LIME [21] have become widely adopted to gain insight into model behavior. Although these methods are useful for general interpretability, they are not explicitly designed to detect or quantify the bias introduced by contextual features. For this reason, they often fail to reveal the impact that context can have on fairness-related aspects of a model [10]. This work addresses this gap by proposing a methodology to evaluate the potential bias introduced by contextual features in ML models. The goal is to ensure that context-aware systems are not only accurate and personalized but also fair and reliable when deployed in real-world scenarios.

To this end, we proposed a novel formalization to quantify the bias associated with contextual features in a ML model. This general metric is used to guide an evaluation algorithm that leverages machine unlearning techniques [9] to obtain a model that discards the influence of the contextual feature under analysis. Specifically, we employ the data removal-enabled (DaRE) forest [9] to simulate the removal of individual contextual features after training, without requiring full retraining of the model. By comparing the predictions of the original model with the new one, we are able to investigate the role of the contextual features and how the biases change.

Lastly, we present a real-world dataset of Points of Interest (PoIs) in the city of Verona, where contextual information plays a key role in shaping tourist behaviors [6]. This dataset, which reflects temporal, spatial, and environmental factors, will serve as the basis for our case study and will support future research on context-aware and bias-aware machine learning in the tourism domain.

The remainder of the paper is organized as follows. In Sect. 2, the related work is discussed. Section 3 introduces the formalization of the problem and presents the new evaluation metric, while Sect. 4 outlines the proposed algorithm. Section 5 discusses a preliminary analysis of the dataset. Finally, Sect. 6 summarizes the findings and outlines the future directions.

2   Related Work

In recent years, the discussion about ethical concerns in the development and deployment of intelligent systems is gaining increasing attention [5,7,8,20,22]. Despite this growing interest, there is still no clear consensus on universally accepted ethical guidelines [23]. A key challenge is that ethical principles are often context-dependent, meaning that their interpretation and application can vary depending on the specific domain or use case. The authors in [13] discuss bias from a different perspective: it is not always necessary to completely remove it. The elimination of one form of bias may unintentionally give rise to another.

Instead of enforcing strict, unbiased situations, a possible approach may involve designing systems that allow users to transparently explore and adjust existing biases.

2.1 Bias in Machine Learning

In the machine learning domain, bias is typically understood as a deviation in model behavior that leads to unfair outcomes, especially for underrepresented groups in the dataset [17]. In classification tasks, for example, a biased model may achieve high overall accuracy while underperforming on specific classes or demographic groups. [10,24] The state of the art on ethical issues related to machine learning is explored in [11], where various approaches aimed at enhancing fairness in machine learning are outlined. The authors conclude by presenting five open dilemmas for the research. This highlights that despite years of exploration of existing methodologies designed to mitigate potential ethical biases and inequities, it remains challenging to define the concept of fairness in a consistent manner [15,19]. Recent studies have also investigated the relationship between human cognitive biases and machine learning models, highlighting how biases present in training data, originating from human behaviors and societal patterns, can be learned and amplified by machine learning systems. For instance, the study in [4] introduces a method to detect bias by swapping sensitive attributes, measuring the impact on predictions. Their results reveal biases, including lower wage predictions for females and Asians, underscoring the importance of bias detection for developing fairer machine learning models.

2.2 Context in Machine Learning

The integration of contextual information, such as time, location, weather, or user situation, has become increasingly common in machine learning, particularly in domains like recommendation systems, smart mobility, and personalized services. Contextual features enrich the input space and allow models to adapt their predictions to external conditions, often beneficial for increasing prediction accuracy, improving personalization, and enabling better interpretability [3]. For instance, in context-aware recommender systems, contextual factors allow the model to learn the user preferences that depend on the situation they are acting in [2]. Similarly, in spatiotemporal forecasting or mobile health, context is essential to model the variability of human behavior or system responses across time and space [1,12]. In [26], the authors provide a systematic literature review describing how context is incorporated at various stages of recommender system development, identifying the most commonly used contextual features across different application domains, and discussing evaluation mechanisms in terms of datasets, metrics, and validation protocols. More recent work has investigated the synergies between context and temporal aspects, such as data aging, to enhance recommendations in dynamic environments like wearable devices and smart TVs [12]. These approaches emphasize the need for models to adapt not only to static contextual variables but also to evolving user behavior and environmental conditions. In a similar direction, the authors in [18] study how contextual models can be leveraged to forecast Point of Interest occupation and users' preferences in the tourism domain, proposing an architecture that leverages both historical and contextual data.

To the best of our knowledge, the work most closely related to ours is presented in [25], where the authors present a system that explains fairness violations in machine learning models. This system identifies training data subsets that most contribute to fairness violations by estimating how model fairness changes when those subsets are removed. However, it primarily focuses on a single protected feature, whereas our approach considers the effects of multiple contextual factors on model behavior. Building upon these foundations, our work explores how contextual factors may influence the behavior of machine learning models, with a particular focus on their potential impact on model bias. This broader perspective allows us to better understand the complex role of context in model behavior and its potential implications in real-world applications.

3   Bias in Contextual Machine Learning

Concerns about fairness and bias have emerged as critical aspects since machine learning systems increasingly incorporate contextual information to enhance performance and personalization [3,27]. Typically, the bias concept in machine learning refers to systematic deviations in model behavior that disadvantage certain groups or data subsets. This section first presents the traditional contextual machine learning models, and then the potential biases emerging from those models when contextual data is introduced.

3.1   Context Definition in Machine Learning

Nowadays, we are flooded by data generated from different sources, like sensors, wearable devices, or other types of platforms. As a result, tools that help users understand data values and trends are becoming an invaluable assistant both for users and service providers. In many real-world applications, supplementary information, such as the geographical location, weather conditions and holidays, plays a crucial role in improving the performance and the interpretability of such tools. In this work, we refer to this kind of additional information as contextual features, which encompass any external or auxiliary data that can describe the circumstances under which user interactions or events occur.

To formally capture the role of contextual features in a machine learning setting, we define the model input as a combination of both intrinsic and contextual data. Let $\mathcal{M}$ be a general machine learning model, $X \in \mathbb{R}^n$ the original input feature vector, and $Y \in \mathbb{R}$ the target variable. Each element $x \in X$ represents an intrinsic attribute of the instance. Let $C \in \mathbb{R}^m$ be the set of contextual features, where each $c \in C$ encodes external information related to the instance, such as temporal or environmental context, i.e. contextual features. The complete input to the model is then defined as the augmented feature vector:

$$\hat{X} = [X \parallel C] \in \mathbb{R}^{n+m}$$

where $\parallel$ denotes the concatenation of the original features and the contextual features. The model is trained to learn a function:

$$\mathcal{M} : \mathbb{R}^{n+m} \rightarrow \mathbb{R}$$

mapping the enriched input $\hat{X}$ to the predicted output Y.

3.2   Bias Definition in Contextual Machine Learning

In recent years, the concept of bias and fairness has emerged as a topic to pay attention to in different areas of computer science. Also, machine learning is becoming a domain in which the implications of these concepts have to be explored [17]. In general, bias in machine learning can be seen as a deviation during the learning process in which some aspects of the datasets are considered more important, resulting in unfair, skewed, or wrong outcomes of the model [14,17]. These problems, in most cases, originate from an unbalanced dataset. For instance, if a dataset is biased towards a particular class, the resulting model will excel for that class, while it may underperform for the less represented group. However, analyzing a well-processed dataset can reveal how each feature influences the outcome and whether it is still biased. Specifically, when contextual features are introduced, they can reveal an unexpected bias due to the provided additional information.

In this work, we investigate the possible presence of bias by studying how contextual features influence the learning process and whether they introduce disparities in model performance across different target classes. Rather than treating context as a neutral input, we propose a novel perspective in which context is analyzed as a potential source of bias that may affect some prediction outcomes. More in detail, we investigate the precision of the model in a classification task, analyzing when there is a significant difference between the model trained with and without each contextual feature. We define contextual bias in a classification model as the disparity in predictive performance across classes of different models. This bias can be quantified by measuring the variation in precision

Let $\mathcal{M}$ be a classification model trained to predict class labels $Y \in \{1, \ldots, K\}$ from input instances $\hat{X} = [X \parallel C]$, where $X \in \mathbb{R}^n$ represents the original feature and $C \in \mathbb{R}^m$ the contextual features. Let $\mathcal{M}_{-c}$ be the model trained without the feature $c \in C$. We define the *Contextual Bias* introduced by the feature c, denoted as $CB(c)$, as the disparity in class-level predictive performance between $\mathcal{M}$ and $\mathcal{M}_{-c}$. Using the precision evaluation metric, the contextual bias is measured as the standard deviation of the change in per-class precision:

$$CB(c) = \sqrt{\frac{1}{K} \sum_{t=1}^{K} \left(\Delta Prec_t(c) - \overline{\Delta Prec(c)} \right)^2}$$

where:

- $\Delta Prec_t(c) = Prec_t(\mathcal{M}) - Prec_t(\mathcal{M}_{-c})$
- $\overline{\Delta Prec(c)} = \frac{1}{K} \sum_{t=1}^{K} \Delta Prec_t(c)$
- $Prec_t(\cdot)$ denotes the precision on class t under the given model.

A high value of $CB(c)$ indicates that the contextual feature c has a dispro-portionate effect on the model's performance across different classes, suggesting it may be a source of contextual bias. This formulation is based on the per-class precision in a multi-class setting, as:

$$Prec_t(\cdot) = \frac{TP_t}{TP_t + FP_t}$$

where TP_t and FP_t denote the number of true positive and false positive pre-dictions for class t. However, the contextual bias measure $CB(c)$ is orthogonal and can be generalized to any class-level evaluation function ϕ, such as recall, F1-score, or calibration error, depending on the desired fairness criteria.

4   Contextual Bias Evaluation

Evaluating the influence of individual features in a trained model typically involves retraining the model multiple times, each time excluding a single fea-ture. Let $\hat{X} = [X \parallel C]$ be the full set of input features, composed of input features $x \in \mathbb{R}^n$ and contextual information $C \in \mathbb{R}^m$. The traditional app-roach trains m separate models $\mathcal{M}_{-c}$, where $c \in C$ and each model is trained on $\hat{X} \setminus \{-c\}$ While this method is straightforward and widely used, it becomes computationally expensive when applied to a model, especially when the number of contextual features is large.

To address this limitation, in this study, we adopt a more efficient alternative based on machine unlearning introduced in [9], which enables efficient removal of training data from Random Forest models through a specialized structure called DaRE (Data Removal-Enabled) forests. DaRE forests support exact and efficient unlearning by modifying only the necessary subtrees when data is removed, avoiding the need for full retraining. This is achieved by strategically introducing random nodes in the upper levels of the trees and caching sufficient statistics at decision and leaf nodes. In other words, retraining the model $\mathcal{M}$ is not required each time feature c needs to be excluded. We refer to the resulting model, in which c has been removed, as $\mathcal{M}_{-c}^{DaRE}$.

Specifically, we apply a feature-level unlearning procedure that allows us to estimate the effect of removing a specific feature from the trained model, without requiring full retraining. This enables a more scalable analysis of feature influ-ence, particularly in scenarios involving high-dimensional or context-rich data, where the traditional approaches based on retraining would be computationally expensive.

To analyze the potential bias introduced by each feature, we adopt an iter-ative approach based on DaRE presented in Algorithm 1. Particularly, the

Algorithm 1.

Input: Dataset $\mathcal{D}$, contextual features C
Output: DaRE Model set $\mathcal{M}_{all}$, contextual bias score set S
1: $\mathcal{M}_{all} \leftarrow \{\ \}$
2: $S \leftarrow \{\ \}$
3: $\mathcal{M} \leftarrow trained(\mathcal{D})$
4: **for** each $c \in C$ **do**
5:     $\mathcal{M}_{all} \leftarrow \mathcal{M}_{all} \cup \mathcal{M}_{-c}^{DaRE}$
6:     $S \leftarrow S \cup eval(\mathcal{M}, \mathcal{M}_{-c}^{DaRE}, \mathcal{D})$
7: **end for**
8: **return** $\mathcal{M}_{all}, S$

unlearning procedure is applied iteratively, removing one feature at a time until all features have been exhausted. For each resulting model, we evaluate its performance on the test set, saving the previously defined contextual bias measure (CB) computed by the *eval* function. By observing the results collected in S, it is possible to identify which contextual feature introduces the most bias by analyzing its corresponding CB value. A value close to zero indicates that the feature has a uniform impact across all classes. Conversely, a value significantly greater than zero suggests that the feature affects the model differently across classes, thus introducing contextual bias. The study focuses not only on the overall accuracy of the model, but also on the outcomes obtained for each target class, in order to capture possible shifts in predictive performance across different classes. Thus, we aim to identify whether the feature affects the prediction quality for specific subsets of the data, which may indicate the presence of feature-specific or context-induced bias.

4.1   Complexity Analysis

As demonstrated by the authors in [9], the complexity of training a DaRE model is equivalent to that of training a standard Random Forest, i.e.,

$$\mathcal{O}(T \cdot n \cdot \tilde{p} \cdot k \cdot d_{\max})$$

where T is the number of trees in the forest, n is the number of training instances, $\tilde{p}$ is the number of attributes randomly selected at each node, k is the number of candidate thresholds per attribute, and $d_{\max}$ is the maximum tree depth. The time complexity of deleting a single instance from a DaRE tree, when no structural modification is required and all attribute thresholds remain valid, is:

$$\mathcal{O}(\tilde{p} \cdot k \cdot d_{\max})$$

However, if a node's attribute thresholds become invalid, an additional cost of $\mathcal{O}(|D| \log |D|)$ is incurred to select a new valid threshold, where $|D|$ is the number of data points reaching that node. Furthermore, if the node at depth d with $|D|$ instances requires retraining, the cost of rebuilding the affected subtree is:

$$\mathcal{O}(\tilde{p} \cdot |D| \cdot (d_{\max} - d))$$

The overall time complexity of Algorithm 1 can be decomposed into three components. The first corresponds to the initial training of the DaRE model in line 3, which takes $\mathcal{O}(T \cdot n \cdot \tilde{p} \cdot k \cdot d_{\max})$. Next, for each contextual feature $c \in C$, the algorithm performs machine unlearning to obtain the model $\mathcal{M}_{-c}^{DaRE}$ at line 5, whose complexity depends on the specific operations required as described above. Finally, the evaluation step of line 6 adds an additional cost of $\mathcal{O}(n)$ per feature. Consequently, the total complexity is the sum of these costs over all contextual features.

5   Early Findings on PoIs Case Study

In this work, we apply the proposed methodology to a case study focused on Points of Interest (PoIs) in the tourism domain. Understanding and incorporating contextual information is crucial in this setting, as tourists' behaviors and visit patterns are strongly influenced by external factors such as time, weather, or events. To validate the importance of context in this domain, we build upon previous work presented in [6], where the authors investigate how contextual factors influence the occupancy of PoIs, focusing on the city of Verona, Italy.

More in detail, the study introduces the notion of a *Touristic Visit* representing the set of input feature vectors X with the corresponding *Visit Context*, which defines the tuple of contextual features C defined in this work. The objective of the study was to forecast the occupation of each PoI at different timeslots of the day, then support the development of a crowd-aware recommendation system capable of suggesting to tourists the less crowded attractions.

The problem was modeled using both traditional machine learning (Random Forest) and deep learning approaches. These models were evaluated in two settings: using raw historical data only, and using data enriched with contextual features. The results obtained shows that including context significantly improves accuracy. Furthermore, models trained with contextual data demonstrated better generalization capabilities when applied to PoIs with limited historical information. That study demonstrates how integrating contextual features improves the forecast of the PoI crowding, benefiting both tourists, who can better plan their visits, and city tourist service providers, who can manage urban crowding more efficiently.

The role of context in the tourism domain is further investigated in this work from a different perspective. While previous studies, including [3,6], have demonstrated the benefits of incorporating contextual information to improve recommendations, the current investigation focuses on understanding when such contextual features may act as sources of bias in the model. The aim is to ensure that context-aware models are not only more accurate but also fair and reliable when applied to real-world tourism scenarios.

5.1   Dataset

To validate our approach, this work focuses on the use case presented in [6] regarding the touristic scenario. More in detail, the real-world dataset tracks touristic activity in the city of Verona in Italy. The dataset spans nine years from 2014 to 2022 and includes approximately 4.4 million visit entries, involving about 1 million unique tourists and seventeen Points of Interest (PoIs). Each entry consists of a timestamped visit linked to an anonymized user ID and a PoI identifier. To enhance the dataset, we integrated contextual variables, including meteorological conditions and calendar-based indicators such as public holidays. Specifically, the augmented feature vector $\vec{X}$ is represented by the concatenation of the input features

$$X = \{u, p, t, lat, long\}$$

where u is the user identifier, p is the PoI identifier, t is the timestamp of the visit, and lat and $long$ are the geographical coordinates of the PoI, and of the contextual features

$$C = \{ts, day, month, year, doy, dow, week, rain, temp, festive, hol\}$$

where ts is the timeslot of the performed visit (e.g. morning, noon, afternoon), $day, month, year$ are the day of the month, the month, the year respectively, doy is the day of the year, dow and $week$ are the day of the week and the week number in a year, $rain$ is a string for the weather condition (e.g., rainy, sunny), $temp$ is the temperature, $festive$ and hol are two boolean indicators for the day is a weekend or a public holiday.

5.2   The Role of the Context

Firstly, it is important to highlight the relationship between the crowding levels at PoIs and the contextual factors added to the original dataset.

As expected, the weather conditions strongly influence the tourists' affluence in different PoI, and the tourist visit patterns. Figure 1 illustrates that the number of visits to certain PoIs on the same calendar day in two different years with different weather conditions can notably vary. Outdoor locations, such as POI 49, show a clear drop in attendance on rainy days. Conversely, indoor venues such as PoIs 62 and 63 may even benefit from adverse weather, attracting more visitors who will prefer indoor activities when outdoor conditions are unfavorable.

Figure 2 presents another illustrative case, showing the number of visits to Juliet's House in February. This PoI, recognized as the symbol of love, attracts more visitors around Valentine's Day. Looking at the plot, the red bars represent the number of visits on February 14th across different years, while the blue bars represent the average number of visits for the same weekday throughout the month. There is a notable spike on Valentine's Day compared to the baseline, highlighting how special events can significantly influence tourist behavior, leading to atypical crowding on typically less congested days.

5.3   The per-Class Evaluation

The Fig. 3 illustrates the class-wise evaluation of Precision, Recall, and F1-score for the Random Forest classifier applied to our dataset enriched with contextual

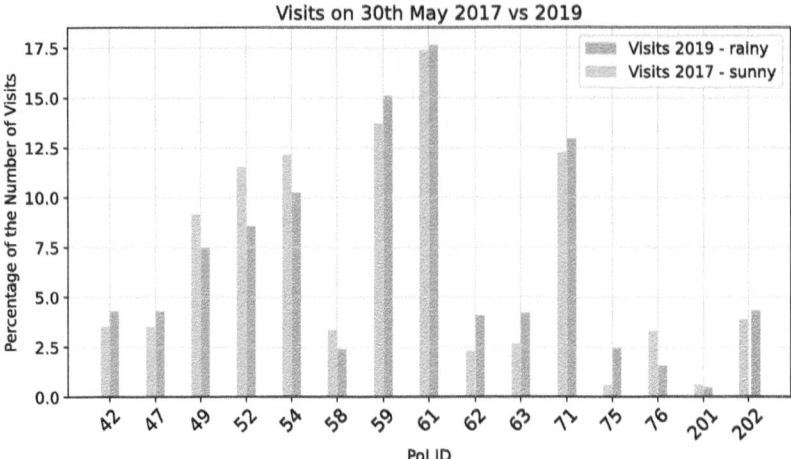

Fig. 1. Number of visits on the 30th of May of two years with different weather conditions.

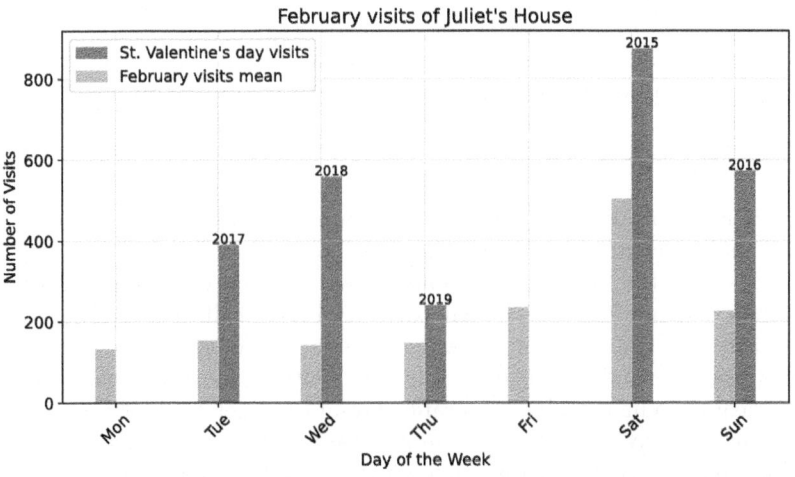

Fig. 2. February weekly visits. (Color figure online)

factors. We trained a Random Forest classifier where occupancy is discretized into three classes (i.e., Low, Medium, High) based on percentile thresholds computed for each PoI. Each subplot represents one metric, with classes denoted on the x-axis and the respective metric on the y-axis. It is clear that the class *Medium* exhibits lower scores across all three metrics compared to the other classes. This indicates that the model systematically penalizes instances from class *Medium*, highlighting a potential bias or imbalance in model performance.

Consequently, a deeper analysis is required to discern whether the observed underperformance of *Medium* is driven by contextual features that influence the

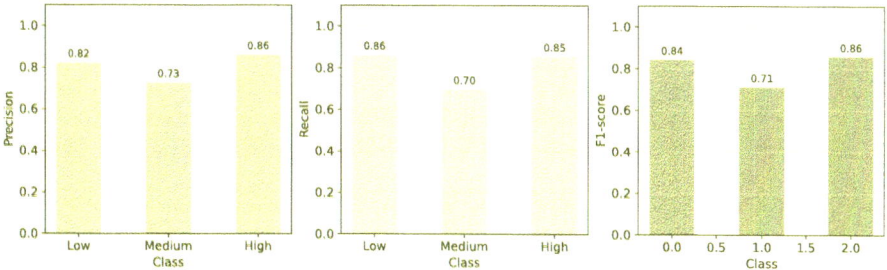

Fig. 3. Precision, Recall, and F1-score of the Random Forest Classifier.

model's predictions or whether it reflects input features. Such an investigation is essential to properly interpret these results and to guide the training in order to obtain an unbiased model.

6   Conclusion and Future Work

This work presents a preliminary investigation into the role of contextual features as potential sources of bias in machine learning models. While the use of contextual information to enhance model performance is widely discussed in the literature, its interactions with class labels and other input features remain insufficiently explored. To address this challenge, we present a general metric to quantify the bias introduced by contextual features and propose a novel methodology based on machine unlearning. Our approach evaluates new models without the contextual feature being analyzed, eliminating the need for full model retraining, thanks to the DaRE forest approach. Additionally, we presented a real-world dataset of tourist visits in the city of Verona, where contextual information plays a crucial role in user behavior and recommendation dynamics. This dataset will serve as a starting point for future studies on context-aware, unbiased, and explainable machine learning models. Beyond its methodological contribution, this study lays the foundation for an alternative explanation of feature importance, based not only on model accuracy but also on its potential to introduce biases across classes.

Future work will focus on conducting a more comprehensive empirical evaluation of the proposed methodology using the presented dataset. Additionally, we plan to investigate its applicability across other domains to further assess the generalizability and orthogonality of the contextual bias problem discussed. Another direction involves integrating our approach with traditional feature importance techniques, such as feature importance scores in tree-based models, permutation-based feature importance, and other model-agnostic approaches.

References

1. Abowd, G.D., Dey, A.K., Brown, P.J., Davies, N., Smith, M., Steggles, P.: Towards a better understanding of context and context-awareness. In: Proceedings of Handheld and Ubiquitous Computing HUC'99. LNCS, vol. 1707, pp. 304–307. Springer (1999)
2. Adomavicius, G., Sankaranarayanan, R., Sen, S., Tuzhilin, A.: Incorporating contextual information in recommender systems using a multidimensional approach **23**(1) (2005)
3. Adomavicius, G., Tuzhilin, A.: Context-aware recommender systems. In: Recommender Systems Handbook, pp. 191–226 (2015)
4. Alelyani, S.: Detection and evaluation of machine learning bias. Appl. Sci. **11**(14), 6271 (2021)
5. Araújo, T.B., Efthymiou, V., Christophides, V., Pitoura, E., Stefanidis, K.: TREATS: fairness-aware entity resolution over streaming data. Inf. Syst. **129**, 102506 (2025)
6. Belussi, A., Cinelli, A., Dalla Vecchia, A., Migliorini, S., Quaresmini, M., Quintarelli, E.: Forecasting POI occupation with contextual machine learning. In: Advances in Databases and Information Systems, pp. 361–376. Springer (2022)
7. Borges, R., Sahlgren, O., Koivunen, S., Stefanidis, K., Olsson, T., Laitinen, A.: Multi-objective fairness in team assembly. In: New Trends in Database and Information Systems - ADBIS 2023 Short Papers, Doctoral Consortium and Workshops: AIDMA, DOING, K-Gals, MADEISD, PeRS, Barcelona, Spain, September 4-7, 2023, Proceedings. Communications in Computer and Information Science, vol. 1850, pp. 106–116 (2023)
8. Borges, R., Stefanidis, K.: Feature-blind fairness in collaborative filtering recommender systems. Knowl. Inf. Syst. **64**(4), 943–962 (2022). https://doi.org/10.1007/s10115-022-01656-x
9. Brophy, J., Lowd, D.: Machine unlearning for random forests. In: International Conference on Machine Learning, pp. 1092–1104. PMLR (2021)
10. Buolamwini, J., Gebru, T.: Gender shades: Intersectional accuracy disparities in commercial gender classification. In: Proceedings of the 1st Conference on Fairness, Accountability and Transparency. Proceedings of Machine Learning Research, vol. 81, pp. 77–91. PMLR (2018)
11. Caton, S., Haas, C.: Fairness in machine learning: a survey. ACM Comput. Surv. **56**(7) (2024)
12. Dalla Vecchia, A., Marastoni, N., Oliboni, B., Quintarelli, E.: The synergies of context and data aging in recommendations. In: Big Data Analytics and Knowledge Discovery, pp. 80–87. Springer (2023)
13. Demartini, G., Roitero, K., Mizzaro, S.: Data bias management. Commun. ACM **67**(1), 28–32 (2023)
14. Friedman, B., Nissenbaum, H.: Bias in computer systems. ACM Trans. Inf. Syst. **14**(3), 330–347 (1996)
15. Hutchinson, B., Mitchell, M.: 50 years of test (un)fairness: Lessons for machine learning. In: Proceedings of the Conference on Fairness, Accountability, and Transparency, pp. 49–58. FAT* '19, Association for Computing Machinery (2019)
16. Lundberg, S.M., Lee, S.I.: A unified approach to interpreting model predictions. In: Proceedings of the 31st International Conference on Neural Information Processing Systems, NIPS'17, pp. 4768–4777. Curran Associates Inc. (2017)

17. Mehrabi, N., Morstatter, F., Saxena, N., Lerman, K., Galstyan, A.: A survey on bias and fairness in machine learning. ACM Comput. Surv. **54**(6) (2021)
18. Migliorini, S., Dalla Vecchia, A., Belussi, A., Quintarelli, E.: ARTEMIS: a context-aware recommendation system with crowding forecaster for the touristic domain. Information Systems Frontiers, pp. 1–27 (2024)
19. Narayanan, A.: Translation tutorial: 21 fairness definitions and their politics. In: Proc. Conf. Fairness Accountability Transp, vol. 1170, p. 3 (2018)
20. Pitoura, E., Stefanidis, K., Koutrika, G.: Fairness in rankings and recommendations: an overview. VLDB J. **31**(3), 431–458 (2022)
21. Ribeiro, M.T., Singh, S., Guestrin, C.: "Why Should I Trust You?": explaining the Predictions of Any Classifier. In: Proceedings of the 22nd ACM SIGKDD International Conference on Knowledge Discovery and Data Mining, KDD '16, pp. 1135–1144. Association for Computing Machinery (2016)
22. Sacharidis, D., Giannopoulos, G., Papastefanatos, G., Stefanidis, K.: Auditing for spatial fairness. In: Proceedings 26th International Conference on Extending Database Technology, EDBT 2023, Ioannina, Greece, March 28-31, 2023, pp. 485–491 (2023)
23. Steen, M.: Ethics as a participatory and iterative process. Commun. ACM **66**(5), 27–29 (2023)
24. Suresh, H., Guttag, J.: A framework for understanding sources of harm throughout the machine learning life cycle. In: Proceedings of the 1st ACM Conference on Equity and Access in Algorithms, Mechanisms, and Optimization. EAAMO '21. Association for Computing Machinery (2021)
25. Surve, T., Pradhan, R.: Explaining fairness violations using machine unlearning (2025)
26. Villegas, N.M., Sánchez, C., Díaz-Cely, J., Tamura, G.: Characterizing context-aware recommender systems: a systematic literature review. Knowl.-Based Syst. **140**, 173–200 (2018)
27. Zhao, Y., Wang, Y., Liu, Y., Cheng, X., Aggarwal, C.C., Derr, T.: Fairness and diversity in recommender systems: a survey. ACM Trans. Intell. Syst. Technol. **16**(1) (2025)

RideLink: Enhancing Route Quality for Urban Multimodal Mobility

Leon Schardin[1]([⊠]), Ulrike Steffens[2], and Annett Ungethüm[1]

[1] Department of Informatics, University of Hamburg, Hamburg, Germany
leon.schardin@studium.uni-hamburg.de, annett.ungethuem@uni-hamburg.de
[2] Department of Computer Science, Hamburg University of Applied Sciences,
Hamburg, Germany
ulrike.steffens@haw-hamburg.de

Abstract. Navigation in unfamiliar cities is conveniently done using an appropriate mobile app. Such apps must account for different methods of transportation, especially when the use of a car is not a viable option, and public transportation in combination with bicycling or a shared e-scooter is used instead. Existing multimodal route planners developed for this task treat response times as first-class citizens. As a consequence, they often fall short in terms of result quality by potentially missing optimal routes due to restrictive assumptions or insufficient exploration of the numerous nearby public transport access points. This paper presents RideLink, a database-centric routing system that deliberately reverses the usual priorities of these optimization goals such that route quality, i.e. the compound travel time, becomes the primary objective. It evaluates all stops reachable within an empirically chosen cycling/scooter radius (5 km) before pruning alternative routes, thereby discovering near-optimal multimodal journeys, which are often multiple minutes faster than those of other widely used approaches, while still responding within a few seconds on commodity hardware. A fully deployed system demonstrates real-world viability and underpins the contribution to smart urban tourism.

Keywords: Tourist Mobility · Comprehensive Routing · Multimodal Routing · Public Transportation · Bicycle · E-Scooter · Smart Tourism · RideLink

1 Introduction

The use of mobile apps to navigate cities has become ubiquitous, especially among tourists who are unfamiliar with the local traffic and methods of transportation, or locals who infrequently travel to different destinations. Moreover, there are multiple reasons why using a (rental) car in an inner city is not always advantageous compared to other modes of transportation, e.g., dense traffic, parking fees, and time-consuming cruising for parking. An example where this is especially prevalent is the city of Hamburg in Germany. With almost 1.78 million

© The Author(s), under exclusive license to Springer Nature Switzerland AG 2026
P. K. Chrysanthis et al. (Eds.): ADBIS 2025, CCIS 2676, pp. 568–582, 2026.
https://doi.org/10.1007/978-3-032-05727-3_46

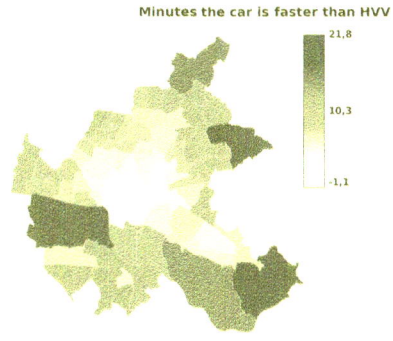

Minutes the car is faster than HVV

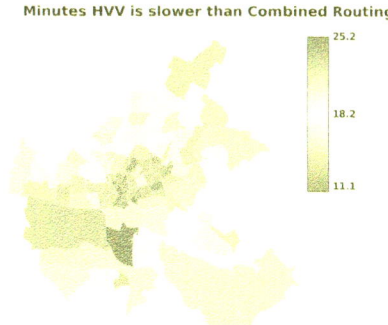

Minutes HVV is slower than Combined Routing

a) Travel time advantage of cars over public transportation (HVV) in Hamburg

b) Travel time disadvantage of public transportation in Hamburg over combined routing (bicycle and public transportation)

Fig. 1. Choropleth map of the city of Hamburg divided into zip code areas, illustrating a) the average travel time difference between using a car and public transportation for any destination from a starting point within the respective area, and b) the travel time disadvantage of public transportation over combined routing using public transportation and a bicycle or e-scooter. Two things can be observed: First, a car does not lead to faster travel times than public transportation in the city center. Second, combined routing is always faster than using public transportation only.

inhabitants[1] it is the second largest city in Germany. Additionally, it hosts 7.54 million overnight stays per year and 100 million day visitors per year and the trend is accelerating[2]. Figure 1a shows a map of Hamburg divided into zip code areas. The colors encode the average time advantage of taking a car over public transportation (PT) with a starting point within the respective area. The dataset to produce this graph consists of 20,000 routes at randomly chosen day times, the start and end points were sampled proportionally to the inhabitant count in each zip code area. While a starting point in the outer areas benefits from using a car, traveling from within the inner city is faster using public transportation. This is one of the reasons why the passenger numbers in public transportation are increasing. In 2024, a new record of almost 1.1 billion passengers was reached by the HVV, the local traffic enterprise.[3] However, getting from the doorstep to a start station and from an end station to the final destination, takes time and can be done in various ways. Besides walking, bicycles are a popular method of transportation for this so-called *first mile* or *last mile*. In 2024, "Stadtrad", the local bicycle sharing company registered an increase of 11%

---

[1] https://www.hamburg.com/visitors/hamburg-by/numbers-22762, accessed 19 jun 2025.

[2] https://hamburg-business.com/en/news/hamburg-tourismus-presents-good-first-half-year-figures, accessed 19 jun 2025.

[3] https://www.hvv.de/de/ueber-uns/der-hvv/zahlen-daten-fakten/nachfrageentwicklung, accessed 19 jun 2025.

in the number of rides[4] amounting to 1.65 million trips per year. This does not include smaller bicycle rental businesses and privately owned bicycles. Shared e-scooters are another popular mode of transportation with 12 million trips in 2023 alone[5]. Thus, using public transportation often leads to multimodal transportation which an app for navigation must account for. Figure 1b shows the potential difference between optimal multimodal routing, which may involve cycling to a farther but better-served stop, and public transportation only, i.e., accounting only for the way to the closest stations.

However, the technological hurdles in delivering truly effective multimodal route planning are twofold. On the one side, the task is computationally complex. While the number of potential start and end stops is limited when considering a walking distance, this range increases when the first or last mile can be covered with a faster method of transportation which is easily available, e.g., a bicycle or an e-scooter. This can lead to unacceptably long response times. Second, existing route planners, e.g., GTFS-Router [9], which overcome the performance challenge tend to struggle with the route quality, i.e., how optimal a route really is. Other systems, e.g., the DB Navigator[6], may overly restrict the search for suitable public transport access points and prefer stops near the user's true origin or destination. By considering only a very limited set of nearby stops or using rudimentary heuristics, they risk missing genuinely better or more convenient multimodal options that a more thorough search might uncover.

In this work, we present an alternative approach for multimodal routing which treats route quality as the primary optimization goal and performance as secondary optimization goal. We reach this by pruning routes as late as possible to not omit alternatives which only become more ideal in a later step, e.g., when accounting for the *last mile* problem. For real-world schedules of public transport, we rely on the data provided by HVV in the open data standard GTFS [8]. This way, we find more optimal routes than HVV's own solution (Geofox)[7], DB Navigator, and even Google Maps[8] which uses the well-established Transfer Patterns approach [1,2]. The remainder of this paper is organized as follows:

- In Sect. 2, we describe our routing system which prioritizes the result quality.
- In Sect. 3, we describe the performance bottlenecks of our routing system and how we address them.
- The practical application as a mobile app is shortly introduced in Sect. 4.
- We evaluate our approach in Sect. 5 with the city of Hamburg as our running example.
- We conclude this paper with a selection of related work in Sect. 6 and a conclusion in Sect. 7.

---

[4] https://www.ndr.de/nachrichten/hamburg/Hamburger-Stadtraeder-werden-deutlich-haeufiger-genutzt,stadtrad254.html, accessed 19 jun 2025.

[5] https://www.statista.com/statistics/1399212/shared-e-scooter-ridership-leading-cities-europe/, accessed 19 jun 2025.

[6] https://bahn.de, accessed 19 jun 2025.

[7] https://www.geofox.de/web/de/connections, accessed 19 jun 2025.

[8] https://www.google.com/maps, accessed 19 jun 2025.

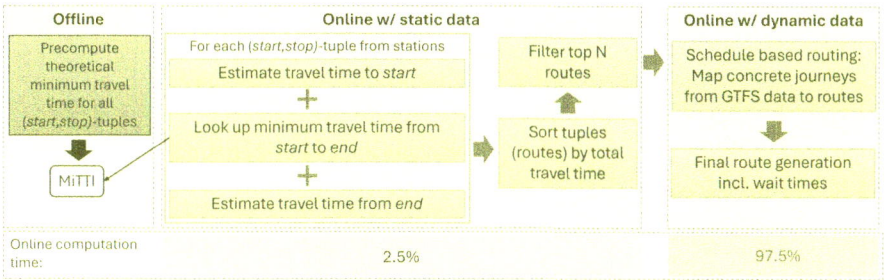

Fig. 2. Overview of our routing approach. By refraining from prefiltering stations, all possible routes are considered in the first online processing step. This increases the chances of finding optimal routes. However, the second online processing steps using dynamic data, creates a performance bottleneck.

2   Multimodal Routing Approach

Our routing system is explicitly designed to prioritize *route quality* over raw computational performance, addressing a key shortcoming of many existing journey planners, such as Google Maps or DB Navigator. In this section, we describe the individual steps of our routing approach as shown in Fig. 2, which consists of three parts: 1) An offline part which is only executed once for a provided network of public transport, 2) An online part which uses static data, e.g., geographical data, and 3) An online part using dynamic data provided by the public transport provider, i.e., timetables.

2.1   Offline Precomputation of the Minimum-Travel-Time Index (MiTTI)

RideLink examines every public-transport stop within a 5 km cycling/e-scooter radius around both origin and destination. In Hamburg this produces up to 500×500 ($\approx 2.5 \times 10^5$) stop pairs per query, each of which expands into thousands of timetable paths. To keep such exhaustive searches feasible at run time, we precompute a *Minimum-Travel-Time Index* (MiTTI).

For every ordered pair of stops (s_o, s_d), the MiTTI stores the fastest possible in-vehicle travel times (including transfer times) but *excluding* additional waiting times. These values form tight lower bounds that enable aggressive yet safe pruning of the candidate set before any schedule-dependent search is invoked.

The index is built once per network using PostgreSQL/PostGIS. A SQL loop iterates over every source stop, performs a depth-limited (≤ 3 trips) recursive search and records up to the 30 shortest journeys to each reachable target stop. To achieve this, a table which lists the time-independent travel times of the routes is created. Static walk and transfer times stem from a GTFS table. For this work, we used the table provided by the local traffic enterprise in Hamburg [5]. Clock-dependent waiting times are deliberately ignored. Thus, the only difference to the actual travel time is the time added by waiting at stations exceeding the transfer time. Since information depending on the day of the week or the time of the day is not considered, this step does not require consulting the live schedule. Consequently, rebuilding the MiTTI is required only when services are added, removed, or permanently rerouted.

2.2   Online Selection of Route Candidates

When a user query is received, the system performs the following for every potential start-stop/end-stop combination:

1. Estimate the first-mile travel time from the user's start coordinate to the potential start-stop, e.g., using PostGIS 'ST_Distance' and an average cycling speed while considering rivers.
2. Retrieve the pre-calculated, best PT travel times for this stop pair using the MiTTI.
3. Estimate the last-mile travel time, analogous to step 1, from the potential end stop to the user's final destination.

The sum of these three values yields a highly effective, albeit optimistic, *total travel time* for that specific stop-pair including the travel time to the start stop and from the end stop to the goal. The system calculates these estimates for up to all 250,000 pairs and sorts them. Then, only the top N candidates with the lowest estimated total travel times are passed to the next step. This crucial filtering stage reduces the (s_o, s_d)-pairs by over 99% while retaining basically all route candidates that are feasible in order to find the optimal final route from start to end. This "late pruning" strategy preserves optimality while reducing the dynamic search space immensely.

2.3   Online Estimation of Optimal Journeys

With the search space now reduced to a small set of high-potential stop-pairs, the last step performs a detailed, time-dependent search, each of them implemented via an individual database query:

1. **Focused, Schedule-Based Routing:** The system executes a recursive Common Table Expression (CTE) in PostgreSQL to traverse the live GTFS timetable. This computationally expensive search is constrained to operate *only* on the routes and stops associated with the promising candidate pairs identified in the previous stage. The search is seeded with the candidate (s_o, s_d)-pairs, factoring in the initial bicycle/scooter travel time to calculate the earliest possible arrival at each stop. From these entry points, the query iteratively expands paths by finding valid, time-dependent connections, thereby considering actual waiting times and service schedules, e.g., holiday or night services.
2. **Final Route Generation:** The query evaluates all feasible paths that successfully reach a candidate destination stop, calculates exact total travel times, and presents the best final routes (journeys) to the user.

While this method leads to a better route quality than the existing popular route planners, as we will show in Sect. 5, the query answer times are too high for a real-world application. Hence, the next Section takes a closer look at the performance bottlenecks and how to relax them.

3   Performance Optimization

As outlined in Sect. 2, our routing system, while effective in finding high-quality routes, is severely hampered by a performance bottleneck: The recursive SQL query (CTE) responsible for traversing the time-dependent public transport schedule. In the remainder of this Section, we take a closer look at our performance bottlenecks and how we relax them by applying traditional optimization techniques in relational databases.

3.1   Quantitative Bottleneck Analysis

To gain a better insight into the performance bottleneck in the last step of our routing approach, we inspected the generated query execution plan and the real costs per (sub)operator using SQL's EXPLAIN ANALYZE command. The analysis was performed on a representative complex query processing approximately 187,000 (s_o, s_d) combinations. However, the query was particularly challenging, because the top-ranked candidate routes were highly diverse, utilizing vastly different paths and transfer points, which significantly increased the computational load of the recursive search. The test system for this initial analysis was an Apple MacBook Pro, equipped with an M1 Max CPU and 32 GB RAM. For the database system, we used PostgreSQL 14.13. The key findings are quantified in Table 1. They confirm that the massive combinatorial space of route segments led to a noticeably increased memory and disk access, and CPU usage. While the disk access could be reduced by installing more main memory, the remaining shortcomings cannot sufficiently be solved by adding or upgrading hardware components.

Table 1. Quantified performance bottlenecks identified in our test query

Identified Bottleneck	Quantitative Manifestation (Example Query)
External Sorting Operations	Sorting 3.6 million intermediate route candidates took 26 s and caused 1.8 GB of temporary disk I/O.
Recursive CTE Explosion	Generates approximately 8 million intermediate results, leading to excessive memory consumption and CPU load.
Inefficient I/O Operations	The whole query generates over 5.3 GB of temporary data on disk due to memory pressure from large intermediate datasets. Our test system had 32 GB of main memory.
Cache Inefficiency	Tables and indexes, approx. 500 MB, are loaded repeatedly from secondary storage due to the sprawling nature of the search.

3.2   The Optimized Query Structure

Based on the bottleneck analysis, we created a revised query. The new structure retains the comprehensive search principle but incorporates several optimization techniques designed to mitigate the identified performance issues. We applied classical best practices for optimizing SQL code as well as some adjustments to our algorithm.

Classical SQL Optimization

- **Consistent CTE Materialization:** All major Common Table Expressions (CTEs), such as those identifying candidate stops and prefiltered routes, are explicitly defined as materialized views. This forces PostgreSQL to compute these potentially large intermediate result sets only once, preventing redundant calculations and ensuring their efficient reuse throughout the query plan.
- **Optimized Data Structures and Indexing:** The original, large MiTTI table is replaced with a highly optimized and compact version. The initial table was created for ease of debugging and included non-essential data such as human-readable stop names, query plans in text-form and redundant geometric information. The new, optimized table contains only the three essential columns for the lookup: `start_stop_id`, `end_stop_id` and `min_travel_time`. To support this, a new compound B-tree index was created on the (`start_stop_id`, `end_stop_id`) columns. This combination of a lean table structure and a targeted index allows for extremely fast lookups, often as efficient index-only scans, drastically accelerating access times.

Adjustments to the Algorithm

- **Aggressive Search Space Pruning:** While the initial stop selection remains broad to ensure comprehensiveness, intelligent pruning is introduced to limit the combinatorial explosion: A `best_time_estimate` CTE establishes a tight upper-bound travel time, e.g., best estimate + 30% buffer. This bound is then used within the main recursive query to aggressively discard any partial path that already exceeds this threshold, pruning entire search branches at the earliest possible stage.
- **Improved Join and Filter Logic:** The query now employs more efficient `EXISTS` subqueries in place of some costly joins. Furthermore, all joins to the core GTFS timetable data, e.g., `stop_times`, are rigorously prefiltered to include only the stop and route IDs present in the remaining set of promising candidates. This dramatically reduces the amount of data processed by the recursive CTE.

4 RideLink: A Proof-of-Concept for Enhanced Tourist Mobility

To validate the practical applicability of the optimized routing system, the "RideLink" mobile application was developed as a proof-of-concept. This application serves to demonstrate how the performance gains and the underlying comprehensive search logic translate into a tangible and useful tool for the target user: the urban tourist.

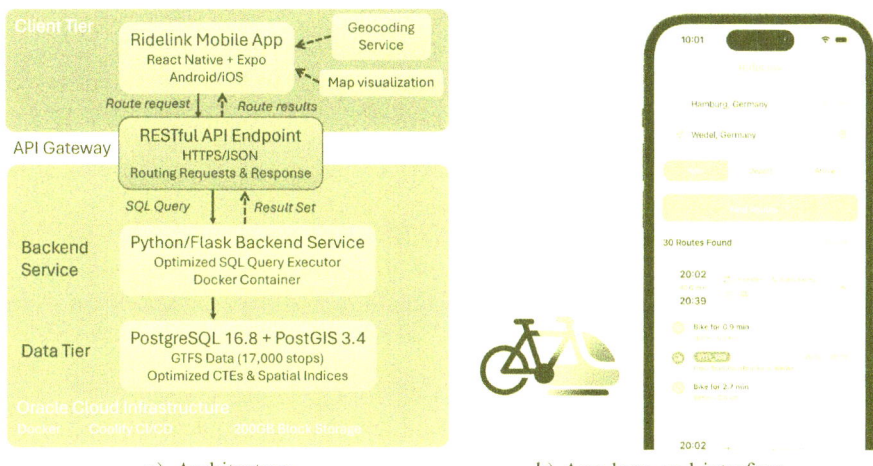

a) Architecture   b) App logo and interface

Fig. 3. a) System Architecture of the *RideLink* application. The diagram illustrates the data flow from the client tier (React Native mobile app) through the backend service (Python/Flask API) to the data tier, where the optimized SQL query is executed on the PostgreSQL/PostGIS database hosted on the OCI. b) The logo of the *RideLink* app and an impression of the user interface.

System Architecture. The system is implemented using a standard three-tier architecture, as shown in Fig. 3a. The service layer is a Python/Flask-based RESTful API that acts as the intermediary between the client and the database. It receives start/destination coordinates and time parameters, executes the optimized SQL query, and returns structured route results. The entire backend is containerized with Docker and deployed on an Virtual Private Server (VPS) running on the Oracle Cloud Infrastructure (OCI)[9] using Coolify[10] for Continuous Integration/Continuous Deployment automation. The client tier is the RideLink mobile application, a cross-platform app built with React Native[11] and Expo[12]. The key features are specifically tailored for tourist usability and include:

---

[9] https://www.oracle.com/cloud/free/, accessed 20 jun 2025.
[10] https://coolify.io, accessed 20 jun 2025.
[11] https://reactnative.dev/, accessed 20 jun 2025.
[12] https://expo.dev/, accessed 20 jun 2025.

- Intuitive address search with geocoding and interactive map-based point selection
- Flexible departure and arrival time settings
- A clear presentation of multimodal route options, detailing public transport segments, transfers, and estimated micro-mobility times for first/last-mile connections as shown in Fig. 3b.
- Visualization of the complete route path on an integrated map.

The application was successfully built and submitted to the Google Play Store[13] as well as the Apple App Store[14], underscoring its readiness as a practical demonstrator.

Practical Evaluation and User Experience. In typical usage, the deployed RideLink application delivers response times for complex multimodal queries, e.g., crossing significant portions of Hamburg with cycling legs, in the range of 3 to 5 s. While not instantaneous, this is a noticeable improvement over the potential minute-long waits of the unoptimized system and is generally acceptable for on-the-go planning by tourists. Crucially, route quality remains high, often saving tens of minutes of travel time thanks to the broader set of access and egress stops evaluated. The system consistently presents more diverse and often more convenient options than simpler planners by virtue of considering a wider array of public transport access points. However, the system is not without limitations. On the resource-constrained VPS, particularly demanding queries can occasionally lead to excessive memory consumption, resulting in query termination in roughly 5–10% of the most demanding test cases, due to limited cloud resources. This highlights the ongoing tension between search comprehensiveness and resource constraints in a real-world deployment on particularly constrained hardware.

Nevertheless, RideLink successfully demonstrates that an optimized, database-centric approach can provide tourists with a powerful and practical tool for discovering and planning efficient, flexible multimodal journeys, thereby enhancing their ability to explore cultural sites and urban environments.

5   Evaluation

The evaluation was designed to empirically validate the route quality and the effectiveness of the optimization strategies detailed in Sect. 3. All benchmarks were conducted on an Apple MacBook Pro (M1 Max, 32GB RAM) running PostgreSQL 14.13 with PostGIS 3.2 in order to run a controlled local benchmark. The test data consisted of the Hamburger Verkehrsverbund (HVV) GTFS dataset from January 2025 [5], comprising approximately 17,000 stop points and over 2 million timetable entries. Performance was measured using PostgreSQL's `EXPLAIN ANALYZE` command on a representative, complex query (Altona to Poppenbüttel, approx. 15 km) that requires a large initial search space.

---

[13] https://play.google.com/store/apps/details?id=com.top3.RideLink, accessed 20 jun 2025.

[14] https://apps.apple.com/de/app/id6745747327, accessed 20 jun 2025.

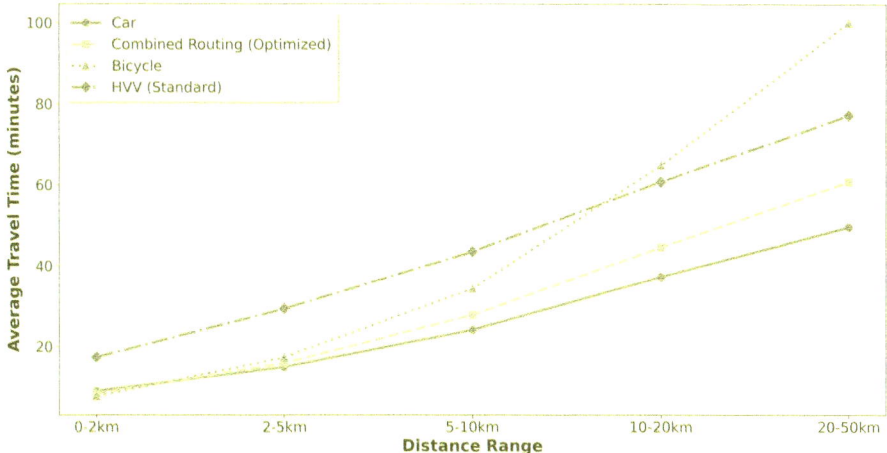

Fig. 4. Comparison of average travel times across different distance ranges based on more than 20,000 evaluated exemplary routes. The proposed Combined Routing system consistently outperforms standard public transportation (HVV) and remains highly competitive with the private car. This demonstrates the effectiveness of strategically integrating micro-mobility for first/last-mile connections to optimize the overall door-to-door journey and bridge the efficiency gap to private vehicles.

5.1 Route Quality

In this work, we use the estimated travel time as a measure of route quality. In that regard, our approach successfully achieves its primary goal: It returns high-quality, often non-obvious multimodal routes by evaluating a wide range of *first/last-mile* options, thereby being able to display the full potential of combining these modes of transportation.

Figure 4 shows the average of the travel times for a collection of 20,000 routes and different methods of transportation for different distances in Hamburg. The times for multimodal transportation were retrieved by our RideLink approach. The results are available as our *Ridelink_Dataset*[15]. For short distances of less than 3 km, our multimodal approach outperforms the car for the most part while all methods, except for public transport, are generally close. For longer journeys, the gap between the different methods of transportation increases. While bicycle and public transportation on their own cannot compete with a car, our combined approach is only 10 min slower than a car at a maximum, even for distances as long as 30 km.

---

[15] https://docs.google.com/spreadsheets/d/1rmCL-0WBKvHGix4im3abCbcYEPDu VFPs, accessed 20 jun 2025.

Route	HVV (Geofox)	DB Navigator (5km)	DB Navigator (10km)	Google Maps	Our Approach
Robert-Blum-Straße 25 A → Röhrigstraße 24 Tuesday 9:10	45	41	51	39	27
Ernst-August-Stieg 7 B → Jenfelder Allee 55 Friday, 13:15	33	63	59	53	37
Richeystraße 6 C → Burgwedelkamp 24 Monday,15:00	60	48	63	48	48
Stader Str. 203 D → Paul-Ehrlich-Straße 1/Haus 1 Tuesday, 10:00	53	33	33	67	33

Fig. 5. Four example scenarios (A-D) and a comparison of the travel times in minutes for the most optimal routes suggested by different route planners.

In the following, we take a closer look at some example routes which consider the specific properties of the city of Hamburg. Hamburg is characterized by its large bodies of water. Besides many small rivers and creeks, there are two main rivers. The river Elbe divides Hamburg into a southern, and a northern part, as well a middle part between the two main branches of the Elbe. With its harbor and sights like the "Elbphilharmonie" it attracts numerous tourists. The Alster, also known as the "Heart of Hamburg" is a river which is partially enlarged to a lake and provides leisure activities for locals and tourists alike. It divides the northern part of Hamburg into an eastern and a western part and has no bridges over its main body of water. These rivers are bottlenecks for transportation. Thus, for our evaluation, we selected example routes on all sides of these waterways:

A) This route is in the north-west of Hamburg and does not require crossing one of the mentioned rivers.
B) This route is in the east of Hamburg and requires crossing one of the main branches of the river Elbe.
C) This route is an east-west connection in the north of Hamburg, north of the enlarged Alster.
D) This route crosses all branches of the river Elbe.

Figure 5 shows the estimated optimal travel times for our approach and for 3 selected multimodal route planners: HVV's own solution, DB Navigator, and Google Maps. In the DB Navigator app, the search radius for stations could be adjusted. We set it to 5 km and to 10 km. The comparison shows that our approach leads to optimal results in 3 out of the 4 cases. In the fourth case, it is only behind by 4 min. This is caused by an inconsistency in the published data for the transfer times regarding Hamburg central station; the found route, regarding lines taken, is the same. Surprisingly, the widely used Google Maps, suggested to not take public transportation at all but to cover the whole distance by bicycle in all cases with one notable exception: In case D, it shows an alternative route breaking up the journey by using a ferry. However, this approach is not the recommended one and only reduces the estimated travel time by 3 min making it still the least optimal choice out of all journeys suggested by the other apps.

In contrast to this, it shows the best result in case A out of the existing apps. However, our approach still outperforms this result by 12 min. Our 27-minute route (Route A) consists of a short bicycle ride (3 min), a buffer (2 min), a single public transport leg (8 min), and a final bicycle segment (14 min). This highlights the ability of our approach to uncover superior, non-obvious journey options that other planners miss.

Table 2. Performance Comparison: Original vs. Optimized Comprehensive Query.

Metric	Original Query	Optimized Query	Improvement
Total execution time (local)	59,538 ms	16,771 ms	3.55×
Generated intermediate results (rows)	16,124,160	2,495,779	6.46×
Final result rows before sorting	1,767,910	737	2,399×
Disk access (temp written)	2,890 MB	415 MB	6.97×

Table 3. Storage Requirement Comparison: Original vs. optimized MiTTI.

Component	Table Size	Index Size	Total Size
Original MiTTI Table	90.1 GB	75.3 GB	165.4 GB
Optimized MiTTI Table	6.3 GB	38.9 GB	45.2 GB
Total Storage Savings	**83.8 GB**	**36.4 GB**	**120.2 GB**

5.2 Performance and Storage Requirements

Since our exhaustive search is significantly slower than the ones we compared with in Sect. 5.1, the primary goal of our performance optimization is to achieve an acceptable performance which is justified by the higher route quality. Hence, in this subsection, we show the effect of the steps described in Sect. 3. These optimizations yielded substantial improvements across all key metrics, as quantified in Table 2, which shows the improvements for our example query in Sect. 3.1.

The optimized query achieves a 3.55× reduction in total execution time on the local test machine. However, the most significant impact was the 2,399× reduction in the number of rows requiring final sorting. This directly demonstrates the effectiveness of the aggressive, but late, pruning and optimized join strategies in managing the combinatorial complexity of the comprehensive search. While a 16-s local execution time on the hardest cases observed still exceeds ideal real-time targets, it represents a critical step towards practicality which is compensated by the time savings caused by the better route quality. In the deployed environment, the end-to-end response time for RideLink on an average query is 3–5 s, a significant improvement that makes interactive tourist planning feasible.

Further, our optimizations not only improve the query response time, but also lead to a significant reduction of the memory footprint. This is a direct result of redesigning the MiTTI data structure. Table 3 compares the storage requirements of the original and optimized routing tables. The optimized system architecture successfully addresses the critical performance bottlenecks, making the comprehensive routing approach viable for practical deployment.

6   Related Work

Optimizing multimodal route queries, particularly those combining scheduled public transportation with flexible *first/last mile* access, presents challenges distinct from static road network routing. The reason for this is a combination of factors, e.g., the time dependency of schedules and the combinatorial complexity of routing. This leads to a trade-off between performance and the comprehensiveness necessary for route quality. While existing approaches prefer performance, our work treats route quality as a primary optimization goal.

The Transfer Patterns (TP) approach used by Google Maps, proposed by Bast et al. [1,2], significantly accelerates queries by precomputing optimal sequences of transfer stops. During a query, the system retrieves these patterns and finds concrete connections for the given time. While TPs can return individual stop-to-stop routes in milliseconds, they lack any mechanism to pre-filter stops for multimodal routing. As a result, executing, for example, 250 000 such queries for a single user request would raise response times to unacceptable levels. Other state-of-the-art algorithms include the Connection Scan Algorithm (CSA) [4] and Round-Based Public Transit Routing (RAPTOR) [3]. However, these approaches do also not explicitly consider the *first/last mile* problem. At the same time, there are approaches tackling the *first/last mile* problem specifically, e.g., the route planning approach by Horstmannshoff et al. which incorporates user preferences [6]. Unfortunately, the test city for that work was Göttingen, which is significantly smaller than Hamburg having less than 10% of the population of Hamburg. Hence, it is not clear if the concept is feasible for larger networks. Moreover, it does not provide a comprehensive approach which includes the actual routing within the PT network. Instead, the classical Dijkstra algorithm is used for routing, which has already been reworked but is still hard to parallelize [7].

7   Conclusion and Future Work

In this paper, we propose RideLink, a routing system for multimodal transportation considering public transportation, and bicycles or e-scooters for the *first/last mile* problem. We developed a comprehensive approach which treats route quality as a primary optimization goal and performance as a secondary optimization goal. Through systematic bottleneck analysis and the application of several optimization techniques, notably CTE materialization, aggressive search space pruning, and refined join strategies, the query performance increased by a factor of 3.55x. This enhancement significantly improves the practicality of a system that, by design, considers a vast array of *first/last mile* options. While sub-second responses for such broad queries remain an ambitious target for a database-centric approach, the achieved speeds transform the system from a research prototype into a usable tool for tourist mobility planning. Further, we created a mobile app as a proof-of-concept.

Several options exist for further enhancement of our approach:

1. **Hybrid Routing Architectures:** Explore a tighter integration where the database handles spatial querying and relevant stop-selection while a dedicated, in-memory engine, e.g., using CSA [4] or RAPTOR [3], performs the transit-leg calculations.
2. **Personalization and Context-Awareness:** Allow users to specify preferences, e.g., scenic routes, minimize cycling, or adapt suggestions based on factors like weather.
3. **Scalability and Deployment:** Test the system in other cities to ensure broader applicability and explore more powerful deployment strategies to mitigate the resource constraints observed on the current VPS.
4. **OLAP-Backend Analytics Pipeline:** Offload large, heavy tasks, such as the periodic MiTTI recomputation and aggregated performance monitoring, to a column, oriented OLAP system. Separating analytical workloads from the operational PostgreSQL/PostGIS instance should allow us to shorten batch windows without impacting real-time query latency.

In conclusion, we demonstrate that the result quality of multimodal routing can be increased at a bearable performance cost. While further optimization of our approach will be required as the user count grows, we are confident that our approach can benefit tourists and adventurous locals alike.

Disclosure of Interests. The authors have no competing interests to declare that are relevant to the content of this article.

References

1. Bast, H., Brodesser, M., Storandt, S.: Result diversity for multi-modal route planning. In: 13th Workshop on Algorithmic Approaches for Transportation Modelling, Optimization, and Systems (ATMOS 2013). OpenAccess Series in Informatics (OASIcs), vol. 33, pp. 123–136. Schloss Dagstuhl–Leibniz-Zentrum für Informatik (2013). https://doi.org/10.4230/OASIcs.ATMOS.2013.123
2. Bast, H., Carlsson, E., Eigenwillig, A., Geisberger, R., Harrelson, C., Raychev, V., Viger, F.: Fast routing in very large public transportation networks using transfer patterns. In: Algorithms – ESA 2010. LNCS, vol. 6346, pp. 290–301. Springer (2010). https://doi.org/10.1007/978-3-642-15775-2_25
3. Delling, D., Pajor, T., Werneck, R.F.: Round-based public transit routing. In: 14th Workshop on Algorithm Engineering and Experiments (ALENEX 2012), pp. 130–140. SIAM (2012). https://doi.org/10.1137/1.9781611972924.13

4. Dibbelt, J., Pajor, T., Strasser, B., Wagner, D.: Intriguingly simple and fast transit routing. In: 12th International Symposium on Experimental Algorithms (SEA 2013). Lecture Notes in Computer Science, vol. 7933, pp. 43–54. Springer (2013). https://doi.org/10.1007/978-3-642-38527-8_6

5. Hamburger Verkehrsverbund (HVV): HVV Fahrplandaten (GTFS) Mai 2025 – Dez 2025. https://suche.transparenz.hamburg.de/dataset/hvv-fahrplandaten-gtfs-mai-2025-bis-dezember-2025 (2025). Accessed 14 June 2025

6. Horstmannshoff, T., Redmond, M.: Identifying alternative stops for first and last-mile urban travel planning. Public Transport **16**, 359–379 (2024). https://doi.org/10.1007/s12469-024-00355-w

7. Madduri, K., Bader, D.A., Berry, J.W., Crobak, J.R.: Parallel shortest path algorithms for solving large-scale instances. In: The Shortest Path Problem, pp. 249–290 (2006)

8. McHugh, B.: Pioneering open data standards: The gtfs story. Beyond transparency: open data and the future of civic innovation, pp. 125–135 (2013)

9. Padgham, M., Stepniak, M.: gtfsrouter: Routing with 'GTFS' (General Transit Feed Specification) Data (2025). https://github.com/UrbanAnalyst/gtfsrouter, r package version 0.1.4

Author Index

P. K. Chrysanthis et al. (Eds.): ADBIS 2025, CCIS 2676, pp. 583–584, 2026.
https://doi.org/10.1007/978-3-032-05727-3

The manufacturer's authorised representative in the EU is Springer
Nature Customer Service Centre GmbH, Europaplatz 3, 69115 Heidelberg,
Germany. If you have any concerns regarding our products, please
contact ProductSafety@springernature.com

Printed and bound by CPI Group (UK) Ltd, Croydon, CR0 4YY

28/04/2026

02098524-0014